IMPORTANCE MEASURES IN RELIABILITY, RISK, AND OPTIMIZATION

IMPORTANCE MEASURES IN RELIABILITY, RISK, AND OPTIMIZATION

PRINCIPLES AND APPLICATIONS

Way Kuo

City University of Hong Kong, Hong Kong

Xiaoyan Zhu

The University of Tennessee, Knoxville, Tennessee, USA

A John Wiley & Sons, Ltd., Publication

Registered office
John Wiley & Sons Ltd, The Atrium, Southern Gate, Chichester, West Sussex, PO19 8SQ, United Kingdom

For details of our global editorial offices, for customer services and for information about how to apply for permission to reuse the copyright material in this book please see our website at www.wiley.com.

Library of Congress Cataloging-in-Publication Data

Kuo, Way, 1951–
    Importance measures in reliability, risk, and optimization : principles and applications / Way Kuo, Xiaoyan Zhu.
      p. cm.
    Includes bibliographical references and index.
    ISBN 978-1-119-99344-5 (hardback)
    1. Reliability (Engineering)   2. Risk assessment.   3. Industrial priorities.   I. Zhu, Xiaoyan.   II. Title.
    TA169.K855 2012
    620′.00452–dc23

                                                                                                    2011052840

A catalogue record for this book is available from the British Library.

Print ISBN: 978-1-119-99344-5

Typeset in 10/12pt Times by Aptara Inc., New Delhi, India
Printed and bound in Malaysia by Vivar Printing Sdn Bhd

1   2012

Contents

List of Figures

List of Tables

Preface

Appearing on the scene in the late 1940s and early 1950s, reliability was first defined and applied to communication and transportation, and it has become a major part of performance measure for military systems. Much of the early work was confined to reliability analysis, and many of the early theories were developed hypothetically without considering the real problems encountered.

Along with reliability design, the concept of importance measures of components in a coherent system was proposed in the 1960s (Birnbaum 1969). At that moment, the system of interest was confined to a binary system of two states, functioning or failed. Strictly, the early version of the importance measures was sensitivity analysis of a probabilistic system. Given the increasing trend of using complex and complicated systems, other types of importance measures have also been investigated, one of which is the Fussell–Vesely importance (Fussell 1975; Vesely 1970), which was used in the design for reliability and safety in nuclear power plants. The precise relationships among various importance measures and their applications in the areas of reliability, risk, mathematical programming, and even broader categories have not yet been investigated thoroughly. This book will be the first to meet the need of addressing these unresolved issues.

In the 1980s, consecutive-k-out-of-n ($Con/k/n$) systems were proposed as a type of complex systems (Chiang and Niu 1981). It was quickly studied that optimal design for such systems is nearly impossible without further investigating other types of importance measures. Similarly, applications such as design for reliable software systems were needed, and with the aid of importance measures, one is able to improve software reliability with limited resources (Fiondella and Gokhale 2008). Still, optimal design has been only remotely possible. In the 1990s, as with other optimization design tools, heuristics were proposed along with importance measures to facilitate optimal reliability design (Kuo and Prasad 2000; Kuo and Wan 2007).

In this book, we provide a comprehensive picture of importance measures in reliability. Furthermore, we generalize the ideas of importance measures to solve some of the problems in reliability design, risk, and mathematical programming. Appendix A provides the proofs of some theorems. The book is divided into five parts.

Part One, "**Introduction and Background**," includes two chapters. The importance measures carry very wide range of applications including both deterministic and stochastic systems. Often it is most economic to quickly identify the improvement for the system performance such as reliability or other measures, instead of searching for a global solution in the crisis management. Chapter 1 gives examples of importance measures in various areas. Chapter 2 presents the fundamentals of systems reliability. In additional to defining reliability and availability, we present the classical definitions of coherent systems, cuts and paths, system signatures, various system configurations, redundancy, reliability optimization, and complexity.

Part Two, "**Principles of Importance Measures**," includes six chapters, introducing different types of importance measures in aspects of mathematical expressions, physical meanings, probabilistic interpretations, computation, and so on. Chapter 3 presents the essence and classifications of importance measures of components in coherent systems defined in Chapter 2. Chapters 4, 5, and 6 present the major importance measures in three different classes—reliability, lifetime, and structure importance measures, respectively. Structure importance measures evaluate the relative importance of positions in the system; thus, the structure importance of components actually represents the importance of their positions in the system.

The importance measures that are discussed in Chapters 4–6 are the index of individual components and evaluate the importance of an individual component with respect to system performance. In contrast, Chapter 7 discusses importance measures of pairs or groups of components, which evaluate the interaction of a pair or a group of components on system performance. The importance measures of pairs or groups of components can provide additional information that the importance measures of individual components cannot. Chapter 8 presents the importance measures and their relationships for $Con/k/n$ systems, including both the F and G systems. In these chapters, we make general assumptions about the system and its components as presented in Chapter 2. Examples of using these importance measures in dealing with network reliability problems and those beyond the context of coherent systems are also given.

Part Three, "Importance Measures for Reliability Design," includes five chapters. Through the many importance measures introduced in Part Two and the importance measures specially designed for their applications in Part Three, we are able to solve problems of redundancy allocation, system upgrading, and component assignment problems (CAP*). Chapters 9, 10, and 11–13 address these three applications, respectively. For example, a reliability analyst would present the system structure and evaluate the system performance, while a systems analyst would like to identify the critical components so that the systems can be improved with minimum additional resources. Importance measures in upgrading systems, along with special examples, are addressed in Chapter 10.

The CAP itself is of great interest in the area of reliability optimization. The importance measures have the close relationships to CAP and can be used to find optimal or approximate solutions. Chapter 11 introduces a CAP, which is to assign components to different positions in a system with the objective of maximizing systems reliability. Chapter 12 focuses on the CAP in $Con/k/n$ systems and some variations of $Con/k/n$ systems, and Chapter 13 presents heuristics for the CAP using the B-importance. For systems that are large and as complex as the various versions of $Con/k/n$ systems, heuristics and sometimes metaheuristics are of great use in the optimal design.

Part Four, "Relations and Generalizations," includes two chapters. Chapter 14 investigates the relationships of importance measures presented in Parts Two and Three and compares

---

* The singular and plural of an acronym are always spelled the same.

them in various aspects. The summary and comparisons center on the essential role of the B-importance, and the issues addressed in Chapter 14 include the computation of importance measures. In Chapter 15, we release one or more assumptions made in Chapter 2 and investigate the corresponding importance measures. In particular, Chapter 15 discusses the importance measures for noncoherent systems, multistate systems, continuum systems, and repairable systems.

Part Five, "Broad Implications to Risk and Mathematical Programming," includes four chapters. Importance measures have been adopted and applied in various disciplines, such as network flows, $\mathbb{K}$-terminal networks, mathematical programming, sensitivity and uncertainty analysis, perturbation analysis, software reliability, fault diagnosis, and probabilistic risk analysis (PRA) and probabilistic safety assessment (PSA). Chapters 16–19 address these applications, showing that importance measures could be used in optimization problems, for example, to illustrate a simplex algorithm for linear programming and to design a branch-and-bound method for integer programming, as in Chapter 17.

A sketch note

As we deal with many modern systems using the fastest computing facilities, we still find it a challenge to create an optimal design for a generic system. It is even more difficult when component reliability values cannot be accurately estimated. This book provides a comprehensive view on modeling and combining importance measures with other design tools. Included are some solution methods that require only orderings or ranges of component reliability without the need to know their exact values; this feature lessens, to some extent, the trouble of estimating component reliability, a task that is normally necessary but hard to perform in addressing various reliability problems. With respect to this feature, some importance measures are designed under the assumption of unknown component reliability; some others are designed for different ranges of component reliability. Theoretical justifications are made in the book by rigorous proofs whenever possible although throughout the writing of this book, we identified many difficult problems that will need to be further solved by readers. This book presents many conclusive results for the first time based on our studies.

In preparing for this book, we have thoroughly reviewed articles that discuss and analyze many importance measures. We found major and minor errors in these articles, and in attempting to streamline the proofs, sometimes we reached different conclusions. After updating the theories, we have adopted the correct versions in this book. As much as we would wish to present the materials in the book as complete, we may inadvertently make errors. In all cases, we acknowledge the contributions made by the authors of those articles, who provided us with many insightful viewpoints although some of them reached erroneous conclusions.

This book is an integral part of reliability modeling (Kuo and Zuo 2003) and reliability design (Kuo et al. 2006). In broader research, importance-measure-based methods for solving various hard problems in the fields of reliability, risk, and mathematical programming deserve investigation, since we believe that the importance-measure-based methods are among the most practical decision tools. The concept of importance measures finds broad implications to mathematical optimization, but the optimization community is unfamiliar of this concept and its further development. Inspired by this book, we hope that readers can develop feasible, effective methods for solving various open-ended problems in their research and practical work. Finally, we hope that you enjoy reading this book, which has been planned for many years over the course of our careers as reliability practitioners and researchers.

References

Birnbaum ZW. 1969. On the importance of different components in a multicomponent system. In *Multivariate Analysis, Vol. 2* (ed. Krishnaiah PR). Academic Press, New York, pp. 581–592.

Chiang DT and Niu SC. 1981. Reliability of consecutive-k-out-of-n:F system. *IEEE Transactions on Reliability* **R-30**, 87–89.

Fiondella L and Gokhale SS. 2008. Importance measures for a modular software system. *Proceedings of the 8th International Conference on Quality Software*, pp. 338–343.

Fussell JB. 1975. How to hand-calculate system reliability and safety characteristics. *IEEE Transactions on Reliability* **R-24**, 169–174.

Kuo W and Prasad V. 2000. An annotated overview of system-reliability optimization. *IEEE Transactions on Reliability* **49**, 176–187.

Kuo W, Prasad VR, Tillman FA, and Hwang CL. 2006. *Optimal Reliability Design: Fundamentals and Applications, 2nd edn*. Cambridge University Press, Cambridge, UK.

Kuo W and Wan R. 2007. Recent advances in optimal reliability allocation. *IEEE Transactions on Systems, Man, and Cybernetics, Series A* **37**, 143–156.

Kuo W and Zuo MJ. 2003. *Optimal Reliability Modeling: Principles and Applications*. John Wiley & Sons, New York.

Vesely WE. 1970. A time dependent methodology for fault tree evaluation. *Nuclear Engineering and Design* **13**, 337–360.

Acknowledgments

We are appreciative of the research funds granted from National Science Foundation during the past 25 consecutive years. We would also like to acknowledge Army Research Office, National Research Council, Nuclear Regulatory Commission, Bell Labs, and IBM. This book grows from the research projects, supported in part by the above agencies and companies. The Weisenbaker Chair fund of Texas A&M University and the Distinguished Professorships of National Tsing Hua University and National Taiwan University are acknowledged for providing us with the platform in conducting the exploratory study that leads to this book.

The early draft of the book has been used in classes and seminars attended by the graduate students and professionals. We are grateful for the input to this manuscript from Larry Yu-Chi Ho of Harvard University, Cambridge, Massachusetts; Min Xie of National University of Singapore and now City University of Hong Kong, Hong Kong; Fan-Chin Meng of Academia Sinica, Taipei; Nozer D. Singpurwalla of George Washington University, Washington, DC; Shu-Cherng Fang of North Carolina State University, Raleigh, North Carolina; Markos V. Koutras of the University of Piraeus, Piraeus; Baoding Liu of Tsinghua University, Beijing; Haijun Li of Washington State University, Pullman, Washington; and Enrico Zio of Ecole Centrale Paris, Châtenay-Malabry.

Acronyms and Notation

The following list the acronyms and notation that are used in this book. Additional illustrations of them have been given if necessary when they are introduced for the first time.

Acronyms

- AGSM: ANOVA-decomposition-based global sensitivity measure
- B-MGSM: Borgonovo moment-independent global sensitivity measure
- B&B: branch-and-bound
- BDD: binary decision diagrams
- BITS: B-importance-based two-stage method
- BIGLS: B-importance-based genetic local search algorithm
- BTMMS: binary-type multistate monotone system
- CAP: component assignment problem
- $Con/k/n$: consecutive-k-out-of-n system
  $Lin/Con/k/n$:F: linear consecutive-k-out-of-n failure system
  $Lin/Con/k/n$:G: linear consecutive-k-out-of-n good system
  $Cir/Con/k/n$:F: circular consecutive-k-out-of-n failure system
  $Cir/Con/k/n$:G: circular consecutive-k-out-of-n good system
- $Con/k/r/n$:F ($Con/k/r/n$:G): consecutive-k-out-of-r-from-n:F (G) system with $k \le r \le n$
- $Con/(k_1, k_2)/(m, n)$: two-dimensional consecutive-$k_1 k_2$-out-of-mn system with $2 \le k_1 \le m$ and $2 \le k_2 \le n$
  $Rec/Con/(k_1, k_2)/(m, n)$:F: rectangular $Con/(k_1, k_2)/(m, n)$:F system
  $Rec/Con/(k_1, k_2)/(m, n)$:G: rectangular $Con/(k_1, k_2)/(m, n)$:G system
  $Cyl/Con/(k_1, k_2)/(m, n)$:F: cylindrical $Con/(k_1, k_2)/(m, n)$:F system
  $Cyl/Con/(k_1, k_2)/(m, n)$:G: cylindrical $Con/(k_1, k_2)/(m, n)$:G system
- DIM: differential importance measure
- DIM^I: first-order differential importance measure
- DIM^{II}: second-order differential importance measure, that is, differential importance measure of a pair of components
- DIM^k: differential importance measure of order k, that is, differential importance measure of a group of k components
- ESA: European Space Agency
- ETS: electricity transmission system
- FIS: firewater injection system

- GGA: a general genetic algorithm
- i.i.d.: statistically independent, identically distributed
- JFI: joint failure importance
- JRI: joint reliability importance
- JRI^{II}: second-order JRI, that is, JRI of a pair of components
- JRI^k: joint reliability importance of order k, that is, JRI of a group of k components
- JSI: joint structure importance
- DGSM: derivative-based global sensitivity measure
- LLOCA: large loss of coolant accident
- MCS: multistate coherent system
- MDSM: multidirectional sensitivity measure
- MINLP: mixed integer nonlinear programming
- MIP: mixed integer programming
- MMS: multistate monotone system
- MTBF: mean time between failures
- MTTR: mean time to repair
- NBU: new better than used
- NWU: new worse than used
- PRA: probabilistic risk analysis
- PSA: probabilistic safety assessment
- PMX: partial matched crossover
- RAW: reliability achievement worth
- RAW^{II}: second-order reliability achievement worth, that is, reliability achievement worth of a pair of components
- RRW: reliability reduction worth
- RRW^{II}: second-order reliability reduction worth, that is, reliability reduction worth of a pair of components
- TOI: total order importance
- TDL: time-dependent lifetime
- TIL: time-independent lifetime

Notation

- $\backslash$: set exclusive
- $\succ$: lexicographic greater
- $\mapsto$: map to
- $\exists$: exist
- $\emptyset$: empty set
- $|\cdot|$: cardinality or size of set $\cdot$
- $(\cdot)^c$: complement of set $\cdot$
- $(\cdot)^T$: transpose of matrix $\cdot$
- $(\cdot)^{-1}$: inverse of matrix $\cdot$
- $\lfloor\cdot\rfloor$: integer part of fractional number $\cdot$
- $\leq^{st}$: stochastic ordering
- $\leq^{hr}$: hazard rate ordering
- $\leq^{lr}$: likelihood ratio ordering

- *s.t.*: such that
- $(\alpha_i, \beta_j, \mathbf{p}^{(ij)})$: vector with elements $p_i = \alpha$, $p_j = \beta$, and $\mathbf{p}^{(ij)}$ for others where $\alpha, \beta \in [0, 1]$
- $(\alpha_i, \beta_j, \mathbf{x}^{(ij)})$: vector with elements $x_i = \alpha$, $x_j = \beta$, and $\mathbf{x}^{(ij)}$ for others where $\alpha, \beta \in \{0, 1\}$
- $(\alpha_i, \overline{\mathbf{F}}(t)^M)$: vector with elements $\overline{F}_j(t)$, $j \in M$ and $\overline{F}_i(t) = \alpha$ for $t \geq 0$, $i \in M$
- $(\alpha_i, \mathbf{p}) = (p_1, p_2, \ldots, p_{i-1}, \alpha, p_{i+1}, \ldots, p_n)$ for $\alpha \in [0, 1]$
- $(\alpha_i, \mathbf{x}) = (x_1, x_2, \ldots, x_{i-1}, \alpha, x_{i+1}, \ldots, x_n)$ for $\alpha = 0$ or 1
- $(\alpha_i, \mathbf{X}) = (X_1, X_2, \ldots, X_{i-1}, \alpha, X_{i+1}, \ldots, X_n)$ for $\alpha = 0$ or 1
- $\Gamma(\alpha)$: Gamma function of α
- Π: set of all $n!$ permutations of integers 1 through n
- π: a permutation $(\pi_1, \pi_2, \ldots, \pi_n)$ of the integers 1 through n, that is, component π_i occupying position i in the system, $1 \leq i \leq n$
- $\pi(i, j)$: a permutation obtained from permutation π by exchanging the integers in positions i and j
- $\tau(\mathbf{T})$: system lifetime function of component lifetimes
- ϕ: state of the system
- $\phi(\mathbf{x})$: structure function of the system
- $\phi^D(\mathbf{x}) = 1 - \phi(\mathbf{1} - \mathbf{x})$: dual structure of $\phi(\mathbf{x})$
- $a_i(\ell, d)$: the number of collections of ℓ distinct minimal cuts such that the union of each collection contains exactly d components and includes component i, $1 \leq \ell \leq u$, $1 \leq d \leq n$
- $a_{(i)}(\ell, d)$: the number of collections of ℓ distinct minimal cuts such that the union of each collection contains exactly d components and does not include component i, $1 \leq \ell \leq u$, $1 \leq d \leq n$
- $\tilde{a}(\ell, d) = a_i(\ell, d) + a_{(i)}(\ell, d)$, $1 \leq \ell \leq u$, $1 \leq d \leq n$
- A_ϕ: steady-state availability of system ϕ
- $A_\phi(t)$: availability function of system ϕ
- $\overline{A}_\phi(t)$: unavailability function of system ϕ
- $A_i(t)$: availability function of component i
- $\overline{A}_i(t)$: unreliability function of component i
- $b_i(d) = \sum_{\ell=1}^{u}(-1)^{\ell-1}a_i(\ell, d)$, $1 \leq d \leq n$
- $\tilde{b}(d) = \sum_{\ell=1}^{u}(-1)^{\ell-1}\tilde{a}(\ell, d)$, $1 \leq d \leq n$
- $\mathbf{b}_i = (b_i(1), b_i(2), \ldots, b_i(n))$
- $B(i) = \{(\cdot_i, \mathbf{x}) : \phi(1_i, \mathbf{x}) - \phi(0_i, \mathbf{x}) = 1\}$: set of state vectors $(\cdot_i, \mathbf{x})$ in which component i is critical for the system
- C_k: the kth minimal cut for $k = 1, 2, \ldots, u$
- $\mathscr{C}$: set of cuts
- $\overline{\mathscr{C}} = \{C_k : 1 \leq k \leq u\}$: set of minimal cuts
- $\mathscr{C}_i$: set of cuts containing component i
- $\overline{\mathscr{C}}_i = \{C_k : i \in C_k, 1 \leq k \leq u\}$: set of minimal cuts containing component i
- $\mathscr{C}_{(i)}$: set of cuts not containing component i
- $\overline{\mathscr{C}}_{(i)} = \{C_k : i \notin C_k, 1 \leq k \leq u\}$: set of minimal cuts not containing component i
- $\overline{\mathscr{C}}_i^{(j)} = \{C_k : i \in C_k, j \notin C_k, 1 \leq k \leq u\}$: set of minimal cuts containing component i but not j
- $\mathscr{C}(d)$: set of cuts of size d, $1 \leq d \leq n$
- $\overline{\mathscr{C}}(d)$: set of minimal cuts of size d, $1 \leq d \leq n$
- $\mathscr{C}_i(d)$: set of cuts of size d that contains component i, $1 \leq d \leq n$

- $\overline{\mathscr{C}}_i(d)$: set of minimal cuts of size d that contains component i, $1 \le d \le n$
- $\mathscr{C}_{(i)}(d)$: set of cuts of size d that does not contain component i, $1 \le d \le n$
- $\overline{\mathscr{C}}_{(i)}(d)$: set of minimal cuts of size d that does not contain component i, $1 \le d \le n$
- $\overline{\mathscr{C}}_M$: set of minimal cuts whose failure results in the failure of modular set M
- $\mathrm{Corr}(\cdot, \cdot)$: correlation coefficient of two random variables
- $\mathrm{Cov}(\cdot, \cdot)$: covariance of two random variables
- $e = \min_{C \in \overline{\mathscr{C}}} |C|$: the smallest size of minimal cuts
- $e' = \min_{P \in \overline{\mathscr{P}}} |P|$: the smallest size of minimal paths
- e_i: the smallest size of a cut containing component i
- $\mathbb{E}(\cdot)$: the expected value of random variable $\cdot$
- $f_\phi(t)$: probability density function of T_ϕ
- $f_i(t)$: probability density function of T_i
- $F_\phi(t) = \Pr\{T_\phi < t\}$: lifetime distribution of system ϕ, that is, the cumulative distribution function of T_ϕ
- $\overline{F}_\phi(t) = 1 - F_\phi(t)$: reliability distribution of system ϕ
- $F_i(t) = \Pr\{T_i < t\}$: lifetime distribution of component i, that is, the cumulative distribution function of T_i
- $\overline{F}_i(t) = 1 - F_i(t) = \Pr\{T_i > t\} = \Pr\{X_i(t) = 1\}$: reliability distribution (function) of component i
- $\mathbf{F}(t) = (\overline{F}_1(t), \overline{F}_2(t), \dots, \overline{F}_n(t))$
- $\mathbf{F}(t)^M$: vector with elements $\overline{F}_i(t)$, $i \in M$
- $\mathbf{F}(t)^{(ij)}$: vector obtained by deleting $\overline{F}_i(t)$ and $\overline{F}_j(t)$ from $\mathbf{F}(t)$
- $g_i(d) = \sum_{\ell=1}^{v}(-1)^{\ell-1} w_i(\ell, d)$, $1 \le d \le n$
- $\mathbf{g}_i = (g_i(1), g_i(2), \dots, g_i(n))$
- G: a graph representing a network
- $G - e$: graph with edge e deleted from G
- G^*e: graph obtained from G with edge e contracted
- $\overline{G}_i(t)$: updated reliability distribution (function) of component i
- $\mathcal{I}(A)$: indicator function of event A, $\mathcal{I}(A) = 1$ if A occurs and 0 otherwise
- $i >_x j$: position i is more important than position j according to importance measure of type x; $i =_x j$ equally important to; $i \ge_x j$ more important than or equally important to. Type x could be a^c, a^p, B^h, B^u, c, cd, cp, ex, in, li, mc, mp, p, and pe, representing the different types of structure importance measures.
- $I_x^y(\cdot; \phi)$: the structure importance measure under type x. Subscript x represents the types of importance measures. Superscript y is optional and could be empty, M, C, P, or others. "Empty" means the importance of components; "M" means the importance of modular sets; "C" means the importance of minimal cuts; and "P" means the importance of minimal paths.
- $I_x^y(\cdot; \phi, \mathbf{F})$ and $I_x^y(\cdot; \phi, \mathbf{F}(t))$: the time-independent lifetime (TIL) and time-dependent lifetime (TDL) importance measures under type x, respectively. For the sake of simplification, the ϕ in notation $I_x^y(\cdot; \phi, \mathbf{F})$ and $I_x^y(\cdot; \phi, \mathbf{F}(t))$ may be ignored. $I_x^y(\cdot; \overline{F})$ and $I_x^y(\cdot; \overline{F}(t))$ denote the corresponding importance under the assumption of $\overline{F}_1(t) = \overline{F}_2(t) = \cdots = \overline{F}_n(t) = \overline{F}(t)$. See above for illustrations of superscript y.
- $I_x^y(\cdot; \phi, \mathbf{p})$: the reliability importance measure under type x. For the sake of simplification, the ϕ in notation $I_x^y(\cdot; \phi, \mathbf{p})$ may be ignored. $I_x^y(\cdot; p)$ denotes the corresponding importance

under the assumption of $p_1 = p_2 = \cdots = p_n = p$. See above for illustrations of superscript y.

- Main types of importance measures:
  - I_A, $I_{A^{\text{var}}}$, $I_{A^{\text{tot}}}$: ANOVA-decomposition-based global sensitivity measures (AGSM)
  - I_B: B-importance
  - I_{Bf}: B-importance for system failure
  - I_{Bs}: B-importance for system functioning (success) (i.e., improvement potential importance)
  - I_{Bs^V}: B-importance for system functioning (success) relative to the variance of systems reliability
  - I_{B^T}: transformed B-importance in ETS
  - I_{B^V}: B-importance relative to the variance of systems reliability
  - I_{Bay}: Bayesian importance
  - I_{BM}: Borgonovo moment-independent global sensitivity measure (B-MGSM)
  - $I_{BM\text{-Fuzzy}}$: B-MGSM for fuzzy input uncertainty
  - I_{BP}: BP importance
  - I_{Cf}: criticality importance for system failure
  - I_{Cf^T}: transformed criticality importance for system failure in ETS
  - I_{CM}: CHT moment-independent global sensitivity measure (CHT-MGSM)
  - I_{Cs}: criticality importance for system functioning (success)
  - I_{DIM}: differential importance measure of a component
  - $I_{DIM^{ll}}$: differential importance measure of a pair of components
  - I_{DIM^k}: differential importance measure of order k
  - $I_{D\mu}$, $I_{D\widetilde{\mu}}$, $I_{D\sigma}$, $I_{D\widetilde{\sigma}}$, I_{D^2}, $I_{D_A^2}$: derivative-based global sensitivity measures (DGSM)
  - I_{E_k}, $k = 1, 2, \ldots, 6$: edge importance measures in a network flow system
  - $I_{E\mu}$, $I_{E\widetilde{\mu}}$, $I_{E\sigma}$: elementary effect sensitivity measures
  - I_{FCI}: simulation-based failure criticality importance
  - I_{Flow}: global flow-based importance of an edge in a network flow system
  - I_{FT}: first-term importance
  - I_{FV^c}: c-type FV (c-FV) importance
  - I_{FV^p}: p-type FV (p-FV) importance
  - I_{FV^T}: transformed FV importance in ETS
  - I_{Impact}: global impact-based importance of an edge in a network flow system
  - I_{IU}: I-uncertainty importance
  - $I_{JF^{ll}}$: joint failure importance of a pair of components
  - I_{JF^k}: joint failure importance of order k
  - $I_{JR^{ll}}$: joint reliability importance of a pair of components
  - I_{JR^k}: joint reliability importance of order k
  - I_{L_1}, I_{L_2}, I_{L_3}, I_{L_4}, I_{L_5}, I_{L_6}: $L_1 - L_6$ importance measures related to improving the expected system lifetime
  - I_{MaxD}, I_{MinD}: maximum and minimum degradation importance measures in B&B for integer programming
  - I_{MDSM}: multidirectional sensitivity measure
  - $I_{\text{NF-B}}$: generalization of B-reliability importance in a network flow system
  - $I_{\text{NF-FV}}$: generalization of FV importance in a network flow system

- $I_{\text{NF-RAW}}$: generalization of reliability achievement worth in a network flow system
- $I_{\text{NF-RRW}}$: generalization of reliability reduction worth in a network flow system
- $I_{\text{NFA-B}}$: alternative generalization of B-reliability importance in a network flow system
- $I_{\text{NFA-FV}}$: alternative generalization of FV importance in a network flow system
- $I_{\text{NFA-RAW}}$: alternative generalization of reliability achievement worth in a network flow system
- $I_{\text{NS}_1}, I_{\text{NS}_2}, I_{\text{NS}_3}$: node selection importance measures in B&B for integer programming
- I_{OCI}: simulation-based operational criticality importance
- I_{PR}: parallel redundancy importance
- I_{RAW}: reliability achievement worth of a component
- $I_{\text{RAW}^{II}}$: reliability achievement worth of a pair of components
- I_{RAW^T}: transformed reliability achievement worth in ETS
- I_{RCI}: simulation-based restore criticality importance
- I_{RE}: rare-event importance
- I_{RRW}: reliability reduction worth of a component
- $I_{\text{RRW}^{II}}$: reliability reduction worth of a pair of components
- I_{RRW^T}: transformed reliability reduction worth in ETS
- I_{SR}: series redundancy importance
- $I_{\text{TOI}^{II}}$: total order importance of order 2
- I_{TOI^k}: total order importance of order k
- $I^{\mu}_{\text{U-B}}$: utility B-reliability importance of state of component with respect to system performance utility
- $I_{\text{U-B}}$: utility B-structure importance with respect to system performance utility
- $I^{\mu}_{\text{U-BP}}$: BP TIL importance of state of component with respect to system performance utility
- $I^{\mu}_{\text{U-D}}$: utility-decomposition reliability importance of state of component with respect to system performance utility
- $I_{\text{U-JR}}$: joint reliability importance with respect to system performance utility
- $I_{\text{U-JS}}$: joint structure importance with respect to system performance utility
- I_Y: yield importance
- K: the number of states in multistate systems
- $\mathbb{K}$: a designated set of target vertices in a $\mathbb{K}$-terminal reliability network
- ℓ_i: the number of distinct minimal cuts of size e_i containing component i
- m: the number of modules in a module decomposition of a system
- m_i: the number of critical cut vectors for component i, $m_i = \sum_{d=1}^{n} m_i(d)$
- $m_i(d)$: the number of critical cut vectors for component i of size d, $1 \le d \le n$
- m_i': the number of critical path vectors for component i, $m_i' = \sum_{d=1}^{n} m_i'(d)$
- $m_i'(d)$: the number of critical path vectors for component i of size d, $1 \le d \le n$
- $\{(M_k, \chi_k)\}_{k=1}^{m}$: a modular decomposition with m modules
- n: the number of components in the system
- $N = \{1, 2, \ldots, n\}$: set of components $1, 2, \ldots, n$ in the system
- (N, ϕ): the system with component set N and structure ϕ
- $N_0(\mathbf{x}) = \{i \in N : x_i = 0\}$. If $\phi(\mathbf{x}) = 0$, $N_0(\mathbf{x})$ is a cut and $\mathbf{x}$ is a cut vector.
- $N_1(\mathbf{x}) = \{i \in N : x_i = 1\}$. If $\phi(\mathbf{x}) = 1$, $N_1(\mathbf{x})$ is a path and $\mathbf{x}$ is a path vector.
- $nc(G, \mathbb{K})$: the number of minimum cutsets whose removal disconnects the target vertices in $\mathbb{K}$ of graph G

- $np(G, \mathbb{K})$: the number of minimum pathsets joining all target vertices in $\mathbb{K}$ of graph G
- p: component reliability when $p_1 = p_2 = \cdots = p_n = p$
- $\mathbf{p} = (p_1, p_2, \ldots, p_n)$: component reliability vector
- $(1_i, p)$: vector having 1 in the ith position and p in all other positions
- $\mathbf{p}_\pi$: vector of reliability specified by permutation π
- p_e: reliability of edge e
- $p_i = \Pr\{X_i = 1\}$: reliability of component i
- $p_{i,j}$: reliability of the component in row i and column j in a two-dimensional consecutive-$k_1 k_2$-out-of-mn
- P_r: the rth minimal path for $r = 1, 2, \ldots, v$
- $\mathbf{p}^M$: vector with elements $p_i, i \in M$
- $\mathbf{p}^{(ij)}$: vector obtained by deleting p_i and p_j from $\mathbf{p}$
- $\mathscr{P}, \overline{\mathscr{P}}, \mathscr{P}_i, \overline{\mathscr{P}}_i, \mathscr{P}_{(i)}, \overline{\mathscr{P}}_{(i)}, \overline{\mathscr{P}}_i^{(j)}, \mathscr{P}(d), \overline{\mathscr{P}}(d), \mathscr{P}_i(d), \overline{\mathscr{P}}_i(d), \mathscr{P}_{(i)}(d)$, and $\overline{\mathscr{P}}_{(i)}(d)$ are defined similarly for (minimal) paths as for (minimal) cuts.
- $q = 1 - p$: component unreliability
- $q_i = \Pr\{X_i = 0\} = 1 - p_i$: unreliability of component i
- $r_\phi(t) = f_\phi(t)/R(\overline{\mathbf{F}}(t))$: failure rate function of system ϕ
- $r_i(t) = f_i(t)/\overline{F}_i(t)$: failure rate function or hazard rate of component i
- $R(p)$: reliability of the system when $p_1 = p_2 = \cdots = p_n = p$
- $R(\mathbf{p})$ or $R_\phi(\mathbf{p})$: reliability of system ϕ under component reliability vector $\mathbf{p}$
- $R(\overline{\mathbf{F}}(t)) = \overline{F}_\phi(t)$: reliability function of system ϕ, which is a function of component reliability distributions
- $R(\mathbf{p}_\pi)$: reliability of the system in which component π_i is in the ith position in the system for $1 \le i \le n$
- R_χ: reliability of module (M, χ)
- $R_C(k, (i, j))$: reliability of a $Cir/Con/k/j - i + 1$:F subsystem consisting of components $i, i + 1, \ldots, j$
- $R_C(k, n)$: reliability of a $Cir/Con/k/n$:F system. Note that $R_C(k, (1, n)) = R_C(k, n)$.
- $R'_C(k, (i, j))$: reliability of a $Cir/Con/k/j - i + 1$:G subsystem consisting of components $i, i + 1, \ldots, j$
- $R'_C(k, n)$: reliability of a $Cir/Con/k/n$:G system. Note that $R'_C(k, (1, n)) = R'_C(k, n)$.
- $R_L(k, (1, i - 1), (i + 1, n))$: reliability of a $Lin/Con/k/n - 1$:F subsystem consisting of components $1, 2, \ldots, i - 1, i + 1, \ldots, n$
- $R_L(k, (i, j))$: reliability of a $Lin/Con/k/j - i + 1$:F subsystem consisting of components $i, i + 1, \ldots, j$
- $R_L(k, n)$: reliability of a $Lin/Con/k/n$:F system. Note that $R_L(k, (1, n)) = R_L(k, n)$.
- $R'_L(k, (1, i - 1), (i + 1, n))$: reliability of a $Lin/Con/k/n - 1$:G subsystem consisting of components $1, 2, \ldots, i - 1, i + 1, \ldots, n$
- $R'_L(k, (i, j))$: reliability of a $Lin/Con/k/j - i + 1$:G subsystem consisting of components $i, i + 1, \ldots, j$
- $R'_L(k, n)$: reliability of a $Lin/Con/k/n$:G system. Note that $R'_L(k, (1, n)) = R'_L(k, n)$.
- $Rel(G, \mathbb{K}, p)$: reliability of graph G subject to a designated set of target vertices $\mathbb{K}$, assuming that all edges have the same reliability p
- $Rel(G, \mathbb{K}, \mathbf{p})$: reliability of graph G subject to a designated set of target vertices $\mathbb{K}$, and the reliability of edges specified by vector $\mathbf{p}$

- $\mathbb{R}_+^m$: set of nonnegative real m-dimensional vectors
- $sc(G, \mathbb{K})$: size of a minimum cutset whose removal disconnects the target vertices in $\mathbb{K}$ of graph G
- $sp(G, \mathbb{K})$: size of a minimum pathset joining all target vertices in $\mathbb{K}$ of graph G
- T_ϕ: random variable representing the lifetime of system ϕ, $\tau(\mathbf{T}) = T_\phi$
- T_i: random variable representing the lifetime of component i
- $\mathbf{T} = (T_1, T_2, \ldots, T_n)$
- u: the number of minimal cuts
- v: the number of minimal paths
- $\mathrm{Var}(\cdot)$: variance of random variable $\cdot$
- $w_i(\ell, d)$: the number of collections of ℓ distinct minimal paths such that the union of each collection contains exactly d components and includes component i, $1 \leq \ell \leq v$, $1 \leq d \leq n$
- $\mathbf{x} = (x_1, x_2, \ldots, x_n)$: component state vector whose element x_i, $1 \leq i \leq n$, is a binary variable

$$x_i = \begin{cases} 1 & \text{if component } i \text{ functions} \\ 0 & \text{if component } i \text{ fails.} \end{cases}$$

- $\mathbf{X} = (X_1, X_2, \ldots, X_n)$: random vector whose element X_i, $1 \leq i \leq n$, is a binary random variable

$$X_i = \begin{cases} 1 & \text{if component } i \text{ functions} \\ 0 & \text{if component } i \text{ fails.} \end{cases}$$

- $\mathbf{X}(t) = (X_1(t), X_2(t), \ldots, X_n(t))$: random vector where for $1 \leq i \leq n$,

$$X_i(t) = \begin{cases} 1 & \text{if component } i \text{ functions at time } t \\ 0 & \text{if component } i \text{ fails at time } t. \end{cases}$$

- $\mathbf{x}^M$: vector with elements x_i, $i \in M$
- $\mathbf{X}(t)^M$: vector with elements $X_i(t)$, $i \in M$
- $\mathbf{x}^{(ij)}$: vector obtained by deleting x_i and x_j from $\mathbf{x}$
- $\mathbf{X}^{(ij)}$: vector obtained by deleting X_i and X_j from $\mathbf{X}$
- $\mathbb{Z}_+^n$: set of nonnegative integer n-dimensional vectors

Part One

Introduction and Background

Introduction

This part includes two chapters. Chapter 1 introduces the concept of importance measures and their applications in broad areas. Chapter 2 provides the background that is necessary to define and apply importance measures, including basic concepts and topics in reliability optimization and complexity, as well as the basic underlying assumptions of this book. Other background knowledge is introduced whenever it is necessary in a particular chapter.

1

Introduction to Importance Measures

Importance measures are used in various fields to evaluate the relative importance of various objects such as components in a system. The absolute values of importance measures may not be as important as their relative rankings. In general, a system is a collection of components performing a specific function. For example, a computer system performs a range of functions, such as computing, data processing, data input and output, playing music and movies, and others. It consists of the following major components: a computer unit, a monitor, a keyboard, a mouse, a printer, and a pair of speakers. The computer unit as a key component of the computer system can also be treated as a system by itself. It consists of one or more central processing units, a motherboard, a display card, disk controller cards, hard and floppy disk drives, CD-ROM drives, a sound card, and possibly other components. This chapter provides a wide range of modern problems that can be dealt with by the various types of importance measures. Some of them are further addressed in the rest of this book.

Example 1.0.1 (Fukushima nuclear accidents)   The recent nuclear accidents in Japan have resulted in the release of heavy doses of radioactive materials from the Fukushima I Nuclear Power Plant following the Tohoku 9-magnitude earthquake and the subsequent 46-foot-high tsunami on March 11, 2011. The plant comprises six boiling water reactors as sketched in Figure 1.1. Three had been shut down prior to the earthquake for maintenance, and the other three shut down automatically after the earthquake. Although the emergency generators were functional to cool down the reactors right after the shutdown of the nuclear reactors, they very quickly were knocked out by the flood, so the reactors and even the spent fuel pools started to overheat. As a consequence, partial core meltdown in the nuclear fuels caused hydrogen explosions that destroyed the upper cladding of some buildings housing reactors and the containment inside one reactor. In addition, fuel rods stored in the spent fuel pools began to overheat as water levels in the pools dropped. All these led to the leaking of radiation of Iodine 131 and Cesium 137 to the surrounding cities, including the greater Tokyo area.

Importance Measures in Reliability, Risk, and Optimization: Principles and Applications, First Edition.
Way Kuo and Xiaoyan Zhu. © 2012 John Wiley & Sons, Ltd. Published 2012 by John Wiley & Sons, Ltd.

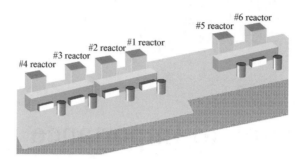

Figure 1.1   Sketch of Fukushima I nuclear power plant

The bottleneck resulting in these accidents was the flooded cable of the emergency generators that supply power to cool down the heated fuel rods in the reactors and the spent fuel pools. Because of the tsunami, the floodwaters prevented assistance from being obtained elsewhere, and apparently no spare parts were put in place for several days. This disaster has highlighted the necessity to know how crucial the emergency generators play in terms of system failure under flooding. The importance measures addressed in this book would provide the most efficient way to identify the bottleneck in the system failure by providing a methodology to follow. In particular, Chapters 3–7 and 19 give an idea and guideline by identifying maintenance methods for practitioners to follow. Smith and Borgonovo (2007) have noticed that importance measures are useful in probabilistic risk analysis (PRA) and probabilistic safety assessment (PSA), which firm a core step in precursor analysis for decision-making during nuclear power plant incidents.

Example 1.0.2 (Random forest variables in bioinformatics)   The notion of importance measures as in Chapter 3 is widely used in multivariate data analysis and ensemble learning methods (e.g., a recently developed method – random forests) that generate many classifiers and aggregate their results. Random forests yield variable importance measures for each candidate predictor in identifying the true predictor(s) among a large number of candidate predictors (Archer and Kimes 2008). The importance of a variable is estimated by the extent of prediction error increase when data for that variable is permuted, while all others are unchanged (Liaw and Wiener 2002). The necessary calculations are carried out tree by tree as the random forest is constructed. The importance measures of random forest variables have been used in bioinformatics for investigating factors to the disease risk (Schwender et al. 2011; Strobl et al. 2007) and microarray studies (Archer and Kimes 2008).

Example 1.0.3 (Transportation and rail industry)   Importance measures have been used to prioritize rail sections in order to effectively decrease delay in the railway industry (Zio et al. 2007). For example, in the winter season, large areas of rail are often blocked by snow, and some trains have to make detours to the limited open rails outside the snowy areas. The priorities for rescheduling trains could be assigned according to the appropriate importance measures of train lines so that the overall quantity to be conveyed is maximized or the overall

Figure 1.2   Lithium-ion polymer electric battery

delay is minimized, as discussed in Chapters 7 and 15. On the other hand, importance measures could also be used to prioritize the rail sections in the snowy areas for clearing snow with the limited labor and equipment so that the increase of the quantity to be conveyed could be maximized.

In addition to the rail industry, importance measures can find great use in identifying and releasing ground traffic congestion in big cities, dispatching the traffic flow after major sport events, planning highway construction in developing countries, and dealing with other transportation-related issues.

The high-speed train crashed in Wenzhou, China, on July 23, 2011, killed at least 40 persons. During the accident, a train ran into the back of another that had stalled on a viaduct after lightning cut its power supply. This accident has highlighted the crucial role of reliability, safety, and scheduling in constructing and operating a complex high-tech transportation network, no matter if the accident was due to a mechanical fault, a management problem, or a manufacturing problem. If a better planning were studied using the concept of importance measures, such an accident would have been greatly minimized.

Example 1.0.4 (Lithium-ion polymer electric battery)   It is expected that more electric buses will, over time, come into operation to make society more environment-friendly. The city of Chattanooga, Tennessee, USA, has operated nine electric buses since 1992; they have carried 11.3 million passengers and covered a distance of $1,930,000$ miles over 9 years. Two of the Chattanooga buses were used for the 1996 Atlanta Olympics. Beginning in the summer of 2000, the Hong Kong Airport Authority also began operating a 16-passenger Mitsubishi Rosa electric shuttle bus. Electric vehicles are dependent on powerful lithium-ion batteries that must last a long time and are replaced only with fully charged ones at the recharging station to allow failure-free and continuous operation of the buses. Normally, a box of battery cell consists of more than 100 battery units in series, as depicted in Figure 1.2, and typically each unit has an expected life of 5–7 years. In such a system, the power of the battery cell is significantly reduced if more than 4 or 5 units fail. The importance measures of each unit are clearly different depending on the locations and conditions of the units within the cell. This can be typically modeled as a consecutive-k-out-of-n system, as indicated in Chapters 8 and 12. A similar description and illustration can be applied to a recently developed wireless charging system (Naoki and Hiroshi 2004).

Example 1.0.5 (Reliability design)   When limited resources are available for upgrading a system (e.g., replacing one or more old components with new ones or adding redundancy),

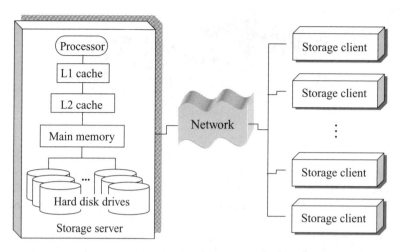

Figure 1.3   Storage system architecture

it is best to put the resources into the components whose upgrade could bring the greatest improvement in system performance (e.g., reliability). The relative importance of components in such a case could be ranked according to the importance measure that is designed to determine which component(s) merits the most additional research and development to improve overall systems reliability at minimum cost or effort as detailed in Chapters 9 and 10. Details on reliability optimization and redundancy allocation and the corresponding mathematical programming techniques can be found in Kuo et al. (2006).

A system may need multiple units that perform the same function, for example, the pump stations in an oil pumping system along a pipeline. Because of a limited budget or different extents of wear, the same functional units that are to be used in distinct positions in the system may have different brands, quality, ages, and/or conditions. Generally, the newer and more expensive ones have higher quality and are more reliable. Allocating these functionally interchangeable units of different reliability into the system is critical. The component assignment problems (CAP*) can maximize the systems reliability by optimally allocating these units into the system with the aid of importance measures, as discussed in Chapters 11–13. Essentially, the CAP is combinatorial optimization.

Example 1.0.6 (Data cache management in cloud computing)   A modern distributed data cache consists of multiple layers, which may correspond to different types of storage media with varying prices and performance characteristics. The distributed data cache helps make data more accessible, rendering cloud computing a more appealing option for computation and analysis. As illustrated in Figure 1.3, a typical computer storage system contains L1 and L2 CPU caches implemented in static random access memory, main memory based on dynamic random access memory, and hard disk drives recording data on rotating platters with magnetic surfaces. Generally speaking, the closer to the CPU, the higher price/capacity ratio but the lower access latency a storage component has. Therefore, it is beneficial to place certain

---

* The singular and plural of an acronym are always spelled the same.

data blocks in fast and expensive storage devices based on predictions of future accesses. In computer science this method is called caching, and the corresponding storage component is a so-called cache. Caches are widely used in computer systems. A data request goes to permanent storage devices such as hard disks only if the data is absent in all caches.

While enhancing the cost-effectiveness of computer systems, this multi-level storage hierarchy also creates opportunities and challenges in data management, including data placement and transfer. Due to the spatial locality of real applications, prefetching is widely used where certain blocks are fetched into a cache from lower levels before they are requested by the applications. The main objective is to minimize the data access latency of the applications.

Given limited cache sizes, two issues that need to be addressed (Zhang et al. 2009) include (i) how to handle the prefetching resource allocation between concurrent sequential access streams with different request rates and (ii) how to coordinate prefetching at multiple levels in the data access path. For sequential prefetching, "what to prefetch" is given and the key problem is to decide "how much to prefetch" and "when to prefetch," two metrics related to prefetching aggressiveness. In many current systems, this is determined at runtime through two prefetching parameters: prefetch degree, which controls how much data to prefetch for each prefetching request, and trigger distance, which controls how early to issue the next prefetch request. Different from a permanent storage device that keeps and deletes data following commands from its users, a cache needs its management mechanism to make the above decisions. In many scenarios, one storage server needs to support many storage clients through local or wide-area networks. This further increases the depth of storage stacks and the complexity of data cache management.

Optimization models and methods are in high demand for data cache management. In a novel view, the problem can be dealt with using supply chain inventory management models and methods, treating data storage system as multiple-echelon, multiple-product, supply chain and each client as a unique product.

However, for such a complex problem, the "optimal" management policy cannot be too complicated due to consideration regarding the practical implementation in the operating system (e.g., the Linux kernel) as well as the issue of consuming and occupying the system memory, which is shared by the operating system, applications, and the memory cache. Over-occupying system resources by calculating and implementing a data management policy will seriously degrade the performance of the whole system and increase the time of operating and data access. It is desirable to use importance-measure-based models and policies so that the decisions could be dynamically made on the basis of a set of index values rather than on other complicated mathematical tools, as shown in Chapter 13. Importance-measure-based methods may potentially become powerful tools in computer science research.

Example 1.0.7 (Survivability in an airline network) Suppose that a flight leg from Houston to New York is down for some reason such as the lateness or breakdown of the airplane or the absence of the crew. If this flight leg is canceled, the entire airplane network will be badly affected. The analysis based on importance measures could enhance the survivability of the entire airline network, for example, by identifying a flight leg that has less effect on the entire airplane network than the flight leg from Houston to New York and reassigning an airplane or crew that was originally assigned to this "less important" flight leg to the one that is down. Therefore, it is logical to enhance the reliability of the link that would add more survivability to the entire network. This is of particular usefulness when the available resources are limited.

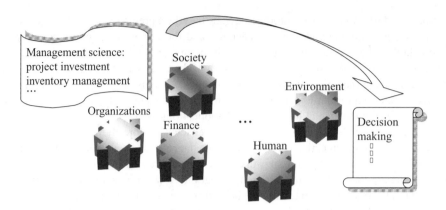

Figure 1.4   Management decision-making

Proper arrangement of resources to each link in the network as a result of careful analysis can enhance the safety and survivability, as addressed in Chapters 16 and 17. In fact, this is a typical mathematical programming problem in operations research. Including the survivability in problem modeling and the importance measures of links can improve the effectiveness of the mathematical programming.

Example 1.0.8 (Software development)   Software development is costly, very labor-intensive, and depending on the applications that are also very knowledge dependent. Unfortunately, software is rarely assured for failure-free operation because of the complexity and special requirements of the different users' profiles that may be involved. It is a common practice that for software that requires high reliability, the same specification set may be given to different developers so that redundant software may be adopted. In all of these processes, one needs to identify a critical module or subprogram in the software that requires more resources in the development phase. Therefore, as pointed out by Simmons et al. (1998), it would be very beneficial to find a formal way to identify the important modules of the software and invest in their development in order to achieve the high software reliability and meanwhile minimize the software development cost. Subsection 18.3.1 presents an example to illustrate the concept. Another case study on optimal design for software reliability can be found in Kuo et al. (2006).

Example 1.0.9 (Management science)   The importance measures are diversely used in management science and making financial decisions (Soofi et al. 2000), which is a complicated process as in Figure 1.4. For example, Borgonovo and Peccati (2004) applied the differential importance measure (DIM) based on local sensitivity analysis for investment project decisions (see Chapter 18). Borgonovo (2008) extended the DIM in combination with comparative statics technique to inventory management, in which the models are implicit, and the problem is to determine the most influential parameter on an inventory policy.

Example 1.0.10 (Risk analysis)   Risk can be analyzed through sensitivity analysis that allows to identify the impact of the variation of individual inputs on the overall decision-making.

The goal is to reduce the risk of a change in the input value and analyze steps to minimize any potential adverse consequences. In a typical sensitivity analysis, the process is repeated for all input variables for which risks and the ranking of the risks are identified. Therefore, one can identify the critical inputs in the analysis process. The commonly used procedures are treated uniformly as importance measures in Chapters 18 and 19.

Importance measures can also be used to quantify the contribution of input parameters on uncertainty over the optimal decision as described in Chapter 18. In this regard, Coyle et al. (2003) conducted a case study on a cost utility analysis of adjunct therapy for the treatment of Parkinson's disease patients.

Similarly, it is strongly recommended that one identifies the critical elements in a simulation process using the concept of importance measures. By incorporating importance measures in determining the input variables, simulation can be conducted with meaning and efficiency as well.

Example 1.0.11 (Fault diagnosis and maintenance)   In a system consisting of multiple components, the system may fail to perform its desired functions due to the failure of one or more components. For example, large pipeline networks are widely used to transport different kinds of fluids from production sites to consumption ones. Transmission pipelines are the most essential components of such networks and used to transport fluids over very long distances. The safe handling of such networks is of great importance due to the serious consequences which may result from any faulty operation. In particular, leaks quite frequently occur due to material ageing, pipes bursting, cracking of the welding seam, hole corrosion, and unauthorized actions by third parties. When a system fails, diagnosis must be efficiently conducted, and maintenance must locate the component(s) that caused the failure of the system and bring the system back to function as soon as possible. Then the maintenance needs to check the component that most likely caused the failure first, then the second most likely component, and so on, until the system is fixed. For this purpose, one might propose an importance measure to indicate how important the different components are in terms of system failure. The importance measures in Section 19.6 may suggest the most efficient way to diagnose system failure by generating a repair checklist for an operator to follow, and indeed, importance-measure-based maintenance methods are used by practitioners (ReliaSoft 2003).

Example 1.0.12 (Nuclear power plant)   Nuclear power plants are designed so that major external pressures, including bombing, will not compromise the safety concern. Special attention has also been paid to seismic factors in the siting, design, and construction of nuclear plants and in identifying other critical elements in order to enhance the safety margin. However, in analyzing all key factors, people need formal procedures to identify the risk factors and to design for reliability in power plants before they are put in operation.

Fussell (1975) and Vesely (1970) may be the first to apply the concept of importance measures, that is, the Fussell–Vesely importance (see Sections 4.2 and 5.2), to evaluating systems reliability and risk based on complete probabilistic information. Since then, various importance measures have been applied to evaluate and assess nuclear power plants. In an early work, Lambert (1975) explicitly proposed importance measures of events and cuts in fault trees and used them in generating a checklist for fault diagnosis of a low-pressure injection system, which is a standby safety system that forms part of the emergency core cooling system at a nuclear power plant.

For a recent example, with regard to safety assessment for nuclear waste disposal, Saltelli and Tarantola (2002) used global sensitivity measures to judge the effect of model parameter uncertainty reductions on the overall model output uncertainty (see Section 18.2). In this respect, PRA and PSA are methodologies that produce numerical estimates of the basic events, components, and parameters with respect to a number of risk metrics for complex technological systems. Core damage frequency and large early release frequency are the common risk metrics of interest in nuclear power plants. PRA and PSA are generally conducted using certain importance measures as explained in Chapter 19.

The concept of importance measures of components in a reliability system was initially proposed in the 1960s. After that, many types of importance measures were proposed with respect to the diverse considerations of system performance. Performance of a system can be measured using systems reliability, availability, maintainability, safety, utility in Equation (15.1), structural function of system efficiency, which is a set of discrete values from zero to a positive maximum value (Finkelstein 1994), and so on. Part Two presents major existing importance measures in reliability, including their definitions, probabilistic interpretations, properties, computation, and comparability. Part Three thoroughly discusses applications of importance measures in the areas of systems reliability design and introduces more types of importance measures that were proposed to address problems of interest. Although importance measures reflect different probabilistic interpretations and potential applications, they are related to each other in some aspects. Part Four presents the relationships among these types of importance measures and generalizes them to nonetheless important situations. Part Five investigates broad implications of importance measures to risk, mathematical programming, and even broader categories.

References

Archer KJ and Kimes RV. 2008. Empirical characterization of random forest variable importance measures. *Computational Statistics and Data Analysis* **52**, 2249–2260.

Borgonovo E. 2008. Differential importance and comparative statics: An application to inventory management. *International Journal of Production Economics* **111**, 170–179.

Borgonovo E and Peccati L. 2004. Sensitivity analysis in investment project evaluation. *International Journal of Production Economics* **90**, 17–25.

Coyle D, Buxton MJ, and O'Brien BJ. 2003. Measures of importance for economic analysis based on decision modeling. *Journal of Clinical Epidemiology* **56**, 989–997.

Finkelstein MS. 1994. Once more on measures of importance of system components. *Microelectronics and Reliability* **34**, 1431–1439.

Fussell JB. 1975. How to hand-calculate system reliability and safety characteristics. *IEEE Transactions on Reliability* **R-24**, 169–174.

Kuo W, Prasad VR, Tillman FA, and Hwang CL. 2006. *Optimal Reliability Design: Fundamentals and Applications*, 2nd edn. Cambridge University Press, Cambridge, UK.

Lambert HE. 1975. *Fault Trees for Decision Making in System Safety and Availability*. PhD Thesis, University of California, Berkeley.

Liaw A and Wiener M. 2002. Classification and regression by ramdomforest. *Resampling Methods in R: The Boot Package* **2**, 18–22.

Naoki S and Hiroshi M. 2004. Wireless charging system by microwave power transmission for electric motor vehicles. *IEICE Transactions on Electronics* **J87-C**, 433–443.

ReliaSoft. 2003. Using reliability importance measures to guide component improvement efforts. Available at http://www.maintenanceworld.com/Articles/reliasoft/usingreliability.html.

Saltelli A and Tarantola S. 2002. On the relative importance of input factors in mathematical models: Safety assessment for nuclear waste disposal. *Journal of American Statistical Association* **97**, 702–709.

Schwender H, Bowers K, Fallin M, and Ruczinski I. 2011. Importance measures for epistatic interactions in case-parent trios. *Annals of Human Genetics* **75**, 122–132.

Simmons D, Ellis N, Fujihara H, and Kuo W. 1998. *Software Measurement: A Visualization Toolkit for Project Control and Process Improvement*. Prentice Hall, New Jersey. 459 pages and CD software.

Smith CL and Borgonovo E. 2007. Decision making during nuclear power plant incidents – a new approach to the evaluation of precursor events. *Risk Analysis* **27**, 1027–1042.

Soofi ES, Retzer JJ, and Yasai-Ardekani M. 2000. A framework for measuring the importance of variables with applications to management research and decision models. *Decision Sciences* **31**, 595–625.

Strobl C, Boulesteix AL, Zeileis A, and Hothorn T. 2007. Bias in random forest variable importance measures: Illustrations, sources and a solution. *BMC Bioinformatics* **8**, 1–25.

Vesely WE. 1970. A time dependent methodology for fault tree evaluation. *Nuclear Engineering and Design* **13**, 337–360.

Zhang Z, Kulkarni A, Ma X, and Zhou Y. 2009. Memory resource allocation for file system prefetching: From a supply chain management perspective. *Proceedings of the EuroSys Conference*, pp. 75–88.

Zio E, Marella M, and Podofillini L. 2007. Importance measures-based prioritization for improving the performance of multi-state systems: Application to the railway industry. *Reliability Engineering and System Safety* **92**, 1303–1314.

2

Fundamentals of Systems Reliability

In the field of reliability, a component occupies a unique position in a system and may fail at any time by chance; thus, each component has two aspects: its position and its reliability. Correspondingly, a system has two properties: its structure and its reliability. The system structure (i.e., configuration) can normally be represented by a block diagram and expressed as a mathematical structure function. Typical system structures, to be introduced in this chapter, include series, parallel, parallel–series, series–parallel, bridge, k-out-of-n, and consecutive-k-out-of-n systems. The reliability of a system, a common performance measure of the system, depends on the system structure and the reliability of its components. Thus, the relative importance of the components for the system performance may differ due to their positions in the system and their reliability.

This chapter introduces background knowledge in systems reliability. This book always uses n to indicate the number of components in the system and N to denote the set of components in the system, $N = \{1, 2, \ldots, n\}$. Unless stated otherwise, assume that the components and the system are binary, that is, they have two possible states: functioning (i.e., working) or failed.

2.1 Block Diagrams

A reliability block diagram is often used to depict the logical relationship between the functioning of a system and the functioning of its components. As an example, a reliability diagram of a bridge structure is given in Figure 2.1. In a reliability block diagram, a circle or a rectangle is often used to represent a component. This book uses a circle to represent a component. Components can be identified by their names or component index $1, 2, \ldots, n$. The name of a component may also be given in the block. More often, however, only the index of each component is given in the block. The reliability block diagram in Figure 2.1 shows the five components in the bridge system (Kuo and Zuo 2003).

A reliability block diagram does not necessarily represent how the components are physically connected in the system. It only indicates how the functioning of the components ensures

Importance Measures in Reliability, Risk, and Optimization: Principles and Applications, First Edition.
Way Kuo and Xiaoyan Zhu. © 2012 John Wiley & Sons, Ltd. Published 2012 by John Wiley & Sons, Ltd.

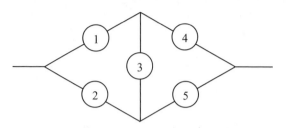

Figure 2.1   Bridge system

the functioning of the system. As shown in Figure 2.1, a reliability block diagram is best interpreted with the signal flowing through the components from left to right. A functioning component allows the signal to flow through it, while a failed one does not.

However, not all systems can be represented by reliability block diagrams. For example, the k-out-of-n system for $1 < k < n$, which functions if and only if at least k of the n components function, cannot be represented by a reliability block diagram without duplicating components. Sometimes, block diagrams become directed networks in which each edge has associated arrow(s) to show the direction(s) along which a functioning component allows the signal to flow through it. The system functions if the signal can flow though the directed networks; otherwise it fails. Another variation of the reliability block diagram is a network where an edge is used to represent a component. This book uses the block diagrams similar to Figure 2.1 and does not use the other variations.

2.2   Structure Functions

Assume that the system and each component are binary, that is, they are only in one of two possible states: functioning or failed. Let binary variable x_i indicate the state of component i for $i = 1, 2, \ldots, n$, and

$$x_i = \begin{cases} 1 & \text{if component } i \text{ functions} \\ 0 & \text{if component } i \text{ fails.} \end{cases}$$

Then, vector $\mathbf{x} = (x_1, x_2, \ldots, x_n)$ represents the states of all components and is known as the component state vector. Let ϕ represent the state of the system, and

$$\phi = \begin{cases} 1 & \text{if the system functions} \\ 0 & \text{if the system fails.} \end{cases}$$

The state of the system is completely determined by and is a deterministic function of the states of the components. Thus, it is often written as

$$\phi = \phi(\mathbf{x}) = \phi(x_1, x_2, \ldots, x_n),$$

where $\phi(\mathbf{x})$ is called the structure function of the system. Each unique system corresponds to a unique structure function $\phi(\mathbf{x})$. Thus, the terms "system" and "structure" are interchangeable.

Figure 2.2 Series system

Notation (N, ϕ) represents a system with the set of components N and the structure function ϕ, or sometimes, simply use "a system (or structure) ϕ."

Example 2.2.1 (Series systems) A series system functions if and only if every component functions. It fails whenever any component fails. The reliability block diagram of a series system is given in Figure 2.2. The structure function of a series system is given by

$$\phi(\mathbf{x}) = \prod_{i=1}^{n} x_i = \min\{x_1, x_2, \ldots, x_n\}.$$

Example 2.2.2 (Parallel systems) A parallel system fails if and only if all components fail. It functions as long as at least one component functions. The reliability block diagram of a parallel system is given in Figure 2.3. The structure function of a parallel system is given by

$$\phi(\mathbf{x}) = 1 - \prod_{i=1}^{n} (1 - x_i) = \max\{x_1, x_2, \ldots, x_n\}.$$

Example 2.2.3 (k-out-of-n systems) A k-out-of-n system functions if and only if at least k of the n components function for $1 \leq k \leq n$. The structure function is given by

$$\phi(\mathbf{x}) = \begin{cases} 1 & \text{if } \sum_{i=1}^{n} x_i \geq k \\ 0 & \text{if } \sum_{i=1}^{n} x_i < k. \end{cases}$$

When $k = 1$, it is a parallel system. When $k = n$, it is a series system.

Pivotal decomposition is a very useful technique in deriving the structure function of a system and in carrying through inductive proofs. This technique relies on the enumeration of

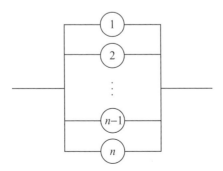

Figure 2.3 Parallel system

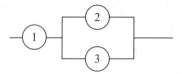

Figure 2.4 Component 1 in series with components 2 and 3

the states of a selected component. For any $i = 1, 2, \ldots, n$, the following equations can be used for any system structure function:

$$\phi(\mathbf{x}) = x_i \phi(1_i, \mathbf{x}) + (1 - x_i)\phi(0_i, \mathbf{x}) \tag{2.1}$$

$$= \phi(0_i, \mathbf{x}) + x_i[\phi(1_i, \mathbf{x}) - \phi(0_i, \mathbf{x})] \tag{2.2}$$

$$= \phi(1_i, \mathbf{x}) - (1 - x_i)[\phi(1_i, \mathbf{x}) - \phi(0_i, \mathbf{x})],$$

where $(\cdot_i, \mathbf{x}) = (x_1, x_2, \ldots, x_{i-1}, \cdot, x_{i+1}, \ldots, x_n)$.

In Equation (2.1), the structure function of a system with n components is expressed in terms of the structure functions of two different subsystems, each with $n-1$ components. In the first (second) subsystem, the state of component i is equal to 1 (0), while the states of the other $n-1$ components are random variables. Through repeated applications of Equation (2.1), the system can be eventually decomposed into subsystems whose structure functions are known (e.g., series or parallel structure). The selection of the components to be decomposed first in this process is critical in order to quickly find the structure function of the system. By repeated applications of Equation (2.1), structure function $\phi(\mathbf{x})$ can be expressed as

$$\phi(\mathbf{x}) = \sum_{\mathbf{y}} \left(\prod_{i=1}^{n} x_i^{y_i} (1 - x_i)^{1-y_i} \right) \phi(\mathbf{y}), \tag{2.3}$$

where the sum is extended over all n-dimensional binary vectors $\mathbf{y}$ and $0^0 \equiv 1$.

Example 2.2.4 For the three-component system in Figure 2.4, applying pivotal decomposition (2.1) on component 1, it is easy to derive the structure function as $\phi(\mathbf{x}) = x_1[1 - (1 - x_2)(1 - x_3)]$. Similarly, for the three-component system in Figure 2.5, its structure function is $\phi(\mathbf{x}) = x_1 + (1 - x_1)x_2x_3$.

For each system structure, Barlow and Proschan (1975b) defined a dual structure function, as in Definition 2.2.5. A structure and its dual have a close relationship.

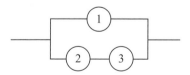

Figure 2.5 Component 1 in parallel with components 2 and 3

Definition 2.2.5 *Given a structure ϕ, define its dual ϕ^D by*

$$\phi^D(\mathbf{x}) = 1 - \phi(\mathbf{1} - \mathbf{x}), \tag{2.4}$$

where $\mathbf{1} - \mathbf{x} = (1 - x_1, 1 - x_2, \ldots, 1 - x_n)$.

Example 2.2.6   Consider the structure function of a series system $\phi(\mathbf{x})$,

$$\phi^D(\mathbf{x}) = 1 - \phi(\mathbf{1} - \mathbf{x}) = 1 - \prod_{i=1}^{n}(1 - x_i),$$

which is exactly the structure function for a parallel system. Thus, the dual structure of a series system of n components is a parallel system of n components. One can confirm that the dual structure of a parallel system is a series system. That is, the series and the parallel systems are dual of each other. More generally, the dual of a k-out-of-n structure is an $(n - k + 1)$-out-of-n structure. A general property for any structure function is shown in the following theorem.

Theorem 2.2.7   *The dual of the dual is the structure itself.*

Proof. Let ϕ be a system structure and ϕ^D be the dual of ϕ. Then

$$\phi^D(\mathbf{x}) = 1 - \phi(\mathbf{1} - \mathbf{x}), \text{ that is, } \phi^D(\mathbf{1} - \mathbf{x}) = 1 - \phi(\mathbf{x}).$$

Let φ be the dual of ϕ^D, then

$$\varphi(\mathbf{x}) = 1 - \phi^D(\mathbf{1} - \mathbf{x}) = 1 - (1 - \phi(\mathbf{x})) = \phi(\mathbf{x}).$$

Hence, the dual of ϕ^D is ϕ.

Components in a system may be structurally symmetric as defined in Definition 2.2.8 (Xie and Lai 1996). Let $\mathbf{x}^{(ij)}$ be the vector obtained by deleting x_i and x_j from $\mathbf{x}$ and $(\alpha_i, \beta_j, \mathbf{x}^{(ij)})$ be the vector with $x_i = \alpha, x_j = \beta$, and $\mathbf{x}^{(ij)}$ for others.

Definition 2.2.8   *Components i and j are structurally symmetric if $\phi(\mathbf{x})$ is permutation symmetric in x_i and x_j, that is, $\phi(\alpha_i, \beta_j, \mathbf{x}^{(ij)}) = \phi(\beta_i, \alpha_j, \mathbf{x}^{(ij)})$ for all $\mathbf{x}^{(ij)}$ and $\alpha, \beta \in \{0, 1\}$.*

Example 2.2.9 (Example 2.2.3 continued)   All components in a k-out-of-n system (including a series and a parallel system) are structurally symmetric.

Example 2.2.10 (Example 2.2.4 continued)   For the two three-component systems in Figures 2.4 and 2.5, it is intuitive and easy to verify through their structure functions in Example 2.2.4 that components 2 and 3 are structurally symmetric.

2.3   Coherent Systems

Definition 2.3.1   *Component i is irrelevant to the structure ϕ if ϕ is constant in x_i, that is, $\phi(1_i, \mathbf{x}) = \phi(0_i, \mathbf{x})$ for all $(\cdot_i, \mathbf{x})$. Otherwise, component i is relevant to the structure ϕ.*

If a component is relevant, it means that there exists at least one component state vector **x** such that the state of component i dictates the state of the system. The irrelevant components can never directly change the state of the system and are useless for the analysis of systems reliability; therefore, assume that all components are relevant.

Definition 2.3.2   *A system of components is coherent if (i) its structure function is nondecreasing and (ii) each component is relevant.*

Condition (i) assumes that the structure function of the system is a nondecreasing function of the state of every component, implying that $\phi(\mathbf{x}) \leq \phi(\mathbf{y})$ for any **x** and **y** such that $\mathbf{x} \leq \mathbf{y}$. This condition reflects the reality that the improvement of any component usually does not degrade the performance of the system. For example, replacing a failed component in a functioning system usually does not make the system fail. Of course, replacing a failed component in a failed system does not necessarily restore the functioning of the system because there may be other failed components in the system. Combining conditions (i) and (ii) in Definition 2.3.2, a coherent structure function must also satisfy $\phi(\mathbf{0}) = 0$ and $\phi(\mathbf{1}) = 1$. In other words, the system fails when all components fail and the system functions when all components function.

According to Definition 2.3.2, it is easy to show the following theorem.

Theorem 2.3.3   *If ϕ is a coherent structure, then its dual ϕ^D is a coherent structure, too.*

Barlow and Proschan (1975b) first defined the relevant components and the coherent systems as described in this section. For the rest of the book, assume that the system is coherent unless stated otherwise. Section 15.1 investigates noncoherent systems specifically.

2.4   Modules within a Coherent System

Intuitively, in a coherent system, a module is a subset of components that behaves like a "supercomponent." It has a single input from and a single output to the rest of the system. In the reliability block diagram, any cluster of components with one wire leading into it and one wire leading out of it can be treated as a module. A module itself is an *embedded system*. Definition 2.4.1 gives a formal definition of a module (Barlow and Proschan 1975b).

Definition 2.4.1   *The coherent system (M, χ) is a module (embedded system) of the coherent system (N, ϕ) if $\phi(\mathbf{x}) = \psi(\chi(\mathbf{x}^M), \mathbf{x}^{M^c})$, where ψ is a coherent structure function, M is a subset of N with complement M^c $(M^c = N \setminus M)$, and $\mathbf{x}^M$ is a vector with elements $x_i, i \in M$. The set $M \subseteq N$ is called a modular set of the system (N, ϕ).*

On the basis of this definition, each component and its indicator function constitute a module. *Modular decomposition* is a technique that can be used to decompose a coherent system into several disjoint modules. Such a decomposition is useful if both the structure function of each module and the structure function relating the system state to the states of these disjoint modules can be easily derived. Definition 2.4.2 gives a formal definition of modular decomposition (Barlow and Proschan 1975b).

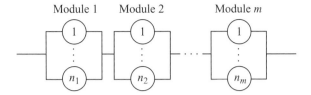

Figure 2.6 Series–parallel system

Definition 2.4.2 *A modular decomposition of a coherent system (N, ϕ) is a set of disjoint modules $\{(M_k, \chi_k)\}_{k=1}^{m}$ together with an organizing structure ψ, such that (i) $N = \bigcup_{k=1}^{m} M_k$, where $M_k \bigcap M_\ell = \emptyset, k \neq \ell$, and (ii)*

$$\phi(\mathbf{x}) = \psi \left[\chi_1(\mathbf{x}^{M_1}), \chi_2(\mathbf{x}^{M_2}), \ldots, \chi_m(\mathbf{x}^{M_m}) \right]. \tag{2.5}$$

Example 2.4.3 (Series–parallel and parallel–series systems) Series–parallel and parallel–series systems are basic examples of systems with modular decomposition, as shown in Figures 2.6 and 2.7, respectively. For a series–parallel system, the organizing structure is a series structure, and each module k is a parallel structure of n_k components such that $\sum_{k=1}^{m} n_k = n$. For a parallel–series system, the organizing structure is a parallel structure, and each module k is a series structure of n_k components such that $\sum_{k=1}^{m} n_k = n$.

In practice, a system may be decomposed into major modules and each major module may also be decomposed into smaller modules or components. The process is continued until each module can be easily dealt with. Modular decomposition is useful in determining the structure and the reliability functions of the system and in obtaining the bounds on the systems reliability. It is also helpful in determining the importance measures of components in the system by means of first determining the importance measures of components in the module and then the importance measure of the module in the system.

The contribution of all components in a module to the performance of the whole system must go through the module and can be represented by the state of the module. Of course, the state of the module is determined by the states of the components within the module. This property is not owned by an arbitrary subset of components. In addition, modular sets in a system are disjoint with each other; thus, their behavior is independent if components in the system are independent. However, a collection of arbitrary subsets of components may be statistically dependent because of replications of components.

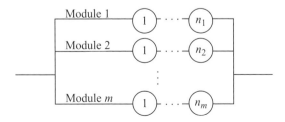

Figure 2.7 Parallel–series system

2.5   Cuts and Paths of a Coherent System

For the set of components $N = \{1, 2, \ldots, n\}$ and a component state vector $\mathbf{x}$, define $N_0(\mathbf{x}) = \{i \in N : x_i = 0\}$ and $N_1(\mathbf{x}) = \{i \in N : x_i = 1\}$. Note that a set S is said to be of size d if its cardinality equals d (i.e., $|S| = d$).

Definition 2.5.1   *A cut vector is a vector $\mathbf{x}$ such that $\phi(\mathbf{x}) = 0$. The corresponding cut is $N_0(\mathbf{x})$. A minimal cut vector is a cut vector $\mathbf{x}$ such that $\phi(\mathbf{y}) = 1$ for all $\mathbf{y} \geq \mathbf{x}$, $\mathbf{y} \neq \mathbf{x}$. The corresponding minimal cut is $N_0(\mathbf{x})$. A path vector is a vector $\mathbf{x}$ such that $\phi(\mathbf{x}) = 1$. The corresponding path is $N_1(\mathbf{x})$. A minimal path vector is a path vector $\mathbf{x}$ such that $\phi(\mathbf{y}) = 0$ for all $\mathbf{y} \leq \mathbf{x}$, $\mathbf{y} \neq \mathbf{x}$. The corresponding minimal path is $N_1(\mathbf{x})$.*

Note that if $\mathbf{x}$ is a cut vector and $\mathbf{y} \leq \mathbf{x}$, then $\phi(\mathbf{y}) = 0$. Similarly, if $\mathbf{x}$ is a path vector and $\mathbf{y} \geq \mathbf{x}$, then $\phi(\mathbf{y}) = 1$. In the view of physical meaning, a minimal cut is a minimal set of components whose failure causes the system to fail; a minimal path is a minimal set of components whose functioning ensures the functioning of the system. That is, a cut (path) set is minimal if it has no proper subset as a cut (path) set. For a series system, each component by itself forms a minimal cut, and the set of all components is a minimal path. For a parallel system, each component is a minimal path, and the set of all components is a minimal cut. Thus, a series system has n distinct minimal cuts and one minimal path, and a parallel system has n minimal paths and one minimal cut.

Let u and v represent the numbers of minimal cuts and minimal paths of structure ϕ, respectively. Let C_k for $k = 1, 2, \ldots, u$ and P_r for $r = 1, 2, \ldots, v$ indicate the kth minimal cut and the rth minimal path, respectively. Hwang (2005) stated some fundamental relationships between cuts and paths, as in Theorem 2.5.2. Before giving these relationships, first introduce the notation related to the cuts. Path-related notation $\mathscr{P}$, $\overline{\mathscr{P}}$, $\mathscr{P}_i$, $\overline{\mathscr{P}}_i$, $\mathscr{P}_{(i)}$, $\overline{\mathscr{P}}_{(i)}$, $\overline{\mathscr{P}}_i^{(j)}$, $\mathscr{P}(d)$, $\overline{\mathscr{P}}(d)$, $\mathscr{P}_i(d)$, $\overline{\mathscr{P}}_i(d)$, $\mathscr{P}_{(i)}(d)$, and $\overline{\mathscr{P}}_{(i)}(d)$ can be similarly defined.

Notation

- $\mathscr{C}$ ($\overline{\mathscr{C}} = \{C_k : k = 1, 2, \ldots, u\}$): set of (minimal) cuts
- $\mathscr{C}_i$ ($\overline{\mathscr{C}}_i = \{C_k : i \in C_k, k = 1, 2, \ldots, u\}$): set of (minimal) cuts containing component i
- $\mathscr{C}_{(i)}$ ($\overline{\mathscr{C}}_{(i)} = \{C_k : i \notin C_k, k = 1, 2, \ldots, u\}$): set of (minimal) cuts not containing component i
- $\overline{\mathscr{C}}_i^{(j)} = \{C_k : i \in C_k, j \notin C_k, k = 1, 2, \ldots, u\}$: set of minimal cuts containing component i but not component j
- $\mathscr{C}(d)$ ($\overline{\mathscr{C}}(d)$): set of (minimal) cuts of size d
- $\mathscr{C}_i(d)$ ($\overline{\mathscr{C}}_i(d)$): set of (minimal) cuts of size d that contain component i
- $\mathscr{C}_{(i)}(d)$ ($\overline{\mathscr{C}}_{(i)}(d)$): set of (minimal) cuts of size d that do not contain component i

Theorem 2.5.2   *The following properties are true for cuts and paths of a coherent system and for components $i = 1, 2, \ldots, n$.*

(i) S is a cut if and only if the complement $S^c = N \setminus S$ is not a path.
(ii) A path for ϕ is a cut for the dual ϕ^D, and vice versa.
(iii) A minimal path for ϕ is a minimal cut for the dual ϕ^D, and vice versa.

(iv) $|\mathscr{C}_i| + |\mathscr{P}_{(i)}| = 2^{n-1}$.
(v) $|\mathscr{P}_i| + |\mathscr{C}_{(i)}| = 2^{n-1}$.
(vi) $|\mathscr{C}_i(d)| + |\mathscr{P}_{(i)}(n-d)| = \binom{n-1}{d-1}$.
(vii) $|\mathscr{P}_i(d)| + |\mathscr{C}_{(i)}(n-d)| = \binom{n-1}{d-1}$.
(viii) $|\mathscr{P}_i| - |\mathscr{P}_{(i)}| = |\mathscr{C}_i| - |\mathscr{C}_{(i)}|$.

Proof. Part (i): Suppose to the contrary that S is a cut, and S^c is a path. Consider the case that a component functions if and only if it is in S^c. Then the system would simultaneously function, and fail, which is absurd.

Parts (ii) and (iii): They are straightforward from the definition of dual structure given in Equation (2.4) and the definitions of the cuts and paths.

Part (iv): For any of the 2^{n-1} subsets containing component i, either it is a cut in $\mathscr{C}_i$ or its complement is a path in $\mathscr{P}_{(i)}$ by part (i).

Part (v): Similar to part (iv).

Parts (vi) and (vii): Note that if a set is of size d, then its complement is of size $n-d$. Then, similar to parts (iv) and (v), part (i) implies the results in parts (vi) and (vii).

Part (viii): The result is obtained directly by parts (iv) and (v).

A system fails if and only if at least one minimal cut fails, and a system functions if and only if at least one minimal path functions. Thus, the system may be considered as a parallel structure with each minimal path as a component or as a series structure with each minimal cut as a component. On the basis of the definitions of minimal paths and minimal cuts, the components in each minimal path have a series structure, and the components in each minimal cut have a parallel structure. Note that different minimal paths (cuts) may have components in common.

Then, the reliability block diagram of any coherent system can be depicted in terms of minimal cuts or minimal paths as in Figure 2.8 or 2.9, respectively. Correspondingly, the structure function can be represented in terms of its minimal cuts or minimal paths (Barlow and Proschan 1975a):

$$\phi(\mathbf{x}) = \prod_{k=1}^{u} \left[1 - \prod_{i\in C_k}(1 - x_i) \right] = \min_{1\leq k\leq u} \max_{i\in C_k} x_i, \tag{2.6}$$

representing the structure as a series arrangement of the minimal cut parallel structures;

$$\phi(\mathbf{x}) = 1 - \prod_{r=1}^{v} \left[1 - \prod_{i\in P_r} x_i \right] = \max_{1\leq r\leq v} \min_{i\in P_r} x_i,$$

representing the structure as a parallel arrangement of the minimal path series structure. Thus, the set of minimal cuts (paths) of a coherent system completely determines its structure function, and vice versa. Kuo and Zuo (2003) presented methods for generating minimal cuts

Figure 2.8 Reliability block diagram in terms of minimal cuts

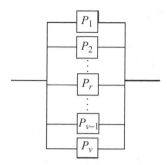

Figure 2.9   Reliability block diagram in terms of minimal paths

and minimal paths. As a result, minimal cuts and minimal paths can be used in deriving the structure of a coherent system and play a crucial role in analyzing the systems reliability and defining the importance measures.

According to Barlow and Iyer (1988) and Iyer (1992), any structure function ϕ has a unique form

$$\phi(\mathbf{x}) = \sum_{S \subseteq N} b_S \prod_{i \in S} x_i, \tag{2.7}$$

where the sum is taken over all subsets of N, and b_S is the signed domination of the structure $\phi(\mathbf{0}^{N \setminus S}, \mathbf{x}^S)$, in which all components not in S are assumed to fail. The signed domination is defined as follows and could be positive, zero, or negative.

Definition 2.5.3   *The signed domination of structure (N, ϕ) is the number of odd formations of N minus the number of even formations of N. A formation of N is a set of minimal paths whose union is N; it is odd (even) if the number of minimal paths in it is odd (even).*

2.6   Critical Cuts and Critical Paths of a Coherent System

Definition 2.6.1   *Component i is critical for coherent structure ϕ at the state vector $(\cdot_i, \mathbf{x})$ when $\phi(1_i, \mathbf{x}) - \phi(0_i, \mathbf{x}) = 1$ (implying $\phi(1_i, \mathbf{x}) = 1$ and $\phi(0_i, \mathbf{x}) = 0$). Furthermore, component i is said to be critical for the functioning of ϕ at $(0_i, \mathbf{x})$ (because $\phi(0_i, \mathbf{x}) = 0$ and the functioning of component i will restore the system to functioning) and to be critical for the failure of ϕ at $(1_i, \mathbf{x})$ (because $\phi(1_i, \mathbf{x}) = 1$ and the failure of component i will result in the failure of the system). Then, $(0_i, \mathbf{x})$ is a critical cut vector for component i, $N_0(0_i, \mathbf{x})$ the corresponding critical cut for component i, $(1_i, \mathbf{x})$ a critical path vector for component i, and $N_1(1_i, \mathbf{x})$ the corresponding critical path for component i.*

Component i is critical for structure ϕ at $(\cdot_i, \mathbf{x})$ in the sense that at the other component states determined by $(\cdot_i, \mathbf{x})$, the functioning of component i makes the system function and the failure of component i results in the failure of the system. That is, at the state $(\cdot_i, \mathbf{x})$, component i

can cause the failure or functioning of the system. Thus, component i is more important in this case than if $\phi(1_i, \mathbf{x}) = \phi(0_i, \mathbf{x}) = 1$ or $\phi(1_i, \mathbf{x}) = \phi(0_i, \mathbf{x}) = 0$. Essentially, component i is critical for system functioning because a minimal path of component i functions, with component i functioning last; component i is critical for system failure because a minimal cut of component i fails, with component i failing last. This concept of "critical" is used in defining several importance measures of components, as illustrated later.

Indeed, critical vectors of components can be found by minimal cuts and minimal paths as shown in Theorem 2.6.2 and Examples 2.6.3 and 2.6.4. Meng (1996) was the first to show that the condition in Theorem 2.6.2 is necessary for a vector to be critical for a component. We show that this condition is indeed both necessary and sufficient.

Theorem 2.6.2 $\phi(1_i, \mathbf{x}) - \phi(0_i, \mathbf{x}) = 1$, *that is, $(\cdot_i, \mathbf{x})$ is a critical vector for component i, if and only if there exists a minimal path P_r and a minimal cut C_k such that (i) $P_r \cap C_k = \{i\}$ and (ii) $x_j = 1$ for all $j \in P_r \backslash \{i\}$, and $x_j = 0$ for all $j \in C_k \backslash \{i\}$.*

Proof. ("Only if")   Suppose that $\phi(1_i, \mathbf{x}) = 1$ and $\phi(0_i, \mathbf{x}) = 0$. Let $S = N_1(1_i, \mathbf{x}) \backslash \{i\}$ and $H = N_0(0_i, \mathbf{x}) \backslash \{i\}$, then $\phi(1_i, \mathbf{1}^S, \mathbf{0}^H) = 1$ and $\phi(0_i, \mathbf{1}^S, \mathbf{0}^H) = 0$. Therefore, $\phi(1_i, \mathbf{1}^{S_1}, \mathbf{0}^{S_2}, \mathbf{0}^H) = 1$ for some minimal path $S_1 \cup \{i\}$, where $S_1 \subseteq S$ and $S_2 = S \backslash S_1$; $\phi(0_i, \mathbf{1}^S, \mathbf{1}^{H_2}, \mathbf{0}^{H_1}) = 0$ for some minimal cut $H_1 \cup \{i\}$, where $H_1 \subseteq H$ and $H_2 = H \backslash H_1$. Here, $(1_i, \mathbf{1}^S, \mathbf{0}^H)$ is a vector with $x_i = 1$, $x_j = 1$ for $j \in S$, and $x_j = 0$ for $j \in H$. Similar definitions are applied to vectors $(0_i, \mathbf{1}^S, \mathbf{0}^H)$, $(1_i, \mathbf{1}^{S_1}, \mathbf{0}^{S_2}, \mathbf{0}^H)$, $(0_i, \mathbf{1}^S, \mathbf{1}^{H_2}, \mathbf{0}^{H_1})$, and so on.

("If")   Now suppose that there exists a minimal path P_r and a minimal cut C_k such that $P_r \cap C_k = \{i\}$. Construct a state vector $(\cdot_i, \mathbf{x})$ such that $x_j = 1$ for all $j \in P_r \backslash \{i\}$ and $x_j = 0$ for all $j \in C_k \backslash \{i\}$, then $\phi(1_i, \mathbf{x}) = 1$ and $\phi(0_i, \mathbf{x}) = 0$.

Theorem 2.6.2 implies that if $(\cdot_i, \mathbf{x})$ is a critical vector for component i, then there exists at least one minimal path containing component i, P_r, such that $P_r \subseteq N_1(1_i, \mathbf{x})$, as well as a minimal cut containing component i, C_k, such that $C_k \subseteq N_0(0_i, \mathbf{x})$; otherwise, the failure of component i will not cause the system to fail and the functioning of component i will not restore the system to functioning.

However, $N_1(1_i, \mathbf{x})$ itself is not necessarily a minimal path because the failure of the components in $N_1(\cdot_i, \mathbf{x})$ may not take the system down. Similarly, $N_0(0_i, \mathbf{x})$ is not necessarily a minimal cut. Thus, the critical paths or critical cuts are not equivalent to minimal paths or minimal cuts. On the other hand, no minimal path not containing component i is in $N_1(\cdot_i, \mathbf{x})$; otherwise, the failure of component i would not cause the failure of the system. Similarly, no minimal cut not containing component i is in $N_0(\cdot_i, \mathbf{x})$; otherwise, the functioning of component i would not restore the system to functioning.

Example 2.6.3   For the system in Figure 2.10, $\phi(\mathbf{x}) = 1 - (1 - x_1 x_2 x_3)(1 - x_1 x_2 x_4)(1 - x_2 x_5)$. The minimal paths and minimal cuts of the system are $P_1 = \{1, 2, 3\}$, $P_2 = \{1, 2, 4\}$, and $P_3 = \{2, 5\}$; $C_1 = \{2\}$, $C_2 = \{1, 5\}$, and $C_3 = \{3, 4, 5\}$. With regard to component 4, the minimal paths and minimal cuts that contain component 4 are P_2 and C_3. Thus, there is exactly one critical vector for component 4: $(1_1, 1_2, 0_3, \cdot_4, 0_5)$.

Example 2.6.4   Consider the bridge structure in Figure 2.1. The minimal paths and minimal cuts of the system are $P_1 = \{1, 4\}$, $P_2 = \{2, 5\}$, $P_3 = \{1, 3, 5\}$, and $P_4 = \{2, 3, 4\}$; $C_1 = \{1, 2\}$,

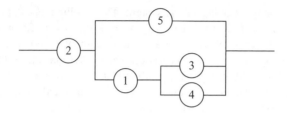

Figure 2.10 System of five components with component 2 in series with others

$C_2 = \{4, 5\}$, $C_3 = \{1, 3, 5\}$, and $C_4 = \{2, 3, 4\}$. The minimal paths and minimal cuts that contain component 3 are P_3, P_4, C_3, and C_4. Hence, there are two critical vectors for component 3: $(1_1, 0_2, \cdot_3, 0_4, 1_5)$ and $(0_1, 1_2, \cdot_3, 1_4, 0_5)$.

Let

$$m_i = \sum_{\{(0_i, \mathbf{x})\}} [\phi(1_i, \mathbf{x}) - \phi(0_i, \mathbf{x})]$$

denote the total number of critical cuts (cut vectors) for component i, and

$$m_i' = \sum_{\{(1_i, \mathbf{x})\}} [\phi(1_i, \mathbf{x}) - \phi(0_i, \mathbf{x})]$$

denote the total number of critical paths (path vectors) for component i, $i = 1, 2, \ldots, n$. Note that $\phi(1_i, \mathbf{x}) - \phi(0_i, \mathbf{x}) = 1$ implies simultaneously that $(0_i, \mathbf{x})$ is a critical cut vector and $(1_i, \mathbf{x})$ is a critical path vector. Thus, apparently,

$$m_i = m_i'. \tag{2.8}$$

A critical cut (path) vector $(0_i, \mathbf{x})$ $((1_i, \mathbf{x}))$ for component i is of size d if $|N_0(0_i, \mathbf{x})| = d$ ($|N_1(1_i, \mathbf{x})| = d$) for $1 \leq d \leq n$. Let $m_i(d)$ and $m_i'(d)$ denote, respectively, the numbers of critical cut vectors and of critical path vectors for component i of size d. Then, $m_i = \sum_{d=1}^{n} m_i(d)$, and $m_i' = \sum_{d=1}^{n} m_i'(d)$. Note that

$$m_i(d) = m_i'(n + 1 - d). \tag{2.9}$$

Theorem 2.6.5 *The number of critical cuts (critical paths) for component i equals the difference in the number of minimal cuts (paths) containing component i and the one not containing component i, and the number of critical cuts equals the number of critical paths. That is,*

$$m_i = |\mathscr{C}_i| - |\mathscr{C}_{(i)}| = |\mathscr{P}_i| - |\mathscr{P}_{(i)}| = m_i'.$$

Proof. First, show that $|\mathscr{P}_{(i)}| + m_i' \leq |\mathscr{P}_i|$. Let an arbitrary minimal path $P \in \mathscr{P}_{(i)}$, then $P \cup \{i\} \in \mathscr{P}_i$. Let $(\cdot_i, \mathbf{x})$ be a critical vector for component i, then $\phi(1_i, \mathbf{x}) = 1$ and $N_1(1_i, \mathbf{x}) \in \mathscr{P}_i$. Note that $N_1(\cdot_i, \mathbf{x}) \notin \mathscr{P}_{(i)}$; otherwise, $\phi(0_i, \mathbf{x}) \neq 0$. Thus, $|\mathscr{P}_{(i)}| + m_i' \leq |\mathscr{P}_i|$.

Now show that $|\mathscr{P}_i| \le |\mathscr{P}_{(i)}| + m_i'$. Let $P \in \mathscr{P}_i$, then $\phi(1_i, \mathbf{1}^{P \setminus \{i\}}, \mathbf{0}^{N \setminus P}) = 1$. If $\phi(0_i, \mathbf{1}^{P \setminus \{i\}}, \mathbf{0}^{N \setminus P}) = 1$, then $P \setminus \{i\} \in \mathscr{P}_{(i)}$. Otherwise, if $\phi(0_i, \mathbf{1}^{P \setminus \{i\}}, \mathbf{0}^{N \setminus P}) = 0$, then $(1_i, \mathbf{1}^{P \setminus \{i\}}, \mathbf{0}^{N \setminus P})$ is a critical path vector for component i. Thus, $|\mathscr{P}_i| \le |\mathscr{P}_{(i)}| + m_i'$.

Hence $m_i' = |\mathscr{P}_i| - |\mathscr{P}_{(i)}|$. A similar argument can prove that $m_i = |\mathscr{C}_i| - |\mathscr{C}_{(i)}|$, and the proof is completed by noting, as in Equation (2.8), that $m_i = m_i'$.

2.7 Measures of Performance

Any component that is expected to function properly may fail at any time by chance. The length of time that a new functioning component continues to function properly is defined as its *lifetime* or *failure time*. Since a system consists of a set of components, it may fail at any time with a certain probability, and its lifetime is also a random variable. This section presents three major measures of system performance. The other measures of system performance are described when they are introduced.

2.7.1 Reliability for a Mission Time

Many systems of interest have a mission time, which is defined to be the period of time during which a system is required to function properly. For example, an airplane making an overseas flight is required to function properly for the entire duration of the flight, including takeoff and landing. For a system with a specified mission time, reliability is often used to measure performance. The reliability of a system, denoted by R, is defined to be the probability that it functions properly during the mission time. Note that the mission time, which is usually implicit and fixed, is not a random variable.

Let X_i be a random variable, and

$$X_i = \begin{cases} 1 & \text{if component } i \text{ functions} \\ 0 & \text{if component } i \text{ fails,} \end{cases}$$

with $\Pr\{X_i = 1\} = p_i = \mathbb{E}(X_i)$, for $i = 1, 2, \ldots, n$, where $\mathbb{E}(\cdot)$ denotes the expected value of a random variable. The probability that component i functions, p_i, is referred to as the reliability of component i, and $q_i = 1 - p_i$ as the unreliability of component i. Assume that random variables X_i, $i = 1, 2, \ldots, n$, are mutually statistically independent unless stated otherwise.

Introducing $\mathbf{X} = (X_1, X_2, \ldots, X_n)$ and $\mathbf{p} = (p_1, p_2, \ldots, p_n)$, reliability of system ϕ is given by

$$R = \Pr\{\phi(\mathbf{X}) = 1\} = \mathbb{E}(\phi(\mathbf{X})).$$

Under the assumption that components are mutually statistically independent, the reliability of system ϕ is a function of the reliability of the components,

$$R = R_\phi(\mathbf{p}) \quad \text{(or simply } R(\mathbf{p})),$$

which is referred to as the reliability function of the structure ϕ. When $p_1 = p_2 = \ldots = p_n = p$, the symbol $R(p)$ is used where p is a scalar. $R(\mathbf{p})$ reflects the unique relation between the reliability of each distinct system structure and the reliability of the components.

Example 2.7.1 (Examples 2.2.1, 2.2.2, and 2.2.3 continued)   The series structure $\phi(\mathbf{x}) = \prod_{i=1}^{n} x_i$ in Example 2.2.1 has the reliability function

$$R(\mathbf{p}) = \prod_{i=1}^{n} p_i.$$

The parallel structure $\phi(\mathbf{x}) = 1 - \prod_{i=1}^{n}(1 - x_i)$ in Example 2.2.2 has the reliability function

$$R(\mathbf{p}) = 1 - \prod_{i=1}^{n}(1 - p_i).$$

Assuming $p_1 = p_2 = \ldots = p_n = p$, the k-out-of-n system, in which $\phi(\mathbf{x}) = 1$ if and only if $\sum_{i=1}^{n} x_i \geq k$ as in Example 2.2.3, has the reliability function

$$R(p) = \sum_{i=k}^{n} \binom{n}{i} p^i (1 - p)^{n-i}.$$

Recall that the structure function of a coherent system is nondecreasing. The corresponding monotonicity property for reliability functions is given in Theorem 2.7.2 (Barlow and Proschan 1975b). It also shows that $R(\mathbf{p})$ is multilinear, that is, linear in each p_i (see also Section 2.10). Moreover, when $p_1 = p_2 = \ldots = p_n = p$, $R(p)$ is a polynomial function in p.

Theorem 2.7.2   *Let $R(\mathbf{p})$ be the reliability function of a coherent structure. Then, for $i = 1, 2, \ldots, n$,*

$$R(\mathbf{p}) = p_i h(1_i, \mathbf{p}) + (1 - p_i)R(0_i, \mathbf{p}), \tag{2.10}$$

where $(\cdot_i, \mathbf{p}) = (p_1, p_2, \ldots, p_{i-1}, \cdot_i, p_{i+1}, \ldots, p_n)$. Moreover, $R(\mathbf{p})$ is strictly increasing in each p_i for $0 < p_i < 1$, $i = 1, 2, \ldots, n$.

Proof. Using pivotal decomposition (2.1) and the independence of components,

$$R(\mathbf{p}) = \mathbb{E}(\phi(\mathbf{X})) = \mathbb{E}(X_i)\mathbb{E}(\phi(1_i, \mathbf{X})) + (1 - \mathbb{E}(X_i))\mathbb{E}(\phi(0_i, \mathbf{X})).$$

Equation (2.10) follows immediately. From Equation (2.10),

$$\frac{\partial R(\mathbf{p})}{\partial p_i} = R(1_i, \mathbf{p}) - R(0_i, \mathbf{p}) \tag{2.11}$$

so that

$$\frac{\partial R(\mathbf{p})}{\partial p_i} = \mathbb{E}(\phi(1_i, \mathbf{X}) - \phi(0_i, \mathbf{X})). \tag{2.12}$$

Because ϕ is nondecreasing, then $\phi(1_i, \mathbf{x}) - \phi(0_i, \mathbf{x}) \geq 0$. In addition, $\phi(1_i, \mathbf{x}_0) - \phi(0_i, \mathbf{x}_0) = 1$ for some $\mathbf{x}_0$ because each component is relevant. Since $0 < p_i < 1$ for all $i = 1, 2, \ldots, n$, $\mathbf{x}_0$ has a positive probability of occurring. Thus, $\mathbb{E}(\phi(1_i, \mathbf{X}) - \phi(0_i, \mathbf{X})) > 0$, and the result follows.

By summing over all 2^n vectors $\mathbf{x}$ with 0 or 1 elements, $R(\mathbf{p})$ can be calculated as

$$R(\mathbf{p}) = \sum_{\mathbf{x}} \left(\prod_{i=1}^{n} p_i^{x_i} (1 - p_i)^{1-x_i} \right) \phi(\mathbf{x}), \tag{2.13}$$

which is similar to Equation (2.3) for determining the system's structure function. When all components have the same reliability p, Equation (2.13) implies the reliability polynomial

$$R(p) = \sum_{\mathbf{x}} \phi(\mathbf{x}) p^{\sum_{i=1}^{n} x_i} (1 - p)^{n - \sum_{i=1}^{n} x_i}.$$

Apart from this direct calculation method, some other methods have been proposed in the literature for calculating the exact value of systems reliability $R(\mathbf{p})$, as well as its lower and upper bounds. The exact calculation of reliability of a complex system is usually a formidable task. For this reason, much research has focused on the bounds on systems reliability. Readers are referred to Barlow and Proschan (1975a) and Kuo and Zuo (2003) for various computation methods and bounds.

If a coherent system (N, ϕ) has a modular decomposition $\{(M_k, \chi_k)\}_{k=1}^{m}$ together with an organizing structure ψ, the reliability function R_ϕ can be expressed in terms of the reliability of modules as

$$R_\phi(\mathbf{p}) = R_\psi \left(R_{\chi_1}(\mathbf{p}^{M_1}), R_{\chi_2}(\mathbf{p}^{M_2}), \ldots, R_{\chi_m}(\mathbf{p}^{M_m}) \right), \tag{2.14}$$

where $\mathbf{p}^{M_k}$ denotes the vector with elements p_i for $i \in M_k$ and $k = 1, 2, \ldots, m$. Equation (2.14) is similar to Equation (2.5) for structure function ϕ.

2.7.2   Reliability Function (of Time t)

The reliability measure defined and illustrated in Subsection 2.7.1 does not deal directly with the time factor, in which the required service time is implicit and fixed. Thus, time t does not appear in the equations of systems reliability evaluation. However, in many practical applications, no finite mission time is specified in advance. For example, a certain kind of service such as supply of electronic power is needed for a long time or indefinitely. In these situations, the lifetime of the system is of interest.

Let

$$X_i(t) = \begin{cases} 1 & \text{if component } i \text{ functions at time } t \\ 0 & \text{if component } i \text{ fails at time } t. \end{cases}$$

Introduce $\mathbf{X}(t) = (X_1(t), X_2(t), \ldots, X_n(t))$. Assuming that the stochastic processes $\{X_i(t), t \geq 0\}$, $i = 1, 2, \ldots, n$, are mutually independent, then the system structure function

$$\phi(\mathbf{X}(t)) = \begin{cases} 1 & \text{if the system functions at time } t \\ 0 & \text{if the system fails at time } t. \end{cases}$$

Let random variable T_i be the lifetime of component i, $i = 1, 2, \ldots, n$. Assuming that component i has an absolutely continuous lifetime distribution $F_i(t)$ with density $f_i(t)$, then the reliability (survival) function of component i at time t is given by

$$\overline{F}_i(t) = 1 - F_i(t) = \Pr\{T_i > t\} = \Pr\{X_i(t) = 1\},$$

which is the probability that component i survives beyond a time duration of length $t, t > 0$. Introduce $\mathbf{T} = (T_1, T_2, \ldots, T_n)$ and $\overline{\mathbf{F}}(t) = (\overline{F}_1(t), \overline{F}_2(t), \ldots, \overline{F}_n(t))$. The *failure rate function* (*hazard rate*) of component i is defined as

$$r_i(t) = \frac{f_i(t)}{\overline{F}_i(t)}.$$

Therefore,

$$\overline{F}_i(t) = \exp\left(-\int_0^t r_i(u)du\right). \tag{2.15}$$

A commonly used model in reliability is to assume that the lifetimes of components have proportional hazards, that is,

$$\overline{F}_i(t) = e^{-\lambda_i h(t)}, \ t \geq 0, \ i = 1, 2, \ldots, n, \tag{2.16}$$

where $\lambda_i > 0$ is a parameter and $h(t)$ is a common function of t. Then, the hazard rate of component i is

$$r_i(t) = \lambda_i h'(t),$$

where $h'(t)$ is the derivative of function $h(t)$. When $h(t) = t$ and so $h'(t) = 1$, then the components have exponential lifetime distributions and constant hazard rates. When $h(t) = t^\alpha$, $\alpha > 0$, the lifetimes of the components are Weibull distributed with the same shape parameter α. The Weibull distribution with shape parameter α and scale parameter λ_i is

$$\overline{F}_i(t) = e^{-(\lambda_i t)^\alpha}, \ \alpha > 0, \lambda_i > 0, t \geq 0, i = 1, 2, \ldots, n. \tag{2.17}$$

Another commonly used lifetime distribution is the Gamma distribution, which has a density with shape parameter α and scale parameter λ_i:

$$f_i(t) = \frac{\lambda_i^\alpha t^{\alpha-1} e^{-\lambda_i t}}{\Gamma(\alpha)}, \ \alpha > 0, \lambda_i > 0, t \geq 0, i = 1, 2, \ldots, n, \tag{2.18}$$

where $\Gamma(\alpha)$ is the Gamma function of α. If α is a positive integer, then

$$\overline{F}_i(t) = \sum_{k=0}^{\alpha-1} \frac{(\lambda_i t)^k}{k!} e^{-\lambda_i t}, \ \lambda_i > 0, t \geq 0, i = 1, 2, \ldots, n. \tag{2.19}$$

Let T_ϕ be a random variable representing the system lifetime and $\tau(\mathbf{T})$ be the function of the system lifetime, which is a function of component lifetimes, and

$$T_\phi = \tau(\mathbf{T}).$$

Let $F_\phi(t)$ be an absolutely continuous lifetime distribution of the system. Then, the reliability (survival) of the system at time t is given by

$$\overline{F}_\phi(t) = 1 - F_\phi(t) = \Pr\{T_\phi > t\} = \Pr\{\phi(\mathbf{X}(t)) = 1\} = \mathbb{E}(\phi(\mathbf{X}(t))) = R(\overline{\mathbf{F}}(t)).$$

The systems reliability distribution $\overline{F}_\phi(t)$ is defined to be the probability that a system survives beyond a time duration of length t, $t > 0$. $R(\overline{\mathbf{F}}(t))$ is the systems reliability function of component lifetime distributions. When the lifetimes of components are known, $R(\overline{\mathbf{F}}(t))$ is a function of t and equivalent to $\overline{F}_\phi(t)$. $R(\overline{\mathbf{F}}(t))$ can be easily obtained from $R(\mathbf{p})$ by replacing $\mathbf{X}$ with $\mathbf{X}(t)$ and $\mathbf{p}$ with $\overline{\mathbf{F}}(t)$ when the components are independent and nonrepairable, and thus the system as well is nonrepairable.

Other functions characterizing the lifetime of a system include its probability density function $f_\phi(t)$, its failure rate function $r_\phi(t)$, and its expected lifetime $\mathbb{E}(T_\phi)$. They can be derived as follows:

$$f_\phi(t) = \frac{\mathrm{d}F_\phi(t)}{\mathrm{d}t} = -\frac{\mathrm{d}R(\overline{\mathbf{F}}(t))}{\mathrm{d}t},$$

$$r_\phi(t) = \frac{f_\phi(t)}{R(\overline{\mathbf{F}}(t))},$$

$$\mathbb{E}(T_\phi) = \int_0^\infty \overline{F}_\phi(t)\mathrm{d}t.$$

2.7.3 Availability Function

For repairable components and systems, availability is often used as a measure of performance. Let $A_i(t)$ be the availability of component i at time t, that is, the probability that component i functions at time t. Then, the unavailability $\overline{A}_i(t) = 1 - A_i(t)$. Consider a repairable system in which some or all components can be repaired after failure and can be reused in the system after repair. A system may function properly for a period of time and eventually fail. Once it fails, repairs and overhauls may be performed to the system to restore it to a functioning state again. Thus, the state of the system may change between 1 and 0 several times before it is scrapped. The probability that the system functions at time t is called its availability function and is given by

$$A_\phi(t) = \Pr\{\phi(\mathbf{X}(t)) = 1\}.$$

For a repairable system, $\{\phi(\mathbf{X}(t)) = 1\}$ indicates that the system is in the functioning state at time t. It does not say anything about the states that the system has experienced before time t. It may go through several functioning and failure cycles before staying in the functioning state at time t. If the system is expected to provide service at a specific time point t, $A_\phi(t)$ is a good indicator of its availability at time t. For a nonrepairable system, the availability function equals the reliability function.

The availability function is usually difficult to derive. Instead, the long-term average availability of the system, namely, the *steady-state availability* of the system is tractable if the steady state of a system exists. For the system in steady state, the mean time between failures (MTBF) is defined to be the average continuous functioning duration of the system and the mean time to repair (MTTR) to be the average amount of time needed to repair a failed system. The steady-state availability of the system, A_ϕ, can then be expressed as

$$A_\phi = \frac{\mathrm{MTBF}}{\mathrm{MTBF} + \mathrm{MTTR}}.$$

A repair can be classified according to its extent. If a repair restores a component to "as good as new" condition, it is a perfect repair; for example, a new replacement is used. Another type of repair is a minimal repair, which simply restores the component to the functioning state that is "as bad as old" condition. In other words, the failure rate of the component right after the repair is the same as that just before the failure; thus, a minimal repair does not change the health condition of a component. An imperfect repair is used to describe a situation that is neither "as good as new" nor "as bad as old"; instead, the component's health condition following the repair is somewhere in the middle. Thus, the lifetime distribution of a repaired system depends on the extent of the repair and the condition of the system before the failure. Several importance measures for investigating the effect of different types of repair on upgrading system performance are introduced later.

2.8   Stochastic Orderings

Three different stochastic orderings are defined as follows. These definitions apply equally when a pair of random variables (X, Y) are discrete or continuous.

Definition 2.8.1   *The random variable X is less than the random variable Y in the stochastic ordering (denoted by $X \leq^{st} Y$) if their respective survival functions satisfy $\overline{F}^X(t) \leq \overline{F}^Y(t)$ for all t.*

Definition 2.8.2   *The random variable X is less than the random variable Y in the hazard rate (or uniform stochastic) ordering (denoted by $X \leq^{hr} Y$) if the ratio of survival functions $\overline{F}^X(t)/\overline{F}^Y(t)$ is nondecreasing in t.*

Definition 2.8.3   *The random variable X is less than the random variable Y in the likelihood ratio ordering (denoted by $X \leq^{lr} Y$) if the ratio $f^Y(t)/f^X(t)$ is nondecreasing in t, where $f^X(t)$ and $f^Y(t)$ represent the density or probability mass functions of X and Y, respectively.*

Suppose that two random variables (X, Y) are discrete and their discrete distributions are $\mathbf{f}^X$ and $\mathbf{f}^Y$ on the set $\{1, 2, \ldots, n\}$. Then, the stochastic ordering of $X \leq^{st} Y$ is also equivalently denoted by $\mathbf{f}^X \leq^{st} \mathbf{f}^Y$; the similar notation is applied for the other two orderings. For this discrete case, the earlier definitions can be specialized as follows:

(i) $\mathbf{f}^X \leq^{st} \mathbf{f}^Y$ if and only if $\sum_{j=i}^{n} f_j^X \leq \sum_{j=i}^{n} f_j^Y$ for all $i = 1, 2, \ldots, n$;

(ii) $\mathbf{f}^X \leq^{hr} \mathbf{f}^Y$ if and only if $\sum_{j=i}^{n} f_j^Y / \sum_{j=i}^{n} f_j^X$ is nondecreasing in i;

(iii) $\mathbf{f}^X \leq^{lr} \mathbf{f}^Y$ if and only if f_j^Y/f_j^X is nondecreasing in j, that is, $f_j^X f_i^Y \leq f_i^X f_j^Y$ for all $1 \leq i < j \leq n$.

Note that they are readily extended to any discrete sets in addition to the set $\{i = 1, 2, \ldots, n\}$.

It is well known that likelihood ratio ordering is the most stringent of these three orderings and that hazard rate ordering is more stringent than the stochastic ordering. It is also well known that if two random variables $X \leq^{st} Y$, then $\mathbb{E}(X) \leq \mathbb{E}(Y)$.

2.9   Signature of Coherent Systems

Structure functions are algebraic expressions in one-to-one correspondence with coherent systems, providing a way of indexing and comparing the systems. However, since the number of n-component coherent systems grows exponentially with n, indexing systems by means of their structure functions tends to be of limited usefulness in problems involving comparisons or optimization among systems. This section introduces an alternative index, system signature, a probability vector that has the virtues of being both quite manageable and easily interpreted but less general than a structure function. Its precise meaning is specified in Definition 2.9.1. Recall that T_ϕ is the lifetime of system ϕ.

Definition 2.9.1   *Assume that the lifetimes of n components in a coherent system ϕ are independent and identically distributed (i.i.d.). The signature of system ϕ, denoted by $\mathbf{s}_\phi$ or simply by $\mathbf{s}$, is an n-dimensional probability vector whose jth element s_j is the probability that the jth component failure causes system failure. In brief,*

$$s_j = \Pr\{T_\phi = X_{(j)}\},$$

where $X_{(j)}$ is the jth order statistic of the n component failure times representing the jth smallest component lifetime, that is, the time of the jth component failure, $j = 1, 2, \ldots, n$.

System signatures are used primarily in comparing the structures (i.e., designs) of the systems being of the same number of components. The system signature, as a certain probability vector, can be well defined without the i.i.d. assumption on component lifetimes, but the assumption is nonetheless made so that there is a fair basis for comparing the differences in system performance attributed to the system structures. It should be noted that a comparison between two systems with quite different component characteristics may be either misleading or inconclusive. From an analytical point of view, the system signatures defined under the i.i.d. assumption have three advantages. They allow one to utilize the tools of combinatorial mathematics for the calculation of system characteristics. Also, the well-known distribution theory for the order statistics of an i.i.d. sample from a continuous component lifetime distribution is available for studying the performance of a system with a given signature. Finally, system signatures depend only on the permutation distribution of the n observed failure times and do not depend on the underlying lifetime distribution. Therefore, the system signature can be viewed as a pure measure of the structure of a system, and is, therefore, also referred to as the structural signature (Navarro et al. 2010). In turn, the signature-based analysis may be inexact in the applications in which components are neither independent nor identically distributed, but the relative performance of systems under the i.i.d. assumption can still provide worthwhile information about system quality.

The computation of system signatures is essentially combinatorial mathematics. In terms of the orderings of the component lifetimes $T_1, T_2, \ldots, T_n$, the system signature can be computed as the probability vector with elements:

$$s_j = \frac{\text{Number of orderings for which the } j\text{th failure causes system failure}}{n!},$$

for $j = 1, 2, \ldots, n$. Since the T_i, $i = 1, 2, \ldots, n$, are assumed to be i.i.d., the $n!$ permutations of these n distinct failure times are equally likely. In addition, $\sum_{j=1}^{n} s_j = 1$. The vector $\mathbf{s}$ is most easily obtained from the cut representation of system lifetime, namely,

$$T_\phi = \min_{1 \le k \le u} \max_{i \in C_k} T_i, \qquad (2.20)$$

where $C_1, C_2, \ldots, C_u$ are the minimal cuts of the system (see also Equation (2.6)). The following example illustrates the computation of the system signatures.

Example 2.9.2 For the system in Figure 2.4, the minimal cuts of this system are $C_1 = \{1\}$ and $C_2 = \{2, 3\}$. By Equation (2.20),

$$T_\phi = \min\{T_1, \max\{T_2, T_3\}\}.$$

The order statistic equivalent to T_ϕ is shown as follows for each of the 3! orderings of the component lifetimes:

Ordering of component failure times	T_ϕ
$T_1 < T_2 < T_3$	$X_{(1)}$
$T_1 < T_3 < T_2$	$X_{(1)}$
$T_2 < T_1 < T_3$	$X_{(2)}$
$T_2 < T_3 < T_1$	$X_{(2)}$
$T_3 < T_1 < T_2$	$X_{(2)}$
$T_3 < T_2 < T_1$	$X_{(2)}$

Since each ordering is equally likely, the signature of the system in Figure 2.4 is $\mathbf{s} = (1/3, 2/3, 0)$.

Theorem 2.9.3 (Kochar et al. 1999; Samaniego 2007) establishes a fundamental property of system signature $\mathbf{s}$ that the distribution of the system lifetime T_ϕ, given i.i.d. components with lifetime distribution $F(t), t \ge 0$, can be expressed as a function of $\mathbf{s}$ and $F(t)$ alone. It is worth mentioning that the representation in Equation (2.21) holds under the less stringent assumption that the lifetimes of n components are interchangeable. Samaniego (2007) and Triantafyllou and Koutras (2008) presented more results between the system lifetime and system signature that can be induced from Theorem 2.9.3.

Theorem 2.9.3 *Let an n-component coherent system ϕ be with signature $\mathbf{s}$ and the components' lifetimes be i.i.d. according to the (continuous) distribution $F(t), t \ge 0$. Then*

$$\overline{F}_\phi(t) = \Pr\{T_\phi > t\} = \sum_{j=1}^{n} s_j \sum_{i=0}^{j-1} \binom{n}{i} (F(t))^i (\overline{F}(t))^{n-i}. \qquad (2.21)$$

Proof. First note that the system fails concurrently with the failure of one of its components so that $T_\phi \in \{X_{(1)}, X_{(2)}, \ldots, X_{(n)}\}$ with probability 1. Then, utilizing the Law of Total Probability and the i.i.d. assumption on component lifetimes,

$$\Pr\{T_\phi > t\} = \sum_{j=1}^{n} \Pr\{T_\phi > t, T_\phi = X_{(j)}\}$$

$$= \sum_{j=1}^{n} \Pr\{T_\phi > t | T_\phi = X_{(j)}\} \Pr\{T_\phi = X_{(j)}\}$$

$$= \sum_{j=1}^{n} s_j \Pr\{X_{(j)} > t\}$$

$$= \sum_{j=1}^{n} s_j \sum_{i=0}^{j-1} \binom{n}{i} (F(t))^i (\overline{F}(t))^{n-i}.$$

As is clear from Equation (2.21), the lifetime of a coherent system with i.i.d. components depends on the structure of the system only through the signature. Indeed, if two systems have the same signature (which is possible), the stochastic behavior of their lifetimes is identical. It is easily verified that the four-component system with minimal cuts $\{\{1, 2\}, \{2, 4\}, \{3, 4\}\}$ has the same signature as the four-component system with minimal cuts $\{\{1, 2\}, \{1, 3\}, \{1, 4\}, \{2, 3, 4\}\}$. The system signature $\mathbf{s}$ serves as a compact but complete summary of the structure function ϕ, and also eliminates the duplication inherent in different structure functions whose impact on the distribution of system lifetime is identical. Theorem 2.9.4 (Kochar et al. 1999) gives a further simplification, showing that the signature of a given system can be obtained from the signature of its dual system without further computation.

Theorem 2.9.4 *Let* $\mathbf{s}$ *be the signature of a coherent system* ϕ, *whose* n *components have i.i.d. lifetimes, and let* $\mathbf{s}^D$ *be the signature of its dual system* ϕ^D. *Then, for* $j = 1, 2, \ldots, n$,

$$s_j = s_{n-j+1}^{D}.$$

Samaniego (2007) has given an up-to-date review on system signatures and presented their applications in various reliability problems in which the primary interest is in comparing the performance of two coherent systems. For this purpose, Theorem 2.9.5 (Kochar et al. 1999) provides three preservation rules that are based on three different stochastic orderings, which are defined in Section 2.8.

Theorem 2.9.5 *Let* $\mathbf{s}_1$ *and* $\mathbf{s}_2$ *be the signatures of two coherent systems of* n *i.i.d. components, and let* $T_{\phi 1}$ *and* $T_{\phi 2}$ *be their respective lifetimes.*

(i) If $\mathbf{s}_1 \leq^{\text{st}} \mathbf{s}_2$, *then* $T_{\phi 1} \leq^{\text{st}} T_{\phi 2}$, *and, in turn,* $\mathbb{E}(T_{\phi 1}) \leq \mathbb{E}(T_{\phi 2})$.
(ii) If $\mathbf{s}_1 \leq^{\text{hr}} \mathbf{s}_2$, *then* $T_{\phi 1} \leq^{\text{hr}} T_{\phi 2}$.
(iii) If $\mathbf{s}_1 \leq^{\text{lr}} \mathbf{s}_2$, *then* $T_{\phi 1} \leq^{\text{lr}} T_{\phi 2}$.

Note that the signatures $\mathbf{s}_1$ and $\mathbf{s}_2$ are two discrete distributions on the set $\{1, 2, \ldots, n\}$. It is obvious that ordered structure functions, that is, $\phi_1(\mathbf{x}) \leq \phi_2(\mathbf{x})$ for all $\mathbf{x} \in \{0, 1\}^n$, imply ordered signatures, that is, $\mathbf{s}_1 \leq^{\text{st}} \mathbf{s}_2$. In addition, when the underlying distributions are absolutely continuous, hazard rate ordering in Definition 2.8.2 is equivalent to the ordering of the failure rates, with $\mathbf{s}_1 \leq^{\text{hr}} \mathbf{s}_2$ if and only if $r_{\phi 1}(t) \geq r_{\phi 2}(t)$ for all t.

2.10 Multilinear Functions and Taylor (Maclaurin) Expansion

In binary systems, multilinear functions play a crucial role; in fact, many reliability results are a consequence of the properties of Boolean functions (structure functions of the systems)

(Agrawal and Barlow 1984; Barlow and Proschan 1975b; Birnbaum 1969; Boros et al. 2000; Finkelstein 1994; and Khachiyan et al. 2007). Borgonovo (2010) summarized the properties of multilinear functions and Taylor (Maclaurin) expansion related to the reliability functions of coherent and noncoherent systems. This section presents these properties because they are essential in studying some importance measures such as those included in Chapter 7.

From a strictly mathematical viewpoint, a function $f : \mathbb{R}^n \mapsto \mathbb{R}$ is multilinear if it is separately affine in each variable (Marinacci and Montrucchio 2005) and can be represented as

$$f(\mathbf{x}) = \sum_{k=1}^{n} \sum_{i_1 < i_2 < \ldots < i_k} \delta_{i_1, i_2, \ldots, i_k} x_{i_1} x_{i_2} \cdots x_{i_k}, \tag{2.22}$$

with constants $\delta_{i_1, i_2, \ldots, i_k}$ and $k = 1, 2, \ldots, n$. The constants $\delta_{i_1}, \delta_{i_1 i_2}, \ldots$ can be determined in a way from the initial data of $\mathbf{x}$ as

$$\delta_{i_1} = f(0_1, 0_2, \ldots, 0_{i_1 - 1}, 1_{i_1}, 0_{i_1 + 1}, \ldots, 0_n),$$

$$\delta_{i_1 i_2} = f(0_1, 0_2, \ldots, 0_{i_1 - 1}, 1_{i_1}, 0_{i_1 + 1}, \ldots, 0_{i_2 - 1}, 1_{i_2}, 0_{i_2 + 1}, \ldots, 0_n) - \delta_{i_1} - \delta_{i_2},$$

and so on.

The multilinear function f in Equation (2.22) is a homogeneous function and satisfies Euler's equation of order 1:

$$\mathbf{x} \nabla f = f,$$

where ∇f denotes the gradient of function f. The following example illustrates Equation (2.22).

Example 2.10.1 Let the multilinear function $f(x_1, x_2, x_3) = 10x_1 x_2 - 5x_1 x_3$. According to Equation (2.22), $\delta_0 = \delta_1 = \delta_2 = \delta_3 = \delta_{2,3} = \delta_{1,2,3} = 0$, $\delta_{1,2} = 10$, and $\delta_{1,3} = -5$.

Note that Equation (2.22) can be regarded as a multilinear function centered at the origin. However, one can translate Equation (2.22) by centering it at $\mathbf{x}_0$ as

$$f(\mathbf{x}; \mathbf{x}^0) = \sum_{k=0}^{n} \sum_{i_1 < i_2 < \ldots < i_k} \delta_{i_1, i_2, \ldots, i_k} (x_{i_1} - x_{i_1}^0)(x_{i_2} - x_{i_2}^0) \cdots (x_{i_k} - x_{i_n}^0), \tag{2.23}$$

with constants $\delta_{i_1, i_2, \ldots, i_k}$ and $k = 0, 1, \ldots, n$. For the sake of notation simplicity, the remainder refers to Equation (2.22). However, all results applicable to Equation (2.22) are readily extended to Equation (2.23).

The following two theorems present the properties of the Maclaurin expansion of f (Taylor expansion if Equation (2.23) is of concern). For a thorough discussion of multilinear functions and the proofs of these results, readers are referred to Borgonovo (2010), Foldes and Hammer (2005), and Grabisch et al. (2000). In particular, Theorem 2.10.2 addresses the coincidence of a multilinear function with its Taylor (Maclaurin) polynomial.

Theorem 2.10.2 *Let $f : \mathbb{R}^n \mapsto \mathbb{R}$ be a multilinear function.*

(i) The Taylor (Maclaurin) expansion of f is exact and at most is of order O with

$$O = \min\{n, r_{\text{largest}}\} \tag{2.24}$$

and r_{largest} is the size of the largest product in Equation (2.22) for f.

(ii) f coincides with its Taylor (Maclaurin) polynomial.

Borgonovo (2010) showed that the reliability function of any coherent (i.e., $R(\mathbf{p})$) or noncoherent system is multilinear as in Theorem 2.10.3(i). Then, by Theorem 2.10.2, the Taylor expansion of any systems reliability function is exact and can be stopped in a finite order O, as defined in Equation (2.24). El-Neweihi (1980) showed the result in Theorem 2.10.3(ii) in the case of coherent systems and Borgonovo (2010) extended it to the case of noncoherent systems.

Theorem 2.10.3 *Consider a coherent or noncoherent n-component system with dependent or independent component failures. Let $\mathbf{x}$ denote the vector of all conditional reliability/unreliability of the components and let G denote the systems reliability or systems unreliability.*

(i) $G(\mathbf{x})$ is a multilinear function of $\mathbf{x}$.

(ii) The Taylor expansion of the reliability/unreliability function of any system is exact, and the highest order of the expansion, O as defined in Equation (2.24), is less than or equal to n.

2.11 Redundancy

In a parallel system, not all components need to function properly for the system as a whole to function properly. In fact, only one component needs to function properly to make the system function properly. The other $n-1$ components in the parallel system are called redundant components. They are included to increase the probability that there is at least one functioning component. Redundancy is commonly used in system design to enhance systems reliability, especially when it is difficult to increase component reliability itself.

There are two types of redundancy used in engineering design. One type is called *active redundancy*, in which all n components in a parallel system ($n \geq 2$) are used simultaneously even though only one functioning component is required for the system to function. Another type of redundancy is called *standby redundancy*. In this case, only one component in the system is active, but one or more additional spare components may be placed in the system solely as standbys. A sensing and switching mechanism is then used to monitor the operation of the active component. Whenever the active component fails, a standby component is immediately switched into active operation. Figure 2.11 depicts a standby system with one standby component.

Furthermore, there are three types of standby redundancy: hot standby, warm standby, and cold standby. A hot standby has the same failure rate as the active component. A cold standby has a zero failure rate. Warm standby implies that inactive components have a failure rate that

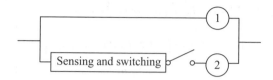

Figure 2.11   Standby system with one standby component

is between 0 and the failure rate of active components. A warm standby and a hot standby may fail while in the standby condition, but a cold standby will not fail.

Active redundancy can be further classified into two types, parallel (often called active redundancy) and series, according to the connection method of the redundant components in the system. The use of parallel active redundancy increases systems reliability, which is desirable in most cases. However, sometimes, decreasing systems reliability through series active redundancy is preferable. For example, adding a monitor in series within a prison security system can decrease the probability of a prisoner escaping. Chapter 9 discusses more examples and the importance measures related to various redundancy.

2.12   Reliability Optimization and Complexity

In general, reliability optimization problems involve maximizing or improving the systems reliability or minimizing the reliability-related system cost. Reliability optimization problems are classified as (Kuo et al. 2006) (i) redundancy allocation problems where the decision variables are the numbers of redundant components, (ii) reliability allocation problems where the decision variables are the reliability values of components, (iii) reliability-redundancy allocation problems where the decision variables are a combination of the numbers of redundant components and the component reliability values, and (iv) CAP, where the assignment of components in the system can make a difference in systems reliability.

Reliability problems are generally NP-hard (Lin and Chen 1997). For example, Chern (1992) showed that a redundancy allocation problem in a simple system—a series system—is NP-hard. This redundancy allocation problem is to determine the numbers of redundant components that are identical to and in parallel with each individual component in the series system. The optimal design is to maximize the systems reliability subject to linear constraints on the amount of available resources and upper-bound constraints on the number of redundant components for each component in the series system. Such a problem is proved to be NP-hard even when only one type of resource is available or when two types of resources are available and all components are identical. Chern (1992) also showed the NP-hardness of another redundancy allocation problem for the case of only one single type of resource, which is to minimize the sum of resources over redundant components (linear objective function) subject to lower-bound requirement on the resulting systems reliability and upper-bound constraints on the number of redundant components for each component in the series system.

In CAP assuming that n physical units (components) of different reliability are functionally interchangeable, then the components can be arbitrarily assigned to any positions in the system. Essentially, CAP are combinatorial optimization problems and generally NP-hard.

Chapters 11–13 investigate CAP in depth, focusing on the application of importance measures in developing solution methods.

In a general setting, the evaluation of network reliability is NP-hard (Ball 1980, 1986; Valiant 1979). Ball (1986) presented an overview related to the computational complexity of the network reliability analysis problem, which is to compute a reliability measure of a stochastic network. To prove the NP-hard complexity of a problem, a common approach is to find a polynomial time reduction from some known NP-hard problem to the problem in question. By reduction from the classically NP-hard combinatorial and optimization problems such as integer programming, the traveling salesman problem and the satisfiability problem (see Garey and Johnson, 1979, for more details), Ball (1980) showed that several systems reliability problems are NP-hard. For the NP-hard reliability problems, the exact solution methods can only solve problems of limited size in a reasonable time. Another approach would be to look for approximate answers. In particular, one popular approach to reliability analysis is Monte Carlo simulation. The general description of complexity can be seen in Kuo and Zuo (2003).

2.13 Consecutive-k-out-of-n Systems

A consecutive-k-out-of-n system, denoted by $Con/k/n$, is a system of n ordered components that are connected either linearly or circularly. A consecutive-k-out-of-n failure system, denoted by $Con/k/n$:F, fails if and only if at least k consecutive components fail. A consecutive-k-out-of-n good system, denoted by $Con/k/n$:G, functions if and only if at least k consecutive components function. Thus, there are four types of $Con/k/n$ systems: linear consecutive-k-out-of-n:F ($Lin/Con/k/n$:F) systems, circular consecutive-k-out-of-n:F ($Cir/Con/k/n$:F) systems, linear consecutive-k-out-of-n:G ($Lin/Con/k/n$:G) systems, and circular consecutive-k-out-of-n:G ($Cir/Con/k/n$:G) systems. For $Con/k/n$:F (G) systems, every k consecutive components forms a minimal cut (path). In a circular case, assume that the components are labeled clockwise from 1 to n.

$Con/k/n$ systems (both linear and circular) include series and parallel systems as special cases. The linear and circular $Con/k/n$:F system becomes a series system when $k = 1$ and a parallel system when $k = n$.

Chiang and Niu (1981) introduced the $Con/k/n$:F system, and a key report can be found in Kuo and Zuo (2003). A $Con/k/n$ system can represent telecommunication and pipeline networks (Chiang and Niu 1981), vacuum systems accelerators (Kao 1982), street lights and microwave towers (Chao and Lin 1984), spacecraft relay stations (Chiang and Chiang 1986), computer ring networks (Hwang 1989), quality control lot acceptance sampling (Shen and Zuo 1994), parking spaces (Kuo et al. 1990), photographing of a nuclear accelerator (Kuo and Zuo 2003), and so on. Example 2.13.1 gives a clear picture of $Con/k/n$ systems.

Example 2.13.1 (a) Consider a pipeline system for transporting oil from one point to another by n pump stations. Each pump station is able to transport oil to a distance including k other pump stations. If one pump station is down, the flow of oil is not interrupted because the subsequent $k - 1$ stations can still carry the load. However, when at least k consecutive pump stations fail, the oil flow is interrupted and the system fails. This is a $Lin/Con/k/n$:F system.

The pump failures are likely to be dependent because the load on neighboring pumps increases when a pump fails.

(b) A sequence of n microwave stations relays signals from one place to another. Each microwave station is able to transmit signals over a distance including to k nearby microwave stations. The transmission of such a system stops if and only if at least k consecutive microwave stations fail. Thus, the system is a $Lin/Con/k/n$:F system. The reliability of the stations may be different because of differences in environmental conditions and operational procedures among the individual microwave stations, and station failures are likely to be independent.

(c) The ring network is one of the most popular architectures for computer networks. The topology of a ring network is a loop consisting of n nodes $1, 2, \ldots, n$ and n links from each node i to node $i+1 \pmod n$ for $i = 1, 2, \ldots, n$. In order to increase the reliability of communication between nodes, a particularly simple and systematic technique is to add links from node i to each of nodes $i+2, i+3, \ldots, i+k \pmod n$ for $i = 1, 2, \ldots, n$ and for some k. Assuming that nodes but not links can fail, the communication between any two functioning nodes cannot be disrupted except when some k consecutive nodes all fail. Thus, this is a $Cir/Con/k/n$:F system.

(d) In quality control lot acceptance sampling, the concepts of both the $Lin/Con/k/n$:F and G systems are applicable. If consecutive k out of n lots are rejected (in other words, a $Lin/Con/k/n$:F system fails) under the normal sampling scheme, the tightened sampling scheme goes into effect. If consecutive k out of n lots are accepted (in other words, a $Lin/Con/k/n$:G system functions) under the tightened sampling scheme, the normal sampling scheme returns. If consecutive k out of n lots are rejected under the tightened sampling scheme, the inspection is discontinued and the products are rejected.

(e) Consider a street parking problem where there are n parallel-parking spaces on a street. Each space is suitable for one car, but if a bus parks on the street, two consecutive spaces are required. Every parking space has a probability that it is available. The probability that the bus can park on this street is equivalent to the reliability of a $Lin/Con/2/n$:G system, although the usual assumption of mutually statistically independent states is probably quite untrue in parallel parking for automobiles.

(f) In a nuclear accelerator, a set of n high-speed cameras is installed around the accelerator to take pictures of the acceleration activities. The photographing system functions properly if and only if at least k consecutive cameras function properly. This is an example of the $Cir/Con/k/n$:G system. One interesting problem is how to arrange cameras of different reliability values to maximize the reliability of the photographing system.

Theorem 2.13.2 and Corollary 2.13.3 (Kuo et al. 1990) indicate the relations between $Con/k/n$:F and G systems. Using these relations, the results of one type of $Con/k/n$ system can be extended to another.

Theorem 2.13.2 *$Con/k/n$:F and $Con/k/n$:G systems are dual of each other.*

Proof. Let $\phi_F(\mathbf{x})$ and $\phi_G(\mathbf{x})$ denote structure functions of $Lin/Con/k/n$:F and G systems, respectively. Supposing there are k consecutive 0 elements in state vector $\mathbf{x}$, then there are k consecutive 1 elements in state vector $\mathbf{1} - \mathbf{x}$. Thus, $\phi_F(\mathbf{x}) = 0$, and $\phi_G(\mathbf{1} - \mathbf{x}) = 1$. If there are no k consecutive 0 elements in state vector $\mathbf{x}$, then there are no k consecutive 1 elements in state vector $\mathbf{1} - \mathbf{x}$. Thus, $\phi_F(\mathbf{x}) = 1$ and $\phi_G(\mathbf{1} - \mathbf{x}) = 0$. Hence, overall, $\phi_G(\mathbf{1} - \mathbf{x}) = 1 - \phi_F(\mathbf{x})$. These arguments can also be applied to $Cir/Con/k/n$:F and G systems.

Corollary 2.13.3 *If two $Con/k/n$ systems have the same values of k and n and arrangement (either linear or circular) and if the reliability of component i in one type of $Con/k/n$ system (e.g., $Con/k/n$:F) is equal to the unreliability of component i in another type of $Con/k/n$ system (e.g., $Con/k/n$:G) for $i = 1, 2, \ldots, n$, then the reliability of one type of system is equal to the unreliability of another type of system, and the $Con/k/n$:F and G systems are mirror images of each other.*

Proof. Let R_F and $\bar{R}_F$ denote the reliability and unreliability of a $Con/k/n$:F system, respectively. Similarly, let R_G and $\bar{R}_G$ denote the reliability and unreliability of a $Con/k/n$:G system, respectively.

$$\bar{R}_F = 1 - R_F = \text{Pr\{at least } k \text{ consecutive components fail in the } Con/k/n\text{:F system\}}$$
$$= \text{Pr\{at least } k \text{ consecutive 0 elements are contained in } \mathbf{x}\}$$
$$R_G = 1 - \bar{R}_G = \text{Pr\{at least } k \text{ consecutive components function in the } Con/k/n\text{:G system\}}$$
$$= \text{Pr\{at least } k \text{ consecutive 1 elements are contained in } \mathbf{x}\}$$

If $\text{Pr}\{X_i = 0\}$ in the $Con/k/n$:F system is the same as $\text{Pr}\{X_i = 1\}$ in the $Con/k/n$:G system for $i = 1, 2, \ldots, n$, then $\bar{R}_F = R_G$ and $R_F = \bar{R}_G$.

Chapters 8, 12, and 13 provide a comprehensive investigation on $Con/k/n$ systems and their extensions, showing their broad applications and mathematical tractability. The covered topics include the expressions and comparisons of various importance measures as well as the importance-measure-based rules and methods for CAP in these systems.

2.14 Assumptions

On the basis of the discussion in this chapter, the following basic assumptions about components and system are made. They hold true throughout the book unless stated otherwise.

(i) The components and the system are binary, that is, having two possible states: functioning or failed.
(ii) The system (N, ϕ) is coherent with $N = \{1, 2, \ldots, n\}$.
(iii) Behaviors of components are mutually statistically independent.
(iv) The components and the system are nonrepairable.
(v) If related, $F_i(t)$ for $i = 1, 2, \ldots, n$ and $F_\phi(t)$ are absolutely continuous with density and failure rate functions. That is, the components and the system have absolutely continuous lifetime distributions.

As discussed in this book, most importance measures were proposed under the aforementioned assumptions. Chapter 15 investigates importance measures under the relaxation of one or more assumptions. In addition to binary systems, research has been done to develop the multistate, continuum, and mixed (continuous and discrete) systems. Chapter 15 presents importance measures in those systems and introduce the fundamentals of those systems accordingly.

References

Agrawal A and Barlow R. 1984. A survey of network reliability and domination theory. *Operations Research* **32**, 478–492.

Ball MO. 1980. Complexity of network reliability computations. *Networks* **10**, 153–165.

Ball MO. 1986. Computational complexity of network reliability analysis: An overview. *IEEE Transactions on Reliability* **R-35**, 230–239.

Barlow R and Iyer S. 1988. Computational complexity of coherent systems and the reliability polynomial. *Probability in the Engineering and Informational Sciences* **2**, 461–469.

Barlow RE and Proschan F. 1975a. Importance of system components and fault tree events. *Statistic Processes and Their Applications* **3**, 153–172.

Barlow RE and Proschan F. 1975b. *Statistical Theory of Reliability and Life Testing Probability Models*. Holt, Rinehart and Winston, New York.

Birnbaum ZW. 1969. On the importance of different components in a multicomponent system. In *Multivariate Analysis, Vol. 2* (ed. Krishnaiah PR). Academic Press, New York, pp. 581–592.

Borgonovo E. 2010. The reliability importance of components and prime implicants in coherent and non-coherent systems including total-order interactions. *European Journal of Operational Research* **204**, 485–495.

Boros E, Crama Y, Ekin O, Hammer P, Ibaraki T, and Kogan A. 2000. Boolean normal forms, shellability, and reliability computations. *SIAM Journal on Discrete Mathematics* **13**, 212–226.

Chao MT and Lin GD. 1984. Economical design of large consecutive-k-out-of-n:F systems. *IEEE Transactions on Reliability* **R-33**, 411–413.

Chern MS. 1992. On the computational complexity of reliability redundancy allocation in a series system. *Operations Research Letters* **11**, 309–315.

Chiang DT and Chiang RF. 1986. Relayed communication via consecutive-k-out-of-n:F system. *IEEE Transactions on Reliability* **35**, 65–67.

Chiang DT and Niu SC. 1981. Reliability of consecutive-k-out-of-n:F system. *IEEE Transactions on Reliability* **R-30**, 87–89.

El-Neweihi E. 1980. A relationship between partial derivatives of the reliability function of a coherent system and its minimal path (cut) sets. *Mathematics of Operations Research* **5**, 553–555.

Finkelstein MS. 1994. Once more on measures of importance of system components. *Microelectronics and Reliability* **34**, 1431–1439.

Foldes S and Hammer PL. 2005. Submodularity, supermodularity, and higher-order monotonicities of pseudo-Boolean functions. *Mathematics of Operations Research* **30**, 453–461.

Garey M and Johnson D. 1979. *Computers and Intractability: A Guide to the Theory of NP-Completeness*. Freeman, San Francisco.

Grabisch M, Marichal JL, and Roubens M. 2000. Equivalent representations of set functions. *Mathematics of Operations Research* **25**, 157–178.

Hwang FK. 1989. Invariant permutations for consecutive-k-out-of-n cycles. *IEEE Transactions on Reliability* **R-38**, 65–67.

Hwang FK. 2005. A hierarchy of importance indices. *IEEE Transactions on Reliability* **54**, 169–172.

Iyer S. 1992. The Barlow-Proschan importance and its generalizations with dependent components. *Stochastic Processes and Their Applications* **42**, 353–359.

Kao SC. 1982. Computing reliability from warranty. *Proceedings of the 1982 American Statistical Association, Section on Statistical Computing*, pp. 309–312.

Khachiyan L, Boros E, Elbassioni K, Gurvich V, and Makino K. 2007. Enumerating disjunctions and conjunctions of paths and cuts in reliability theory. *Discrete Applied Mathematics* **155**, 137–149.

Kochar S, Mukerjee H, and Samaniego FJ. 1999. The "signature" of a coherent system and its application to comparisons among systems. *Naval Research Logistics* **46**, 507–523.

Kuo W, Prasad VR, Tillman FA, and Hwang CL. 2006. *Optimal Reliability Design: Fundamentals and Applications, 2nd edn*. Cambridge University Press, Cambridge, UK.

Kuo W, Zhang W, and Zuo MJ. 1990. A consecutive k-out-of-n:G: The mirror image of a consecutive k-out-of-n:F system. *IEEE Transactions on Reliability* **R-39**, 244–253.

Kuo W and Zuo MJ. 2003. *Optimal Reliability Modeling: Principles and Applications.* John Wiley & Sons, New York.

Lin MS and Chen DJ. 1997. The computational complexity of the reliability problem on distributed systems. *Information Processing Letters* **64**, 143–147.

Marinacci M and Montrucchio L. 2005. Ultramodular functions. *Mathematics of Operations Research* **30**, 311–332.

Meng FC. 1996. Comparing the importance of system components by some structural characteristics. *IEEE Transactions on Reliability* **45**, 59–65.

Navarro J, Spizzichino F, and Balakrishnan N. 2010. Average systems and their role in the study of coherent systems. *Journal of Multivariate Analysis* **101**, 1471–1482.

Samaniego FJ. 2007. *System Signatures and Their Applications in Engineering Reliability.* Springer, New York.

Shen J and Zuo MJ. 1994. Optimal design of series consecutive-k-out-of-n:G system with age-dependent minimal repair. *Reliability Engineering and System Safety* **45**, 277–283.

Triantafyllou IS and Koutras MV. 2008. On the signature of coherent systems and applications. *Probability in the Engineering and Informational Sciences* **22**, 19–35.

Valiant LG. 1979. The complexity of enumeration and reliability problems. *SIAM Journal on Computing* **8**, 410–421.

Xie M and Lai CD. 1996. Exploiting symmetry in the reliability analysis of coherent systems. *Naval Research Logistics* **43**, 1025–1034.

Part Two

Principles of Importance Measures

Introduction

This part includes Chapters 3–8. Chapter 3 presents the classifications of importance measures as well as the terminology used to describe and to differentiate among the types of importance measures. For each of the three classes of importance measures, Chapters 4–6 introduce the most commonly used importance measures of individual components in a coherent system, and Chapter 7 concentrates on the importance measures of pairs or groups of components. Chapter 8 studies various importance measures for $Con/k/n$ systems, which are of broad applications as illustrated in Section 2.13. Because of their mathematical tractability, the importance measures for $Con/k/n$ systems are further specified and deeply investigated, focusing on the B-importance and some structure importance measures that have close relations to the B-importance. These results are used in solving the CAP in Chapters 12 and 13. According to the review in Kuo and Zhu (2012a, 2012b) on the importance measures in reliability, they are adapted and extended in this part.

References

Kuo W and Zhu X. 2012a. Relations and generalizations of importance measures in reliability. *IEEE Transactions on Reliability* **61**.

Kuo W and Zhu X. 2012b. Some recent advances on importance measures in reliability. *IEEE Transactions on Reliability* **61**.

3

The Essence of Importance Measures

A system normally consists of multiple components, which are not necessarily equally important for the performance of the system. Such a system often needs to be designed, enhanced, or maintained efficiently using limited resources. However, for highly complex systems, it may be too tedious, or not even possible, to develop a formal optimal strategy. In these situations, it is desirable to allocate resources according to how important the components are to the system and to concentrate the resources on the small subset of components that are most important to the system. Thus follows the notion of importance measures of components.

For tackling different problems (e.g., Examples 1.0.5 and 1.0.11), the distinct importance measures and the associated algorithms should be appropriately designed. Various importance measures have been proposed to judge the relative strength of a component in a system with respect to different criteria. As stated in Griffith and Govindarajulu (1985), no single type of importance measure is universal, since different perspectives on the same system can lead to different views about which factors make one component more important than another. As in Example 1.0.5, a design engineer may think of a component as being more important if a given reliability improvement in it does more to improve systems reliability than the same reliability improvement in another component. In contrast, in Example 1.0.11, an engineer may devise a checklist of components for finding the cause of system failure based on his or her assessment of the conditional probability that a particular component has contributed to system failure. The probabilistic interpretation describing the relation of a component to the functioning or failure of the system is different in each case.

3.1 Importance Measures in Reliability

In reliability, an importance measure evaluates the relative importance of individual components or groups of components in a system and can be determined on the basis of the system structure, component reliability, and/or component lifetime distributions. It might be a measure that can be determined for each individual component or set of components. Or, it might take the form of a relative ranking (i.e., an importance measure being defined implicitly by means

Importance Measures in Reliability, Risk, and Optimization: Principles and Applications, First Edition.
Way Kuo and Xiaoyan Zhu. © 2012 John Wiley & Sons, Ltd. Published 2012 by John Wiley & Sons, Ltd.

of a comparison) among components, where the ranking might be partial (i.e., some of components are not comparable) or complete (i.e., any two of the components can be compared).

In general, when using importance measures, the relative rankings of the importance measures of components are more essential than their absolute values. The rankings are possibly determined with the knowledge of the orderings or ranges of component reliability but not the exact values. This is attractive since the orderings and ranges of component reliability are much easier to determine than their exact values. For example, in the case that the ages of the components are known and their common lifetime distribution has an unknown but increasing failure rate, then with fixed mission time, the reliability of components can be inversely ordered according to their ages. This idea is similar to ordinal optimization (Ho et al. 2007), which works on orders rather than values and involves compromise for the "good enough" solutions, taking advantages of computational efficiency. Properties of importance measures relative to ranking the importance of two or more components are presented in the book.

3.2   Classifications

Birnbaum (1969) categorized importance measures into three classes according to the knowledge needed for determining the importance measures. The first class is the *structure importance measure*, which measures the relative importance of various components with respect to their positions in a system. The structure importance measures are based on knowledge of the system structure only and do not involve the reliability of components. It can be determined completely by the design of the system, that is, ϕ. Note that structure importance measures of components actually represent the importance of the positions in the system that the components occupy.

The second class is the *reliability importance measure*, which is considered when the mission time of a system is implicit and fixed and, consequently, the components are evaluated by their reliability at a fixed time point (i.e., the probability that a component functions properly during the mission time). Reliability importance measures depend on both the system structure ϕ and the reliability of components; thus, to calculate reliability importance measures, the mission time and the reliability of components must be determined in advance.

The third class is the *lifetime importance measure*, which is considered when a system and components have long-term or infinite service missions. Lifetime importance measures depend on both the positions of components within the system and the component lifetime distributions. They can be further divided into two subclasses: time-dependent lifetime (TDL) importance and time-independent lifetime (TIL) importance, depending on whether they are a function of time. A TDL importance measure evaluates the importance of components at any time and the induced rankings of component importance may vary with time. During the system development phase, a TIL importance measure, which takes into account the component lifetime over the long term, as such, perhaps gives a more global view of component importance, for example, in addressing the expected system lifetime.

When applying TDL importance measures during the system development phase, an analyst needs to determine which points of time are important. Similarly, for reliability importance measures, the mission time has to be determined. For this reason, the reliability and TDL importance measures are often criticized. Reliability importance measures are essentially time dependent, since the implicit mission time associated with them has to be determined in advance. For different mission times, the values of p_i, $i = 1, 2, \ldots, n$, may be different. A reliability importance measure can be transformed to a corresponding TDL importance measure as stated in Theorem 3.2.1.

Theorem 3.2.1 *Every type of reliability importance measure can be transformed to a corresponding TDL importance measure by substituting $\overline{F}_i(t)$ for p_i, $i = 1, 2, \ldots, n$, without changing the probabilistic interpretation of the importance measure.*

Proof. A reliability importance measure is based on reliability at a fixed time point of t; therefore, the reliability of component i (at time t) is $p_i = \overline{F}_i(t)$. If multiple time points are of concern or time t is not fixed, $\overline{F}_i(t)$ needs to be used instead of p_i, which results in the corresponding TDL importance measure.

According to Theorem 3.2.1, the definitions and results for a reliability importance measure can be extended to the corresponding TDL importance measure by replacing p_i with $\overline{F}_i(t)$, $i = 1, 2, \ldots, n$. In this book, sometimes this kind of extension is applied implicitly without special mention. However, a TDL importance measure may not be transformed to a corresponding reliability importance measure by replacing $\overline{F}_i(t)$ with p_i because complex operations on $\overline{F}_i(t)$, such as differential or integration, may be involved in the definition of the TDL importance measure.

Importance measures that involve more information might (at first) seem to provide better evaluation about component importance. However, at the same time, their calculations might be more extensive, especially for large complex systems. Another issue is that the probabilistic information required for their calculations might not be available in practice. Thus, it is not true that the more complicated the importance measure, the better. For example, when the probabilistic information of components is not available or the calculation involved is prohibitively extensive, structure importance measures should be used and could provide a fair basis to compare relative importance among components.

Additionally, in regard to repairable components and system, availability, which involves information of repair operations, is often used in defining importance measures. The importance measures involving availability of repairable components are specifically discussed in Section 15.7.

3.3 *c*-Type and *p*-Type Importance Measures

Many importance measures can be defined through (minimal) cuts and/or (minimal) paths. According to Hwang (2001), an importance measure is said to be *c*-type if it can be defined by cuts and is *p*-type if it can be defined by paths. If the definition of an importance measure based on cuts is equivalent to the one based on paths, the importance measure does not need to be identified by *c*-type and *p*-type and is simply referred without any such prefix.

3.4 Importance Measures of a Minimal Cut and a Minimal Path

Minimal cuts and minimal paths play a basic role in the analysis of coherent systems and of fault trees. For example, the reliability of a coherent system or a lower and upper bound on it can be obtained from knowledge of the reliability of the minimal cuts and minimal paths. The failure of a single component may not cause the failure of the system, but the failure of any single minimal cut must cause the failure of the system. Thus, it is of concern which minimal cut is more important than the others with respect to the performance of the system. Some importance measures have been proposed for minimal cuts extended from their definitions for components. Normally, independent component failures are assumed, but minimal cuts are generally dependent because they may have some components in common. The similar arguments are applied to minimal paths.

Fault tree analysis can generate minimal cuts and minimal paths. Fault trees are logical block diagrams that display the state of a system (top event) in terms of the states of its components (basic events). They are presented in terms of events rather than components. A fault tree traces the top failure (system failure) to the basic (primary) failure (component failure) and generates the minimal cuts. The dependence of the top event on the basic events is analogous to the dependence of a system state to the states of the components. Fault trees have been used by engineers and reliability analysts to represent schematically basic events and their various logical combinations that may result in the so-called top event. The most commonly used logical operations in a fault tree are AND, OR, and NOT. If a fault tree does not contain a NOT operation, the two fundamental conditions for coherence, that is, nondecreasing and relevancy, will be met, and the fault tree is coherent (Lu and Jiang 2007). On the other hand, if a fault tree contains at least one NOT operation, it is noncoherent because the first condition of the coherence, nondecreasing, is no longer valid for the basic event that is associated with the NOT operation. In this book, fault trees are assumed to be coherent unless stated otherwise. Kuo and Zuo (2003) presented more methods for generating minimal cuts and minimal paths.

This book uses the terminology of coherent structure theory, even though fault trees are more general than coherent structures in the sense that basic events in a fault tree need not necessarily correspond to component failures in a coherent system. Meanwhile all results also apply to fault trees when proper identifications are made.

3.5   Terminology

Because the structure importance measure of components actually represents the importance of the positions in the system that the components occupy, the terms "structure importance of components" and "structure importance of positions" are used interchangeably in this book, depending on the context. For simplicity, in some situations (e.g., in Chapter 13), the reliability (importance measure) of the component in position i is also referred to as the reliability (importance measure) of position i. In addition, a type of importance measure is often simply referred to as a type of importance or a type of measure without ambiguity. Furthermore, a type of structure importance measure that is defined by means of a comparison may also be referred to as a type of importance ranking.

For convenience in referring to importance measures, this book denotes the structure importance measures as $I_x^y(\cdot; \phi)$ since their calculations depend only on the structure function ϕ; the reliability importance measures as $I_x^y(\cdot; \phi, \mathbf{p})$ since their calculations need, in addition to the structure function ϕ, the knowledge of component reliability; and the TIL and TDL importance measures as $I_x^y(\cdot; \phi, \overline{\mathbf{F}})$ and $I_x^y(\cdot; \phi, \overline{\mathbf{F}}(t))$, respectively, since their calculations need, in addition to the structure function ϕ, the knowledge of component lifetime distributions. The notation $I_x^y(\cdot; \phi, \overline{\mathbf{F}})$ differs with $I_x^y(\cdot; \phi, \overline{\mathbf{F}}(t))$ because the latter is a function of time t. For the sake of simplification, the ϕ in notation $I_x^y(\cdot; \phi, \mathbf{p})$, $I_x^y(\cdot; \phi, \overline{\mathbf{F}})$, and $I_x^y(\cdot; \phi, \overline{\mathbf{F}}(t))$ may be ignored. Additionally, $I_x^y(\cdot; p)$, $I_x^y(\cdot; \overline{F})$, and $I_x^y(\cdot; \overline{F}(t))$ with scalar p and real function $\overline{F}$ represent the case where $p_1 = p_2 = \ldots = p_n = p$ and $\overline{F}_1(t) = \overline{F}_2(t) = \ldots = \overline{F}_n(t) = \overline{F}(t)$, and thus, the corresponding vectors are reduced to the scalars.

Subscript x represents the types of importance measures, perhaps being the abbreviations of the names of the people who first proposed them. Superscript y is optional and could be empty, M, C, or P. "Empty" means the importance of components; "M" means the importance of modular sets; "C" means the importance of minimal cuts; and "P" means the importance of minimal paths. For the structure importance measures that are defined by means of comparisons, notation $i >_x j$ is used to denote that component i is more important than component j according to the importance measure of type x.

References

Birnbaum ZW. 1969. On the importance of different components in a multicomponent system. In *Multivariate Analysis, Vol. 2* (ed. Krishnaiah PR). Academic Press, New York, pp. 581–592.

Griffith WS and Govindarajulu Z. 1985. Consecutive k-out-of-n failure systems: reliability, availability, component importance, and multistate extensions. *American Journal of Mathematical and Management Sciences* **5**, 125–160.

Ho YC, Zhao QC, and Jia QS. 2007. *Ordinal Optimization: Soft Optimization for Hard Problems*. Springer, New York.

Hwang FK. 2001. A new index of component importance. *Operations Research Letters* **28**, 75–79.

Kuo W and Zuo MJ. 2003. *Optimal Reliability Modeling: Principles and Applications*. John Wiley & Sons, New York.

Lu L and Jiang J. 2007. Joint failure importance for noncoherent fault trees. *IEEE Transactions on Reliability* **56**, 435–443.

4

Reliability Importance Measures

This chapter introduces two major types of reliability importance measures: the B-importance in Section 4.1 and the FV importance in Section 4.2. Both the B-reliability importance and FV reliability importance can be extended to the TDL type and simplified to the structure type of importance measures. Their TDL versions and structure versions are discussed in Chapters 5 and 6, respectively.

4.1 The B-reliability Importance

The B-importance (Birnbaum 1969) considers the probability that a component is critical for the system. This section first introduces the B-reliability importance of components and then discusses its extension to the modular sets. Subsections 4.1.1, 4.1.2, and 4.1.3 introduce the improvement potential importance, criticality importance, and Bayesian importance, respectively, noting that they are derived from the B-reliability importance.

Definition 4.1.1 *The B-reliability importance of component i for the functioning (i.e., success as denoted in subscript) of system ϕ, denoted by $I_{Bs}(i; \mathbf{p})$, is defined as*

$$I_{Bs}(i; \mathbf{p}) = \Pr\{\phi(\mathbf{X}) = 1 | X_i = 1\} - \Pr\{\phi(\mathbf{X}) = 1\}. \tag{4.1}$$

The B-reliability importance of component i for the failure of ϕ, denoted by $I_{Bf}(i; \mathbf{p})$, is defined as

$$I_{Bf}(i; \mathbf{p}) = \Pr\{\phi(\mathbf{X}) = 0 | X_i = 0\} - \Pr\{\phi(\mathbf{X}) = 0\}.$$

The B-reliability importance of component i for ϕ, denoted by $I_B(i; \mathbf{p})$, is defined as

$$
\begin{aligned}
I_B(i; \mathbf{p}) &= I_{Bs}(i; \mathbf{p}) + I_{Bf}(i; \mathbf{p}) \\
&= \Pr\{\phi(\mathbf{X}) = 0 | X_i = 0\} - \Pr\{\phi(\mathbf{X}) = 0 | X_i = 1\} \tag{4.2} \\
&= \frac{1}{q_i} \Pr\{\mathbf{X} \in \mathscr{C}_i\} - \frac{1}{p_i} \Pr\{\mathbf{X} \in \mathscr{C}_{(i)}\} \tag{4.3}
\end{aligned}
$$

Importance Measures in Reliability, Risk, and Optimization: Principles and Applications, First Edition.
Way Kuo and Xiaoyan Zhu. © 2012 John Wiley & Sons, Ltd. Published 2012 by John Wiley & Sons, Ltd.

$$= \Pr\{\phi(\mathbf{X}) = 1 | X_i = 1\} - \Pr\{\phi(\mathbf{X}) = 1 | X_i = 0\} \tag{4.4}$$

$$= \frac{1}{p_i} \Pr\{\mathbf{X} \in \mathscr{P}_i\} - \frac{1}{q_i} \Pr\{\mathbf{X} \in \mathscr{P}_{(i)}\}. \tag{4.5}$$

Theorem 4.1.2 *For a coherent system with independent components,*

$$I_{\mathrm{Bs}}(i; \mathbf{p}) = q_i \mathbb{E}\left(\phi(1_i, \mathbf{X}) - \phi(0_i, \mathbf{X})\right) = q_i \frac{\partial R(\mathbf{p})}{\partial p_i}, \tag{4.6}$$

$$I_{\mathrm{Bf}}(i; \mathbf{p}) = p_i \mathbb{E}\left(\phi(1_i, \mathbf{X}) - \phi(0_i, \mathbf{X})\right) = p_i \frac{\partial R(\mathbf{p})}{\partial p_i}, \tag{4.7}$$

and

$$I_{\mathrm{B}}(i; \mathbf{p}) = \mathbb{E}\left(\phi(1_i, \mathbf{X}) - \phi(0_i, \mathbf{X})\right) \tag{4.8}$$

$$= \Pr\{\phi(1_i, \mathbf{X}) - \phi(0_i, \mathbf{X}) = 1\} = \Pr\{\phi(1_i, \mathbf{X}) > \phi(0_i, \mathbf{X})\} \tag{4.9}$$

$$= \frac{\partial R(\mathbf{p})}{\partial p_i} \tag{4.10}$$

$$= R(1_i, \mathbf{p}) - R(0_i, \mathbf{p}). \tag{4.11}$$

Proof. First by Equation (2.2), the covariance of X_i and $\phi(\mathbf{X})$ is

$$\mathrm{Cov}(X_i, \phi(\mathbf{X})) = \mathrm{Cov}\left(X_i, X_i \left(\phi(1_i, \mathbf{X}) - \phi(0_i, \mathbf{X})\right) + \phi(0_i, \mathbf{X})\right)$$

$$= \mathrm{Cov}\left(X_i, X_i \left(\phi(1_i, \mathbf{X}) - \phi(0_i, \mathbf{X})\right)\right)$$

$$= \mathbb{E}\left(X_i^2 \left(\phi(1_i, \mathbf{X}) - \phi(0_i, \mathbf{X})\right)\right) - \mathbb{E}(X_i)\mathbb{E}\left(X_i \left(\phi(1_i, \mathbf{X}) - \phi(0_i, \mathbf{X})\right)\right)$$

$$= \mathbb{E}\left(X_i \left(\phi(1_i, \mathbf{X}) - \phi(0_i, \mathbf{X})\right)\right) - p_i \mathbb{E}\left(X_i \left(\phi(1_i, \mathbf{X}) - \phi(0_i, \mathbf{X})\right)\right)$$

$$= p_i q_i \mathbb{E}\left(\phi(1_i, \mathbf{X}) - \phi(0_i, \mathbf{X})\right). \tag{4.12}$$

Note that X_i is independent of $\phi(1_i, \mathbf{X})$ and $\phi(0_i, \mathbf{X})$.

Now by Equation (4.1),

$$I_{\mathrm{Bs}}(i; \mathbf{p}) = \frac{\Pr\{X_i = \phi(\mathbf{X}) = 1\}}{\Pr\{X_i = 1\}} - \mathbb{E}(\phi(\mathbf{X})) = \frac{1}{p_i}\mathbb{E}(X_j\phi(\mathbf{X})) - \mathbb{E}(\phi(\mathbf{X}))$$

$$= \frac{1}{p_i}[\mathbb{E}(X_j)\mathbb{E}(\phi(\mathbf{X})) + \mathrm{Cov}(X_i, \phi(\mathbf{X}))] - \mathbb{E}(\phi(\mathbf{X})) = \frac{1}{p_i}\mathrm{Cov}(X_i, \phi(\mathbf{X}))$$

$$= q_i \mathbb{E}\left(\phi(1_i, \mathbf{X}) - \phi(0_i, \mathbf{X})\right).$$

The last equality holds due to Equation (4.12). Equation (4.6) follows from Equation (2.12). A similar argument yields Equation (4.7). Equations (4.8)–(4.10) follow by adding Equations (4.6) and (4.7), and then Equation (4.11) follows from Equation (2.11).

Part of Theorem 4.1.2 has been presented by Birnbaum (1969) and Barlow and Proschan (1975). Equations (4.8)–(4.11) are more frequently used than the original definitions (4.2)–(4.5). In the case of dependent components, the B-reliability importance of component i could be defined using Equation (4.8) (Barlow and Proschan 1975).

According to Equation (4.9), $I_B(i; \mathbf{p})$ is the probability that component i is critical for the system, that is, the probability that the failure and functioning of component i coincide with system failure and functioning. By Equation (4.9), the range of the B-reliability importance is specified in Corollary 4.1.3.

Corollary 4.1.3 *For a coherent system with $n \geq 2$ and each component reliability in $(0, 1)$, then $0 < I_B(i; \mathbf{p}) < 1$ for $i = 1, 2, \ldots, n$.*

According to Equation (4.10), $I_B(i; \mathbf{p})$ measures the significance of component i to systems reliability by the rate at which systems reliability improves with the reliability of component i. Thus, the B-reliability importance is also known as the *marginal reliability importance* (Armstrong 1995; Hong and Lie 1993; Hsu and Yuang 1999; Lu and Jiang 2007). Note that

$$\frac{dR}{dz} = \sum_{i=1}^{n} \frac{\partial R}{\partial p_i} \frac{dp_i}{dz} = \sum_{i=1}^{n} I_B(i; \mathbf{p}) \frac{dp_i}{dz}, \tag{4.13}$$

where z is a common parameter. Thus, the rate at which systems reliability grows is a weighted combination of the rates at which component reliability values grow, where the weights are the values of the B-reliability importance. Approximately,

$$\Delta R \simeq \sum_{i=1}^{n} I_B(i; \mathbf{p}) \Delta p_i, \tag{4.14}$$

where ΔR is the perturbation in systems reliability corresponding to perturbations Δp_i in the reliability of component i for $i = 1, 2, \ldots, n$. Thus, the system improvement is *monotone* in the order of the B-reliability importance of components in accordance with Equation (4.14). This *monotone property* is desirable and useful in upgrading input parameters for systems reliability improvement, for example, determining the component(s) on which additional research and development effort can be most profitably expended.

As shown in Equation (4.11), the B-reliability importance of component i *does not depend on p_i itself*. This is a drawback of the B-importance. The improvement of a highly reliable component (i.e., high value of p_i) is very costly, but the B-reliability importance does not reflect this issue since it is independent of p_i.

Computation

With the known expression of $R(\mathbf{p})$, $I_B(i; \mathbf{p})$ can be obtained by taking the differentiation of $R(\mathbf{p})$ with respect to p_i as Equation (4.10). Alternatively, $I_B(i; \mathbf{p})$ of an independent component can be calculated according to Equation (4.11); Hsu and Yuang (1999) presented such an efficient algorithm for evaluating the B-reliability importance of an edge (corresponding to a component) in a reducible network (corresponding to a system). Lee et al. (1997) computed the B-reliability importance of gates in a fault tree by a conventional fault-tree algorithm.

Another method for calculating $I_B(i; \mathbf{p})$ is to find the critical vectors and then use Equation (4.9) directly. Theorem 2.6.2 shows that critical vectors of components can be found through minimal cuts and minimal paths. Thus, it is possible to calculate the B-reliability importance without calculating $R(\mathbf{p})$ (see Examples 4.1.8 and 4.1.9). Let $B(i) = \{(\cdot_i, \mathbf{x}) : \phi(1_i, \mathbf{x}) - \phi(0_i, \mathbf{x}) = 1\}$ be the set of state vectors $(\cdot_i, \mathbf{x})$ in which component i is critical for the system. Then,

$$I_B(i; \mathbf{p}) = \Pr\{(\cdot_i, \mathbf{X}) \in B(i)\}. \tag{4.15}$$

Example 4.1.4 (Example 2.2.1 continued) For a series system, $I_B(i; \mathbf{p}) = \prod_{j=1, j \neq i}^{n} p_j = \prod_{j=1}^{n} p_j / p_i$; $I_{Bs}(i; \mathbf{p}) = p_i^{-1} q_i \prod_{j=1}^{n} p_j$; $I_{Bf}(i; \mathbf{p}) = \prod_{j=1}^{n} p_j$. That is, all components have the same B-reliability importance for system failure and the most reliable component has the smallest B-reliability importance and the smallest B-reliability importance for system functioning.

Example 4.1.5 (Example 2.2.2 continued) For a parallel system, $I_B(i; \mathbf{p}) = \prod_{j=1, j \neq i}^{n} q_j = \prod_{j=1}^{n} q_j / q_i$; $I_{Bs}(i; \mathbf{p}) = \prod_{j=1}^{n} q_j$; $I_{Bf}(i; \mathbf{p}) = p_i q_i^{-1} \prod_{j=1}^{n} q_j$. That is, the most reliable component has the largest B-reliability importance, all components are equally B-reliability important for system functioning, and the most reliable component also has the largest B-reliability importance for system failure.

Example 4.1.6 (Example 2.2.3 continued) For a k-out-of-n system, $\phi(1_i, \mathbf{x}) - \phi(0_i, \mathbf{x}) = 1$ if and only if exactly $k - 1$ of the $n - 1$ components (excluding component i) function. Therefore, $I_B(i; \mathbf{p}) = \sum p_{i_1} p_{i_2} \cdots p_{i_{k-1}} (1 - p_{i_k})(1 - p_{i_{k+1}}) \cdots (1 - p_{i_{n-1}})$, where the sum is taken over all permutations $(i_1, i_2, \ldots, i_{n-1})$ of the integers $\{1, 2, \ldots, i-1, i+1, \ldots, n\}$.

Furthermore, without loss of generality, assuming $p_1 \leq p_2 \leq \cdots \leq p_n$, $1 \leq k < n$, and $n > 2$, Boland and Proschan (1983) and Chadjiconstantinidis and Koutras (1999) showed that if $p_i \geq (\leq)(k-1)/(n-1)$ for $i = 1, 2, \ldots, n$, then $I_B(1; \mathbf{p}) \leq I_B(2; \mathbf{p}) \leq \cdots \leq I_B(n; \mathbf{p})$ ($I_B(1; \mathbf{p}) \geq I_B(2; \mathbf{p}) \geq \cdots \geq I_B(n; \mathbf{p})$).

Example 4.1.7 For a system with k components in series, in series with $n-k$ components in parallel as shown in Figure 4.1, $I_B(i; \mathbf{p}) = p_i^{-1} \prod_{j=1}^{k} p_j [1 - \prod_{j=k+1}^{n} (1 - p_j)]$ for $i = 1, 2, \ldots, k$ and $I_B(i; \mathbf{p}) = (1 - p_i)^{-1} \prod_{j=1}^{k} p_j \prod_{j=k+1}^{n} (1 - p_j)$ for $i = k+1, k+2, \ldots, n$, and the corresponding expressions are immediately obtained for $I_{Bs}(i; \mathbf{p})$ and $I_{Bf}(i; \mathbf{p})$.

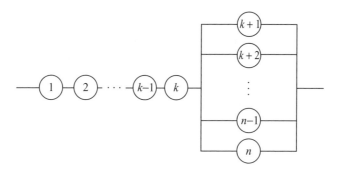

Figure 4.1 System with k components in series, in series with $n - k$ components in parallel

Example 4.1.8 (Example 2.6.3 continued) For the system in Figure 2.10, there is one critical vector for component 4: $B(4) = \{(\cdot_4, 1_1, 1_2, 0_3, 0_5)\}$. Then, $I_B(4; \mathbf{p}) = p_1 p_2 q_3 q_5$.

Example 4.1.9 (Example 2.6.4 continued) Consider the bridge system in Figure 2.1. There are two critical vectors for component 3: $B(3) = \{(\cdot_3, 1_1, 0_2, 0_4, 1_5), (\cdot_3, 0_1, 1_2, 1_4, 0_5)\}$. Then, $I_B(3; \mathbf{p}) = p_1 p_5 q_2 q_4 + p_2 p_4 q_1 q_5$.

The B-reliability importance in a dual system

Theorem 4.1.10 studies the B-reliability importance in a dual system.

Theorem 4.1.10 *For a coherent system ϕ with component reliability $\mathbf{p} = (p_1, p_2, \ldots, p_n)$ and its dual system ϕ^D with component reliability $\mathbf{q} = (q_1, q_2, \ldots, q_n)$, where $q_i = 1 - p_i$ for $i = 1, 2, \ldots, n$, let $I_B(i; \phi, \mathbf{p})$ and $I_B(i; \phi^D, \mathbf{q})$ denote the B-reliability importance of component i in system ϕ and its dual system ϕ^D, respectively. Then $I_B(i; \phi, \mathbf{p}) = I_B(i; \phi^D, \mathbf{q})$.*

Proof. Let $R(\mathbf{p})$ and $R^D(\mathbf{q})$ denote the reliability of system ϕ and its dual system ϕ^D, respectively, then $R^D(\mathbf{q}) = 1 - R(\mathbf{p})$. Thus,

$$\frac{\partial R(\mathbf{p})}{\partial p_i} = \frac{\partial(1 - R^D(\mathbf{q}))}{\partial q_i} \frac{\partial q_i}{\partial p_i} = \frac{\partial R^D(\mathbf{q})}{\partial q_i}.$$

By Equation (4.10), then $I_B(i; \phi, \mathbf{p}) = I_B(i; \phi^D, \mathbf{q})$.

Relations to systems reliability

Let $(\alpha_i, \beta_j, \mathbf{p}^{(ij)})$ denote the vector for which $p_i = \alpha$ and $p_j = \beta$, where $\alpha, \beta \in [0, 1]$ and $\mathbf{p}^{(ij)}$ is the vector obtained by deleting p_i and p_j from $\mathbf{p}$. Similarly, $(\alpha_j, \mathbf{p}^{(ij)})$ denotes the $(n - 1)$-dimensional vector specifically with $p_j = \alpha$. Recall that $I_B(i; \mathbf{p})$ is independent of p_i, and thus, can also be denoted as $I_B(i; \alpha_j, \mathbf{p}^{(ij)})$ when $p_j = \alpha$, $j \neq i$. Using this notation, Meng (1995) presented Theorem 4.1.11.

Theorem 4.1.11 *Given $i, j \in \{1, 2, \ldots, n\}, i \neq j$, and $0 \leq \mathbf{p}^{(ij)} \leq 1$, if $I_B(i; \alpha_j, \mathbf{p}^{(ij)}) > I_B(j; \alpha_i, \mathbf{p}^{(ij)})$ for some $0 \leq \alpha \leq 1$, then $R(\beta_i, \gamma_j, \mathbf{p}^{(ij)}) > R(\gamma_i, \beta_j, \mathbf{p}^{(ij)})$ for all $0 \leq \gamma < \beta \leq 1$.*

Proof. Using pivotal decomposition (2.10) on component j,

$$I_B(i; \alpha_j, \mathbf{p}^{(ij)}) = R(1_i, \alpha_j, \mathbf{p}^{(ij)}) - R(0_i, \alpha_j, \mathbf{p}^{(ij)})$$
$$= \alpha R(1_i, 1_j, \mathbf{p}^{(ij)}) + (1 - \alpha)R(1_i, 0_j, \mathbf{p}^{(ij)}) - \alpha R(0_i, 1_j, \mathbf{p}^{(ij)})$$
$$- (1 - \alpha)R(0_i, 0_j, \mathbf{p}^{(ij)}).$$

Thus, for all $0 \leq \alpha \leq 1$,

$$I_B(i; \alpha_j, \mathbf{p}^{(ij)}) - I_B(j; \alpha_i, \mathbf{p}^{(ij)}) = R(1_i, 0_j, \mathbf{p}^{(ij)}) - R(0_i, 1_j, \mathbf{p}^{(ij)}) > 0.$$

Similarly, the following can be obtained:

$$R(\beta_i, \gamma_j, \mathbf{p}^{(ij)}) - R(\gamma_i, \beta_j, \mathbf{p}^{(ij)}) = (\beta - \gamma)[R(1_i, 0_j, \mathbf{p}^{(ij)}) - R(0_i, 1_j, \mathbf{p}^{(ij)})].$$

Thus, the result follows immediately.

Given two positions i and j and $\mathbf{0} \leq \mathbf{p}^{(ij)} \leq \mathbf{1}$, if $I_{\mathrm{B}}(i; \alpha_j, \mathbf{p}^{(ij)}) > I_{\mathrm{B}}(j; \alpha_i, \mathbf{p}^{(ij)})$ for some $0 \leq \alpha \leq 1$, then it is true for any number α in $[0, 1]$. Thus, in practical applications, α could be set to 0 or 1 for simplicity. This is useful for investigating redundancy allocation and other reliability operations. As an example, suppose that a single spare can be put in redundancy with one of two identical components in the system. It then follows from Theorem 4.1.11 that the spare should be allocated to position i rather than position j if and only if $I_{\mathrm{B}}(i; \alpha_j, \mathbf{p}^{(ij)}) > I_{\mathrm{B}}(j; \alpha_i, \mathbf{p}^{(ij)})$ for some $0 \leq \alpha \leq 1$ (Meng 1995).

The B-reliability importance of a modular set

Let a coherent system (N, ϕ) have a module (M, χ).

Definition 4.1.12 *The B-reliability importance of modular set M, denoted by $I_{\mathrm{B}}^M(M; \mathbf{p})$, is defined as the probability that the module is critical for the system, that is, the probability that the failure and functioning of the module coincide with the failure and functioning of the system. Then*

$$I_{\mathrm{B}}^M(M; \mathbf{p}) = R(1_M, \mathbf{p}) - R(0_M, \mathbf{p}),$$

where $R(\alpha_M, \mathbf{p})$ denotes the reliability of the system when module (M, χ) is either functioning ($\alpha = 1$) or failed ($\alpha = 0$).

Note that $I_{\mathrm{B}}^M(M; \mathbf{p}) \neq \sum_{i \in M} I_{\mathrm{B}}(i; \phi, \mathbf{p})$. Theorem 4.1.13 (Birnbaum 1969) gives the relationship between $I_{\mathrm{B}}^M(M; \mathbf{p})$ and $I_{\mathrm{B}}(i; \phi, \mathbf{p})$, which is referred to as the *chain rule property* because it can be obtained by the chain rule for differentiation using Equations (2.14) and (4.10). Theorem 4.1.13 in conjunction with modular decomposition can simplify the calculation of the B-reliability importance of a component in a module of a system by computing the importance of the component in a module and of the module in the system and by repeating this process step-by-step as modules are substituted for components.

Theorem 4.1.13 *For $i \in M$,*

$$I_{\mathrm{B}}(i; \phi, \mathbf{p}) = (R(1_M, \mathbf{p}) - R(0_M, \mathbf{p})) \left(R_\chi \left(1_i, \mathbf{p}^M \right) - R_\chi \left(0_i, \mathbf{p}^M \right) \right), \quad (4.16)$$

$$= I_{\mathrm{B}}^M(M; \mathbf{p}) I_{\mathrm{B}}(i; \chi, \mathbf{p}^M), \quad (4.17)$$

where $\mathbf{p}^M$ denotes the vector with elements p_i, $i \in M$, $R_\chi\left(\alpha_i, \mathbf{p}^M\right)$ denotes the reliability of module (M, χ) when component i has reliability α, and $I_{\mathrm{B}}(i; \chi, \mathbf{p}^M)$ denotes the B-reliability importance of component i in module (M, χ). Similarly,

$$I_{\mathrm{Bs}}(i; \phi, \mathbf{p}) = I_{\mathrm{B}}^M(M; \mathbf{p}) I_{\mathrm{Bs}}(i; \chi, \mathbf{p}^M),$$

$$I_{\mathrm{Bf}}(i; \phi, \mathbf{p}) = I_{\mathrm{B}}^M(M; \mathbf{p}) I_{\mathrm{Bf}}(i; \chi, \mathbf{p}^M).$$

As discussed in Section 2.4, Equation (4.16) verifies again that the failure of component i causes the module to fail and that module failure causes the system to fail. That is, component i cannot cause system failure except through module failure.

4.1.1 The B-reliability Importance for System Functioning and for System Failure

According to Equations (4.6)–(4.11), the B-reliability importance measures of component i for the functioning and for the failure of ϕ are proportional to the B-reliability importance, that is,

$$I_{\mathrm{Bs}}(i; \mathbf{p}) = R(1_i, \mathbf{p}) - R(\mathbf{p}) = q_i I_{\mathrm{B}}(i; \mathbf{p}), \tag{4.18}$$

$$I_{\mathrm{Bf}}(i; \mathbf{p}) = R(\mathbf{p}) - R(0_i, \mathbf{p}) = p_i I_{\mathrm{B}}(i; \mathbf{p}). \tag{4.19}$$

The B-reliability importance for system functioning is also known as the *improvement potential importance* (Aven and Jensen 1999; Freixas and Pons 2008) or *risk achievement importance* (van der Borst and Schoonakker 2001). It is commonly used in reliability and risk design (Aven and Nøkland 2010). $I_{\mathrm{Bs}}(i; \mathbf{p})$ is equal to the increase of systems reliability when component i is perfect, representing the maximum potential improvement in systems reliability that can be obtained by improving the reliability of component i. A component is a *perfect component* if its reliability is 1. $I_{\mathrm{Bf}}(i; \mathbf{p})$ represents the significance of the failure of component i to the system failure.

According to Equations (4.18) and (4.19) and Theorem 4.1.10, the B-reliability importance measures for system functioning and the one for system failure are counterparts for a system and its dual system, as stated in Corollary 4.1.14.

Corollary 4.1.14 *For a coherent system ϕ with component reliability vector $\mathbf{p}$ and its dual system ϕ^D with component reliability vector $\mathbf{q} = \mathbf{1} - \mathbf{p}$, $I_{\mathrm{Bs}}(i; \phi, \mathbf{p}) = I_{\mathrm{Bf}}(i; \phi^D, \mathbf{q})$, where $I_{\mathrm{Bs}}(i; \phi, \mathbf{p})$ and $I_{\mathrm{Bf}}(i; \phi^D, \mathbf{q})$ denote the B-reliability importance of component i for the functioning of ϕ and that for the failure of ϕ^D, respectively. Similarly, $I_{\mathrm{Bf}}(i; \phi, \mathbf{p}) = I_{\mathrm{Bs}}(i; \phi^D, \mathbf{q})$.*

4.1.2 The Criticality Reliability Importance

Kuo and Zuo (2003) proposed two criticality reliability importance measures, which overcome a disadvantage of the B-reliability importance, that the B-reliability importance of a component is independent of its own reliability value.

Definition 4.1.15 *The criticality reliability importance of component i for system functioning, denoted by $I_{\mathrm{Cs}}(i; \mathbf{p})$, is defined as the probability that component i functions and is critical for system functioning, given that the system functions. Mathematically,*

$$I_{\mathrm{Cs}}(i; \mathbf{p}) = \frac{p_i \left(R(1_i, \mathbf{p}) - R(0_i, \mathbf{p}) \right)}{R(\mathbf{p})} = \frac{p_i}{R(\mathbf{p})} I_{\mathrm{B}}(i; \mathbf{p}). \tag{4.20}$$

The criticality reliability importance of component i for system failure, denoted by $I_{\mathrm{Cf}}(i; \mathbf{p})$, is defined as the probability that component i fails and is critical for system failure, given that the system fails. Mathematically,

$$I_{\mathrm{Cf}}(i; \mathbf{p}) = \frac{q_i \left(R(1_i, \mathbf{p}) - R(0_i, \mathbf{p}) \right)}{1 - R(\mathbf{p})} = \frac{q_i}{1 - R(\mathbf{p})} I_{\mathrm{B}}(i; \mathbf{p}). \tag{4.21}$$

$I_{\text{Cs}}(i; \mathbf{p})$ represents the significance of component i to system functioning given that the system functions. Note that $I_{\text{Cs}}(i; \mathbf{p}) = I_{\text{Bf}}(i; \mathbf{p})/R(\mathbf{p})$. That is, $I_{\text{Cs}}(\cdot; \mathbf{p})$ and $I_{\text{Bf}}(\cdot; \mathbf{p})$ produce the same ranking of components, evaluating two sides (functioning and failure) to the problem of judging the importance of components.

The criticality reliability importance, defined in terms of system failure, can be used in fault diagnosis. When a system fails, the component with the largest $I_{\text{Cf}}(i; \mathbf{p})$ value is most likely to have caused the failure and thus should be checked first. In fact, $I_{\text{Cf}}(i; \mathbf{p}) = I_{\text{Bs}}(i; \mathbf{p})/(1 - R(\mathbf{p}))$; thus, these two importance measures are equivalent in use.

Similar to Corollary 4.1.14, the criticality reliability importance for system functioning and the one for system failure are counterparts for a system and its dual system, as stated in Corollary 4.1.16.

Corollary 4.1.16 *For a coherent system ϕ with component reliability vector $\mathbf{p}$ and its dual system ϕ^D with component reliability vector $\mathbf{q} = \mathbf{1} - \mathbf{p}$, $I_{\text{Cs}}(i; \phi, \mathbf{p}) = I_{\text{Cf}}(i; \phi^D, \mathbf{q})$, where $I_{\text{Cs}}(i; \phi, \mathbf{p})$ and $I_{\text{Cf}}(i; \phi^D, \mathbf{q})$ denote the criticality reliability importance of component i for the functioning of ϕ and that for the failure of ϕ^D, respectively. Similarly, $I_{\text{Cf}}(i; \phi, \mathbf{p}) = I_{\text{Cs}}(i; \phi^D, \mathbf{q})$.*

4.1.3 The Bayesian Reliability Importance

When a system fails, it is of interest to figure out which component has caused the failure and how important the different components are for the system failure. For this purpose, Birnbaum (1969) proposed the Bayesian reliability importance, which is defined as follows.

Definition 4.1.17 *The Bayesian reliability importance of component i, denoted by $I_{\text{Bay}}(i; \mathbf{p})$, is defined as the probability that component i fails, given that the system fails. Mathematically,*

$$I_{\text{Bay}}(i; \mathbf{p}) = \Pr\{X_i = 0 | \phi(\mathbf{X}) = 0\}.$$

From the proof of Theorem 4.1.2, it follows that

$$\Pr\{\phi(\mathbf{X}) = 0 | X_i = 0\} = 1 - R(\mathbf{p}) + p_i I_{\text{B}}(i; \mathbf{p}).$$

Thus, by Bayesian formulation (Singpurwalla 2006), for $i = 1, 2, \ldots, n$,

$$
\begin{aligned}
I_{\text{Bay}}(i; \mathbf{p}) &= \Pr\{X_i = 0 | \phi(\mathbf{X}) = 0\} \\
&= \frac{\Pr\{\phi(\mathbf{X}) = 0 | X_i = 0\} \Pr\{X_i = 0\}}{\Pr\{\phi(\mathbf{X}) = 0\}} \\
&= q_i \left(1 + \frac{p_i}{1 - R(\mathbf{p})} I_{\text{B}}(i; \mathbf{p})\right).
\end{aligned}
$$

Thus, the Bayesian reliability importance is a linear function of the B-reliability importance. Unlike $I_{\text{B}}(i; \mathbf{p})$, which does not depend on the reliability of component i, p_i, $I_{\text{Bay}}(i; \mathbf{p})$ is related to p_i.

4.2 The FV Reliability Importance

The FV importance is classified into c-type and p-type. The c-type FV importance, referred to as c-FV importance, takes into account the contribution of a component to system failure, and its definition is based on cuts. The p-type FV importance, referred to as p-FV importance, takes into account the contribution of a component to system functioning and is derived on the basis of paths. Youngblood (2001) proposed the *prevention worth measure* of basic events in the context of risk and safety analysis, which is based on paths and is similar to the p-FV importance. Subsections 4.2.1 and 4.2.2 define the c-FV and p-FV reliability importance measures, respectively. Subsection 4.2.3 introduces a method for calculating the FV importance based on the decomposition of state vectors. Subsection 4.2.4 presents the properties of the FV importance, especially when a component is in series or parallel with the rest of the system.

4.2.1 The c-FV Reliability Importance

The FV importance was initially proposed (Fussell 1975; Vesely 1970) in the context of fault trees; however, this book illustrates the FV importance using the terminology of coherent system for consistency with other importance measures. For both repairable and nonrepairable systems, Vesely (1970) presented exact expressions for the lifetime distributions of the system, minimal cuts, and components using a time-dependent methodology, indicating that with these probabilities, the components and the minimal cuts, which are most likely to cause the system to fail, can be determined. On that basis, Fussell (1975) proposed the time-dependent unreliability importance of component i, which is defined as the probability (a function of time) that failure of component i contributes to system failure given that the system fails. Assuming that a mission time is implicit and fixed, then the c-FV reliability importance can be defined as follows.

Definition 4.2.1 *The c-FV reliability importance of component i, denoted by $I_{FV^c}(i; \mathbf{p})$, is defined as the probability that a component state vector has a corresponding cut that causes system failure and contains a minimal cut in $\overline{\mathscr{C}}_i$ (i.e., containing component i). Mathematically,*

$$I_{FV^c}(i; \mathbf{p}) = \Pr\{\exists C \in \overline{\mathscr{C}}_i \ s.t. \ C \subseteq N_0(\mathbf{X}) | \phi(\mathbf{X}) = 0\}$$

$$= \frac{q_i \Pr\{(0_i, \mathbf{X}) : \exists C \in \overline{\mathscr{C}}_i \ s.t. \ C \subseteq N_0(0_i, \mathbf{X})\}}{1 - R(\mathbf{p})}, \qquad (4.22)$$

where $C \in \overline{\mathscr{C}}_i$ denotes the minimal cut containing component i.

4.2.2 The p-FV Reliability Importance

Essentially, the p-FV importance (Meng 1996) is the probability that at least one minimal path containing component i functions, given that the system functions.

Definition 4.2.2 *The p-FV reliability importance of component i, denoted by $I_{FV^p}(i; \mathbf{p})$, is defined as the probability that a component state vector has the corresponding path that* *makes system function and contains a minimal path in $\overline{\mathscr{P}}_i$ (i.e., containing component i). Mathematically,*

$$I_{\text{FV}^p}(i; \mathbf{p}) = \Pr\{\exists P \in \overline{\mathscr{P}}_i \text{ s.t. } P \subseteq N_1(\mathbf{X}) | \phi(\mathbf{X}) = 1\}$$

$$= \frac{p_i \Pr\{(1_i, \mathbf{X}) : \exists P \in \overline{\mathscr{P}}_i \text{ s.t. } P \subseteq N_1(1_i, \mathbf{X})\}}{R(\mathbf{p})}, \tag{4.23}$$

where $P \in \overline{\mathscr{P}}_i$ denotes the minimal path containing component i.

As shown in Theorem 4.2.3, the p-FV importance is closely related to the c-FV importance. Theorem 4.2.3 is easily proven by Definition 2.2.5 of duality and by noting that the minimal cuts in a system are the minimal paths in its dual system, and vice versa.

Theorem 4.2.3 *For a coherent system ϕ with component reliability vector $\mathbf{p}$ and its dual system ϕ^D with component reliability vector $\mathbf{q} = \mathbf{1} - \mathbf{p}$, $I_{\text{FV}^c}(i; \phi, \mathbf{p}) = I_{\text{FV}^p}(i; \phi^D, \mathbf{q})$, where $I_{\text{FV}^c}(i; \phi, \mathbf{p})$ denotes the c-FV importance of component i in system ϕ, and $I_{\text{FV}^p}(i; \phi^D, \mathbf{q})$ denotes the p-FV importance of component i in its dual system ϕ^D. Similarly, $I_{\text{FV}^p}(i; \phi, \mathbf{p}) = I_{\text{FV}^c}(i; \phi^D, \mathbf{q})$.*

4.2.3 Decomposition of State Vectors

Decomposition of state vectors for a component, which was initially presented by Meng (1996), further illustrates the FV importance and is useful in computing and deriving the properties of the FV importance. For $i = 1, 2, \ldots, n$, let sets

$$U_1(i) = \{(\cdot_i, \mathbf{x}) : \exists C \in \overline{\mathscr{C}}_i \text{ s.t. } C \subseteq N_0(0_i, \mathbf{x})\}, \tag{4.24}$$

$$U_2(i) = \{(\cdot_i, \mathbf{x}) : \exists P \in \overline{\mathscr{P}}_i \text{ s.t. } P \subseteq N_1(1_i, \mathbf{x})\}. \tag{4.25}$$

Then, $\phi(0_i, \mathbf{x}) = 0$ if $(\cdot_i, \mathbf{x}) \in U_1(i)$, and $\phi(1_i, \mathbf{x}) = 1$ if $(\cdot_i, \mathbf{x}) \in U_2(i)$. Recall that $B(i)$ is the set of state vectors in which component i is critical for the system:

$$B(i) = \{(\cdot_i, \mathbf{x}) : \phi(1_i, \mathbf{x}) - \phi(0_i, \mathbf{x}) = 1\} = U_1(i) \cap U_2(i). \tag{4.26}$$

Define sets $V_1(i)$ and $V_2(i)$ as

$$V_1(i) = U_1(i) \setminus B(i) = \{(\cdot_i, \mathbf{x}) : \phi(1_i, \mathbf{x}) = 0; \exists C \in \overline{\mathscr{C}}_i \text{ s.t. } C \subseteq N_0(0_i, \mathbf{x})\}, \tag{4.27}$$

$$V_2(i) = U_2(i) \setminus B(i) = \{(\cdot_i, \mathbf{x}) : \phi(0_i, \mathbf{x}) = 1; \exists P \in \overline{\mathscr{P}}_i \text{ s.t. } P \subseteq N_1(1_i, \mathbf{x})\}. \tag{4.28}$$

If a vector $(\cdot_i, \mathbf{x})$ is not in one of the three *disjoint* sets $B(i)$, $V_1(i)$, or $V_2(i)$, then $(\cdot_i, \mathbf{x})$ must be in either $D_1(i)$ or $D_2(i)$, which are defined as

$$D_1(i) = \{(\cdot_i, \mathbf{x}) : \phi(1_i, \mathbf{x}) = 0; C \not\subseteq N_0(0_i, \mathbf{x}) \text{ for all } C \in \overline{\mathscr{C}}_i\},$$

$$D_2(i) = \{(\cdot_i, \mathbf{x}) : \phi(0_i, \mathbf{x}) = 1; P \not\subseteq N_1(1_i, \mathbf{x}) \text{ for all } P \in \overline{\mathscr{P}}_i\}.$$

$D_1(i)$ and $D_2(i)$ are the set of unessential vectors for component i because

(i) If $(\cdot_i, \mathbf{x}) \in D_1(i)$, then $\phi(0_i, \mathbf{x}) = \phi(1_i, \mathbf{x}) = 0$, and with respect to $(0_i, \mathbf{x})$, there is no minimal cut containing component i such that all components in the minimal cuts fail. Thus, in this instance, repairing component i alone cannot restore the system to functioning.
(ii) If $(\cdot_i, \mathbf{x}) \in D_2(i)$, then, similarly, cutting component i off cannot take the system down.

Let $\{(\cdot_i, \mathbf{x})\}$ denote the set of all state vectors of $(\cdot_i, \mathbf{x})$, then

$$\{(\cdot_i, \mathbf{x})\} = B(i) \cup V_1(i) \cup V_2(i) \cup D_1(i) \cup D_2(i).$$

Note that $B(i)$, $V_1(i)$, $V_2(i)$, $D_1(i)$, and $D_2(i)$ are mutually disjoint.

Equations (4.29)–(4.32) are obtained from Equations (4.22)–(4.25), (4.27), and (4.28):

$$I_{\text{FV}^c}(i; \mathbf{p}) = \frac{q_i \Pr\{U_1(i); \mathbf{p}\}}{\Pr\{\phi(\mathbf{X}) = 0\}} = \frac{q_i \left[\Pr\{B(i); \mathbf{p}\} + \Pr\{V_1(i); \mathbf{p}\}\right]}{\Pr\{\phi(\mathbf{X}) = 0\}}, \tag{4.29}$$

$$\Pr\{\phi(\mathbf{X}) = 0\} = \Pr\{V_1(i); \mathbf{p}\} + \Pr\{D_1(i); \mathbf{p}\} + q_i \Pr\{B(i); \mathbf{p}\}, \tag{4.30}$$

$$I_{\text{FV}^p}(i; \mathbf{p}) = \frac{p_i \Pr\{U_2(i); \mathbf{p}\}}{\Pr\{\phi(\mathbf{X}) = 1\}} = \frac{p_i \left[\Pr\{B(i); \mathbf{p}\} + \Pr\{V_2(i); \mathbf{p}\}\right]}{\Pr\{\phi(\mathbf{X}) = 1\}}, \tag{4.31}$$

$$\Pr\{\phi(\mathbf{X}) = 1\} = \Pr\{V_2(i); \mathbf{p}\} + \Pr\{D_2(i); \mathbf{p}\} + p_i \Pr\{B(i); \mathbf{p}\}, \tag{4.32}$$

where $\Pr\{B(i); \mathbf{p}\} = \Pr\{(\cdot_i, \mathbf{X}) \in B(i)\}$ and so on. Example 4.2.4 illustrates how to use these equations to compute the FV importance.

Example 4.2.4 (Example 2.2.3 continued) For 2-out-of-3 system, $\phi(x_1, x_2, x_3) = 1 - (1 - x_1 x_2)(1 - x_1 x_3)(1 - x_2 x_3)$. The minimal cuts are $\{1, 2\}$, $\{2, 3\}$, and $\{1, 3\}$, and $B(1) = \{(\cdot_1, 1_2, 0_3), (\cdot_1, 0_2, 1_3)\}$, $V_1(1) = \{(\cdot_1, 0_2, 0_3)\}$, and $V_2(1) = \{(\cdot_1, 1_2, 1_3)\}$. Thus, $\Pr\{B(1); \mathbf{p}\} + \Pr\{V_1(1); \mathbf{p}\} = 1 - p_2 p_3$. Hence, from Equations (4.29) and (4.31), respectively,

$$I_{\text{FV}^c}(1; \mathbf{p}) = \frac{q_1(1 - p_2 p_3)}{1 - R(\mathbf{p})},$$

$$I_{\text{FV}^p}(1; \mathbf{p}) = \frac{p_1(1 - q_2 q_3)}{R(\mathbf{p})},$$

where $R(\mathbf{p}) = p_1 p_2 + p_2 p_3 + p_1 p_3 - 2 p_1 p_2 p_3$.

4.2.4 Properties

Lemma 4.2.5 (Meng 1996) follows from Equations (4.29)–(4.32). Different from the B-reliability importance, the FV reliability importance of component i is dependent on p_i.

Lemma 4.2.5 $I_{\text{FV}^c}(i; \mathbf{p})$ $(I_{\text{FV}^p}(i; \mathbf{p}))$ *is decreasing (increasing) in* p_i.

Using the decomposition of state vectors, Meng (1996) proved Theorem 4.2.6.

Theorem 4.2.6 *Let component i be in series (parallel) with the rest of the system. Then $I_{FV^c}(i; \mathbf{p}) \geq I_{FV^c}(j; \mathbf{p})$ $(I_{FV^p}(i; \mathbf{p}) \leq I_{FV^p}(j; \mathbf{p}))$ for all $0 < \mathbf{p} < 1$ satisfying $p_i \leq p_j, j \neq i$. The strict inequality holds for some $0 < \mathbf{p} < 1$ satisfying $p_i \leq p_j$ unless component j is also a one-component cut (path).*

Proof. Suppose that component i is in series with the rest of the system. Then, $\phi(0_i, \mathbf{x}) = 0$ for all $(\cdot_i, \mathbf{x})$,

$$\{\mathbf{x}^{(ij)} : (\cdot_j, 0_i, \mathbf{x}^{(ij)}) \in U_1(j)\} \subseteq \{\mathbf{x}^{(ij)} : (\cdot_i, 0_j, \mathbf{x}^{(ij)}) \in U_1(i)\}, \text{ and} \quad (4.33)$$

$$\{\mathbf{x}^{(ij)} : (\cdot_j, 1_i, \mathbf{x}^{(ij)}) \in U_1(j)\} \subseteq \{\mathbf{x}^{(ij)} : (\cdot_i, 1_j, \mathbf{x}^{(ij)}) \in U_1(i)\}. \quad (4.34)$$

Thus, for vector $\mathbf{1}^{(ij)}$, $(\cdot_i, 0_j, \mathbf{1}^{(ij)}) \in U_1(i)$ and $(\cdot_i, 1_j, \mathbf{1}^{(ij)}) \in U_1(i)$ because component i itself is a cut. Meantime, $(\cdot_j, 0_i, \mathbf{1}^{(ij)}) \in U_1(j)$ and $(\cdot_j, 1_i, \mathbf{1}^{(ij)}) \in U_1(j)$ if and only if component j itself is a cut. Therefore, the strict containing relations (4.33) and (4.34) hold unless component j is also a one-component cut.

Using relations (4.33) and (4.34) and noting that $p_i \leq p_j$, then

$$q_i \Pr\{U_1(i); \mathbf{p}\} - q_j \Pr\{U_1(j); \mathbf{p}\}$$

$$= q_i p_j \Pr\{(\cdot_i, 1_j, \mathbf{X}^{(ij)}) \in U_1(i)\} + q_i q_j \Pr\{(\cdot_i, 0_j, \mathbf{X}^{(ij)}) \in U_1(i)\}$$

$$- q_j p_i \Pr\{(\cdot_j, 1_i, \mathbf{X}^{(ij)}) \in U_1(j)\} - q_j q_i \Pr\{(\cdot_j, 0_i, \mathbf{X}^{(ij)}) \in U_1(j)\}$$

$$\geq 0.$$

Thus, according to Equation (4.29), $I_{FV^c}(i; \mathbf{p}) \geq I_{FV^c}(j; \mathbf{p})$. Furthermore, the strict inequality holds for some $0 < \mathbf{p} < 1$ satisfying $p_i \leq p_j$ unless component j is also a one-component cut.

The part of the p-FV importance for the parallel connection case can be similarly proved.

If component i is in parallel with the rest of the system, then $V_1(i) = D_1(i) = \emptyset$. Thus, from Equations (4.29) and (4.30), $I_{FV^c}(i; \mathbf{p}) = 1$ irrespective of the values of $p_1, p_2, \ldots, p_n$. Furthermore,

$$\Pr\{B(i); \mathbf{p}\} = \Pr\{\phi(0_i, \mathbf{X}) = 0\},$$

and by putting $P = \{i\}$ in Equation (4.28), then

$$\Pr\{V_2(i); \mathbf{p}\} = \Pr\{\phi(0_i, \mathbf{X}) = 1\}.$$

Hence, $\Pr\{B(i)\} + \Pr\{V_2(i)\} = 1$, and from Equation (4.31),

$$I_{FV^p}(i; \mathbf{p}) = \frac{p_i}{\Pr\{\phi(\mathbf{X}) = 1\}}.$$

Then, in a parallel system, the most reliable component is the most important according to the p-FV importance, which is different from the result obtained from the c-FV importance that all components are equally important irrespective of their reliability values.

If component i is in series with the rest of the system, then $V_2(i) = D_2(i) = \emptyset$ and $I_{FV^p}(i; \mathbf{p}) = 1$ irrespective of the values of $p_1, p_2, \ldots, p_n$; thus, in a series system, $I_{FV^p}(i; \mathbf{p}) = 1$ for all $i = 1, 2, \ldots, n$. On the other hand, it is easy to show that the c-FV reliability importance assigns the largest value to the least reliable component in a series system.

Hence, the c-FV importance and p-FV importance are complementary to each other. The choice of the c-FV or p-FV importance then is related to the system structure.

References

Armstrong MJ. 1995. Joint reliability-importance of components. *IEEE Transactions on Reliability* **44**, 408–412.

Aven T and Jensen U. 1999. *Stochastic Models in Reliability*. Springer, New York.

Aven T and Nøkland TE. 2010. On the use of uncertainty importance measures in reliability and risk analysis. *Reliability Engineering and System Safety* **95**, 127–133.

Barlow RE and Proschan F. 1975. *Statistical Theory of Reliability and Life Testing Probability Models*. Holt, Rinehart and Winston, New York.

Birnbaum ZW. 1969. On the importance of different components in a multicomponent system. In *Multivariate Analysis, Vol. 2* (ed. Krishnaiah PR). Academic Press, New York, pp. 581–592.

Boland PJ and Proschan F. 1983. The reliability of k out of n systems. *Annals of Probability* **11**, 760–764.

Chadjiconstantinidis S and Koutras MV. 1999. Measures of component importance for Markov chain imbeddable reliability structures. *Naval Research Logistics* **46**, 613–639.

Freixas J and Pons M. 2008. The influence of the node criticality relation on some measures of component importance. *Operations Research Letters* **36**, 557–560.

Fussell JB. 1975. How to hand-calculate system reliability and safety characteristics. *IEEE Transactions on Reliability* **R-24**, 169–174.

Hong JS and Lie CH. 1993. Joint reliability-importance of two edges in an undirected network. *IEEE Transactions on Reliability* **42**, 17–33.

Hsu SJ and Yuang MC. 1999. Efficient computation of marginal reliability importance for reducible networks in network management. *Proceedings of the 1999 IEEE International Conference on Communications*, pp. 1039–1045.

Kuo W and Zuo MJ. 2003. *Optimal Reliability Modeling: Principles and Applications*. John Wiley & Sons, New York.

Lee HS, Lie CH, and Hong JS. 1997. A computation method for evaluating importance-measures of gates in a fault tree. *IEEE Transactions on Reliability* **46**, 360–365.

Lu L and Jiang J. 2007. Joint failure importance for noncoherent fault trees. *IEEE Transactions on Reliability* **56**, 435–443.

Meng FC. 1995. Some further results on ranking the importance of system components. *Reliability Engineering and System Safety* **47**, 97–101.

Meng FC. 1996. Comparing the importance of system components by some structural characteristics. *IEEE Transactions on Reliability* **45**, 59–65.

Singpurwalla ND. 2006. *Reliability and Risk: A Bayesian Perspective*. Wiley Series in Probability and Statistics. John Wiley & Sons, New York.

van der Borst M and Schoonakker H. 2001. An overview of PSA importance measures. *Reliability Engineering and System Safety* **72**, 241–245.

Vesely WE. 1970. A time dependent methodology for fault tree evaluation. *Nuclear Engineering and Design* **13**, 337–360.

Youngblood RW. 2001. Risk significance and safety significance. *Reliability Engineering and System Safety* **73**, 121–136.

5

Lifetime Importance Measures

This chapter introduces three lifetime importance measures, including the B-TDL importance, the FV TDL importance, and the BP TIL importance and the BP TDL importance in Sections 5.1, 5.2, 5.3, and 5.4, respectively. The BP importance is defined on the basis of the B-TDL importance. Similar to the B-importance and FV importance, the BP importance is also of the structure type (Section 6.3) in addition to the TIL and TDL types. Section 5.5 compares the TDL importance measures numerically. Finally, Section 5.6 gives a summary on the three lifetime importance measures.

5.1 The B-time-dependent-lifetime Importance

Lambert (1975) and Natvig (1979) extended the B-reliability importance to the B-TDL importance by using component lifetime distributions, as in Theorem 3.2.1. The properties and results for the B-reliability importance in Section 4.1 and the ones for the B-TDL importance in this section are interapplicable by appropriately using p_i and $\overline{F}_i(t)$.

Definition 5.1.1 *The B-TDL importance of component i at time t, denoted by $I_B(i; \overline{\mathbf{F}}(t))$, is defined as the probability that the system is in a state at time t in which component i is critical for the system, that is, the probability that at time t the failure and functioning of component i coincide with system failure and functioning, respectively. Mathematically,*

$$I_B(i; \overline{\mathbf{F}}(t)) = \Pr\{\phi(1_i, \mathbf{X}(t)) - \phi(0_i, \mathbf{X}(t)) = 1\} = R(1_i, \overline{\mathbf{F}}(t)) - R(0_i, \overline{\mathbf{F}}(t)). \qquad (5.1)$$

Xie (1988) gave an upper bound on the B-TDL importance as in inequality (5.2), of which only one of the two terms in the right-hand side can be less than one. The result is useful in deriving Theorem 10.2.7 for the L_1 TIL importance measure introduced in Subsection 10.2.2.

Lemma 5.1.2 *The B-TDL importance of component i is bounded from above as*

$$I_B(i; \overline{\mathbf{F}}(t)) \leq \min \left\{ \frac{\overline{F}_\phi(t)}{\overline{F}_i(t)}, \frac{F_\phi(t)}{F_i(t)} \right\}. \qquad (5.2)$$

Importance Measures in Reliability, Risk, and Optimization: Principles and Applications, First Edition.
Way Kuo and Xiaoyan Zhu. © 2012 John Wiley & Sons, Ltd. Published 2012 by John Wiley & Sons, Ltd.

Proof. Note that $I_B(i; \overline{\mathbf{F}}(t)) = R(1_i, \overline{\mathbf{F}}(t)) - R(0_i, \overline{\mathbf{F}}(t))$ by Equation (4.11). First, $I_B(i; \overline{\mathbf{F}}(t)) \leq \overline{F}_\phi(t)/\overline{F}_i(t)$ because

$$\overline{F}_\phi(t) = \left(R(1_i, \overline{\mathbf{F}}(t)) - R(0_i, \overline{\mathbf{F}}(t))\right)\overline{F}_i(t) + R(0_i, \overline{\mathbf{F}}(t)) \qquad (5.3)$$

$$\geq I_B(i; \overline{\mathbf{F}}(t))\overline{F}_i(t).$$

Then, $I_B(i; \overline{\mathbf{F}}(t)) \leq F_\phi(t)/F_i(t)$ follows from

$$\overline{F}_\phi(t) = R(1_i, \overline{\mathbf{F}}(t)) - \left(R(1_i, \overline{\mathbf{F}}(t)) - R(0_i, \overline{\mathbf{F}}(t))\right)F_i(t) \qquad (5.4)$$

$$\leq 1 - I_B(i; \overline{\mathbf{F}}(t))F_i(t).$$

Following Theorem 4.1.2,

$$I_B(i; \overline{\mathbf{F}}(t)) = \frac{\partial R(\overline{\mathbf{F}}(t))}{\partial \overline{F}_i(t)},$$

and consequently, by Equation (5.2),

$$\frac{\partial \overline{F}_\phi(t)}{\partial \overline{F}_i(t)} \leq \min\left\{\frac{\overline{F}_\phi(t)}{\overline{F}_i(t)}, \frac{F_\phi(t)}{F_i(t)}\right\}. \qquad (5.5)$$

This result can be extended to any module (embedded system) with a lifetime distribution $G(t)$ and a reliability function $\overline{G}(t) = 1 - G(t)$. That is,

$$\frac{\partial \overline{G}(t)}{\partial \overline{F}_i(t)} \leq \min\left\{\frac{\overline{G}(t)}{\overline{F}_i(t)}, \frac{G(t)}{F_i(t)}\right\}. \qquad (5.6)$$

In addition, similar to Equation (4.13),

$$\frac{d\overline{F}_\phi(t)}{dt} = \frac{dR(\overline{\mathbf{F}}(t))}{dt} = \sum_{i=1}^n \frac{\partial R(\overline{\mathbf{F}}(t))}{\partial \overline{F}_i(t)}\frac{d\overline{F}_i(t)}{dt} = \sum_{i=1}^n I_B(i; \overline{\mathbf{F}}(t))\frac{d\overline{F}_i(t)}{dt}, \qquad (5.7)$$

which is used in deriving the properties of some B-TDL importance related measures later.

Properties

The following two theorems are for the special cases in which a component is in series or parallel with the rest of a system (or *more general, with a module*) (Xie 1987). Theorem 5.1.3 is the direct result of Equations (5.3) and (5.4).

Theorem 5.1.3   *If component i is in series with the rest of the system, then*

$$I_B(i; \overline{\mathbf{F}}(t)) = \frac{\overline{F}_\phi(t)}{\overline{F}_i(t)}, \qquad (5.8)$$

and if component i is in parallel with the rest of the system, then

$$I_{\text{B}}(i; \overline{\mathbf{F}}(t)) = \frac{F_{\phi}(t)}{F_i(t)}. \tag{5.9}$$

According to Theorem 5.1.3, when component i is in series or in parallel with the rest of the system, $I_{\text{B}}(i; \overline{\mathbf{F}}(t))$ may be easily computed if the distributions of component i and the whole system are known. No other information about the rest of the system is needed; the component may even be dependent. Generally, $I_{\text{B}}(i; \overline{\mathbf{F}}(t))$ cannot be computed when only $F_{\phi}(t)$ and $F_i(t)$ are known. Theorem 5.1.3 is used in proving Theorem 10.3.8 for the yield TIL importance measure introduced in Subsection 10.3.1.

Theorem 5.1.4 *Let component i be in series (parallel) with a module containing component j. If $F_i(t) \geq F_j(t)$ $(\overline{F}_i(t) \geq \overline{F}_j(t))$ for $t \geq 0$, then $I_{\text{B}}(i; \overline{\mathbf{F}}(t)) \geq I_{\text{B}}(j; \overline{\mathbf{F}}(t))$.*

Proof. First suppose that component i is in series with the rest of the system. Then

$$
\begin{aligned}
I_{\text{B}}(i; \overline{\mathbf{F}}(t)) &= R(1_i, \overline{\mathbf{F}}(t)) \\
&= \overline{F}_j(t)R(1_i, 1_j, \overline{\mathbf{F}}(t)) + (1 - \overline{F}_j(t))R(1_i, 0_j, \overline{\mathbf{F}}(t)) \\
&= R(1_i, 0_j, \overline{\mathbf{F}}(t)) + \overline{F}_j(t)\left(R(1_i, 1_j, \overline{\mathbf{F}}(t)) - R(1_i, 0_j, \overline{\mathbf{F}}(t))\right).
\end{aligned}
$$

Letting $j \neq i$, then

$$
\begin{aligned}
I_{\text{B}}(j; \overline{\mathbf{F}}(t)) &= R(1_j, \overline{\mathbf{F}}(t)) - R(0_j, \overline{\mathbf{F}}(t)) \\
&= \overline{F}_i(t)R(1_i, 1_j, \overline{\mathbf{F}}(t)) - \overline{F}_i(t)R(1_i, 0_j, \overline{\mathbf{F}}(t)) \\
&= \overline{F}_i(t)\left(R(1_i, 1_j, \overline{\mathbf{F}}(t)) - R(1_i, 0_j, \overline{\mathbf{F}}(t))\right).
\end{aligned}
$$

Thus, if $F_i(t) \geq F_j(t)$, the result for the case of component i in series with the rest of the system follows immediately. Similar arguments can be applied to the case of component i in parallel with the rest of the system.

Furthermore, using the chain rule in Theorem 4.1.13, the results can be easily extended to the general case of component i in series or in parallel with a module.

5.1.1   The Criticality Time-dependent Lifetime Importance

Lambert (1975) defined the criticality TDL importance for system failure, which is relevant to failure diagnosis.

Definition 5.1.5   *The criticality TDL importance for system failure of component i at time t, denoted by $I_{\text{Cf}}(i; \overline{\mathbf{F}}(t))$, is defined as the probability that component i has failed by time t and that component i is critical for the system at time t, given that the system has failed by time t. Mathematically,*

$$I_{\text{Cf}}(i; \overline{\mathbf{F}}(t)) = \Pr\{\phi(1_i, \mathbf{X}(t)) - \phi(0_i, \mathbf{X}(t)) = 1 \text{ and } X_i(t) = 0 | \phi(\mathbf{X}(t)) = 0\}$$

$$= \frac{F_i(t)}{1 - R(\overline{\mathbf{F}}(t))} I_{\text{B}}(i; \overline{\mathbf{F}}(t)). \tag{5.10}$$

As shown in Equation (5.10), this criticality TDL importance is proportional to the B-TDL importance. Substituting $\overline{F}_i(t)$ for p_i, the criticality reliability importance for system failure in Equation (4.21) becomes the corresponding criticality TDL importance in Equation (5.10), as stated in Theorem 3.2.1. The criticality TDL importance for system functioning can be similarly defined.

5.2 The FV Time-dependent Lifetime Importance

5.2.1 The c-FV Time-dependent Lifetime Importance

Similar to Definition 4.2.1, the c-FV TDL importance can be defined as follows.

Definition 5.2.1 *The c-FV TDL importance of component i at time t, denoted by $I_{\mathrm{FV}^c}(i; \overline{\mathbf{F}}(t))$, is defined as the probability that at least one minimal cut containing component i fails at time t, given that the system fails at time t. Mathematically,*

$$I_{\mathrm{FV}^c}(i; \overline{\mathbf{F}}(t)) = \Pr\{\exists C \in \overline{\mathscr{C}}_i \ s.t. \ X_j(t) = 0 \ for \ all \ j \in C|\phi(\mathbf{X}(t)) = 0\}$$

$$= \frac{\Pr\{\exists C \in \overline{\mathscr{C}}_i \ s.t. \ X_j(t) = 0 \ for \ all \ j \in C\}}{1 - R(\overline{\mathbf{F}}(t))}$$

$$= \frac{\Pr\left\{1 - \prod_{C \in \overline{\mathscr{C}}_i}\left[1 - \prod_{j \in C}\left(1 - X_j(t)\right)\right]\right\}}{1 - R(\overline{\mathbf{F}}(t))},$$

where $C \in \overline{\mathscr{C}}_i$ denotes the minimal cut containing component i. Note that the term $1 - \prod_{C \in \overline{\mathscr{C}}_i}\left[1 - \prod_{j \in C}\left(1 - X_j(t)\right)\right]$ equals 1 if at least one minimal cut containing component i fails at time t; 0 otherwise.

The c-FV TDL importance of a modular set

Let a coherent system (N, ϕ) have a module (M, χ). Let $\overline{\mathscr{C}}_M$ denote the set of minimal cuts of module (M, χ), that is, the failure of each minimal cut in $\overline{\mathscr{C}}_M$ results in the failure of modular set M.

Definition 5.2.2 *The c-FV TDL importance of modular set M at time t, denoted by $I_{\mathrm{FV}^c}^M(M; \overline{\mathbf{F}}(t))$, is defined as the probability that at least one minimal cut in $\overline{\mathscr{C}}_M$ fails at time t, given that the system fails at time t. Mathematically,*

$$I_{\mathrm{FV}^c}^M(M; \overline{\mathbf{F}}(t)) = \Pr\{\exists C \in \overline{\mathscr{C}}_M \ s.t. \ \chi(\mathbf{X}(t)^M) = 0 \ and \ X_j(t) = 0 \ for \ all \ j \in C|\phi(\mathbf{X}(t)) = 0\}$$

$$= \frac{\Pr\{\exists C \in \overline{\mathscr{C}}_M \ s.t. \ X_j(t) = 0 \ for \ all \ j \in C\}}{1 - R(\overline{\mathbf{F}}(t))}$$

$$= \frac{\Pr\left\{1 - \prod_{C \in \overline{\mathscr{C}}_M}\left[1 - \prod_{j \in C}\left(1 - X_j(t)\right)\right]\right\}}{1 - R(\overline{\mathbf{F}}(t))},$$

where $\mathbf{X}(t)^M$ is a vector with elements $X_i(t)$, $i \in M$. The illustrations on these mathematical expressions are similar and refer to Definition 5.2.1.

On the basis of this definition, the c-FV importance of a modular set is independent of the modular decomposition of the system. Theorem 5.2.3 gives the relations of the c-FV importance of a modular set to that of the components in the module.

Theorem 5.2.3 *The following is true and holds the equality if and only if a module (M, χ) is a series structure:*

$$I_{FV^c}^M(M; \overline{\mathbf{F}}(t)) \leq \sum_{i \in M} I_{FV^c}(i; \overline{\mathbf{F}}(t)).$$

Proof. For any component $i \in M$, $\overline{\mathcal{C}}_i \subseteq \overline{\mathcal{C}}_M$. To illustrate, supposing there exists a minimal cut $C \in \overline{\mathcal{C}}_i$ but $C \notin \overline{\mathcal{C}}_M$, then $C \setminus \{i\}$ is still a cut, which contradicts the fact that C is a minimal cut. Thus, any minimal cut containing at least one component in M must be a minimal cut such that $\chi(\mathbf{X}(t)^M) = 0$. On the other hand, any minimal cut such that $\chi(\mathbf{X}(t)^M) = 0$ must contain at least one component in modular set M. Therefore,

$$\overline{\mathcal{C}}_M = \bigcup_{i \in M} \overline{\mathcal{C}}_i.$$

Then, starting from its definition,

$$I_{FV^c}^M(M; \overline{\mathbf{F}}(t)) = \Pr\{\exists C \in \bigcup_{i \in M} \overline{\mathcal{C}}_i \ s.t. \ X_j(t) = 0 \ \text{for all} \ j \in C | \phi(\mathbf{X}(t)) = 0\}$$

$$\leq \sum_{i \in M} \Pr\{\exists C \in \overline{\mathcal{C}}_i \ s.t. \ X_j(t) = 0 \ \text{for all} \ j \in C | \phi(\mathbf{X}(t)) = 0\} \quad (5.11)$$

$$= \sum_{i \in M} I_{FV^c}(i; \overline{\mathbf{F}}(t)).$$

The inequality in (5.11) holds because it is possible that $\overline{\mathcal{C}}_{i_1} \cap \overline{\mathcal{C}}_{i_2} \neq \emptyset$ for $i_1, i_2 \in M$ and $i_1 \neq i_2$. $\overline{\mathcal{C}}_{i_1} \cap \overline{\mathcal{C}}_{i_2} = \emptyset$ holds for any $i_1, i_2 \in M$ and $i_1 \neq i_2$ if and only if (M, χ) is a series structure. Thus, when (M, χ) is a series structure, the equality in (5.11) holds.

The c-FV TDL importance of a minimal cut

Let u be the number of minimal cuts in the system. According to Lambert (1975), the c-FV TDL importance of a minimal cut is defined as follows.

Definition 5.2.4 *The c-FV TDL importance of minimal cut C_k at time t, $k = 1, 2, \ldots, u$, denoted by $I_{FV^c}^C(C_k; \overline{\mathbf{F}}(t))$, is defined as the probability that minimal cut C_k contributes to system failure, given that the system fails at time t. Mathematically,*

$$I_{FV^c}^C(C_k; \overline{\mathbf{F}}(t)) = \frac{\prod_{j \in C_k} F_j(t)}{1 - R(\overline{\mathbf{F}}(t))}.$$

The c-FV TDL importance of minimal cuts always assigns more value to a minimal cut of smaller size than to one of larger size when reliability values of components are equal.

5.2.2   The p-FV Time-dependent Lifetime Importance

Similar to Definition 4.2.2 for the p-FV reliability importance, the p-FV TDL importance can be defined as follows (Kuo and Zhu 2012).

Definition 5.2.5   *The p-FV TDL importance of component i, denoted by $I_{\mathrm{FV}^p}(i; \overline{\mathbf{F}}(t))$, is defined as*

$$I_{\mathrm{FV}^p}(i; \overline{\mathbf{F}}(t)) = \Pr\{\exists P \in \overline{\mathscr{P}}_i \ s.t. \ X_j(t) = 1 \ for \ all \ j \in P | \phi(\mathbf{X}(t)) = 1\}$$

$$= \frac{\Pr\{\exists P \in \overline{\mathscr{P}}_i \ s.t. \ X_j(t) = 1 \ for \ all \ j \in P\}}{R(\overline{\mathbf{F}}(t))}$$

$$= \frac{\Pr\left\{1 - \prod_{P \in \overline{\mathscr{P}}_i}\left[1 - \prod_{j \in P} X_j(t)\right]\right\}}{R(\overline{\mathbf{F}}(t))},$$

where $P \in \overline{\mathscr{P}}_i$ denotes the minimal path containing component i and the term $1 - \prod_{P \in \overline{\mathscr{P}}_i}\left[1 - \prod_{j \in P} X_j(t)\right]$ equals 1 if at least one minimal path containing component i functions at time t; 0 otherwise.

The p-FV TDL importance of a modular set

Let a coherent system (N, ϕ) have a module (M, χ). Let $\overline{\mathscr{P}}_M$ denote the set of minimal paths of module (M, χ), that is, the functioning of each minimal path in $\overline{\mathscr{P}}_M$ makes modular set M function.

Definition 5.2.6   *The p-FV TDL importance of modular set M at time t, denoted by $I_{\mathrm{FV}^p}^M(M; \overline{\mathbf{F}}(t))$, is defined as the probability that at least one minimal path in $\overline{\mathscr{P}}_M$ functions at time t, given that the system functions at time t. Mathematically,*

$$I_{\mathrm{FV}^p}^M(M; \overline{\mathbf{F}}(t)) = \Pr\{\exists P \in \overline{\mathscr{P}}_M \ s.t. \ \chi(\mathbf{X}(t)^M) = 1 \ and \ X_j(t) = 1 \ for \ all \ j \in P | \phi(\mathbf{X}(t)) = 1\}$$

$$= \frac{\Pr\{\exists P \in \overline{\mathscr{P}}_M \ s.t. \ X_j(t) = 1 \ for \ all \ j \in P\}}{R(\overline{\mathbf{F}}(t))}$$

$$= \frac{\Pr\left\{1 - \prod_{P \in \overline{\mathscr{P}}_M}\left[1 - \prod_{j \in P} X_j(t)\right]\right\}}{R(\overline{\mathbf{F}}(t))}.$$

On the basis of this definition, the p-FV importance of a modular set is also independent of the modular decomposition of the system. Theorem 5.2.7 for the p-FV importance is analogous to Theorem 5.2.3 for the c-FV importance and has a similar proof, which is thus omitted.

Theorem 5.2.7   *The following is true and holds the equality if and only if a module (M, χ) is a parallel structure:*

$$I_{\mathrm{FV}^p}^M(M; \overline{\mathbf{F}}(t)) \leq \sum_{i \in M} I_{\mathrm{FV}^p}(i; \overline{\mathbf{F}}(t)).$$

The p-FV TDL importance of a minimal path

Let v be the number of minimal paths in the system. The p-FV TDL importance of a minimal path can be defined as follows (Kuo and Zhu 2012).

Definition 5.2.8 *The p-FV TDL importance of minimal path P_r at time t, $r = 1, 2, \ldots, v$, denoted by $I^P_{\mathrm{FV}^p}(P_r; \overline{\mathbf{F}}(t))$, is defined as the probability that minimal path P_r contributes to system functioning, given that the system functions at time t. Mathematically,*

$$I^P_{\mathrm{FV}^p}(P_r; \overline{\mathbf{F}}(t)) = \frac{\prod_{j \in P_r} \overline{F}_j(t)}{R(\overline{\mathbf{F}}(t))}.$$

Similar to the c-FV TDL importance of minimal cuts, the p-FV TDL importance of minimal paths always assigns more value to a minimal path of smaller size than one of larger size when reliability values of components are equal.

5.2.3   Decomposition of State Vectors

Note that all results in Subsection 4.2.3 can be straightforwardly extended to the FV TDL importance by replacing p_i with $\overline{F}_i(t)$ as shown in the following equations:

$$I_{\mathrm{FV}^c}(i; \overline{\mathbf{F}}(t)) = \frac{F_i(t) \left[\Pr\{B(i); \overline{\mathbf{F}}(t)\} + \Pr\{V_1(i); \overline{\mathbf{F}}(t)\}\right]}{\Pr\{\phi(\mathbf{X}) = 0\}},$$

$$\Pr\{\phi(\mathbf{X}(t)) = 0\} = \Pr\{V_1(i); \overline{\mathbf{F}}(t)\} + \Pr\{D_1(i); \overline{\mathbf{F}}(t)\} + F_i(t)\Pr\{B(i); \overline{\mathbf{F}}(t)\},$$

$$I_{\mathrm{FV}^p}(i; \overline{\mathbf{F}}(t)) = \frac{\overline{F}_i(t) \left[\Pr\{B(i); \overline{\mathbf{F}}(t)\} + \Pr\{V_2(i); \overline{\mathbf{F}}(t)\}\right]}{\Pr\{\phi(\mathbf{X}(t)) = 1\}},$$

$$\Pr\{\phi(\mathbf{X}(t)) = 1\} = \Pr\{V_2(i); \overline{\mathbf{F}}(t)\} + \Pr\{D_2(i); \overline{\mathbf{F}}(t)\} + \overline{F}_i(t)\Pr\{B(i); \overline{\mathbf{F}}(t)\}.$$

5.3   The BP Time-independent Lifetime Importance

The BP importance of a component (Barlow and Proschan 1975) reveals the relative extent to which each component is critical for a nonrepairable or repairable system. Assume that components fail sequentially in time and that two or more components have a vanishingly small probability to fail at the same instant; therefore, one component must cause the system to fail. Under these assumptions, the BP TIL importance is defined as follows.

Definition 5.3.1 *The BP TIL importance of component i, denoted by $I_{\mathrm{BP}}(i; \overline{\mathbf{F}})$, is defined as the probability that component i is critical for the system (i.e., the system functions if component i functions and fails otherwise, or, in other words, component i causes system failure) over an infinite mission time. Mathematically,*

$$I_{\mathrm{BP}}(i; \overline{\mathbf{F}}) = \int_0^\infty \left(R(1_i, \overline{\mathbf{F}}(t)) - R(0_i, \overline{\mathbf{F}}(t))\right) \mathrm{d}F_i(t). \tag{5.12}$$

By Equation (5.1),

$$I_{BP}(i; \overline{\mathbf{F}}) = \int_0^\infty I_B(i; \overline{\mathbf{F}}(t)) dF_i(t). \tag{5.13}$$

That is, $I_{BP}(i; \overline{\mathbf{F}})$ is a weighted average of the B-TDL importance, $I_B(i; \overline{\mathbf{F}}(t))$, over an infinite mission time. Thus, both the BP importance and B-importance take into account that component i is critical for the system.

Recall that $\tau(\mathbf{T})$ is the system lifetime function of component lifetimes, $T_i, i = 1, 2, \ldots, n$. Boland and El-Neweihi (1995) gave a precise justification that $I_{BP}(i; \overline{\mathbf{F}})$ is exactly the probability that the system lifetime coincides with the lifetime of component i, as follows:

$$\begin{aligned}
\Pr\{\tau(\mathbf{T}) = T_i\} &= \int_0^\infty \Pr\{\tau((t)_i, \mathbf{T}) = t\} \, dF_i(t) \\
&= \int_0^\infty \left(\lim_{n \to \infty} \Pr\left\{ t - \frac{1}{n} < \tau((t)_i, \mathbf{T}) \le t \right\} \right) dF_i(t) \\
&= \int_0^\infty \left(\lim_{n \to \infty} \left[R\left(1_i, \overline{\mathbf{F}}\left(t - \frac{1}{n}\right)\right) - R\left(0_i, \overline{\mathbf{F}}(t)\right) \right] \right) dF_i(t) \\
&= \int_0^\infty \left(R\left(1_i, \overline{\mathbf{F}}(t)\right) - R\left(0_i, \overline{\mathbf{F}}(t)\right) \right) dF_i(t) \\
&= I_{BP}(i; \overline{\mathbf{F}}),
\end{aligned}$$

where $\tau((t)_i, \mathbf{T}) = \tau(T_1, T_2, \ldots, T_{i-1}, t, T_{i+1}, \ldots, T_n)$. In other words, the BP importance of a component is essentially the probability that the failure of the component causes the system failure, given that the system eventually fails.

When components are dependent, Iyer (1992) showed that the BP TIL importance can be equivalently defined and calculated as in Equations (5.14) and (5.15).

Theorem 5.3.2 *Assume that the joint distribution of T_i's is absolutely continuous. Then*

$$I_{BP}(i; \overline{\mathbf{F}}) = \sum_{S \subseteq N \setminus \{i\}} \int_0^\infty b_{S \cup \{i\}} \Pr\left\{ \cap_{j \in S} T_j > t | T_i = t \right\} dF_i(t) \tag{5.14}$$

$$= \sum_{S \subseteq N \setminus \{i\}} \int_0^\infty b_{S \cup \{i\}} \left. \frac{\partial \Pr\left\{ \cap_{j \in S} T_j > t \cap T_i \le t_i \right\}}{\partial t_i} \right|_{t_i = t} dt, \tag{5.15}$$

where $b_{S \cup \{i\}}$ is the signed domination as in Equation (2.7) of the structure $\phi(\mathbf{0}^{N \setminus (S \cup \{i\})}, \mathbf{x}^{S \cup \{i\}})$, in which all components not in $S \cup \{i\}$ are assumed to fail.

The following theorem from Barlow and Proschan (1975) specifies the range of the BP TIL importance.

Theorem 5.3.3

(i) $0 \le I_{\mathrm{BP}}(i; \overline{\mathbf{F}}) \le 1$.
(ii) *If $n \ge 2$ and the intersection of supports of F_j, $j = 1, 2, \ldots, n$, has positive probability with respect to the product distribution $\prod_{j=1}^{n} F_j(t)$, then $0 < I_{\mathrm{BP}}(i; \overline{\mathbf{F}}) < 1$.*

(iii)
$$\sum_{i=1}^{n} I_{\mathrm{BP}}(i; \overline{\mathbf{F}}) = 1. \tag{5.16}$$

Proof. Part (i): The integrand in Equation (5.12) is between 0 and 1, implying part (i).

Part (ii): $I_{\mathrm{BP}}(i; \overline{\mathbf{F}}) = 0$ implies that for t in the support of F_i, $R(1_i, \overline{\mathbf{F}}(t)) - R(0_i, \overline{\mathbf{F}}(t)) = 0$ since component distributions are continuous. This, in turn, implies that component i is irrelevant to system ϕ with positive probability, which contradicts the fact that ϕ is coherent. Likewise, $I_{\mathrm{BP}}(i; \overline{\mathbf{F}}) = 1$ implies that $R(1_i, \overline{\mathbf{F}}(t)) - R(0_i, \overline{\mathbf{F}}(t)) = 1$. Then, all other components are irrelevant with positive probability, again contradicting the fact that ϕ is coherent. Thus, $0 < I_{\mathrm{BP}}(i; \overline{\mathbf{F}}) < 1$.

Part (iii): Equation (5.16) follows the fact that system failure coincides with the failure of exactly one component.

Note that Theorem 5.3.3(ii) is false if the supports do not intersect with positive probability with respect to $\prod_{j=1}^{n} F_j(t)$. Considering two components in parallel, if component 1 has a lifetime distribution with support on $[0, 1]$ and component 2 has lifetime distribution with support on $[2, 3]$, then $I_{\mathrm{BP}}(1; \overline{\mathbf{F}}) = 0$ and $I_{\mathrm{BP}}(2; \overline{\mathbf{F}}) = 1$.

For series and parallel systems with proportional hazard components, Barlow and Proschan (1975) gave the expressions of the BP TIL importance as follows.

Theorem 5.3.4 *Assume that the lifetimes of components have proportional hazards as in Equation (2.16), that is, $\overline{F}_i(t) = e^{-\lambda_i h(t)}$ for $i = 1, 2, \ldots, n$, where $h(t)$ is the common function of t. Then, for a series system,*
$$I_{\mathrm{BP}}(i; \overline{\mathbf{F}}) = \frac{\lambda_i}{\sum_{j=1}^{n} \lambda_j};$$
for a parallel system,
$$I_{\mathrm{BP}}(i; \overline{\mathbf{F}}) = \lambda_i \left[\lambda_i^{-1} - \sum_{j \ne i} (\lambda_i + \lambda_j)^{-1} + \sum_{\substack{j < k \\ j,k \ne i}} (\lambda_i + \lambda_j + \lambda_k)^{-1} \right.$$
$$\left. - \cdots + (-1)^{n-1} (\lambda_1 + \lambda_2 + \cdots + \lambda_n)^{-1} \right].$$

Proof. From the definition in Equation (5.12), for a series system
$$I_{\mathrm{BP}}(i; \overline{\mathbf{F}}) = \int_0^{\infty} \exp\left(-(\sum_{j \ne i} \lambda_j) h(t) \right) \lambda_i \exp(-\lambda_i h(t)) \mathrm{d}h(t) = \frac{\lambda_i}{\sum_{j=1}^{n} \lambda_j}.$$

For a parallel system, the result follows from

$$I_{\mathrm{BP}}(i; \overline{\mathbf{F}}) = \int_0^\infty \left[\prod_{j \neq i} \left(1 - \exp(-\lambda_j h(t)) \right) \right] \lambda_i \exp(-\lambda_i h(t)) \mathrm{d}h(t).$$

Remark 5.3.5 (Monte Carlo Methods)   In general, it is hard to calculate the BP importance for arbitrary lifetime distributions of components. Barlow and Proschan (1975) suggested the Monte Carlo methods as follows for estimating the BP importance for complex systems with large numbers of components, even in the case of proportional hazards. Without loss of generality, assume in Equation (2.16), $\sum_{i=1}^n \lambda_i = 1$. To simulate the sequence of successive component failures that ultimately result in system failure, draw successive independent uniform random variables $U_1, U_2, \ldots$ on [0, 1]. If U_1 falls between $\lambda_1 + \lambda_2 + \cdots + \lambda_{i_1-1}$ and $\lambda_1 + \lambda_2 + \cdots + \lambda_{i_1-1} + \lambda_{i_1}$ (where λ_0 is defined to be 0), then conclude that component i_1 has failed first. Repeat $+\lambda_{i1-1}$ process, using U_2 to determine which component fails second. If U_2 should call for the failure of component i_1 again, simply discard U_2 and use U_3 instead. Continue this process until the system fails. The component causing system failure is then recorded. If component i causes system failure m_i times in m trails, then m_i/m estimates $I_{\mathrm{BP}}(i; \overline{\mathbf{F}})$.

The BP TIL importance of a modular set

Let a coherent system (N, ϕ) have a module (M, χ).

Definition 5.3.6   *The BP TIL importance of modular set M, denoted by $I_{\mathrm{BP}}^M(M; \overline{\mathbf{F}})$, is defined as the probability that the failure of modular set M coincides with the failure of the system. Let $R\left(1_M, \overline{\mathbf{F}}(t)\right) - R\left(0_M, \overline{\mathbf{F}}(t)\right)$ denote the probability that at time t the system functions if the module functions but fails otherwise. Then,*

$$I_{\mathrm{BP}}^M(M; \overline{\mathbf{F}}) = -\int_0^\infty \left[R\left(1_M, \overline{\mathbf{F}}(t)\right) - R\left(0_M, \overline{\mathbf{F}}(t)\right) \right] \mathrm{d}R_\chi(\overline{\mathbf{F}}(t)^M),$$

where $\overline{\mathbf{F}}(t)^M$ is a vector with elements $\overline{F}_i(t)$, $i \in M$, and $R_\chi(\overline{\mathbf{F}}(t)^M)$ is the reliability function of the module.

The BP TIL importance of a modular set can simplify the calculation of the BP importance of the components in the module as shown in Theorem 5.3.7 (Barlow and Proschan 1975).

Theorem 5.3.7

(i) For $i \in M$,

$$I_{\mathrm{BP}}(i; \overline{\mathbf{F}}) = \int_0^\infty \left(R\left(1_M, \overline{\mathbf{F}}(t)\right) - R\left(0_M, \overline{\mathbf{F}}(t)\right) \right) \left(R_\chi\left(1_i, \overline{\mathbf{F}}(t)^M\right) - R_\chi\left(0_i, \overline{\mathbf{F}}(t)^M\right) \right) \mathrm{d}F_i(t).$$

$$(5.17)$$

(ii)
$$I_{BP}^M(M; \overline{\mathbf{F}}) = \sum_{i \in M} I_{BP}(i; \overline{\mathbf{F}}). \tag{5.18}$$

(iii) Let a system (N, ϕ) have a modular decomposition $\{(M_k, \chi_k)\}_{k=1}^m$, then $\sum_{k=1}^{m} I_{BP}^M(M_k; \overline{\mathbf{F}})$
$= 1.$

Proof. Part (i): By the chain rule in Equation (4.16) for the B-importance and letting $p_i = \overline{F}_i(t)$, the result of part (i) is straightforward.

Part (ii):

$$\sum_{i \in M} I_{BP}(i; \overline{\mathbf{F}}) = \int_0^\infty \left(R\left(1_M, \overline{\mathbf{F}}(t)\right) - R\left(0_M, \overline{\mathbf{F}}(t)\right) \right) \sum_{i \in M} \left(R_\chi\left(1_i, \overline{\mathbf{F}}(t)^M\right) - R_\chi\left(0_i, \overline{\mathbf{F}}(t)^M\right) \right) \mathrm{d}F_i(t)$$

$$= -\int_0^\infty \left(R\left(1_M, \overline{\mathbf{F}}(t)\right) - R\left(0_M, \overline{\mathbf{F}}(t)\right) \right) \mathrm{d}R_\chi\left(\overline{\mathbf{F}}(t)^M\right) = I_{BP}^M(M; \overline{\mathbf{F}}).$$

Part (iii): By Equations (5.16) and (5.18), the sum of the BP TIL importance over all disjoint modular sets is unity.

Note that Equation (5.18) holds true when components are dependent (Iyer 1992). Theorem 5.3.7(ii) is not valid for the B-importance, while the chain rule for the B-importance in Theorem 4.1.13 is not valid for the BP importance. Also note the difference between Theorems 5.3.7 and 5.2.3; Theorem 5.2.3 pertains to the FV importance.

Properties

Lemma 5.3.8 *Let component i be in series (parallel) with a module containing component j.*

(i) Then $I_{BP}(i; \overline{\mathbf{F}})$ is increasing (decreasing) in $F_i(t)$ and in $\overline{F}_j(t)$.
(ii) If $F_i(t) = F_j(t)$ for all $t \geq 0$, then $I_{BP}(i; \overline{\mathbf{F}}) \geq I_{BP}(j; \overline{\mathbf{F}})$.

Proof. Part (i): First assume that component i is in series with a module containing component j. Let (M, χ) be a module consisting of component i and the module containing component j. Let $I_B^M(M, \overline{\mathbf{F}}(t))$ denote the B-TDL importance of modular set M. Then, by Equation (5.17) and noting $R_\chi(0_i, \overline{\mathbf{F}}(t)^M) = 0$ by hypothesis,

$$I_{BP}(i; \overline{\mathbf{F}}) = \int_0^\infty I_B^M(M, \overline{\mathbf{F}}(t)) R_\chi(1_i, \overline{\mathbf{F}}(t)^M) \mathrm{d}F_i(t),$$

where $(\alpha_i, \overline{\mathbf{F}}(t)^M)$ is the vector with $\overline{F}_i(t) = \alpha$ for element $i \in M$ and $\overline{\mathbf{F}}(t)^M$ for the remaining elements in M. Note that $I_B^M(M, \overline{\mathbf{F}}(t))$ does not depend on R_χ; thus, the changes of $F_i(t)$ and $F_j(t)$ do not affect $I_B^M(M, \overline{\mathbf{F}}(t))$. $R_\chi(1_i, \overline{\mathbf{F}}(t)^M)$ is increasing in $\overline{F}_j(t)$; thus, $I_{BP}(i; \overline{\mathbf{F}})$ is increasing in $\overline{F}_j(t)$. Also $R_\chi(1_i, \overline{\mathbf{F}}(t)^M)$ is decreasing in t; thus, $I_{BP}(i; \overline{\mathbf{F}})$ is increasing in $F_i(t)$. A similar proof applies when component i is in parallel with a module containing component j.

Part (ii): By the proof of Theorem 5.1.4, $I_B(i; \overline{F}(t)) \geq I_B(j; \overline{F}(t))$ under the conditions of part (ii). By Equation (5.13) and noticing that $F_i(t) = F_j(t)$ for all $t \geq 0$, then $I_{BP}(i; \overline{F}) \geq I_{BP}(j; \overline{F})$.

By the application of Lemma 5.3.8, Theorem 5.3.9 follows immediately.

Theorem 5.3.9 *Let component i be in series (parallel) with a module containing component j. If $F_i(t) \geq F_j(t)$ $(\overline{F}_i(t) \geq \overline{F}_j(t))$ for $t \geq 0$, then $I_{BP}(i; \overline{F}) \geq I_{BP}(j; \overline{F})$.*

Note that Theorem 5.3.9 is analogous to Theorem 5.1.4 for the B-importance and Theorem 4.2.6 for the FV importance. Barlow and Proschan (1975) showed the similar results as Lemma 5.3.8 and Theorem 5.3.9 when component i is in series or parallel with the rest of the system, which is a special case since all components excluding i can be treated as a module. By Theorem 5.3.9, the following corollary is obtained, stating that the least (most) reliable component in a series (parallel) system is the most important according to the BP TIL importance.

Corollary 5.3.10 *In a series (parallel) system, if $F_i(t) \geq F_j(t)$ $(\overline{F}_i(t) \geq \overline{F}_j(t))$ for $t \geq 0$, then $I_{BP}(i; \overline{F}) \geq I_{BP}(j; \overline{F})$.*

Natvig and Gåsemyr (2009) showed the following theorem, which gives lower bounds for how much larger $I_{BP}(i; \overline{F})$ is than $I_{BP}(j; \overline{F})$, not just that it is larger. Note that the right-hand-side terms of the following inequalities are nonnegative and are zero when component j is, respectively, in series and in parallel with the rest of the system.

Theorem 5.3.11 *Let component i be in series (parallel) with the rest of the system. If $F_i(t) \geq F_j(t) > 0$ $(\overline{F}_i(t) \geq \overline{F}_j(t) > 0)$ for $t \geq 0$, $j \neq i$, then*

$$I_{BP}(i; \overline{F}) - I_{BP}(j; \overline{F}) \geq \int_0^\infty \frac{f_j(t)}{\overline{F}_j(t)} R(0_j, \overline{F}(t)) dt$$

$$\left(I_{BP}(i; \overline{F}) - I_{BP}(j; \overline{F}) \geq \int_0^\infty \frac{f_j(t)}{F_j(t)} (1 - R(1_j, \overline{F}(t))) dt \right).$$

The BP TIL importance of a minimal cut

Let u be the number of minimal cuts in the system. Compared to Definition 5.3.1 for the BP TIL importance of components, the BP TIL importance of minimal cuts is similarly defined as follows (Barlow and Proschan 1975).

Definition 5.3.12 *The BP TIL importance of minimal cut C_k, $k = 1, 2, \ldots, u$, denoted by $I_{BP}^C(C_k; \overline{F})$, is defined as the probability that minimal cut C_k causes system failure, that is, the probability that the failure of C_k coincides with system failure.*

It is immediately clear that the sum of the importance over all minimal cuts of a coherent system is at least unity. This is a consequence of the fact that system failure may coincide with the failure of more than one minimal cut in the system (see Examples 5.6.1, 6.3.7, and 6.3.8).

The following theorem gives an explicit expression for the BP TIL importance of a minimal cut in a coherent system (Barlow and Proschan 1975).

Theorem 5.3.13 *Letting $C_k, k = 1, 2, \ldots, u$, be a minimal cut in a coherent system (N, ϕ), then*

$$I_{BP}^C(C_k; \overline{\mathbf{F}}) = \sum_{i \in C_k} \int_0^\infty \left[R\left(1_i, \mathbf{0}^{C_k \setminus \{i\}}, (\overline{\mathbf{F}}(t))^{N \setminus C_k}\right) \prod_{j \in C_k \setminus \{i\}} F_j(t) \right] dF_i(t). \qquad (5.19)$$

Proof. First note that $\prod_{j \in C_k \setminus \{i\}} F_j(t) dF_i(t)$ represents the probability of the joint event that component i fails at time t and that the remaining components in minimal cut C_k have failed by time t. Next note that $R\left(1_i, \mathbf{0}^{C_k \setminus \{i\}}, (\overline{\mathbf{F}}(t))^{N \setminus C_k}\right)$ represents the probability that component i is critical for the system at time t (i.e., the system functions at time t if component i functions but fails otherwise) because $R\left(0_i, \mathbf{0}^{C_k \setminus \{i\}}, (\overline{\mathbf{F}}(t))^{N \setminus C_k}\right) = 0$. Thus, the product yields the probability that component i causes system failure. Summing over $i \in C_k$ (corresponding to the mutually exclusive ways in which minimal cut C_k can fail) gives the probability that C_k causes system failure.

The BP TIL importance of a minimal path

let v be the number of minimal paths in the system. The BP TIL importance of minimal paths can be defined analogous to that of minimal cuts, as shown in Definition 5.3.14 (Kuo and Zhu 2012). The illustration of Equation (5.20) is referred to the proof of Equation (5.19).

Definition 5.3.14 *The BP TIL importance of minimal path $P_r, r = 1, 2, \ldots, v$, denoted by $I_{BP}^P(P_r; \overline{\mathbf{F}})$ is defined as the probability that minimal path P_r restores the system to functioning, that is, the probability that the functioning of P_r coincides with system functioning. Mathematically,*

$$I_{BP}^P(P_r; \overline{\mathbf{F}}) = \sum_{i \in P_r} \int_0^\infty \left[\left(1 - R\left(0_i, \mathbf{1}^{P_r \setminus \{i\}}, (\overline{\mathbf{F}}(t))^{N \setminus P_r}\right)\right) \prod_{j \in P_r \setminus \{i\}} \overline{F}_j(t) \right] dF_i(t). \qquad (5.20)$$

5.4   The BP Time-dependent Lifetime Importance

Assuming that the mission time of the system, t, is a variable, the BP TDL importance is defined as follows (Lambert 1975).

Definition 5.4.1 *The BP TDL importance of component i, denoted by $I_{BP}(i; \overline{\mathbf{F}}(t))$, is defined as the probability that component i has caused the system to fail by time t, given that the system has failed by time t. Mathematically,*

$$I_{BP}(i; \overline{\mathbf{F}}(t)) = \frac{\int_0^t \left[R(1_i, \overline{\mathbf{F}}(u)) - R(0_i, \overline{\mathbf{F}}(u))\right] dF_i(u)}{\sum_{i=1}^n \int_0^t \left[R(1_i, \overline{\mathbf{F}}(u)) - R(0_i, \overline{\mathbf{F}}(u))\right] dF_i(u)} \qquad (5.21)$$

$$= \frac{\int_0^t I_B(i; \overline{\mathbf{F}}(u)) dF_i(u)}{1 - R(\overline{\mathbf{F}}(t))}. \qquad (5.22)$$

Equation (5.22) is obtained from the B-TDL importance in Equation (5.1). The denominators in Equations (5.21) and (5.22) are the probability that the system has failed by time t; the numerators are the probability that component i has caused the system to fail by time t. It is clear that the sum of the BP TDL importance over all components is unity.

Definition 5.4.2 *The BP TDL importance of minimal cut* C_k, $k = 1, 2, \ldots, u$, *denoted by* $I_{BP}^C(C_k; \overline{\mathbf{F}}(t))$, *is defined as the probability that minimal cut* C_k *has caused the system to fail by time t, given that the system has failed by time t. Mathematically,*

$$I_{BP}^C(C_k; \overline{\mathbf{F}}(t)) = \frac{\sum_{i \in C_k} \int_0^t \left[R \left(1_i, \mathbf{0}^{C_k \setminus \{i\}}, (\overline{\mathbf{F}}(u))^{N \setminus C_k} \right) \prod_{j \in C_k \setminus \{i\}} F_j(u) \right] dF_i(u)}{1 - R(\overline{\mathbf{F}}(t))}.$$

Similarly, the BP TDL importance of a minimal path could be defined as the probability that it restores the system to functioning, that is, the probability that its functioning coincides with that of the system.

Taking time t to infinity, the BP TDL importance in Equations (5.21) and (5.22) becomes the BP TIL importance in Equations (5.12) and (5.13). As discussed following Theorem 3.2.1, it is not easy to transform the BP TDL importance to a corresponding reliability importance, since the integration of time t is involved in the definition of the BP TDL importance. But the BP structure importance can be well defined, as in Section 6.3.

5.5 Numerical Comparisons of Time-dependent Lifetime Importance Measures

This section compares the behaviors of the TDL importance measures for the two three-component systems in Figures 2.4 and 2.5. Assume that the lifetimes of components have proportional hazards as in Equation (2.16), that is, $\overline{F}_i(t) = e^{-\lambda_i h(t)}$ for $i = 1, 2, 3$, and simply treat $h(t) = t$ as it does not matter what the common function $h(t)$ is. Furthermore, assume that $\lambda_1 = 0.5\lambda_2$ and $\lambda_2 = \lambda_3$ so that $\overline{F}_1(t) = (\overline{F}_2(t))^{0.5}$ and $\overline{F}_2(t) = \overline{F}_3(t)$. In each of the systems in Figures 2.4 and 2.5, components 2 and 3 have the same value for any lifetime importance measure because these two positions are structurally symmetric and $\overline{F}_2(t) = \overline{F}_3(t)$. Thus, only the importance measures of components 1 and 2 are compared.

Figures 5.1 and 5.2 present the plots of the TDL importance measures of components 1 and 2 as a function of the probability of system failure, $1 - R(\overline{\mathbf{F}}(t))$. For the system in Figure 2.4,

$$1 - R(\overline{\mathbf{F}}(t)) = F_1(t) + \overline{F}_1(t)F_2(t)F_3(t) = 1 - (\overline{F}_2(t))^{0.5} + (\overline{F}_2(t))^{0.5} (1 - \overline{F}_2(t))^2.$$

For the system in Figure 2.5,

$$1 - R(\overline{\mathbf{F}}(t)) = F_1(t) \left(1 - \overline{F}_2(t)\overline{F}_3(t) \right) = \left(1 - (\overline{F}_2(t))^{0.5} \right) \left(1 - (\overline{F}_2(t))^2 \right).$$

In Figures 5.1 and 5.2, labels $B(i)$, $Cf(i)$, $BP(i)$, c-$FV(i)$, and p-$FV(i)$ denote the B-importance, criticality importance for system failure, BP importance, c-FV importance, and p-FV importance of component i for $i = 1, 2$, respectively.

For the system in Figure 2.4, Figure 5.1 indicates that, according to the B-importance, criticality importance, and c-FV importance, component 1 is more important for small values

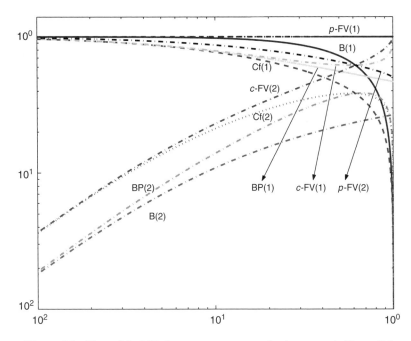

Figure 5.1 Plots of the TDL importance measures for the system in Figure 2.4

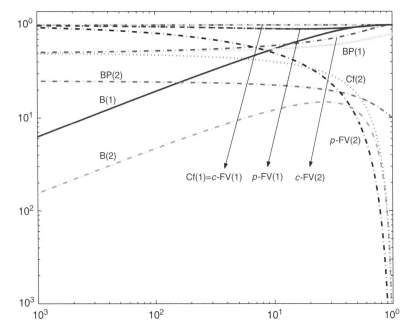

Figure 5.2 Plots of the TDL importance measures for the system in Figure 2.5

of $1 - R(\overline{\mathbf{F}}(t))$ or t (t can be thought of as mission time), while components 2 and 3 are more important for large values of $1 - R(\overline{\mathbf{F}}(t))$ or time t. However, there is disagreement among these three importance measures as to which value of $1 - R(\overline{\mathbf{F}}(t))$ components 2 and 3 would be more important than component 1. In addition, the p-FV importance of component 1 equals 1 for all $t \geq 0$; thus, according to the p-FV importance, component 1 is always more important than components 2 and 3. According to the BP importance, component 1 is also always more important than components 2 and 3. But if set $\lambda_1 = 0.2\lambda_2$ and remain the other assumptions, the BP importance of components 2 and 3 is larger than component 1 after some value of $1 - R(\overline{\mathbf{F}}(t))$ around 0.35.

For the system in Figure 2.5, Figure 5.2 indicates that component 1 is always more important than components 2 and 3; the criticality importance for system failure and the c-FV importance of component 1 equal 1 for all $t \geq 0$; and the p-FV of component 1 approaches 1.

It can be seen from Figures 5.1 and 5.2 that each importance measure has a different time-dependent behavior, that is, there is disagreement in the assessment of importance measures. At the same time point, different importance measures can give different rankings of the importance of components because these importance measures are defined differently and take into account the different aspects of the system and components. Meanwhile, for a given importance measure, the ranking of the importance of components may vary with time. Thus, an analyst should carefully define the probabilistic information regarding the system and components and apply the appropriate importance measure.

5.6 Summary

The c-FV importance takes into account the fact that the failure of a component can *contribute* to system failure without being critical, that is, component failure does not necessarily coincide with system failure. Component i *contributes* to system failure if at least one minimal cut containing component i fails without component i necessarily failing last, while component i *causes* system failure if component i fails last. The B-importance and BP importance consider the case in which a component *causes* system failure, that is, component i is critical for the system (see also the discussion following Definition 2.6.1). Similarly, Carot and Sanz (2000) evaluated the relative importance of nonrepairable components using the marginal improvements of system mean lifetime with respect to mean lifetimes of components. The measure of a component is calculated by means of simulations as the ratio of the sum of lifetimes of the component when it is critical for the system failure to the sum of lifetimes of the component in all cases.

Note that a component can contribute to but not exactly cause system failure because it is possible that when a system failure is observed, two or more minimal cuts could have failed simultaneously. In this case, restoring a failed component to functioning does not necessarily restore the system to functioning, as shown in the following example. Indeed, when component i contributes to system failure, it is always critical for the structure $\prod_{C \in \overline{\mathscr{C}}_i}[1 - \prod_{j \in C}(1 - X_j(t))]$.

Example 5.6.1 For the system in Figure 2.5, if components 2 and 3 have already failed, minimal cuts $\{1, 2\}$ and $\{1, 3\}$ will fail simultaneously when component 1 fails. Then, restoring component 3 to functioning will not restore the system to functioning. The failure of component 3 contributes to the system failure because minimal cut $\{1, 3\}$ has failed, but the failure of component 3 does not cause the system failure because component 1 fails last.

References

Barlow RE and Proschan F. 1975. Importance of system components and fault tree events. *Statistic Processes and Their Applications* **3**, 153–172.

Boland PJ and El-Neweihi E. 1995. Measures of component importance in reliability theory. *Computers and Operations Research* **22**, 455–463.

Carot V and Sanz J. 2000. Criticality and sensitivity analysis of the components of system. *Reliability Engineering and System Safety* **68**, 1147–1152.

Iyer S. 1992. The Barlow-Proschan importance and its generalizations with dependent components. *Stochastic Processes and Their Applications* **42**, 353–359.

Kuo W and Zhu X. 2012. Some recent advances on importance measures in reliability. *IEEE Transactions on Reliability* **61**.

Lambert HE. 1975. Measure of importance of events and cut sets in fault trees. In *Reliability and Fault Tree Analysis* (eds. Barlow RE, Fussell JB, and Singpurwalla ND). Society for Industrial and Applied Mathematics, Philadelphia, pp. 77–100.

Natvig B. 1979. A suggestion of a new measure of importance of system component. *Stochastic Processes and Their Applications* **9**, 319–330.

Natvig B and Gåsemyr J. 2009. New results on the Barlow-Proschan and Natvig measures of component importance in nonrepairable and repairable systems. *Methodology and Computing in Applied Probability* **11**, 603–620.

Xie M. 1987. On some importance measures of system components. *Stochastic Processes and Their Applications* **25**, 273–280.

Xie M. 1988. A note on the Natvig measure. *Scandinavian Journal of Statistics* **15**, 211–214.

6

Structure Importance Measures

This chapter presents structure importance measures, which actually evaluate the relative strength of the positions of the components in a system. Related to the importance measures in Chapters 4 and 5, Sections 6.1, 6.2, and 6.3 propose the B-i.i.d. importance and B-structure importance, the FV structure importance, and the BP structure importance, respectively. The B-i.i.d. importance is the B-reliability importance when all components have the same reliability of p, and it is special due to being a function of p. Section 6.4 proposes a method for defining structure importance measures using the B-i.i.d. importance.

Section 6.5 introduces a permutation importance. Section 6.6 presents an internal and an external domination importance, both of which are related to the permutation importance. Section 6.7 introduces a cut and a path importance, which have close relations to the B-i.i.d. importance as component reliability p approaches 1 and 0, respectively.

Hwang (2001, 2005) proposed four structure importance measures, namely, the absoluteness importance, cut-path importance, min-cut importance, and min-path importance. They are defined in terms of (minimal) cuts and/or (minimal) paths and are related to each other. Section 6.8 discusses the absoluteness importance, and Section 6.9 discusses the other three.

The first-term importance and rare-event importance are another two types of structure importance measures and are intuitively related to the B-i.i.d. importance as p approaches 1 and 0, respectively. They are the counterparts of each other, and Section 6.10 discusses them.

Note that the importance measures in Sections 6.5–6.9 are defined by means of comparisons between two components. Thus, they are also appropriately referred to as the importance rankings. Section 6.11 summarizes and compares the class of structure importance measures in terms of (minimal) cuts and (minimal) paths. Section 6.12 presents the structure importance measures for dual systems. Section 6.13 investigates dominant relations among various importance measures, especially structure importance measures. These three sections elaborate the key results of the review in Kuo and Zhu (2012a).

6.1 The B-i.i.d. Importance and B-structure Importance

When the failures of components are statistically i.i.d. according to the Bernoulli distribution with a same reliability p (i.e., $p_1 = p_2 = \cdots = p_n = p$), the B-reliability importance is referred to as the B-i.i.d. importance and denoted by $I_B(i; p)$ with a scalar p. The B-reliability importance can be used to define other importance measures and is often used as a basis for comparisons among importance measures, as discussed in the preceding and following chapters. However, the B-reliability importance reflects the probabilistic aspect of the system and is hard to compare with structure importance measures, which reflect only the structural aspect. By the use of the i.i.d. model, the B-i.i.d. importance reflects only the structural aspect even though it is a function of p. Thus, this section first investigates the B-i.i.d. importance.

Example 6.1.1 (Examples 2.2.3 and 4.1.6 continued) For a k-out-of-n system, $I_B(i; p) = \binom{n-1}{k-1}p^{k-1}(1-p)^{n-k}$, which is a function of p.

Using Equation (4.15) and noting that $p_1 = p_2 = \cdots = p_n = p$, Lin and Kuo (2002) explored the relations of the B-i.i.d. importance to critical cuts and critical paths, as shown in Theorem 6.1.2. Recall that $m_i(d)$ ($m_i'(d)$) is the number of critical cuts (paths) for component i of size d.

Theorem 6.1.2

$$I_B(i; p) = \sum_{d=1}^{n} m_i(d)p^{n-d}q^{d-1}, \tag{6.1}$$

$$= \sum_{d=1}^{n} m_i'(d)p^{d-1}q^{n-d}. \tag{6.2}$$

Note Equation (2.9) that $m_i(d) = m_i'(n-d+1)$.

The B-i.i.d. importance can also be expressed in terms of sets of cuts and paths on the basis of Equations (4.3) and (4.5).

Theorem 6.1.3

$$I_B(i; p) = \sum_{d=1}^{n} |\mathscr{C}_i(d)|p^{n-d}q^{d-1} - \sum_{d=1}^{n-1} |\mathscr{C}_{(i)}(d)|p^{n-d-1}q^{d}, \tag{6.3}$$

$$= \sum_{d=1}^{n} |\mathscr{P}_i(d)|p^{d-1}q^{n-d} - \sum_{d=1}^{n-1} |\mathscr{P}_{(i)}(d)|p^{d}q^{n-d-1}. \tag{6.4}$$

Since the B-i.i.d. importance is a function of p for $0 < p < 1$, Chang et al. (2002) and Chang and Hwang (2002) investigated the B-i.i.d. importance for different values or ranges of p.

Definition 6.1.4 *Component i is more uniformly B-i.i.d. important than component j if $I_B(i; p) \geq I_B(j; p)$ for all $0 < p < 1$, more half-line B-i.i.d. important than component j if $I_B(i; p) \geq I_B(j; p)$ for all $1/2 \leq p < 1$.*

Compared with the uniform B-i.i.d. importance, the half-line B-i.i.d. importance is easier to handle mathematically and is also practical in applications since the condition $p \geq 1/2$ is easily met in practice.

The B-structure importance, which is initially defined as follows based on the concept that component i is critical for the system (see Section 2.6), is essentially the B-i.i.d. importance with $p = 1/2$. Recall that m_i (m_i') is the number of critical cut (path) vectors for component i.

Definition 6.1.5 *The B-structure importance of component i for the functioning (i.e., success as denoted in subscript) of system ϕ, denoted by $I_{Bs}(i; \phi)$, is defined as*

$$I_{Bs}(i; \phi) = 2^{-n} \sum_{\{(0_i, \mathbf{x})\}} (\phi(1_i, \mathbf{x}) - \phi(0_i, \mathbf{x})) = 2^{-n} m_i. \qquad (6.5)$$

The B-structure importance of component i for the failure of ϕ, denoted by $I_{Bf}(i; \phi)$, is defined as

$$I_{Bf}(i; \phi) = 2^{-n} \sum_{\{(1_i, \mathbf{x})\}} (\phi(1_i, \mathbf{x}) - \phi(0_i, \mathbf{x})) = 2^{-n} m_i'. \qquad (6.6)$$

The B-structure importance of component i for ϕ, denoted by $I_B(i; \phi)$, is defined as

$$I_B(i; \phi) = I_{Bs}(i; \phi) + I_{Bf}(i; \phi) = 2^{-n} \sum_{\{\mathbf{x}\}} (\phi(1_i, \mathbf{x}) - \phi(0_i, \mathbf{x})).$$

From Equation (6.5) (Equation (6.6)), $I_{Bs}(i; \phi)$ ($I_{Bf}(i; \phi)$) is the ratio of the number of state vectors $(0_i, \mathbf{x})$ $((1_i, \mathbf{x}))$ at which component i is critical for the functioning (failure) of system ϕ to all 2^n state vectors, that is, the portion of the critical cut (path) vectors for component i over all 2^n state vectors. As in Equation (2.8), there is a one-to-one correspondence between critical cut vectors and critical path vectors for component i. Hence,

$$I_{Bs}(i; \phi) = I_{Bf}(i; \phi) = \tfrac{1}{2} I_B(i; \phi). \qquad (6.7)$$

Thus, there is no purpose in distinguishing between the B-structure importance for system functioning and that for system failure, and it is equivalent to investigating the B-structure importance $I_B(i; \phi)$, which is the ratio of the number of state vectors $\mathbf{x}$ at which component i is critical for the system to all 2^n state vectors.

From Equations (6.5), (6.6), and (6.7), then

$$I_B(i; \phi) = 2^{-(n-1)} m_i = 2^{-(n-1)} m_i'. \qquad (6.8)$$

Equation (6.8) means that the definition of $I_B(i; \phi)$ in terms of critical cuts is equivalent to that in terms of critical paths. Thus, the c-type and p-type B-structure importance measures are equivalent. In fact, the B-structure importance can be expressed in terms of both paths and cuts, as shown in Corollary 6.1.6 (Hwang 2001). Note that Corollary 6.1.6(i) is a special case of Theorem 6.1.3 when $p = 1/2$. From part (i), it is easy to show parts (ii) and (iii) by noting $|\mathscr{C}_{(i)}| = |\mathscr{C}| - |\mathscr{C}_i|$ and $|\mathscr{P}_{(i)}| = |\mathscr{P}| - |\mathscr{P}_i|$.

Corollary 6.1.6
  (i) $I_B(i; \phi) = 2^{-(n-1)} \left(|\mathscr{C}_i| - |\mathscr{C}_{(i)}| \right) = 2^{-(n-1)} \left(|\mathscr{P}_i| - |\mathscr{P}_{(i)}| \right).$
  (ii) $I_B(i; \phi) \geq I_B(j; \phi)$ *if and only if* $|\mathscr{C}_i| \geq |\mathscr{C}_j|$, *that is, the number of cuts containing component i is greater than or equal to the number of cuts containing component j.*
  (iii) $I_B(i; \phi) \geq I_B(j; \phi)$ *if and only if* $|\mathscr{P}_i| \geq |\mathscr{P}_j|$.

Example 6.1.7 (Examples 2.2.3 and 6.1.1 continued)   For a k-out-of-n system , all components have the same B-structure importance, $I_B(i; \phi) = 2^{-n+1} \binom{n-1}{k-1}$, $i = 1, 2, \ldots, n$. For a fixed n, this importance is largest for $k = n/2$ if n is even, and for $k = (n+1)/2$ and $k = (n+3)/2$ if n is odd. In the case of n components in series (i.e., $k = n$) or in parallel (i.e., $k = 1$), the importance is smallest at the value of $I_B(i; \phi) = 2^{-n+1}$.

Example 6.1.8 (Example 4.1.7 continued)   For the system with k components in series, in series with $n-k$ components in parallel, as shown in Figure 4.1,

$$\phi(\mathbf{x}) = \prod_{i=1}^{k} x_i \left[1 - \prod_{i=k+1}^{n} (1 - x_i) \right].$$

Then,

$$\phi(1_i, \mathbf{x}) - \phi(0_i, \mathbf{x}) = \prod_{\substack{j=1 \\ j\neq i}}^{k} x_j \left[1 - \prod_{j=k+1}^{n} (1 - x_j) \right]$$

for $i = 1, 2, \ldots, k$, and

$$\phi(1_i, \mathbf{x}) - \phi(0_i, \mathbf{x}) = \prod_{j=1}^{k} x_j \prod_{\substack{j=k+1 \\ j\neq i}}^{n} (1 - x_j)$$

for $i = k+1, k+2, \ldots, n$. Thus, $I_B(i; \phi) = 2^{-n+1}(2^{n-k} - 1) = 2^{-k+1} - 2^{-n+1}$ for $i = 1, 2, \ldots, k$, and $I_B(i; \phi) = 2^{-n+1}$ for $i = k+1, k+2, \ldots, n$. That is, components $1, 2, \ldots, k$ each have the B-structure importance $2^{-k+1} - 2^{-n+1}$, much greater than the B-structure importance of each component $k+1, k+2, \ldots, n$, 2^{-n+1}.

If nothing is known about the reliability of the components and, for lack of better knowledge, it is assumed that all state vectors $\mathbf{x}$ are equally probable, that is, each has probability 2^{-n}, then the B-reliability importance in Section 4.1 and the B-i.i.d. importance reduce to the B-structure importance. The B-structure importance is a special case of the B-reliability importance and B-i.i.d. importance when all components have the same reliability of $p = 1/2$.

6.2   The FV Structure Importance

Similar to the relation of the B-structure importance to the B-reliability importance, the FV structure importance is a special case of the FV reliability importance when all components have the same reliability $p_1 = p_2 = \cdots = p_n = 1/2$. Thus, the c-FV structure importance can be defined on the basis of Definition 4.2.1, and the p-FV structure importance can be defined on the basis of Definition 4.2.2.

Definition 6.2.1 *The c-FV structure importance of component i , denoted by $I_{\mathrm{FV}^c}(i; \phi)$, is defined as*

$$I_{\mathrm{FV}^c}(i; \phi) = \frac{\left|\left\{\mathbf{x} : \exists C \in \overline{\mathscr{C}}_i \ s.t. \ C \subseteq N_0(\mathbf{x}); \ \phi(\mathbf{x}) = 0\right\}\right|}{|\{\mathbf{x} : \phi(\mathbf{x}) = 0\}|}$$

$$= \frac{\left|\left\{(0_i, \mathbf{x}) : \exists C \in \overline{\mathscr{C}}_i \ s.t. \ C \subseteq N_0(0_i, \mathbf{x})\right\}\right|}{|\mathscr{C}|}.$$

Definition 6.2.2 *The p-FV structure importance of component i, denoted by $I_{\mathrm{FV}^p}(i; \phi)$, is defined as*

$$I_{\mathrm{FV}^p}(i; \phi) = \frac{\left|\left\{\mathbf{x} : \exists P \in \overline{\mathscr{P}}_i \ s.t. \ P \subseteq N_1(\mathbf{x}); \ \phi(\mathbf{x}) = 1\right\}\right|}{|\{\mathbf{x} : \phi(\mathbf{x}) = 1\}|}$$

$$= \frac{\left|\left\{(1_i, \mathbf{x}) : \exists P \in \overline{\mathscr{P}}_i \ s.t. \ P \subseteq N_1(1_i, \mathbf{x})\right\}\right|}{|\mathscr{P}|}.$$

Then, according to decomposition of state vectors in Subsection 4.2.3,

$$I_{\mathrm{FV}^c}(i; \phi) = \frac{|B(i)| + |V_1(i)|}{|\{\mathbf{x} : \phi(\mathbf{x}) = 0\}|},$$

$$I_{\mathrm{FV}^p}(i; \phi) = \frac{|B(i)| + |V_2(i)|}{|\{\mathbf{x} : \phi(\mathbf{x}) = 1\}|}.$$

6.3   The BP Structure Importance

The BP structure importance is a special case of the BP TIL importance described in Section 5.3 when lifetime distributions are assumed to be the same for all components.

Definition 6.3.1 *The BP structure importance of component i, denoted by $I_{\mathrm{BP}}(i; \phi)$, is defined as*

$$I_{\mathrm{BP}}(i; \phi) = \int_0^1 (R(1_i, p) - R(0_i, p)) \, \mathrm{d}p = \int_0^1 I_{\mathrm{B}}(i; p) \mathrm{d}p, \qquad (6.9)$$

where (α_i, p) is the vector having α in the ith position and p in all other positions for $\alpha = 0$ or 1.

Equation (6.9) is obtained from Equation (5.12) by assuming that all $F_i(t)$ are the same and by making the change of variable $p = \overline{F}_i(t)$ for $i = 1, 2, \ldots, n$. According to Equation (6.9), the BP structure importance is the average (integral) of the B-i.i.d. importance as p ranges over $[0, 1]$.

Example 6.3.2 (Example 2.2.3 continued)   For a k-out-of-n system , the systems reliability is a symmetric function of component reliability $p_1 = p_2 = \cdots = p_n = p$. It follows from

Equation (6.9) that all components have equal value of the BP structure importance. Because $\sum_{i=1}^{n} I_{BP}(i; \phi) = 1$ by Equation (5.16), $I_{BP}(i; \phi) = 1/n$ for $i = 1, 2, \ldots, n$, regardless of value of k. The result can be generalized to the system structure ϕ that is symmetric of component states $x_1, x_2, \ldots, x_n$. Then, similarly, $I_{BP}(i; \phi) = 1/n$ for $i = 1, 2, \ldots, n$. Examples of symmetric ϕ are compositions of k-out-of-n structures, that is, $\phi = \phi_{k|n} \circ \phi_{k|n} \circ \ldots \circ \phi_{k|n}$, where $\phi_{k|n}$ is the structure function for a k-out-of-n system. Another example is the $Cir/Con/k/n$ system.

By Theorem 6.1.2 for the B-i.i.d. importance, the BP structure importance of component i can be computed in terms of the number of critical cuts (paths) for component i.

Theorem 6.3.3

$$I_{BP}(i; \phi) = \frac{1}{n} \sum_{d=1}^{n} m_i(d) \binom{n-1}{d-1}^{-1}, \tag{6.10}$$

$$= \frac{1}{n} \sum_{d=1}^{n} m_i'(d) \binom{n-1}{d-1}^{-1}. \tag{6.11}$$

Proof. Starting from Equation (6.9) and using Equation (6.1),

$$I_{BP}(i; \phi) = \int_0^1 \sum_{d=1}^{n} m_i(d) p^{n-d} (1-p)^{d-1} dp \tag{6.12}$$

$$= \sum_{d=1}^{n} \frac{(d-1)!(n-d)!}{n!} m_i(d) \tag{6.13}$$

$$= \frac{1}{n} \sum_{d=1}^{n} m_i(d) \binom{n-1}{d-1}^{-1}.$$

Similarly, Equation (6.11) follows from Equation (6.2).

Barlow and Proschan (1975) first presented Equation (6.11), in which the term $\binom{n-1}{d-1}$ represents the number of outcomes in which exactly $d-1$ components function among the $n-1$ components excluding component i. Equation (6.11) states that the BP structure importance of component i is the average probability of a vector being a critical path vector for component i. The average is taken over the n different possible sizes $d = 1, 2, \ldots, n$ of a critical path vector, where the probability of a vector being a critical path vector for component i of size d is computed as $m_i'(d) \binom{n-1}{d-1}^{-1}$. Similarly, Equation (6.10) means that the BP structure importance of component i is the average probability of a vector being a critical cut vector for component i.

Similar to Equation (6.12), $I_{BP}(i; \phi) = \int_0^1 \left[\sum_{d=1}^{n} m_i'(d) \binom{n-1}{d-1}^{-1} \binom{n-1}{d-1} p^{d-1} (1-p)^{n-d} \right] dp$. Note that $\binom{n-1}{d-1} p^{d-1} (1-p)^{n-d}$ represents the probability (of binomial $(n-1, p)$ distribution) that among the $n-1$ components excluding component i, $d-1$ components function, while $m_i'(d) \binom{n-1}{d-1}^{-1}$ represents the probability that the $d-1$ functioning components together with component i constitute a critical path for component i. Thus, the integrand represents the probability that component i causes system failure. Integrating this probability over component reliability p is equivalent to assuming that a prior, p is uniformly distributed on $[0, 1]$. A similar interpretation can be performed in terms of critical cuts.

By Theorem 6.1.3 for the B-i.i.d. importance, the BP structure importance can be equivalently expressed in terms of paths and cuts as follows. Thus, the BP structure importance has equivalent c-type and p-type.

Theorem 6.3.4

$$I_{\mathrm{BP}}(i; \phi) = \frac{1}{n} \left(\sum_{d=1}^{n} |\mathscr{C}_i(d)| \binom{n-1}{d-1}^{-1} - \sum_{d=1}^{n-1} |\mathscr{C}_{(i)}(d)| \binom{n-1}{d}^{-1} \right)$$

$$= \frac{1}{n} \left(\sum_{d=1}^{n} |\mathscr{P}_i(d)| \binom{n-1}{d-1}^{-1} - \sum_{d=1}^{n-1} |\mathscr{P}_{(i)}(d)| \binom{n-1}{d}^{-1} \right).$$

Proof. By Equations (6.9) and (6.3),

$$I_{\mathrm{BP}}(i; \phi) = \int_0^1 \left[\sum_{d=1}^{n} |\mathscr{C}_i(d)| p^{n-d} q^{d-1} - \sum_{d=1}^{n-1} |\mathscr{C}_{(i)}(d)| p^{n-d-1} q^d \right] \mathrm{d}p$$

$$= \sum_{d=1}^{n} |\mathscr{C}_i(d)| \int_0^1 p^{n-d} q^{d-1} \mathrm{d}p - \sum_{d=1}^{n-1} |\mathscr{C}_{(i)}(d)| \int_0^1 p^{n-d-1} q^d \mathrm{d}p$$

$$= \frac{1}{n} \left(\sum_{d=1}^{n} |\mathscr{C}_i(d)| \binom{n-1}{d-1}^{-1} - \sum_{d=1}^{n-1} |\mathscr{C}_{(i)}(d)| \binom{n-1}{d}^{-1} \right).$$

Similarly, the second expression of $I_{\mathrm{BP}}(i; \phi)$ in terms of paths follows from Equation (6.4).

The BP structure importance of a minimal cut and a minimal path

Definition 6.3.5 *The BP structure importance of minimal cut C_k, $k = 1, 2, \ldots, u$, denoted by $I_{\mathrm{BP}}^C(C_k; \phi)$, is a special case of the corresponding BP TIL importance when all $F_i(t)$ are the same. After a change of variable $p = \bar{F}_i(t)$ for $i = 1, 2, \ldots, n$ in Equation (5.19), $I_{\mathrm{BP}}^C(C_k; \phi)$ can be expressed as*

$$I_{\mathrm{BP}}^C(C_k; \phi) = \sum_{i \in C_k} \int_0^1 R(1_i, \mathbf{0}^{C_k \setminus i}, p^{N \setminus C_k})(1 - p)^{|C_k|-1} \mathrm{d}p. \tag{6.14}$$

An alternative method for computing $I_{\mathrm{BP}}^C(C_k; \phi)$ is to list the $n!$ permutations, each of which represents a sequence of n component failures, to find the number $\eta(C_k)$ of sequences in which C_k causes system failure, and then to compute $I_{\mathrm{BP}}^C(C_k; \phi) = \eta(C_k)/n!$. This method of computing the structure importance of a minimal cut is valid, since all sequences of component failures are equally likely. However, this method is not efficient.

Theorem 6.3.6 shows that under certain conditions, minimal cuts of smaller size are more BP structure important than those of larger size (Barlow and Proschan 1975). Analogous to this, the c-FV TDL importance of minimal cuts (Definition 5.2.4) always assigns more value to a minimal cut of smaller size than one of larger size when reliability values of components are equal.

Theorem 6.3.6   *Let C_1 and C_2 be two minimal cuts with $|C_1| < |C_2|$. Let components in C_1 not appear in any other minimal cuts. Then, $I_{BP}^C(C_1; \phi) > I_{BP}^C(C_2; \phi)$.*

Proof. Given a sequence of n component failures for which minimal cut C_2 causes system failure, it is always possible to construct a corresponding sequence of component failures for which minimal cut C_1 causes system failure. To do that, simply exchange the first $|C_1|$ failures of the components in minimal cut C_2 with the $|C_1|$ failures of the components in minimal cut C_1. Since the components in minimal cut C_1 do not occur in any other minimal cut, for the failure sequence generated, minimal cut C_1 causes system failure. Moreover, it is obvious that failure sequences so generated are mutually distinct because original failure sequences are mutually distinct.

By the alternative method of the BP importance of minimal cuts following Equation (6.14), $I_{BP}^C(C_1; \phi) \geq I_{BP}^C(C_2; \phi)$. Since $|C_1| < |C_2|$, there exists at least one failure sequence for which minimal cut C_1 causes system failure but minimal cut C_2 does not cause system failure. Therefore, $I_{BP}^C(C_1; \phi) > I_{BP}^C(C_2; \phi)$.

It is easy to show by example that the condition that no components in C_1 appear in other minimal cuts is necessary for the validity of Theorem 6.3.6. Note that the fact that components in C_1 do not appear in any other minimal cuts implies that the components in C_1 are connected in parallel and that they are in series with the rest of the system.

Example 6.3.7 (Example 5.6.1 continued)   For the system in Figure 2.5, the minimal cuts are $C_1 = \{1, 2\}$ and $C_2 = \{1, 3\}$. Using Equation (6.14), $I_{BP}^C(C_1; \phi) = I_{BP}^C(C_2; \phi) = 2/3$. Note that minimal cuts C_1 and C_2 simultaneously fail and cause system failure when the sequence of component failures is 2-3-1 or 3-2-1. Thus, $I_{BP}^C(C_1; \phi) + I_{BP}^C(C_2; \phi) > 1$.

Example 6.3.8 (Example 2.2.3 continued)   For a k-out-of-n system , there are $u = \binom{n}{n-k+1}$ minimal cuts and $\sum_{k=1}^{u} I_{BP}^C(C_k; \phi) = 1$ because two or more minimal cuts cannot simultaneously cause system failure. Since all minimal cuts have equal structure importance and their BP importance values sum to unity, $I_{BP}^C(C_k; \phi) = \binom{n}{n-k+1}^{-1}$ for $k = 1, 2, \ldots, u$.

As a special case of the BP TIL importance in Equation (5.20), the BP structure importance of a minimal path is defined as the probability that it restores the system to functioning under the assumption that lifetime distributions are the same for all components.

6.4   Structure Importance Measures Based on the B-i.i.d. Importance

On the basis of the analysis of the B-i.i.d. importance and the BP structure importance, Barlow and Proschan (1975) proposed an approach for defining structure importance measures using the B-i.i.d. importance. It assumes p having a prior distribution $P(p)$ and integrates the B-i.i.d. importance over p according to distribution $P(p)$, yielding an average value as a structure importance $I(i; \phi)$. That is,

$$I(i; \phi) = \int_0^1 I_B(i; p)\mathrm{d}P(p) = \int_0^1 (R(1_i, p) - R(0_i, p))\mathrm{d}P(p). \tag{6.15}$$

If $P(p) = 0$ for $0 \leq p < 1/2$ and $P(p) = 1$ for $1/2 \leq p \leq 1$, then $I(i; \phi)$ induces the B-structure importance. If $P(p) = p$ (uniform distribution), then $I(i; \phi)$ induces the BP structure importance. The B-structure importance computes the difference $R(1_i, p) - R(0_i, p)$ with p equal to $1/2$, while the BP structure importance averages this difference as p has a uniform prior over $[0, 1]$. In another view, comparing Equations (6.8) and (6.13), the B-structure importance attaches the same weight $2^{-(n-1)}$ to each of terms $m_i'(d)$, while the BP structure importance attaches weight $(d-1)!(n-d)!/n!$ to the $m_i'(d)$. Since $(d-1)!(n-d)!/n!$ is decreasing in d for $d \leq n/2$ and increasing in d for $d \geq n/2$, it implies that the BP structure importance attaches the greatest weight to critical paths of either very small or very large sizes.

Theorem 6.4.1 shows that if $I(i; \phi)$ given in Equation (6.15) satisfies the normalization property that the importance values of all components in a series system sum to unity, then $I(i; \phi)$ must coincide with $I_{BP}(i; \phi)$ (Barlow and Proschan 1975).

Theorem 6.4.1   *Let $I(i; \phi)$ given in Equation (6.15) satisfy $\sum_{i=1}^n I(i; \phi) = 1$ for a series system of n components, then $I(i; \phi) = I_{BP}(i; \phi)$.*

Proof. For the series system, $R(p) = \prod_{i=1}^n p_i$ and $I_B(i; p) = p^{n-1}$. It follows that

$$1 = \sum_{i=1}^n I(i; \phi) = n \int_0^1 p^{n-1}\mathrm{d}P(p). \tag{6.16}$$

By the solution to the Hamburger moment problem (Shohat and Tamarkin, 1943, p. 19), the distribution $P(p) = p$ uniquely satisfies Equation (6.16) and so $I(i; \phi)$ coincides with $I_{BP}(i; \phi)$.

6.5   The Permutation Importance and Permutation Equivalence

This section describes the permutation importance and permutation equivalence that were proposed by Boland et al. (1989). Subsection 6.5.1 demonstrates the relations of the permutation importance to minimal cuts and minimal paths. Subsection 6.5.2 presents the theorems on using the permutation importance for comparing the relative magnitude of systems reliability.

Definition 6.5.1   *Component i is more permutation important than component j, denoted by $i >_{pe} j$, for structure function ϕ if*

$$\phi(1_i, 0_j, \mathbf{x}^{(ij)}) \geq \phi(0_i, 1_j, \mathbf{x}^{(ij)}) \tag{6.17}$$

holds for all $\mathbf{x}^{(ij)}$ and strict inequality holds for some $\mathbf{x}^{(ij)}$. If equality holds for all $\mathbf{x}^{(ij)}$, components i and j are said to be permutation equivalent, denoted by $i =_{pe} j$.

Recall that $(\alpha_i, \beta_j, \mathbf{x}^{(ij)})$ is the vector with $x_i = \alpha$, $x_j = \beta$, and $\mathbf{x}^{(ij)}$ for others, where $\alpha, \beta \in \{0, 1\}$ and $\mathbf{x}^{(ij)}$ is the vector obtained by deleting x_i and x_j from $\mathbf{x}$. Notation $i \geq_{\text{pe}} j$ means that component i is more permutation important than or permutation equivalent to component j. As in Theorem 6.5.2, it is easy to verify the equivalence of the structural symmetry in Definition 2.2.8 and the permutation equivalence in Definition 6.5.1.

Theorem 6.5.2 *Components i and j are permutation equivalent if and only if they are structurally symmetric, that is, $\phi(\mathbf{x})$ is permutation symmetric in x_i and x_j.*

The next three theorems show the transitivity property (Koutras et al. 1994), a dual relation, and a property in a special case of the permutation importance, respectively. Theorem 6.5.4 is straightforward from Definitions 6.5.1 and 2.2.5, and Theorem 6.5.5 is trivial.

Theorem 6.5.3 *If $i >_{\text{pe}} j$ and $j >_{\text{pe}} k$, then $i >_{\text{pe}} k$.*

Proof. It needs to be shown that $\phi(1_i, 0_k, \mathbf{x}^{(ik)}) \geq \phi(0_i, 1_k, \mathbf{x}^{(ik)})$ holds for all $\mathbf{x}^{(ik)}$, that is, all $\mathbf{x}^{(ik)}$ no matter whether $x_j = 1$ or 0.

First, $j >_{\text{pe}} k$ implies $\phi(1_j, 0_k, \mathbf{x}^{(jk)}) \geq \phi(0_j, 1_k, \mathbf{x}^{(jk)})$; thus, $\phi(1_i, 1_j, 0_k, \mathbf{x}^{(ijk)}) \geq \phi(1_i, 0_j, 1_k, \mathbf{x}^{(ijk)})$ when $x_i = 1$. $i >_{\text{pe}} j$ implies $\phi(1_i, 0_j, \mathbf{x}^{(ij)}) \geq \phi(0_i, 1_j, \mathbf{x}^{(ij)})$; thus, $\phi(1_i, 0_j, 1_k, \mathbf{x}^{(ijk)}) \geq \phi(0_i, 1_j, 1_k, \mathbf{x}^{(ijk)})$ when $x_k = 1$. Therefore, $\phi(1_i, 1_j, 0_k, \mathbf{x}^{(ijk)}) \geq \phi(0_i, 1_j, 1_k, \mathbf{x}^{(ijk)})$ for all $\mathbf{x}^{(ijk)}$, that is, $\phi(1_i, 0_k, \mathbf{x}^{(ik)}) \geq \phi(0_i, 1_k, \mathbf{x}^{(ik)})$ for all $\mathbf{x}^{(ik)}$ with $x_j = 1$.

Similarly, it can be shown that $\phi(1_i, 0_k, \mathbf{x}^{(ik)}) \geq \phi(0_i, 1_k, \mathbf{x}^{(ik)})$ for all $\mathbf{x}^{(ik)}$ with $x_j = 0$.

Hence, $\phi(1_i, 0_k, \mathbf{x}^{(ik)}) \geq \phi(0_i, 1_k, \mathbf{x}^{(ik)})$ for all $\mathbf{x}^{(ik)}$. Since $\phi(1_j, 0_k, \mathbf{x}^{(jk)}) > \phi(0_j, 1_k, \mathbf{x}^{(jk)})$ for some $\mathbf{x}^{(jk)}$ and $\phi(1_i, 0_j, \mathbf{x}^{(ij)}) > \phi(0_i, 1_j, \mathbf{x}^{(ij)})$ for some $\mathbf{x}^{(ij)}$, $\phi(1_i, 0_k, \mathbf{x}^{(ik)}) > \phi(0_i, 1_k, \mathbf{x}^{(ik)})$ holds for some $\mathbf{x}^{(ik)}$. The proof is complete.

Theorem 6.5.4 *If $i >_{\text{pe}} j$ for system ϕ, then $i >_{\text{pe}} j$ for its dual system ϕ^D, and vice versa.*

Theorem 6.5.5 *If component i is in series (parallel) with the rest of the system, then $\phi(0_i, 1_j, \mathbf{x}^{(ij)}) = 0$ $(\phi(1_i, 0_j, \mathbf{x}^{(ij)}) = 1)$ for all $\mathbf{x}^{(ij)}$, and thus $i \geq_{\text{pe}} j$ for all $j \neq i$.*

6.5.1 Relations to Minimal Cuts and Minimal Paths

The examination of the permutation importance of two components by the use of Definition 6.5.1 is, in general, laborious, especially for large or structurally complicated systems. For certain structures, however, the minimal cuts and minimal paths are easy to find. The theorems in this subsection provide simple criteria, in terms of minimal cuts and minimal paths, for comparing the permutation importance of components in a coherent system. This subsection uses the notation of various sets of cuts and paths, which is defined in Section 2.5.

Meng (1994) characterized the permutation importance of components in terms of minimal cuts (paths) as in Theorem 6.5.6. Partial results in Theorem 6.5.6 have also been presented by Boland et al. (1989).

Theorem 6.5.6 $i >_{pe} j$ *if and only if one of the following conditions holds: (i) $\overline{\mathscr{C}}_j$ is a proper subset of $\overline{\mathscr{C}}_i$ ($\overline{\mathscr{C}}_j \subset \overline{\mathscr{C}}_i$); (ii) for each $C_k \in \overline{\mathscr{C}}_j^{(i)}$, there exists a $C_{k'} \in \overline{\mathscr{C}}_i^{(j)}$ such that $C_{k'} \setminus \{i\} \subseteq C_k \setminus \{j\}$ and $C_{k'} \setminus \{i\} \subset C_k \setminus \{j\}$ for at least one such pair $(C_k, C_{k'})$; (iii) if in condition (ii) $C_{k'} \setminus \{i\} = C_k \setminus \{j\}$ holds for all such pairs, then $|\overline{\mathscr{C}}_j^{(i)}| < |\overline{\mathscr{C}}_i^{(j)}|$.*

Proof. ("Only if") Suppose that $i >_{pe} j$. First, assume that $\overline{\mathscr{C}}_j^{(i)}$ is empty, then $\overline{\mathscr{C}}_j \subseteq \overline{\mathscr{C}}_i$. Note that if $\overline{\mathscr{C}}_j = \overline{\mathscr{C}}_i$, then $\phi(1_i, 0_j, \mathbf{x}^{(ij)}) = \phi(0_i, 1_j, \mathbf{x}^{(ij)})$ for all $\mathbf{x}^{(ij)}$, which is a contradiction to the assumption that $i >_{pe} j$. Thus, $\overline{\mathscr{C}}_j \subset \overline{\mathscr{C}}_i$.

Now assume that $\overline{\mathscr{C}}_j^{(i)}$ is nonempty. Let C_k be a given minimal cut in $\overline{\mathscr{C}}_j^{(i)}$, that is, $\phi(0^{C_k}, 1) = 0$ and $\phi(\mathbf{x}) = 1$ for all $\mathbf{x} > (0^{C_k}, 1)$ where $(0^{C_k}, 1)$ denotes the vector in which the elements in C_k equal 0, otherwise 1. Hence, $\phi(1_i, 1_j, 0^{C_k \setminus \{j\}}, 1) = 1$ and $\phi(0_i, 1_j, 0^{C_k \setminus \{j\}}, 1) \leq \phi(1_i, 0_j, 0^{C_k \setminus \{j\}}, 1) = 0$ (because $i >_{pe} j$), where $(\alpha_i, \beta_j, 0^{C_k \setminus \{j\}}, 1)$ is the vector with α in position i, β in position j, 0 for $k \in C_k \setminus \{j\}$, and 1 otherwise. This implies that there exists a $C_{k'} \in \overline{\mathscr{C}}_i^{(j)}$ and $C_{k'} \setminus \{i\} \subseteq C_k \setminus \{j\}$.

Thus, either condition (ii) holds or $C_{k'} \setminus \{i\} = C_k \setminus \{j\}$ for all such pairs. For the latter case, there is a one-to-one mapping from $\overline{\mathscr{C}}_j^{(i)}$ to $\overline{\mathscr{C}}_i^{(j)}$, which, in turn, implies that $|\overline{\mathscr{C}}_j^{(i)}| \leq |\overline{\mathscr{C}}_i^{(j)}|$. Supposing that $|\overline{\mathscr{C}}_j^{(i)}| = |\overline{\mathscr{C}}_i^{(j)}|$, then, for a given $C_{k'} \in \overline{\mathscr{C}}_i^{(j)}$, there must exist a $C_k \in \overline{\mathscr{C}}_j^{(i)}$ such that $C_k \setminus \{j\} = C_{k'} \setminus \{i\}$. But then $\phi(0_i, 1_j, \mathbf{x}^{(ij)}) = 0$ implies $\phi(1_i, 0_j, \mathbf{x}^{(ij)}) = 0$, which contradicts the assumption that $i >_{pe} j$. Thus, $|\overline{\mathscr{C}}_j^{(i)}| < |\overline{\mathscr{C}}_i^{(j)}|$.

("If") Let $\mathbf{x}^{(ij)}$ be any vector. If $\phi(1_i, 0_j, \mathbf{x}^{(ij)}) = 1$, then Equation (6.17) is satisfied. If $\phi(1_i, 0_j, \mathbf{x}^{(ij)}) = 0$, then $S \cup \{j\}$ is a cut where $S = \{k : x_k = 0, k \neq i, j\}$.

Condition (i): Because $i \notin S$ and there is no minimal cut containing components j not i (because $\overline{\mathscr{C}}_j \subset \overline{\mathscr{C}}_i$), $S \cup \{j\}$ does not contain any minimal cut containing component j, and thus S must contain a minimal cut (containing neither component i nor j). Hence, $\phi(0_i, 1_j, \mathbf{x}^{(ij)}) = 0$, and Equation (6.17) is satisfied.

Moreover, because $\overline{\mathscr{C}}_j \subset \overline{\mathscr{C}}_i$, there exists a minimal cut C_k containing components i but not j such that $\phi(1_i, 0_j, 0^{C_k \setminus \{i\}}, 1) = 1 > \phi(0_i, 1_j, 0^{C_k \setminus \{i\}}, 1) = 0$. Hence, $i >_{pe} j$.

Condition (ii): If S contains a minimal cut, then $\phi(0_i, 1_j, \mathbf{x}^{(ij)}) = 0$ and Equation (6.17) is satisfied. If S does not contain a minimal cut, then there must exist a minimal cut $C_k \in \overline{\mathscr{C}}_j^{(i)}$ such that $x_\ell = 0$ for all $\ell \in C_k \setminus \{j\}$. By assumption, there is a $C_{k'} \in \overline{\mathscr{C}}_i^{(j)}$ such that $C_{k'} \setminus \{i\} \subseteq C_k \setminus \{j\}$, which in turn implies that $x_\ell = 0$ for all $\ell \in C_{k'} \setminus \{i\}$. Thus, by the monotonicity of ϕ, $\phi(0_i, 1_j, \mathbf{x}^{(ij)}) \leq \phi(0_i, 1_j, 0^{C_{k'} \setminus \{i\}}, 1) = 0$. Therefore, Equation (6.17) holds for all $\mathbf{x}^{(ij)}$.

Moreover, let $C_{k'}$ be a minimal cut in $\overline{\mathscr{C}}_i^{(j)}$ such that $C_{k'} \setminus \{i\} \subset C_k^0 \setminus \{j\}$ for some $C_k^0 \in \overline{\mathscr{C}}_j^{(i)}$. Then, $\phi(0_i, 1_j, 0^{C_{k'} \setminus \{i\}}, 1) = 0$ and $\phi(1_i, 1_j, 0^{C_{k'} \setminus \{i\}}, 1) = 1$. Note that $\phi(1_i, 0_j, 0^{C_{k'} \setminus \{i\}}, 1) = 1 > \phi(0_i, 1_j, 0^{C_{k'} \setminus \{i\}}, 1)$. Otherwise, $\phi(1_i, 0_j, 0^{C_{k'} \setminus \{i\}}, 1) = 0$ and $\phi(1_i, 1_j, 0^{C_{k'} \setminus \{i\}}, 1) = 1$ would imply that there is a $C_k \in \overline{\mathscr{C}}_j^{(i)}$ such that $C_k \setminus \{j\} \subseteq C_{k'} \setminus \{i\}$. But then, $C_k \setminus \{j\} \subset C_k^0 \setminus \{j\}$, which contradicts that C_k^0 is a minimal cut. Hence, $i >_{pe} j$.

Condition (iii): As shown in condition (ii), $\phi(1_i, 0_j, \mathbf{x}^{(ij)}) \geq \phi(0_i, 1_j, \mathbf{x}^{(ij)})$ for all $\mathbf{x}^{(ij)}$. Now, suppose that for each $C_k \in \overline{\mathscr{C}}_j^{(i)}$ there exists a $C_{k'} \in \overline{\mathscr{C}}_i^{(j)}$ such that $C_k \setminus \{j\} = C_{k'} \setminus \{i\}$. From the assumption that $|\overline{\mathscr{C}}_j^{(i)}| < |\overline{\mathscr{C}}_i^{(j)}|$, it then follows that there exists a $C_{k'}^0 \in \overline{\mathscr{C}}_i^{(j)}$ such that $\phi(0_i, 1_j, 0^{C_{k'}^0 \setminus \{i\}}, 1) = 0$ and $\phi(1_i, 0_j, 0^{C_{k'}^0 \setminus \{i\}}, 1) = 1$. Hence, $i >_{pe} j$.

Note that in conditions (i) and (ii), if $\overline{\mathscr{C}}_j \subset \overline{\mathscr{C}}_i$, then $\overline{\mathscr{C}}_j^{(i)} = \emptyset$. Condition (ii) can be stated as follows. For every $S \subseteq N \setminus \{i, j\}$ such that $S \cup \{j\}$ is a minimal cut, the set $S \cup \{i\}$ is a cut as well and there exists $S_0 \subseteq N \setminus \{i, j\}$ such that $S_0 \cup \{i\}$ is a cut while $S_0 \cup \{j\}$ is not. Koutras et al. (1994) independently proved Theorem 6.5.6 using this statement. Moreover, they showed a similar result in terms of cuts, as shown in Theorem 6.5.7, which is weaker than Theorem 6.5.6 since the number of cuts is no less than the number of minimal cuts.

Theorem 6.5.7 $i >_{\mathrm{pe}} j$ *if and only if (i) $\mathscr{C}_j$ is a proper subset of $\mathscr{C}_i$ ($\mathscr{C}_j \subset \mathscr{C}_i$); (ii) for any $S \subseteq N \setminus \{i, j\}$ such that $S \cup \{j\}$ is a cut, the set $S \cup \{i\}$ is a cut too, and moreover, there exists $S_0 \subseteq N \setminus \{i, j\}$ such that $S_0 \cup \{i\}$ is a cut while $S_0 \cup \{j\}$ is not.*

Proof. ("Only if") Suppose that $i >_{\mathrm{pe}} j$. If there is no set $S \subseteq N \setminus \{i, j\}$ such that $S \cup \{j\}$ is a cut, then $\mathscr{C}_j \subset \mathscr{C}_i$. Now assume that there exists a set $S \subseteq N \setminus \{i, j\}$ such that $S \cup \{j\}$ is a cut. By introducing a state vector $\mathbf{x}$ such that $x_k = 0$ for all $k \in S$, it is obtained that $\phi(1_i, 0_j, \mathbf{x}^{(ij)}) = 0$, and, from Equation (6.17), it follows that $\phi(0_i, 1_j, \mathbf{x}^{(ij)}) = 0$. This proves that $S \cup \{i\}$ is a cut. In addition, because $i >_{\mathrm{pe}} j$, there exists a state vector $\mathbf{y}$ such that $\phi(1_i, 0_j, \mathbf{y}^{(ij)}) = 1 > \phi(0_i, 1_j, \mathbf{y}^{(ij)}) = 0$. Let $S_0 = \{k; y_k = 0, k \neq i, j\}$. Then $S_0 \cup \{i\}$ is a cut and $S_0 \cup \{j\}$ is not. This part of the proof is complete.

("If") Condition (i): Suppose that $\mathscr{C}_j \subset \mathscr{C}_i$, then $\overline{\mathscr{C}}_j \subset \mathscr{C}_i$, that is, any minimal cut containing component j contains component i. Thus, $\overline{\mathscr{C}}_j \subset \overline{\mathscr{C}}_i$. By Theorem 6.5.6, then $i >_{\mathrm{pe}} j$.

Condition (ii): Let $\mathbf{x}^{ij}$ be any state vector. If $\phi(1_i, 0_j, \mathbf{x}^{(ij)}) = 1$, then Equation (6.17) is valid. If $\phi(1_i, 0_j, \mathbf{x}^{(ij)}) = 0$, then let $S = \{k; x_k = 0, k \neq i, j\}$, and $S \cup \{j\}$ is a cut. Therefore, $S \cup \{i\}$ is also a cut, which implies that $\phi(0_i, 1_j, \mathbf{x}^{(ij)}) = 0$. Hence, Equation (6.17) is again valid. Moreover, if S_0 is a subset of $N \setminus \{i, j\}$ such that $S_0 \cup \{i\}$ is a cut and $S_0 \cup \{j\}$ is not, by introducing a state vector $\mathbf{y}$ with $y_k = 0$ for all $k \in S_0$, it is deduced that $\phi(1_i, 0_j, \mathbf{y}^{(ij)}) = 1 > \phi(0_i, 1_j, \mathbf{y}^{(ij)}) = 0$. The proof is complete.

From the proof of Theorem 6.5.6, the following characteristic of the permutation equivalence of two components is immediate; the proof is, therefore, omitted and readers are referred to Meng (1994).

Theorem 6.5.8 $i =_{\mathrm{pe}} j$ *if and only if (i) $\overline{\mathscr{C}}_i = \overline{\mathscr{C}}_j$; (ii) $\overline{\mathscr{C}}_i^{(j)}$ can be obtained by replacing component j with component i for each minimal cut in $\overline{\mathscr{C}}_j^{(i)}$, and vice versa. That is, for each $C_k \in \overline{\mathscr{C}}_i^{(j)}, C_k \cup \{j\} \setminus \{i\} \in \overline{\mathscr{C}}_j^{(i)}$, and for each $C_k \in \overline{\mathscr{C}}_j^{(i)}, C_k \cup \{i\} \setminus \{j\} \in \overline{\mathscr{C}}_i^{(j)}$.*

According to Theorems 6.5.7 and 6.5.8, $i >_{\mathrm{pe}} j$, $i =_{\mathrm{pe}} j$, and $i \geq_{\mathrm{pe}} j$ can alternatively be defined in terms of cuts as in Definition 6.5.9.

Definition 6.5.9 (Alternative definition) $i >_{\mathrm{pe}} j$ *if $S \cup \{j\} \in \mathscr{C}_j$ implies $S \cup \{i\} \in \mathscr{C}_i$ for any subset S containing neither component i nor j and there exists a subset S_0 such that $S_0 \cup \{i\} \in \mathscr{C}_i$, while $S_0 \cup \{j\} \notin \mathscr{C}_j$. $i =_{\mathrm{pe}} j$ if $S \cup \{j\} \in \mathscr{C}_j$ implies $S \cup \{i\} \in \mathscr{C}_i$ for any subset S containing neither component i nor j, and vice versa. Together, $i \geq_{\mathrm{pe}} j$ if $S \cup \{j\} \in \mathscr{C}_j$ implies $S \cup \{i\} \in \mathscr{C}_i$ for any subset S containing neither component i nor j.*

Recall from Theorem 6.5.4 that if component i is more permutation important than component j for system ϕ, then component i is more permutation important than component j for its dual system ϕ^D. Furthermore, the (minimal) cuts for ϕ are precisely the (minimal) paths for ϕ^D, and vice versa (Theorem 2.5.2). Thus, *Theorems 6.5.6, 6.5.7, and 6.5.8 remain true if the term "cuts" in these theorems is replaced with "paths." Consequently, Definition 6.5.9 can be restated in terms of paths. Then, the c-type and p-type of the permutation importance are equivalent.*

The following three examples illustrate these results.

Example 6.5.10 (Example 2.6.3 continued)   For the system in Figure 2.10, Example 2.6.3 shows that $\overline{\mathscr{P}}_i \subset \overline{\mathscr{P}}_2$ for $i = 1, 3, 4, 5$; hence, $2 >_{pe} 1, 3, 4, 5$. Condition (ii) in Theorem 6.5.8 applies for components 3 and 4; thus, $3 =_{pe} 4$. Furthermore, $\overline{\mathscr{P}}_5^{(1)} = \{\{2, 5\}\}$ and $\overline{\mathscr{P}}_1^{(5)} = \{\{1, 2, 3\}, \{1, 2, 4\}\}$. Condition (ii) in Theorem 6.5.6 is satisfied by components 5 and 1, so it follows that $5 >_{pe} 1$.

Example 6.5.11   Considering a system of six components connected as shown in Figure 6.1, then $\phi(\mathbf{x}) = 1 - (1 - x_1 x_2 x_4)(1 - x_1 x_2 x_5)(1 - x_1 x_3 x_4)(1 - x_1 x_3 x_5)(1 - x_1 x_3 x_6)$. It is easy to see that component 1 is more permutation important than the other components. Furthermore, $\overline{\mathscr{P}}_2^{(3)} = \{\{1, 2, 4\}, \{1, 2, 5\}\}$ and $\overline{\mathscr{P}}_3^{(2)} = \{\{1, 3, 4\}, \{1, 3, 5\}, \{1, 3, 6\}\}$. Condition (ii) in Theorem 6.5.6 applies for components 2 and 3; hence, $3 >_{pe} 2$. Condition (iii) in Theorem 6.5.6 applies for components 4 and 5, where $\overline{\mathscr{P}}_4^{(5)} = \{\{1, 2, 4\}, \{1, 3, 4\}\}$ and $\overline{\mathscr{P}}_5^{(4)} = \{\{1, 2, 5\}, \{1, 3, 5\}\}$; hence, $4 =_{pe} 5$.

However, considering components $i = 3$ and $j = 4$, for $\mathbf{x}^{(ij)} = (1, 0, \cdot, \cdot, 1, 0)$, $\phi(1_3, 0_4, \mathbf{x}^{(ij)}) = 1 > \phi(0_3, 1_4, \mathbf{x}^{(ij)}) = 0$, and for $\mathbf{x}^{(ij)} = (1, 1, \cdot, \cdot, 0, 0)$, $\phi(1_3, 0_4, \mathbf{x}^{(ij)}) = 0 < \phi(0_3, 1_4, \mathbf{x}^{(ij)}) = 1$. Thus, components 3 and 4 cannot be compared according to the permutation importance.

As shown in Example 6.5.11, the permutation importance is a *partial ranking* in which some of the components may not be compared by means of the permutation importance because the required condition is too strong. This is a shortcoming of the permutation importance.

Note that $i >_{pe} j$ does not necessarily imply $\overline{\mathscr{C}}_j \subset \overline{\mathscr{C}}_i$ (i.e., condition (i) in Theorem 6.5.6), as shown in the next example.

Example 6.5.12 (Example 6.5.11 continued)   Considering the system in Figure 6.1, by Example 6.5.11, $3 >_{pe} 2$, and minimal path $P_1 = \{1, 2, 4\}$ does not contain component 3. Also, $1 >_{pe} 2$ and minimal cut $C_1 = \{2, 3\}$ does not contain component 1.

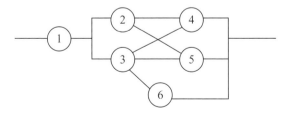

Figure 6.1   Component 1 in series with the rest of the system

6.5.2   Relations to Systems Reliability

With the permutation importance, under the conditions in the following theorems, the reliability of a coherent system can be compared when two components take respectively different
reliability values. Theorems 6.5.13 and 6.5.14 are used in an elimination procedure for generating the optimal arrangements for CAP in Section 11.3, when only the order of component
reliability is known. Theorem 6.5.15 can be used for dealing with redundancy allocation problem in Chapter 9. Boland et al. (1989) and Mi (2003) presented Theorems 6.5.13 and 6.5.15,
respectively. Recall that $(\alpha_i, \beta_j, \mathbf{p}^{(ij)})$ is the vector with $p_i = \alpha$, $p_j = \beta$, and $\mathbf{p}^{(ij)}$ for others,
where $0 \le \alpha, \beta \le 1$ and $\mathbf{p}^{(ij)}$ is the vector obtained by deleting p_i and p_j from $\mathbf{p}$.

Theorem 6.5.13 $i >_{\text{pe}} j$ *if and only if* $R(\beta_i, \alpha_j, \mathbf{p}^{(ij)}) \ge R(\alpha_i, \beta_j, \mathbf{p}^{(ij)})$ *for all* $\mathbf{p}^{(ij)}$ *and all*
$0 < \alpha < \beta < 1$, *with strict inequality for some* $\mathbf{p}^{(ij)}$ *and* α *and* β.

Proof. First note that

$$R(\alpha_i, \beta_j, \mathbf{p}^{(ij)}) = \alpha\beta R(1_i, 1_j, \mathbf{p}^{(ij)}) + (1 - \alpha)(1 - \beta)R(0_i, 0_j, \mathbf{p}^{(ij)})$$

$$+\alpha(1 - \beta)R(1_i, 0_j, \mathbf{p}^{(ij)}) + \beta(1 - \alpha)R(0_i, 1_j, \mathbf{p}^{(ij)}). \qquad (6.18)$$

("Only if") Supposing that $i >_{\text{pe}} j$, then $\phi(1_i, 0_j, \mathbf{x}^{(ij)}) \ge \phi(0_i, 1_j, \mathbf{x}^{(ij)})$ for all $\mathbf{x}^{(ij)}$ and,
consequently, $R(1_i, 0_j, \mathbf{p}^{(ij)}) \ge R(0_i, 1_j, \mathbf{p}^{(ij)})$ for all $\mathbf{p}^{(ij)}$. Thus,

$$R(\beta_i, \alpha_j, \mathbf{p}^{(ij)}) - R(\alpha_i, \beta_j, \mathbf{p}^{(ij)}) = (\beta - \alpha)[R(1_i, 0_j, \mathbf{p}^{(ij)}) - R(0_i, 1_j, \mathbf{p}^{(ij)})] \ge 0.$$

Furthermore,   if $\phi(1_i, 0_j, \mathbf{x}_0^{(ij)}) = 1 > \phi(0_i, 1_j, \mathbf{x}_0^{(ij)}) = 0$ for   some $\mathbf{x}_0^{(ij)}$,   then
$R(1_i, 0_j, \mathbf{p}^{(ij)}) = 1 > R(0_i, 1_j, \mathbf{p}^{(ij)}) = 0$, where $\mathbf{p}^{(ij)}$ is a vector of 0s and 1s such that
$\Pr\{\mathbf{X}^{(ij)} = \mathbf{x}_0^{(ij)} | \mathbf{p}^{(ij)}\} = 1$ and vector $\mathbf{X}^{(ij)}$ is obtained by deleting X_i and X_j from $\mathbf{X}$.
Since $R(\mathbf{p})$ is a continuous function of $\mathbf{p}$ for $\mathbf{0} \le \mathbf{p} \le \mathbf{1}$, there exists a $\mathbf{p}' = (\alpha_i', \beta_j', \mathbf{p}'^{(ij)})$,
$\mathbf{0} < \mathbf{p}' < \mathbf{1}$, such that $R(\alpha_i', \beta_j', \mathbf{p}'^{(ij)}) < R(\beta_i', \alpha_j', \mathbf{p}'^{(ij)})$.

("If") If $i >_{\text{pe}} j$ does not hold, then either $\phi(1_i, 0_j, \mathbf{x}^{(ij)}) = \phi(0_i, 1_j, \mathbf{x}^{(ij)})$ for all $\mathbf{x}^{(ij)}$ (in
this case $R(\beta_i, \alpha_j, \mathbf{p}^{(ij)}) = R(\alpha_i, \beta_j, \mathbf{p}^{(ij)})$ for all $\mathbf{p}^{(ij)}$ and all $0 < \alpha < \beta < 1$) or there exists
an $\mathbf{x}_0^{(ij)}$ such that $\phi(1_i, 0_j, \mathbf{x}_0^{(ij)}) = 0$ and $\phi(0_i, 1_j, \mathbf{x}_0^{(ij)}) = 1$. In the latter case, by an argument
of continuity similar to that used earlier, there exists a $\mathbf{p}' = (\alpha_i', \beta_j', \mathbf{p}'^{(ij)})$, $\mathbf{0} < \mathbf{p}' < \mathbf{1}$, such
that $R(\beta_i', \alpha_j', \mathbf{p}'^{(ij)}) < R(\alpha_i', \beta_j', \mathbf{p}'^{(ij)})$ holds. This completes the proof.

Theorem 6.5.14 $i =_{\text{pe}} j$ *if and only if* $R(\beta_i, \alpha_j, \mathbf{p}^{(ij)}) = R(\alpha_i, \beta_j, \mathbf{p}^{(ij)})$ *for all* $\mathbf{p}^{(ij)}$ *and all*
$0 < \alpha, \beta < 1$.

Proof. ("Only if") If components i and j are permutation equivalent, then $\phi(\mathbf{x})$ is permutation
symmetric in x_i and x_j by Theorem 6.5.2. Then,

$$R(\alpha_i, \beta_j, \mathbf{p}^{(ij)}) = \mathbb{E}(\phi(X_i, X_j, \mathbf{X}^{(ij)}) | (\alpha_i, \beta_j, \mathbf{p}^{(ij)})) = \mathbb{E}(\phi(X_j, X_i, \mathbf{X}^{(ij)}) | (\alpha_i, \beta_j, \mathbf{p}^{(ij)}))$$

$$= \mathbb{E}(\phi(X_i, X_j, \mathbf{X}^{(ij)}) | (\beta_i, \alpha_j, \mathbf{p}^{(ij)})) = R(\beta_i, \alpha_j, \mathbf{p}^{(ij)}).$$

("If") If $i =_{\text{pe}} j$ does not hold, suppose that there exists some $\mathbf{x}_0^{(ij)}$ such that $\phi(1_i, 0_j, \mathbf{x}_0^{(ij)}) =$
1 and $\phi(0_i, 1_j, \mathbf{x}_0^{(ij)}) = 0$. Consequently, for the vector $\mathbf{p}^{(ij)}$ with elements 0s and 1s such that

$\Pr\{\mathbf{X}^{(ij)} = \mathbf{x}_0^{(ij)} | \mathbf{p}^{(ij)}\} = 1$, $R(1_i, 0_j, \mathbf{p}^{(ij)}) = 1$, and $R(0_i, 1_j, \mathbf{p}^{(ij)}) = 0$. By continuity, there exists a $\mathbf{p}' = (\alpha'_i, \beta'_j, \mathbf{p}'^{(ij)})$, $\mathbf{0} < \mathbf{p}' < \mathbf{1}$, such that $R(\beta'_i, \alpha'_j, \mathbf{p}'^{(ij)}) > R(\alpha'_i, \beta'_j, \mathbf{p}'^{(ij)})$, which is a contradiction.

Theorem 6.5.15   *Assume that vectors* $(p_i, p_j, \mathbf{p}^{(ij)})$ *and* $(\tilde{p}_i, \tilde{p}_j, \mathbf{p}^{(ij)})$ *satisfy   (i)* $p_i p_j \leq \tilde{p}_i \tilde{p}_j$; * (ii)* $q_i q_j \geq \tilde{q}_i \tilde{q}_j$ $(q_i = 1 - p_i,$ *and so on); and   (iii)* $p_i \leq \tilde{p}_i$. *If* $i \geq_{\mathrm{pe}} j$, *then* $R(p_i, p_j, \mathbf{p}^{(ij)}) \leq R(\tilde{p}_i, \tilde{p}_j, \mathbf{p}^{(ij)})$.

Proof. According to Equation (6.18),

$$R(p_i, p_j, \mathbf{p}^{(ij)}) - R(\tilde{p}_i, \tilde{p}_j, \mathbf{p}^{(ij)})$$
$$= (\tilde{p}_i - p_i)\tilde{p}_j[R(1_i, 1_j, \mathbf{p}^{(ij)}) - R(0_i, 1_j, \mathbf{p}^{(ij)})]$$
$$+ (\tilde{p}_j - p_j)p_i[R(1_i, 1_j, \mathbf{p}^{(ij)}) - R(1_i, 0_j, \mathbf{p}^{(ij)})]$$
$$+ (\tilde{p}_i - p_i)\tilde{q}_j[R(1_i, 0_j, \mathbf{p}^{(ij)}) - R(0_i, 0_j, \mathbf{p}^{(ij)})]$$
$$+ (\tilde{p}_j - p_j)q_i[R(0_i, 1_j, \mathbf{p}^{(ij)}) - R(0_i, 0_j, \mathbf{p}^{(ij)})].$$

If $i \geq_{\mathrm{pe}} j$, then $R(1_i, 0_j, \mathbf{p}^{(ij)}) \geq R(0_i, 1_j, \mathbf{p}^{(ij)})$ for all $\mathbf{p}^{(ij)}$. Thus,

$$R(p_i, p_j, \mathbf{p}^{(ij)}) - R(\tilde{p}_i, \tilde{p}_j, \mathbf{p}^{(ij)})$$
$$\geq [(\tilde{p}_i - p_i)\tilde{p}_j + (\tilde{p}_j - p_j)p_i][R(1_i, 1_j, \mathbf{p}^{(ij)}) - R(1_i, 0_j, \mathbf{p}^{(ij)})] \qquad (6.19)$$
$$+ [(\tilde{p}_i - p_i)\tilde{q}_j + (\tilde{p}_j - p_j)q_i][R(0_i, 1_j, \mathbf{p}^{(ij)}) - R(0_i, 0_j, \mathbf{p}^{(ij)})]$$

because $\tilde{p}_i - p_i \geq 0$ by assumption (iii). Furthermore, by assumption (i), $(\tilde{p}_i - p_i)\tilde{p}_j + (\tilde{p}_j - p_j)p_i = \tilde{p}_i \tilde{p}_j - p_i p_j \geq 0$; by assumption (ii), $(\tilde{p}_i - p_i)\tilde{q}_j + (\tilde{p}_j - p_j)q_i = q_i q_j - \tilde{q}_i \tilde{q}_j \geq 0$. These two inequalities and inequality (6.19) together imply that $R(p_i, p_j, \mathbf{p}^{(ij)}) \leq R(\tilde{p}_i, \tilde{p}_j, \mathbf{p}^{(ij)})$.

Note that for $i =_{\mathrm{pe}} j$, assumption (iii) in Theorem 6.5.15 is unnecessary because assumptions (i) and (ii) together imply either $p_i \leq \tilde{p}_i$ or $p_j \leq \tilde{p}_j$. Mi (2003) further showed that for a k-out-of-n system, assumptions (i) and (ii) in Theorem 6.5.15 are both necessary and sufficient because all components are permutation equivalent. That is, for a k-out-of-n system, $R(p_i, p_j, \mathbf{p}^{(ij)}) \leq R(\tilde{p}_i, \tilde{p}_j, \mathbf{p}^{(ij)})$ if and only if assumptions (i) and (ii) hold.

6.6   The Domination Importance

Recall from Section 2.5 that $\mathscr{C}$ ($\mathscr{P}$) denotes the set of cuts (paths) of the system, $\overline{\mathscr{C}}_i$ ($\overline{\mathscr{P}}_i$) the set of minimal cuts (paths) containing component i, and $\mathscr{C}_i^{(j)}$ ($\mathscr{P}_i^{(j)}$) the set of cuts (paths) containing components i but not j.

Definition 6.6.1   *Component i internally dominates component j, denoted by $i \geq_{\mathrm{in}} j$ (internal domination importance), if for any $S \in \mathscr{C}_j^{(i)}$, $S \setminus \{j\} \in \mathscr{C}$. Component i externally dominates component j, denoted by $i \geq_{\mathrm{ex}} j$ (external domination importance), if for any $S \in \mathscr{P}_j^{(i)}$, $S \setminus \{j\} \in \mathscr{P}$.*

Theorem 6.6.2 *(i) $i \geq_{in} j$ if and only if for any $S \in \mathscr{P}$ containing components i and j, $S \setminus \{j\} \in \mathscr{P}$. (ii) $i \geq_{ex} j$ if and only if for any $S \in \mathscr{C}$ containing components i and j, $S \setminus \{j\} \in \mathscr{C}$.*

Proof. To show part (i), first assume that $i \geq_{in} j$ and $S \in \mathscr{P}$ such that $i, j \in S$. Letting $S' = S \setminus \{j\}$, it needs to be shown that $S' \in \mathscr{P}$. Note that $S'^c = N \setminus S'$ (complement) contains components j but not i. Because $S \in \mathscr{P}$, by Theorem 2.5.2(i), $S^c = N \setminus S = S'^c \setminus \{j\}$ is not a cut. Because $i \geq_{in} j$, S'^c is not a cut. Thus, by Theorem 2.5.2(i), S' is a path.

Conversely, assume that for any $H \in \mathscr{P}$ containing components i and j, $H \setminus \{j\} \in \mathscr{P}$. To show $i \geq_{in} j$, assuming $S \in \mathscr{C}_j^{(i)}$ and letting $S' = S \setminus \{j\}$, it needs to be shown that S' is a cut. Note that $S'^c = N \setminus S'$ contains both components i and j. Because S is a cut, by Theorem 2.5.2(i), $S^c = N \setminus S = S'^c \setminus \{j\}$ is not a path. Consequently, by the assumption, S'^c is not a path. Thus, by Theorem 2.5.2(i), S' is a cut.

For part (ii), the proof can be done by exchanging "cut" and "path" in the earlier statements.

Freixas and Pons (2008a) proposed the external domination importance as in Definition 6.6.1 and the internal domination importance as in Theorem 6.6.2. They did not show the essential relations among them as Kuo and Zhu (2012b) showed in Theorem 6.6.2. As shown in Definition 6.6.1 and Theorem 6.6.2, the internal and external domination importance measures are c-type and p-type of importance measures with each other. Note that both the internal and external domination importance measures are partial rankings; some of the components may not be compared according to them. In addition, note the similarity and difference among the definitions of the internal and external domination importance measures and of the permutation importance (Definition 6.5.9).

These two domination importance measures can be characterized using minimal cuts and minimal paths as shown in Theorem 6.6.3, partially based on Freixas and Pons (2008a).

Theorem 6.6.3 *Let i and j be two distinct components.*

(i) $i \geq_{in} j$ if and only if for any $S \in \overline{\mathscr{C}}_j$, $i \in S$.
(ii) $i \geq_{in} j$ if and only if for any $S \in \overline{\mathscr{P}}_j$, $(S \cup \{i\}) \setminus \{j\} \in \mathscr{P}$.
(iii) $i \geq_{ex} j$ if and only if for any $S \in \overline{\mathscr{P}}_j$, $i \in S$.
(iv) $i \geq_{ex} j$ if and only if for any $S \in \overline{\mathscr{C}}_j$, $(S \cup \{i\}) \setminus \{j\} \in \mathscr{C}$.

Proof. To show part (i), assume that $i \geq_{in} j$ and $S \in \overline{\mathscr{C}}_j$. Then, if $i \notin S$, it would be $S \setminus \{j\} \in \mathscr{C}$, but this contradicts the minimal character of the cut. Conversely, assuming the hypothesis, let $H \in \mathscr{C}_j^{(i)}$ and then it needs to be shown that $H \setminus \{j\} \in \mathscr{C}$. Let $S \in \overline{\mathscr{C}}$ be such that $S \subseteq H$. If $j \in S$, it would be $i \in S$, and thus, $i \in H$ (contradiction). Thus, $j \notin S$, and as a consequence, $S \subseteq H \setminus \{j\}$ and $H \setminus \{j\} \in \mathscr{C}$.

To show part (ii), assume that $i \geq_{in} j$ and $S \in \overline{\mathscr{P}}_j$. Then $H = S \cup \{i\} \in \mathscr{P}$ and $i, j \in H$; thus, $H \setminus \{j\} = (S \cup \{i\}) \setminus \{j\} \in \mathscr{P}$. Conversely, assume the hypothesis and let $H \in \mathscr{P}$ be such that $i, j \in H$. Then, it needs to be shown that $H \setminus \{j\} \in \mathscr{P}$. Let $S \in \overline{\mathscr{P}}$ be such that $S \subseteq H$. If $j \in S$, then $(S \cup \{i\}) \setminus \{j\} \subseteq (H \cup \{i\}) \setminus \{j\} = H \setminus \{j\}$, and thus, $H \setminus \{j\} \in \mathscr{P}$. On the other hand, if $j \notin S$, then $S \subseteq H \setminus \{j\}$, and thus, $H \setminus \{j\} \in \mathscr{P}$.

Similar arguments can prove parts (iii) and (iv).

Theorem 6.6.4   *For two distinct components i and j in a system, $i \geq_{\text{in}} j$ and $i \geq_{\text{ex}} j$ cannot coexist.*

Proof. Suppose that $i \geq_{\text{in}} j$ and $i \geq_{\text{ex}} j$ coexist. Then, by Theorem 6.6.3(iii), $i \geq_{\text{ex}} j$ implies that for any $S \in \overline{\mathscr{P}}_j$, $i \in S$. Because $S \in \overline{\mathscr{P}}_j$, $S \setminus \{j\}$ is not a path, and thus, $S^c \cup \{j\}$ ($S^c = N \setminus S$) is a cut by Theorem 2.5.2(i). Furthermore, because $i \notin S^c$ and $i \geq_{\text{in}} j$, by Definition 6.6.1, S^c is a cut, which contradicts that S is a path.

Theorem 6.6.5 (Freixas and Pons 2008a) shows the transitivity property.

Theorem 6.6.5   *If $i \geq_{\text{in}} (\geq_{\text{ex}}) j$ and $j \geq_{\text{in}} (\geq_{\text{ex}}) k$, then $i \geq_{\text{in}} (\geq_{\text{ex}}) k$.*

Proof. To prove the result for the external domination importance, suppose that $S \in \mathscr{P}_k^{(i)}$. Then, it needs to be shown that $S \setminus \{k\} \in \mathscr{P}$. There are two possibilities. If $j \notin S$, then $S \setminus \{k\} \in \mathscr{P}$ because $j \geq_{\text{ex}} k$. If $j \in S$, then $S \setminus \{j\} \in \mathscr{P}$ because $i \geq_{\text{ex}} j$ and $i \notin S$. But $k \in S \setminus \{j\}$, and $j \geq_{\text{ex}} k$; thus, $S \setminus \{j, k\} \in \mathscr{P}$, and so $S \setminus \{k\} \in \mathscr{P}$.

By replacing the path with the cut, the result for the internal domination importance can be proven.

Freixas and Pons (2008a) showed the following characteristics for the two domination importance measures, which correspond to Theorems 6.5.13 and 6.5.14 for the permutation importance. The proof of Theorem 6.6.6 is similar to those of Theorems 6.5.13 and 6.5.14.

Theorem 6.6.6   *(i) $i \geq_{\text{in}} j$ if and only if $R(1_i, 0_j, \mathbf{p}^{(ij)}) = R(1_i, 1_j, \mathbf{p}^{(ij)})$ for all $\mathbf{p}^{(ij)}$. (ii) $i \geq_{\text{ex}} j$ if and only if $R(0_i, 0_j, \mathbf{p}^{(ij)}) = R(0_i, 1_j, \mathbf{p}^{(ij)})$ for all $\mathbf{p}^{(ij)}$.*

6.7   The Cut Importance and Path Importance

This section presents the cut importance and path importance. Subsection 6.7.1 gives nice relations of the cut importance and path importance to the B-i.i.d. importance as component reliability approaches one and zero, respectively. Subsection 6.7.2 addresses the determination of the cut importance and path importance, which can be simplified under certain conditions.

Butler (1977) proposed a structure importance measure that defines component i to be more important than component j if $(-e_i, \ell_i) \succ (-e_j, \ell_j)$ in the strict sense of lexicographic order, where e_i is the smallest size of minimal cuts containing component i, ℓ_i is the number of distinct minimal cuts of size e_i containing component i, and $\succ$ denotes lexicographic order. Vectors $\mathbf{x} \succ \mathbf{y}$ if the first nonzero element of $\mathbf{x} - \mathbf{y}$ is positive and $\mathbf{x} \neq \mathbf{y}$. Apparently, this importance measure is a partial ranking only, according to which some components in a system may not be compared.

Extending this partial importance ranking, Butler (1979) proposed another structure importance, namely, the cut importance, which is a complete ranking of all components relative to their importance to the system. The cut importance is defined on the basis of minimal cuts as follows.

Definition 6.7.1   *For each component i in a system (N, ϕ) with u minimal cuts, let $a_i(\ell, d)$ denote the number of collections of ℓ distinct minimal cuts such that the union of each collection contains exactly d components and includes component i, $1 \leq \ell \leq u$, $1 \leq d \leq n$. Let $b_i(d) = \sum_{\ell=1}^{u}(-1)^{\ell-1}a_i(\ell, d)$ and $\mathbf{b}_i = (b_i(1), b_i(2), \ldots, b_i(n))$. Component i is more cut important than component j, denoted by $i >_c j$, if $\mathbf{b}_i \succ \mathbf{b}_j$. Components i and j are equally cut important, denoted by $i =_c j$, if $\mathbf{b}_i = \mathbf{b}_j$.*

Meng (1995) presented Lemma 6.7.2 that refines the aforementioned definition of the cut importance. Let $a^{(i\backslash j)}(\ell, d)$ $(a^{(ij)}(\ell, d))$ denote the number of collections of ℓ distinct minimal cuts such that the union of each collection contains exactly d components and the union includes components i but not j (includes both components i and j). Furthermore, let $y_i(d) = \sum_{\ell=1}^{u}(-1)^{\ell-1}a^{(i\backslash j)}(\ell, d)$ and $\mathbf{y}_i = (y_i(1), y_i(2), \ldots, y_i(n))$. Then, the following lemma is immediate, and the proof is omitted.

Lemma 6.7.2   *The following hold for all $i, j \in \{1, 2, \ldots, n\}$, $1 \leq \ell \leq u$, and $1 \leq d \leq n$:*

$$a_i(\ell, d) = a^{(i\backslash j)}(\ell, d) + a^{(ij)}(\ell, d); \qquad (6.20)$$

$$b_i(d) - b_j(d) = \sum_{\ell=1}^{u}(-1)^{\ell-1}[a^{(i\backslash j)}(\ell, d) - a^{(j\backslash i)}(\ell, d)] = y_i(d) - y_j(d); \quad (6.21)$$

$$\mathbf{b}_i \succ \mathbf{b}_j \;\; \text{if and only if} \;\; \mathbf{y}_i \succ \mathbf{y}_j.$$

Equation (6.20) states that among the $a_i(\ell, d)$ and $a_j(\ell, d)$ collections of ℓ distinct minimal cuts, many of those collections are such that the union of the ℓ minimal cuts contains both components i and j. From Equation (6.21), those collections containing both components i and j can be neglected when comparing the cut importance of components i and j. This observation can significantly reduce the amount of computational work involved in determining the relative cut importance of components. Note that the number of such collections containing both components i and j increases rapidly in d (in particular when $d = n$, $a_i(\ell, d) = a_j(\ell, d)$).

Analogous to the cut importance, the path importance is defined on the basis of minimal paths other than the minimal cuts. The path importance is a complete importance ranking as well.

Definition 6.7.3   *For each component i in a system (N, ϕ) with v minimal paths, let $w_i(\ell, d)$ denote the number of collections of ℓ distinct minimal paths such that the union of each collection contains exactly d components and includes component i, $1 \leq \ell \leq v$, $1 \leq d \leq n$. Let $g_i(d) = \sum_{\ell=1}^{v}(-1)^{\ell-1}w_i(\ell, d)$ and $\mathbf{g}_i = (g_i(1), g_i(2), \ldots, g_i(n))$. Component i is more path important than component j, denoted by $i >_p j$, if $\mathbf{g}_i \succ \mathbf{g}_j$. Components i and j are equally path important, denoted by $i =_p j$, if $\mathbf{g}_i = \mathbf{g}_j$.*

6.7.1   Relations to the B-i.i.d. Importance

One motivation for studying the cut importance is that it has a clear relation to the B-i.i.d. importance for high values of p, as shown in Theorem 6.7.4 (Butler 1979).

Theorem 6.7.4 *The cut importance is equivalent to the B-i.i.d. importance as p approaches 1, that is, $i >_c j$ if and only if $I_B(i; p) > I_B(j; p)$ for all p sufficiently close to 1. Note that*

$$I_B(i; p) = \sum_{d=1}^{n} b_i(d)(1 - p)^{d-1}. \tag{6.22}$$

Proof. To show the results, first derive Equation (6.22). Let A_ℓ denote the event that at least one component in the ℓth minimal cut, C_ℓ, functions, $\ell = 1, 2, \ldots, u$. Then,

$$R(\mathbf{p}) = \Pr\left\{ \bigcap_{\ell=1}^{u} A_\ell \right\} = 1 - \Pr\left\{ \bigcup_{\ell=1}^{u} A_\ell^c \right\}.$$

By the inclusion–exclusion principle (Kuo and Zuo, 2003, pp. 153–157),

$$R(\mathbf{p}) = 1 - \sum_{\ell=1}^{u} (-1)^{\ell-1} S_\ell,$$

where

$$S_\ell = \sum_{1 \le k_1 < k_2 < \cdots < k_\ell \le u} \Pr\left\{ A_{k_1}^c \cap A_{k_2}^c \cap \cdots \cap A_{k_\ell}^c \right\},$$

and $A_{k_1}^c \cap A_{k_2}^c \cap \cdots \cap A_{k_\ell}^c$ is the event that all components $j \in C_{k_1} \cup C_{k_2} \cup \cdots \cup C_{k_\ell}$ fail. Then, using the independent assumption,

$$S_\ell = \sum_{1 \le k_1 < k_2 < \cdots < k_\ell \le u} \left[\prod_{j \in C_{k_1} \cup C_{k_2} \cup \cdots \cup C_{k_\ell}} (1 - p_j) \right].$$

Thus,

$$I_B(i; \mathbf{p}) = \frac{\partial R(\mathbf{p})}{\partial p_i} = \sum_{\ell=1}^{u} (-1)^{\ell-1} \sum_{\substack{1 \le k_1 < k_2 < \cdots < k_\ell \le u \\ i \in C_{k_1} \cup C_{k_2} \cup \cdots \cup C_{k_\ell}}} \left[\prod_{\substack{j \in C_{k_1} \cup C_{k_2} \cup \cdots \cup C_{k_\ell} \\ j \ne i}} (1 - p_j) \right].$$

Recalling the definition of $a_i(\ell, d)$ and noting $p_1 = p_2 = \cdots = p_n = p$,

$$I_B(i; p) = \sum_{\ell=1}^{u} \sum_{d=1}^{n} (-1)^{\ell-1} (1 - p)^{d-1} a_i(\ell, d) = \sum_{d=1}^{n} b_i(d)(1 - p)^{d-1}.$$

Now, using Equation (6.22), it is clear that $i >_c j$ if and only if $\mathbf{b}_i - \mathbf{b}_j \succ \mathbf{0}$, and $\mathbf{b}_i - \mathbf{b}_j \succ \mathbf{0}$ if and only if $I_B(i; p) > I_B(j; p)$ for all p sufficiently close to 1.

Theorem 6.7.4 establishes the relation of the cut importance to the B-i.i.d. importance in the case of high and equal component reliability. Consider the case where the component reliability values are high but unequal. Let $\mathbf{p}(\varepsilon)$ be a vector-valued function of the positive scalar ε for which $0 < p_i(\varepsilon) < 1$ for all $\varepsilon \in (0, \infty)$ and $i = 1, 2, \ldots, n$, and $\lim_{\varepsilon \to 0} \mathbf{p}(\varepsilon) = \mathbf{1}$.

In general, the cut importance ranking of components does not concide with the ranking of components induced by $I_B(\cdot; \mathbf{p}(\varepsilon))$ for all ε sufficiently close to zero. Although Butler (1977, 1979) was able to obtain some limited results with some additional assumptions of $\mathbf{p}(\varepsilon)$, their value is questionable because the hypotheses become too complex. Thus, these limited results are not presented here to conserve space. From a practical standpoint, the cut importance can be useful even when component reliability values are high but unequal; however, it may be misleading if the differences in the orders of magnitude of component reliability are too great.

As shown in Theorem 6.7.5, Meng (1995) gave additional characteristics of the cut importance in terms of the B-reliability importance, which strengthens Theorem 6.7.4. Theorem 6.7.5 differs from Theorem 6.7.4 in that the reliability of the two components being compared can assume an arbitrary same value, not necessarily close to one.

Theorem 6.7.5 *Assuming that the scalar p is sufficiently close to 1, then $i >_c j$ if and only if $I_B(i; \alpha_j, p^{(ij)}) > I_B(j; \alpha_i, p^{(ij)})$, where α is the same reliability of components i and j, $0 < \alpha < 1$, and p is the same reliability of the other $n-2$ components in the system.*

Proof. By Equation (6.22) and Lemma 6.7.2,

$$
I_B(i; \alpha_j, p^{(ij)}) = \sum_{\ell=1}^{u} \sum_{d=2}^{n} (-1)^{\ell-1} (1-\alpha)(1-p)^{d-2} a^{(ij)}(\ell, d)
$$

$$
+ \sum_{\ell=1}^{u} \sum_{d=1}^{n-1} (-1)^{\ell-1} (1-p)^{d-1} a^{(i \backslash j)}(\ell, d).
$$

Hence, $I_B(i; \alpha_j, p^{(ij)}) - I_B(j; \alpha_i, p^{(ij)}) = \sum_{\ell=1}^{u} \sum_{d=1}^{n-1} (-1)^{\ell-1} (1-p)^{d-1} [a^{(i \backslash j)}(\ell, d) - a^{(j \backslash i)}(\ell, d)]$. Then, the result follows as p approaches 1.

Similar to Theorem 6.7.4 for the cut importance, Theorem 6.7.6 (Kuo and Zhu 2012b) provides the results for the path importance. These results are useful for a system consisting of very unreliable components.

Theorem 6.7.6 *The path importance is equivalent to the B-i.i.d. importance as p approaches 0, that is, $i >_p j$ if and only if $I_B(i; p) > I_B(j; p)$ for all p sufficiently close to 0. Note that*

$$
I_B(i; p) = \sum_{d=1}^{n} g_i(d) p^{d-1}. \tag{6.23}
$$

Proof. To show Equation (6.23), consider a coherent system ϕ in which all components have the same reliability p and its dual system ϕ^D in which all components have the same reliability $q = 1 - p$. Then, by Theorem 4.1.10, $I_B(i; \phi, p) = I_B(i; \phi^D, q)$, where $I_B(i; \phi, p)$ and $I_B(i; \phi^D, q)$ represent the B-i.i.d. importance of component i in system ϕ and its dual system ϕ^D, respectively.

Considering the cut importance in dual system ϕ^D, by Equation (6.22), it follows that $I_B(i; \phi^D, q) = \sum_{d=1}^{n} b_i^D(d)(1 - q)^{d-1}$, where $b_i^D(d)$ is defined in Definition 6.7.1 for dual system ϕ^D. Because a minimal path in system ϕ is a minimal cut in its dual system ϕ^D, then $g_i(d) = b_i^D(d)$, $1 \le d \le n$, where $g_i(d)$ is defined in Definition 6.7.3 for system ϕ. Thus, $I_B(i; \phi, p) = I_B(i; \phi^D, q) = \sum_{d=1}^{n} g_i(d)(1 - q)^{d-1}$.

Using Equation (6.23), it is clear that $i >_p j$ if and only if $\mathbf{g}_i - \mathbf{g}_j \succ \mathbf{0}$, and $\mathbf{g}_i - \mathbf{g}_j \succ \mathbf{0}$ if and only if $I_B(i; p) > I_B(j; p)$ for all p sufficiently close to 0.

Corollary 6.7.7 *The following are equivalent: (i) $i =_c j$, (ii) $i =_p j$, and (iii) $I_B(i; p) = I_B(j; p)$ for all p.*

Proof. By Equations (6.22) and (6.23), $i =_c j$ if and only if $I_B(i; p) = I_B(j; p)$ for all p, which is true if and only if $i =_p j$.

6.7.2 Computation

The computation of the cut importance requires only the knowledge of minimal cuts. In the following example, the cut importance ranking of components is determined directly from Definition 6.7.1.

Example 6.7.8 For the system in Figure 6.2, the minimal cuts are $C_1 = \{1, 5\}, C_2 = \{2, 3\}$, $C_3 = \{2, 6\}, C_4 = \{4, 5, 6\}, C_5 = \{3, 4, 5\}$, and $C_6 = \{1, 2\}$. For $i = 2$, the nonzero $a_i(\ell, d)$ are as follows: $a_2(1, 2) = 3$, $a_2(2, 3) = 4$, $a_2(2, 4) = 4$, $a_2(2, 5) = 4$, $a_2(3, 4) = 3$, $a_2(3, 5) = 11$, $a_2(3, 6) = 5$, $a_2(4, 5) = 4$, $a_2(4, 6) = 11$, $a_2(5, 6) = 6$, and $a_2(6, 6) = 1$. Thus, $\mathbf{b}_2 = (0, 3, -4, -1, 3, -1)$. Similarly, $\mathbf{b}_1 = (0, 2, -3, -1, 3, -1)$, $\mathbf{b}_3 = (0, 1, -1, -2, 3, -1)$, $\mathbf{b}_4 = (0, 0, 2, -5, 4, -1)$, $\mathbf{b}_5 = (0, 1, 1, -5, 4, -1)$, and $\mathbf{b}_6 = (0, 1, -1, -2, 3, -1)$. Therefore, $2 >_c 1 >_c 5 >_c 3 =_c 6 >_c 4$.

Because the cut importance depends on lexicographic order, most components can be ranked by determining only the first few elements of $\mathbf{b}_i$. Theorem 6.7.9 (Butler 1979) establishes a simple and computationally convenient formula for determining the first nonzero element in any vector $\mathbf{b}_i$. Other elements in $\mathbf{b}_i$ are computed only as necessary.

Theorem 6.7.9 *For each component i, let e_i be the smallest size of the minimal cuts containing component i and ℓ_i be the number of the minimal cuts of size e_i containing component i. Then, (i) $e_i = \min\{d : b_i(d) \ne 0\}$ and $\ell_i = b_i(e_i)$; (ii) If $e_i < e_j$, then $i >_c j$; (iii) If $e_i = e_j$ and $\ell_i > \ell_j$, then $i >_c j$.*

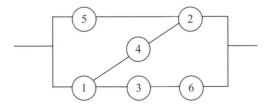

Figure 6.2 System of six components

Proof. By definition, $a_i(1, e_i) = \ell_i$, and any union of two or more minimal cuts at least one of which contains component i must have at least $e_i + 1$ components. Thus, $a_i(\ell, e_i) = 0$ for all $\ell \geq 2$. Therefore,

$$b_i(e_i) = \sum_{\ell=1}^{u}(-1)^{\ell-1}a_i(\ell, e_i) = \ell_i.$$

Also, since component i is contained in no cuts of size smaller than e_i, $a_i(\ell, d) = 0$ for all $d < e_i$, $1 \leq \ell \leq u$. Thus, $b_i(d) = 0$ for all $d < e_i$. Part (i) is proved. Then, parts (ii) and (iii) follow directly.

Corollary 6.7.10   *If component i is in series with the rest of the system, then $i \geq_c j$ for all $j \neq i$. The strict inequality holds unless component j is also a one-component cut.*

The computation can also be simplified when the system under consideration contains modules. Theorem 6.7.11 (Butler 1979) explores the modular structure to calculate the cut importance of components based on the minimal cuts of the module and of the system organizing structure with respect to the module decomposition.

Theorem 6.7.11   *Let a coherent system (N, ϕ) have a module (M, χ) and let $\phi(\mathbf{x}) = \psi\left(\chi(\mathbf{x}^M), \mathbf{x}^{M^c}\right)$. Let $\mathbf{b}_i^\phi$, $\mathbf{b}_i^\chi$, and $\mathbf{b}_M^\psi$ be the vectors corresponding to structures ϕ, χ, and ψ, respectively. Then for $i \in M$,*

$$b_i^\phi(d) = \sum_{j=1}^{d} b_M^\psi(j)b_i^\chi(d - j + 1),$$

where the definition of $\mathbf{b}_i^\chi$ is extended to include zero elements for $j > |M|$, and $\mathbf{b}_M^\psi$ is extended similarly. (This equation is an expression of the fact that $\mathbf{b}_i^\phi$ is the convolution of the finite sequences $\mathbf{b}_M^\psi$ and $\mathbf{b}_i^\chi$.)

Proof. By Equation (6.22) and the chain rule for the B-reliability importance in Equation (4.17), for $i \in M$,

$$\sum_{d=1}^{n} b_i^\phi(d)(1-p)^{d-1} = \left[\sum_{d=1}^{|M^c|+1} b_M^\psi(d)(1-p)^{d-1}\right]\left[\sum_{d=1}^{|M|} b_i^\chi(d)(1-p)^{d-1}\right]$$

$$= \sum_{d=1}^{n}\left[\sum_{j=1}^{d} b_M^\psi(j)b_i^\chi(d - j + 1)\right](1-p)^{d-1}.$$

Since this equality holds for all $0 \leq p \leq 1$, each pair of coefficients of the two polynomials must be identical. This completes the proof.

Example 6.7.12 (Example 6.7.8 continued)   For the system in Figure 6.2, components 3 and 6 form a module with $M = \{3, 6\}$. Then, $z = \chi(\mathbf{x}^M) = x_3x_6$, $\psi(z, \mathbf{x}^{M^c}) = 1 - (1 - x_1x_2x_4)(1 - x_1z)(1 - x_2x_5)$, $\mathbf{b}_3^\chi = (1, -1)$, $\mathbf{b}_6^\chi = (1, -1)$, and $\mathbf{b}_M^\psi = (0, 1, 0, -2, 1)$.

Similar to Theorems 6.7.9 and 6.7.11 and Corollary 6.7.10 and by duality, the results for simplifying the computation of the path importance can be straightforwardly derived. For example, Corollary 6.7.13 is analogous to Corollary 6.7.10.

Corollary 6.7.13 *If component i is in parallel with the rest of the system, then* $i \geq_p j$ *for all* $j \neq i$. *The strict inequality holds unless component j is also a one-component path.*

6.8   The Absoluteness Importance

Recall from Section 2.5 that $\mathscr{C}_i$ ($\mathscr{P}_i$) is the set of cuts (paths) containing component i.

Definition 6.8.1 *Component i is more c-absolutely important than component j, denoted by* $i >_{ac} j$, *if* $\mathscr{C}_j \subset \mathscr{C}_i$ *and more p-absolutely important, denoted by* $i >_{ap} j$, *if* $\mathscr{P}_j \subset \mathscr{P}_i$. *Components i and j are c-absolutely equivalent* ($i =_{ac} j$) *if* $\mathscr{C}_j = \mathscr{C}_i$ *and p-absolutely equivalent* ($i =_{ap} j$) *if* $\mathscr{P}_j = \mathscr{P}_i$.

The c-absoluteness and p-absoluteness importance measures are not equivalent; in fact, no component can be both c-absolutely and p-absolutely more important than another component in a system, as shown in Theorem 6.8.2 (Hwang 2005).

Theorem 6.8.2 *For two distinct components i and j in a system,* $i \geq_{ac} j$ *and* $i \geq_{ap} j$ *cannot coexist.*

Proof. If $\{i\}$ is a cut, then $\{i\}$ is not a path in a nontrivial system with $n \geq 2$. Hence, $N \setminus \{i\}$ is a cut containing components j but not i. If $\{i\}$ is not a cut, then $N \setminus \{i\}$ is a path containing components j but not i.

Example 6.8.3 (Example 2.2.4 continued)   For the system in Figure 2.4, $1 >_{ap} 2$, but $1 \not>_{ac} 2$. For the system in Figure 2.5, $1 >_{ac} 2$, but $1 \not>_{ap} 2$.

Theorem 6.8.4 *If component i is in series (parallel) with the rest of the system, then* $i \geq_{ap} j$ ($i \geq_{ac} j$) *for all* $j \neq i$.

Proof. In the series connection case, $i \in P$ for any path P. Thus, $\mathscr{P}_i = \mathscr{P}$, and consequently, $\mathscr{P}_j \subseteq \mathscr{P}_i$ for any $j \neq i$. In the parallel connection case, $i \in C$ for any cut C. Thus, $\mathscr{C}_i = \mathscr{C}$, and consequently, $\mathscr{C}_j \subseteq \mathscr{C}_i$ for any $j \neq i$.

The absoluteness importance is the strongest structure importance (see Section 6.14). Few components can be compared according to it, seriously limiting its application.

6.9   The Cut-path Importance, Min-cut Importance, and Min-path Importance

Recall from Section 2.5 that $\mathscr{C}_i(d)$ ($\mathscr{P}_i(d)$) is the set of cuts (paths) containing component i of size d and that $\overline{\mathscr{C}}_i(d)$ ($\overline{\mathscr{P}}_i(d)$) is the set of minimal cuts (paths) containing component i of size d.

Definition 6.9.1   *Component i is more cut-path important than component j, denoted by $i >_{cp}$ j, if $|\mathcal{C}_i(d)| \geq |\mathcal{C}_j(d)|$ for all $1 \leq d \leq n$ and the strict inequality holds for some d. Components i and j are cut-path importance equivalent, denoted by $i =_{cp} j$, if $|\mathcal{C}_i(d)| = |\mathcal{C}_j(d)|$ for all d.*

Definition 6.9.2   *Component i is more min-cut important than component j, denoted by $i >_{mc} j$, if $|\overline{\mathcal{C}}_i(d)| \geq |\overline{\mathcal{C}}_j(d)|$ for all $1 \leq d \leq n$ and the strict inequality holds for some d. Components i and j are min-cut importance equivalent, denoted by $i =_{mc} j$, if $|\overline{\mathcal{C}}_i(d)| = |\overline{\mathcal{C}}_j(d)|$ for all d. A min-path importance is defined similarly in terms of minimal paths and denoted by $i >_{mp} j$.*

Different from the permutation importance, domination importance, and absoluteness importance, a comparison of two components according to the cut-path importance, min-cut importance, or min-path importance does not require a containing relation, only a numerical dominance which has to hold for every d. Theorem 6.9.3 (Hwang 2005) shows that it is equivalent to define the cut-path importance using either cuts or paths in Definition 6.9.1, hence the name cut-path importance with no identification of c-type or p-type.

Theorem 6.9.3 $i \geq_{cp} j$ if and only if $|\mathcal{P}_i(d)| \geq |\mathcal{P}_j(d)|$ for all $1 \leq d \leq n$.

Proof. By Theorem 2.5.2(vi),

$$|\mathcal{C}_i(d)| - |\mathcal{C}_j(d)| = \binom{n-1}{d-1} - |\mathcal{P}_{(i)}(n-d)| - \left[\binom{n-1}{d-1} - |\mathcal{P}_{(j)}(n-d)|\right]$$

$$= |\mathcal{P}_{(j)}(n-d)| - |\mathcal{P}_{(i)}(n-d)|$$

$$= (|\mathcal{P}(n-d)| - |\mathcal{P}_j(n-d)|) - (|\mathcal{P}(n-d)| - |\mathcal{P}_i(n-d)|)$$

$$= |\mathcal{P}_i(n-d)| - |\mathcal{P}_j(n-d)|,$$

where $|\mathcal{P}(n-d)|$ is the number of paths of size $n-d$.

The cut-path importance and the min-cut and min-path importance measures are defined similarly in that the cut-path importance is defined in terms of cuts or paths but the min-cut and min-path importance measures are defined in terms of minimal cuts and minimal paths, respectively. Note that the min-path importance is not equivalent to the min-cut importance, although the cut-path importance has its p-type equivalent to its c-type.

Example 6.9.4 (Example 6.8.3 continued)   For the system in Figure 2.4, $1 >_{mp} 2$, but $1 \not>_{mc}$ 2. For the system in Figure 2.5, $1 >_{mc} 2$, but $1 \not>_{mp} 2$.

Theorem 6.9.5 shows a property of the cut-path importance.

Theorem 6.9.5   *If component i is in series (parallel) with the rest of the system, then $i \geq_{cp} j$ for all $j \neq i$.*

Proof. If component i is in series with the rest of the system, that is, $\{i\}$ is a cut, then for any cut $\{j\} \cup S$ where $i \notin S$, $\{i\} \cup S$ is also a cut. Therefore, $|\mathcal{C}_i(d)| \geq |\mathcal{C}_j(d)|$ for all d and then

$i \geq_{\text{cp}} j$ for any $j \neq i$. A similar argument on the basis of paths can be applied to the case that component i is in parallel with the rest of the system.

6.10   The First-term Importance and Rare-event Importance

Inspired by the notion of first-term invariance in the domain of a consecutive-2 system in Santha and Zhang (1987), Chang and Hwang (2002) introduced the first-term importance of component i for the $Lin/Con/k/n$:F system, which is defined as the number of distinct sets of k consecutive components containing component i. Thus, the first-term importance of component i in the $Lin/Con/k/n$:F system is i for $i < k - 1$, is $n - i + 1$ for $i \geq n - k + 2$, and is k otherwise. Extending this importance measure to a general coherent system, the first-term importance is defined as follows.

Definition 6.10.1   *Let* $e = \min_{C \in \mathscr{C}} |C|$ *be the smallest size of minimal cuts. The first-term importance of component i is the number of minimal cuts of size e containing component i:*

$$I_{\text{FT}}(i; \phi) = |\overline{\mathscr{C}}_i(e)|.$$

Recall that $e_i = \min_{C \in \overline{\mathscr{C}}_i} |C|$ is the smallest size of minimal cuts containing component i. Then, $e_i \geq e$. For a general system, most components may hold $e_i > e$ and thus have a zero value for the first-term importance. The first-term importance may only make real sense for a well-structured system such as the $Lin/Con/k/n$ system. In the case that the component reliability p approaches 1, a cut of size d is much more likely to fail than a cut of size d' for $d' > d$. It justifies the use of the first-term importance in this case.

Now compare the first-term importance with the cut importance. If $e_i > e$, then $I_{\text{FT}}(i; \phi) = 0$; otherwise (i.e., $e_i = e$), $I_{\text{FT}}(i; \phi) = |\overline{\mathscr{C}}_i(e)|$. By Definition 6.7.1 of the cut importance, if $I_{\text{FT}}(i; \phi) > I_{\text{FT}}(j; \phi)$, then $i >_c j$. By Theorem 6.7.4 that the cut importance is equivalent to the B-i.i.d. importance as p approaches 1, if $I_{\text{FT}}(i; \phi) > I_{\text{FT}}(j; \phi)$, then $I_B(i; p) > I_B(j; p)$ as p approaches 1. However, if $I_{\text{FT}}(i; \phi) = I_{\text{FT}}(j; \phi)$, it is inconclusive about the relative cut importance ranking of components i and j. On the other hand, if $i =_c j$, then $I_{\text{FT}}(i; \phi) = I_{\text{FT}}(j; \phi)$. But, if $i >_c j$, it may be either $I_{\text{FT}}(i; \phi) = I_{\text{FT}}(j; \phi)$ or $I_{\text{FT}}(i; \phi) > I_{\text{FT}}(j; \phi)$.

Chang and Hwang (2002) defined the rare-event importance of component i for the $Lin/Con/k/n$:F system to be the number of paths containing component i of size $\lfloor n/k \rfloor$, noting that $\lfloor n/k \rfloor$ is the smallest size of minimal paths. The rare-event importance can be defined similarly for a general coherent system.

Definition 6.10.2   *Let* $e' = \min_{P \in \mathscr{P}} |P|$ *be the smallest size of minimal paths. The rare-event importance of component i is the number of minimal paths of size e' containing component i:*

$$I_{\text{RE}}(i; \phi) = |\overline{\mathscr{P}}_i(e')|.$$

The rare-event importance is the counterpart to the first-term importance and relevant as p approaches 0, under which a path with fewer components dominates probability-wise a path with more components. Similarly, the comparison of the rare-event importance and the path importance shows that if $I_{\text{RE}}(i; \phi) > I_{\text{RE}}(j; \phi)$, then $i >_p j$, and $I_B(i; p) > I_B(j; p)$ as p approaches 0.

Although a small value of p is not a practical assumption for components in most real problems, the rare-event importance serves as a useful tool to disprove the uniform B-i.i.d. importance pattern (i.e., the relative ordering of the importance values of the individual components) by providing a counterexample under the rare-event importance. Furthermore, if a pattern is proven under the half-line B-i.i.d. importance or under both the first-term importance and B-structure importance, then proving it further under the rare-event importance is a strong indication that the pattern might hold under the uniform B-i.i.d. importance (i.e., for all p) because p is now covered from both ends. Along this line, Chapter 8 presents more results about the rare-event importance for $Lin/Con/k/n$ systems.

6.11 c-type and p-type of Structure Importance Measures

Sections 6.1–6.10 present many types of structure importance measures, all of which can be defined in terms of (minimal) cuts and/or (minimal) paths. Therefore, the structure importance measure may be identified as c-type if it is defined on (minimal) cuts or as p-type if on (minimal) paths; for some structure importance measures, their c-type and p-type are equivalent. This section summarizes and compares the structure importance measures according to their relations to (minimal) cuts and (minimal) paths.

Table 6.1 presents the expression of each structure importance measure in terms of cuts and/or paths (Kuo and Zhu 2012b). These expressions and their detailed illustrations have been introduced in the preceding sections. For full definitions of the structure importance measures, please refer to the corresponding sections.

As shown in Table 6.1, each of the B-i.i.d. importance, B-structure importance, BP structure importance, permutation importance, and cut-path importance has equivalent c-type and p-type. That is, component i is more important than component j according to a c-type structure importance if and only if it is so according to the corresponding p-type importance. Thus, for these structure importance measures, there is no need to identify them by c-type and p-type, and their properties in terms of (minimal) cuts also hold in terms of (minimal) paths, and vice versa.

On the other hand, the pairs of c-type and p-type of importance measures, including the c-FV importance and p-FV importance, internal and external domination importance measures, cut importance and path importance, min-cut importance and min-path importance, c-absoluteness importance and p-absoluteness importance, and first-term importance and rare-event importance, are different. Particularly, a component cannot both internally and externally dominate another one (Theorem 6.6.4), and it is not possible for a component to be both more c-absolutely and p-absolutely important than another one (Theorem 6.8.2). However, each pair of the c-FV importance and p-FV importance, cut importance and path importance, and first-term importance and rare-event importance can coexist. For example, considering the system in Figure 2.4, component 1 is more important than component 2 according to both c-type and p-type of these importance measures.

6.12   Structure Importance Measures for Dual Systems

This section summarizes structure importance measures for dual systems, which include the entire spectrum of structure importance measures presented in this book. With these dual relations, the importance measure in one system can induce the respective importance measure in the corresponding dual system.

Table 6.1 Structure importance measures

Importance measure	c-type	p-type
B-i.i.d. $I_B(\cdot; p)$	$\sum_{d=1}^n \left(\lvert\mathscr{C}_i(d)\rvert p - \lvert\mathscr{C}_{(i)}(d)\rvert q\right) p^{n-d-1} q^{d-1}$ c-type and p-type are equivalent.	$\sum_{d=1}^n \left(\lvert\mathscr{P}_i(d)\rvert q - \lvert\mathscr{P}_{(i)}(d)\rvert p\right) p^{d-1} q^{n-d-1}$
B-structure $I_B(\cdot; \phi)$	$2^{-n}\left(\lvert\mathscr{C}_i\rvert - \lvert\mathscr{C}_{(i)}\rvert\right)$ c-type and p-type are equivalent.	$2^{-n}\left(\lvert\mathscr{P}_i\rvert - \lvert\mathscr{P}_{(i)}\rvert\right)$
BP-structure $I_{BP}(\cdot; \phi)$	$\left(\sum_{d=1}^n \lvert\mathscr{C}_i(d)\rvert\binom{n-1}{d-1}^{-1} - \sum_{d=1}^{n-1} \lvert\mathscr{C}_{(i)}(d)\rvert\binom{n-1}{d}^{-1}\right)/n$ c-type and p-type are equivalent.	$\left(\sum_{d=1}^n \lvert\mathscr{P}_i(d)\rvert\binom{n-1}{d-1}^{-1} - \sum_{d=1}^{n-1} \lvert\mathscr{P}_{(i)}(d)\rvert\binom{n-1}{d}^{-1}\right)/n$
c-FV-structure $I_{FV^c}(\cdot; \phi)$	$\lvert\{(0_i, \mathbf{x}) : \exists C_k \subseteq \overline{\mathscr{C}}_i \ s.t. \ C_k \subseteq N_0(0_i, \mathbf{x})\}\rvert/\lvert\mathscr{C}\rvert$	$\lvert\{(1_i, \mathbf{x}) : \exists P_r \in \overline{\mathscr{P}}_i \ s.t. \ P_r \subseteq N_1(1_i, \mathbf{x})\}\rvert/\lvert\mathscr{P}\rvert$
p-FV-structure $I_{FV^p}(\cdot; \phi)$	— c-type and p-type are different.	—
Permutation $i \succeq_{pe} j$	If $\{j\} \cup S \in \mathscr{C}_j$, then $\{i\} \cup S \in \mathscr{C}_i$. c-type and p-type are equivalent.	If $\{j\} \cup S \in \mathscr{P}_j$, then $\{i\} \cup S \in \mathscr{P}_i$.
Internal domination $i \succeq_{in} j$	If $S \in \mathscr{C}_j^{(i)}$, then $S \setminus \{j\} \in \mathscr{C}$.	If $S \in \mathscr{P}$ and $i, j \in S$, then $S \setminus \{j\} \in \mathscr{P}$.
External domination $i \succeq_{ex} j$	If $S \in \mathscr{C}$ and $i, j \in S$, then $S \setminus \{j\} \in \mathscr{C}$. Internal domination importance cannot coexist with external domination importance.	If $S \in \mathscr{P}_j^{(i)}$, then $S \setminus \{j\} \in \mathscr{P}$.
Cut $i \succeq_c j$	$(b_i(1), b_i(2), \ldots, b_i(n)) \succeq (b_j(1), b_j(2), \ldots, b_j(n))$	—
Path $i \succeq_p j$	— c-type and p-type are different.	$(g_i(1), g_i(2), \ldots, g_i(n)) \succeq (g_j(1), g_j(2), \ldots, g_j(n))$
c-absoluteness $i \succeq_{a^c} j$	$\mathscr{C}_i \supseteq \mathscr{C}_j$	—
p-absoluteness $i \succeq_{a^p} j$	— c-absoluteness importance cannot coexist with p-absoluteness importance.	$\mathscr{P}_i \supseteq \mathscr{P}_j$
Cut-path $i \succeq_{cp} j$	$\lvert\mathscr{C}_i(d)\rvert \geq \lvert\mathscr{C}_j(d)\rvert$ for all d c-type and p-type are equivalent.	$\lvert\mathscr{P}_i(d)\rvert \geq \lvert\mathscr{P}_j(d)\rvert$ for all d
Min-cut $i \succeq_{mc} j$	$\lvert\overline{\mathscr{C}}_i(d)\rvert \geq \lvert\overline{\mathscr{C}}_j(d)\rvert$ for all d	—
Min-path $i \succeq_{mp} j$	— c-type and p-type are different.	$\lvert\overline{\mathscr{P}}_i(d)\rvert \geq \lvert\overline{\mathscr{P}}_j(d)\rvert$ for all d
First-term $I_{FT}(\cdot; \phi)$	$\lvert\overline{\mathscr{C}}_i(e)\rvert$ with $e = \min_{C \in \overline{\mathscr{C}}} \lvert C\rvert$	—
Rare-event $I_{RE}(\cdot; \phi)$	— c-type and p-type are different.	$\lvert\overline{\mathscr{P}}_i(e')\rvert$ with $e' = \min_{P \in \overline{\mathscr{P}}} \lvert P\rvert$

Theorem 6.12.1 *The following indicate the relations of the structure importance measures for a system ϕ and its dual system ϕ^D.*

    (i) $I_B(i; \phi) = I_B(i; \phi^D)$.
   (ii) $I_{BP}(i; \phi) = I_{BP}(i; \phi^D)$.
  (iii) $I_{FV^c}(i; \phi) = I_{FV^p}(i; \phi^D)$ and $I_{FV^p}(i; \phi) = I_{FV^c}(i; \phi^D)$.
  (iv) If $i >_{pe} j$ for ϕ, then $i >_{pe} j$ for ϕ^D, and vice versa.
   (v) If $i \geq_{in} j$ for ϕ, then $i \geq_{ex} j$ for ϕ^D, and vice versa.
  (vi) If $i >_c j$ for ϕ, then $i >_p j$ for ϕ^D, and vice versa.
 (vii) If $i >_{a^c} j$ for ϕ, then $i >_{a^p} j$ for ϕ^D, and vice versa.
(viii) If $i >_{mc} j$ for ϕ, then $i >_{mp} j$ for ϕ^D, and vice versa.
  (ix) If $i >_{cp} j$ for ϕ, then $i >_{cp} j$ for ϕ^D, and vice versa.
   (x) $I_{FT}(i; \phi) = I_{RE}(i; \phi^D)$ and $I_{RE}(i; \phi) = I_{FT}(i; \phi^D)$.

These dual relations can be normally proved by the definitions of the corresponding importance measures as well as Definition 2.2.5 for dual systems and/or Theorem 2.5.2. By Theorem 4.1.10 for the B-reliability importance with $p_1 = p_2 = \cdots = p_n = p$, parts (i) ($p = 1/2$) and (ii) hold true. Because (minimal) cuts for ϕ are precisely the (minimal) paths for ϕ^D, and vice versa (Theorem 2.5.2(ii) and (iii)), parts (iii)–(x) can be easily verified (see also Theorem 4.2.3 for the FV reliability importance and Theorem 6.5.4 for the permutation importance).

6.13 Dominant Relations among Importance Measures

This section presents various dominant relations among structure importance measures. Suppose that there are two types of importance measures—*A* and *B*. We say that being *A*-type more important *implies* being *B*-type more important if for any two components *i* and *j* such that component *i* is more important than component *j* according to the *A*-type importance, component *i* is also more important than component *j* according to the *B*-type importance.

6.13.1 The Absoluteness Importance with the Domination Importance

Theorem 6.13.1 *(i) $i >_{a^c} j$ implies $i >_{in} j$, and $i >_{a^p}$ implies $i >_{ex} j$. (ii) $i =_{a^c} j$ implies $i =_{in} j$, and $i =_{a^p}$ implies $i =_{ex} j$.*

Proof. $i >_{a^c} j$ implies $\mathscr{C}_j \subset \mathscr{C}_i$, which, in turn, means $\mathscr{C}_j^{(i)} = \emptyset$. Thus, $i >_{in} j$. Similarly, $i >_{a^p} j$ implies $\mathscr{P}_j \subset \mathscr{P}_i$, which means $\mathscr{P}_j^{(i)} = \emptyset$. Thus, $i >_{ex} j$. Part (ii) follows similarly.

6.13.2 The Domination Importance with the Permutation Importance

Theorem 6.13.2 *(i) Both $i >_{in} j$ and $i >_{ex} j$ imply $i >_{pe} j$. (ii) Both $i =_{in} j$ and $i =_{ex} j$ imply $i =_{pe} j$.*

Proof. If $i >_{in} j$, because $j \not>_{in} i$, there exists $S \in \overline{\mathscr{C}}_i^{(j)}$ by Theorem 6.6.3(i). Now $H = (S \cup \{j\}) \setminus \{i\} \notin \mathscr{C}$ because if $H \in \mathscr{C}$, it would be $H \setminus \{j\} = S \setminus \{i\} \in \mathscr{C}$ (because $i >_{in} j$), which is a contradiction. Thus, $j \not>_{pe} i$, and this proves that $i >_{in} j$ implies $i >_{pe} j$. A similar argument based on paths can prove that $i >_{ex} j$ implies $i >_{pe} j$. Part (ii) follows similarly.

6.13.3 The Domination Importance with the Min-cut Importance and Min-path Importance

Theorem 6.13.3 *(i)* $i >_{in} j$ *implies* $i >_{mc} j$, *and* $i >_{ex} j$ *implies* $i >_{mp} j$. *(ii)* $i =_{in} j$ *implies* $i =_{mc} j$, *and* $i =_{ex} j$ *implies* $i =_{mp} j$.

Proof. Part (i) directly follows from Theorem 6.6.3(i) and (iii). That is, $i \geq_{in} j$ if and only if for any $S \in \overline{\mathscr{C}}_j$, $i \in S$, and $i \geq_{ex} j$ if and only if for any $S \in \overline{\mathscr{P}}_j$, $i \in S$. Part (ii) follows similarly.

6.13.4 The Permutation Importance with the FV Importance

Meng (1996, 2000) showed that the permutation importance of two components can be used to determine their relative c-FV importance and p-FV importance rankings by imposing some minor conditions if necessary. The main results are presented in Theorems 6.13.5 and 6.13.7 for the c-FV importance and Theorems 6.13.10 and 6.13.11 for the p-FV importance. The following lemma is used in the proof of Theorem 6.13.5.

Lemma 6.13.4 *Letting* $i >_{pe} j$, *then the following are true.*

(i) If $(\cdot_j, 0_i, \mathbf{x}^{(ij)}) \in B(j)$, then $(\cdot_i, 0_j, \mathbf{x}^{(ij)}) \in B(i)$.
(ii) If $(\cdot_j, 1_i, \mathbf{x}^{(ij)}) \in B(j)$, then $(\cdot_i, 1_j, \mathbf{x}^{(ij)}) \in B(i)$.
(iii) If $(\cdot_j, 1_i, \mathbf{x}^{(ij)}) \in V_1(j)$, then $(\cdot_i, 1_j, \mathbf{x}^{(ij)}) \in V_1(i)$.
(iv) If $(\cdot_j, 0_i, \mathbf{x}^{(ij)}) \in V_1(j)$, then $(\cdot_i, 0_j, \mathbf{x}^{(ij)}) \in V_1(i) \cup B(i)$.
(v) If $(\cdot_j, 1_i, \mathbf{x}^{(ij)}) \in V_2(j)$, then $(\cdot_i, 1_j, \mathbf{x}^{(ij)}) \in V_2(i) \cup B(i)$.
(vi) If $(\cdot_j, 0_i, \mathbf{x}^{(ij)}) \in V_2(j)$, then $(\cdot_i, 0_j, \mathbf{x}^{(ij)}) \in V_2(i)$.

The subsets of state vectors $B(i), V_1(i)$, *and* $V_2(i)$ *are defined by Equations* (4.26), (4.27), *and* (4.28), *respectively.*

Proof. Part (i) follows from that if $\phi(1_j, 0_i, \mathbf{x}^{(ij)}) > \phi(0_j, 0_i, \mathbf{x}^{(ij)})$ and $i >_{pe} j$, then $\phi(1_i, 0_j, \mathbf{x}^{(ij)}) > \phi(0_i, 0_j, \mathbf{x}^{(ij)})$, and part (ii) follows from that if $\phi(1_j, 1_i, \mathbf{x}^{(ij)}) > \phi(0_j, 1_i, \mathbf{x}^{(ij)})$ and $i >_{pe} j$, then $\phi(1_i, 1_j, \mathbf{x}^{(ij)}) > \phi(0_i, 1_j, \mathbf{x}^{(ij)})$.

For part (iii), $(\cdot_j, 1_i, \mathbf{x}^{(ij)}) \in V_1(j)$ means that $\phi(0_j, 1_i, \mathbf{x}^{(ij)}) = \phi(1_j, 1_i, \mathbf{x}^{(ij)}) = 0$ and that there exists a $C_k \in \mathscr{C}_j$ such that $C_k \subseteq N_0(0_j, 1_i, \mathbf{x}^{(ij)})$. Thus, $i \notin C_k$. Hence, either condition (ii) or (iii) in Theorem 6.5.6 holds. So, $\phi(0_i, 1_j, \mathbf{0}^{C_k \setminus \{j\}}, \mathbf{1}) = 0$. This together with $\phi(1_i, 1_j, \mathbf{0}^{C_k \setminus \{j\}}, \mathbf{1}) = 1$ implies that there exists a minimal cut $C_{k'} \in \mathscr{C}_i$ such that $C_{k'} \subseteq C_k \cup \{i\} \setminus \{j\} \subseteq N_0(0_i, 1_j, \mathbf{x}^{(ij)})$. Hence, $(\cdot_i, 1_j, \mathbf{x}^{(ij)}) \in V_1(i)$.

For part (iv), by assumption, $\phi(0_j, 0_i, \mathbf{x}^{(ij)}) = \phi(1_j, 0_i, \mathbf{x}^{(ij)}) = 0$, and there exists a $C_k \in \mathscr{C}_j$ such that $C_k \subseteq N_0(0_j, 0_i, \mathbf{x}^{(ij)})$. If $\phi(1_i, 0_j, \mathbf{x}^{(ij)}) = 1$, then $(\cdot_i, 0_j, \mathbf{x}^{(ij)}) \in B(i)$. If

$\phi(1_i, 0_j, \mathbf{x}^{(ij)}) = 0$, it suffices to show $(\cdot_i, 0_j, \mathbf{x}^{(ij)}) \in V_1(i)$ as follows. If condition (i) in Theorem 6.5.6 holds, then $i \in C_k$. Hence, $(\cdot_i, 0_j, \mathbf{x}^{(ij)}) \in V_1(i)$. If either condition (ii) or (iii) in Theorem 6.5.6 holds, then $\phi(0_i, 1_j, \mathbf{0}^{C_k \setminus \{j\}}, \mathbf{1}) = 0$. Because C_k is a minimal cut, there must exist a $C_{k'} \in \overline{\mathscr{C}}_i$ such that $C_{k'} \subseteq C_k \cup \{i\} \setminus \{j\} \subseteq N_0(0_i, 0_j, \mathbf{x}^{(ij)})$. Thus, $(\cdot_i, 0_j, \mathbf{x}^{(ij)}) \in V_1(i)$.

Part (v) can be proved by duality. By Theorem 6.5.4, if $i >_{\text{pe}} j$ for ϕ, then $i >_{\text{pe}} j$ for ϕ^D. Thus, if $(\cdot_j, 1_i, \mathbf{x}^{(ij)}) \in V_2(j)$ for ϕ, then $(\cdot_j, 0_i, (\mathbf{1} - \mathbf{x})^{(ij)}) \in V_1(j)$ for ϕ^D. This implies $(\cdot_i, 0_j, (\mathbf{1} - \mathbf{x})^{(ij)}) \in V_1(i) \cup B(i)$ for ϕ^D. Hence, $(\cdot_i, 1_j, (\mathbf{1} - \mathbf{x})^{(ij)}) \in V_2(i) \cup B(i)$ for ϕ.

The proof of part (vi) is similar to that of part (v).

Theorem 6.13.5 $i >_{\text{pe}} j$ *if and only if either (i) or (ii) holds:*

(i) *(if $\overline{\mathscr{C}}_j \subset \overline{\mathscr{C}}_i$) $I_{\text{FV}^c}(i; \mathbf{p}) > I_{\text{FV}^c}(j; \mathbf{p})$ for all $\mathbf{0} < \mathbf{p} < \mathbf{1}$;*
(ii) *(if $\overline{\mathscr{C}}_j \not\subset \overline{\mathscr{C}}_i$) $I_{\text{FV}^c}(i; \mathbf{p}) > I_{\text{FV}^c}(j; \mathbf{p})$ for all $\mathbf{0} < \mathbf{p} < \mathbf{1}$ satisfying $p_i \leq p_j$.*

Proof. ("Only if") Part (i): By definition,

$$
\begin{aligned}
I_{\text{FV}^c}(i; \mathbf{p}) &= (1 - R(\mathbf{p}))^{-1} \Pr\{\exists C_k \in \overline{\mathscr{C}}_i \ s.t. \ C_k \subseteq N_0(\mathbf{X})\} \\
&= (1 - R(\mathbf{p}))^{-1} [\Pr\{\exists C_k \in \overline{\mathscr{C}}_i \ s.t. \ C_k \subseteq N_0(\mathbf{X}) \text{ and } \exists C_s \in \overline{\mathscr{C}}_j \ s.t. \ C_s \subseteq N_0(\mathbf{X})\} \\
&\quad + \Pr\{\exists C_k \in \overline{\mathscr{C}}_i^{(j)} \ s.t. \ C_k \subseteq N_0(\mathbf{X}) \text{ and } \nexists C_s \in \overline{\mathscr{C}}_j^{(i)} \ s.t. \ C_s \subseteq N_0(\mathbf{X})\}].
\end{aligned}
$$

Thus, comparing $I_{\text{FV}^c}(i; \mathbf{p})$ and $I_{\text{FV}^c}(j; \mathbf{p})$ is equivalent to comparing the quantities s_i and s_j:

$$
\begin{aligned}
s_i &= \Pr\{\exists C_k \in \overline{\mathscr{C}}_i^{(j)} \ s.t. \ C_k \subseteq N_0(\mathbf{X}) \text{ and } \nexists C_s \in \overline{\mathscr{C}}_j^{(i)} \ s.t. \ C_s \subseteq N_0(\mathbf{X})\} \\
&= q_i \Pr\{(\cdot_i, \mathbf{X}) = (\cdot_i, \mathbf{0}^{C_k \setminus \{i\}}, \mathbf{X}^{N \setminus C_k}) : \exists C_k \in \overline{\mathscr{C}}_i^{(j)} \ s.t. \ C_k \subseteq N_0(0_i, \mathbf{X}) \\
&\quad \text{and } \nexists C_s \in \overline{\mathscr{C}}_j^{(i)} \ s.t. \ C_s \subseteq N_0(0_i, \mathbf{X})\}
\end{aligned} \tag{6.24}
$$

and

$$
\begin{aligned}
s_j &= \Pr\{\exists C_k \in \overline{\mathscr{C}}_j^{(i)} \ s.t. \ C_k \subseteq N_0(\mathbf{X}) \text{ and } \nexists C_s \in \overline{\mathscr{C}}_i^{(j)} \ s.t. \ C_s \subseteq N_0(\mathbf{X})\} \\
&= q_j \Pr\{(\cdot_j, \mathbf{X}) = (\cdot_j, \mathbf{0}^{C_k \setminus \{j\}}, \mathbf{X}^{N \setminus C_k}) : \exists C_k \in \overline{\mathscr{C}}_j^{(i)} \ s.t. \ C_k \subseteq N_0(0_j, \mathbf{X}) \\
&\quad \text{and } \nexists C_s \in \overline{\mathscr{C}}_i^{(j)} \ s.t. \ C_s \subseteq N_0(0_j, \mathbf{X})\}.
\end{aligned} \tag{6.25}
$$

Because the quantity s_j in Equation (6.25) is zero if $\overline{\mathscr{C}}_j \subset \overline{\mathscr{C}}_i$, it suffices to show that $s_i > 0$ for all $\mathbf{0} < \mathbf{p} < \mathbf{1}$ in order to prove part (i). Then $s_i > 0$ simply follows from the existence of a minimal cut $C_k \in \overline{\mathscr{C}}_i^{(j)}$ and the assumption $\mathbf{0} < \mathbf{p} < \mathbf{1}$.

Part (ii): From Lemma 6.13.4 (i)–(iv), if $i >_{\text{pe}} j$, then for $\alpha = 0, 1$,

$$
\{\mathbf{x}^{(ij)} : (\cdot_j, \alpha_i, \mathbf{x}^{(ij)}) \in B(j) \cup V_1(j)\} \subseteq \{\mathbf{x}^{(ij)} : (\cdot_i, \alpha_j, \mathbf{x}^{(ij)}) \in B(i) \cup V_1(i)\};
$$

hence,

$$
\Pr\{(\cdot_j, \alpha_i, \mathbf{x}^{(ij)}) \in B(j) \cup V_1(j)\} \leq \Pr\{(\cdot_i, \alpha_j, \mathbf{x}^{(ij)}) \in B(i) \cup V_1(i)\}. \tag{6.26}
$$

By Theorem 6.5.6, there exists a minimal cut $C_k \in \overline{\mathscr{C}}_i \backslash \overline{\mathscr{C}}_j$, and $C_k \cup \{j\} \backslash \{i\}$ is not a cut. So, $\phi(0_j, 1_i, \mathbf{0}^{C_k \backslash \{i\}}, \mathbf{1}) = \phi(1_j, 1_i, \mathbf{0}^{C_k \backslash \{i\}}, \mathbf{1}) = 1$; hence, $(\cdot_j, 1_i, \mathbf{0}^{C_k \backslash \{i\}}, \mathbf{1}) \notin B(j) \cup V_i(j)$. Clearly, $(\cdot_i, 1_j, \mathbf{0}^{C_k \backslash \{i\}}, \mathbf{1}) \in B(i)$. Thus, it is shown that

$$B(j) \cup V_1(j) \subset B(i) \cup V_1(i). \tag{6.27}$$

Next, from Equation (4.29),

$$I_{\mathrm{FV}^c}(j; \mathbf{p}) = \frac{q_j p_i \Pr\{(\cdot_j, 1_i, \mathbf{x}^{(ij)}) \in B(j) \cup V_1(j)\} + q_j q_i \Pr\{(\cdot_j, 0_i, \mathbf{x}^{(ij)}) \in B(j) \cup V_1(j)\}}{\Pr\{\phi(\mathbf{X}) = 0\}}.$$

$$\tag{6.28}$$

Thus, when $\mathbf{0} < \mathbf{p} < \mathbf{1}$ and $p_i \le p_j$, by Equations (6.26) and (6.27),

$$I_{\mathrm{FV}^c}(j; \mathbf{p}) < \frac{q_i p_j \Pr\{(\cdot_i, 1_j, \mathbf{x}^{(ij)}) \in B(i) \cup V_1(i)\} + q_i q_j \Pr\{(\cdot_i, 0_j, \mathbf{x}^{(ij)}) \in B(i) \cup V_1(i)\}}{\Pr\{\phi(\mathbf{X}) = 0\}}$$

$$= I_{\mathrm{FV}^c}(i; \mathbf{p}).$$

("If") Because part (i) implies part (ii) to prove this direction, it suffices to prove that part (ii) implies $i >_{\mathrm{pe}} j$. It also suffices to show that $I_{\mathrm{FV}^c}(i; \mathbf{p}) > I_{\mathrm{FV}^c}(j; \mathbf{p})$ when $p_i = p_j$ implies $i >_{\mathrm{pe}} j$.

Let $p_i = p_j$ in Equations (6.24) and (6.25), then for all $\mathbf{0} < \mathbf{p}^{(ij)} < \mathbf{1}$,

$$\begin{aligned}
\Pr\{(\cdot_i, \mathbf{X}) = (\cdot_i, \mathbf{0}^{C_k \backslash \{i\}}, \mathbf{X}^{N \backslash C_k}) : \exists C_k \in \overline{\mathscr{C}}_i^{(j)} \text{ s.t. } C_k \subseteq N_0(0_i, \mathbf{X}) \\
\text{and } \nexists C_s \in \overline{\mathscr{C}}_j^{(i)} \text{ s.t. } C_s \subseteq N_0(0_i, \mathbf{X})\} \\
> \Pr\{(\cdot_j, \mathbf{X}) = (\cdot_j, \mathbf{0}^{C_k \backslash \{j\}}, \mathbf{X}^{N \backslash C_k}) : \exists C_k \in \overline{\mathscr{C}}_j^{(i)} \text{ s.t. } C_k \subseteq N_0(0_j, \mathbf{X}) \\
\text{and } \nexists C_s \in \overline{\mathscr{C}}_i^{(j)} \text{ s.t. } C_s \subseteq N_0(0_j, \mathbf{X})\}.
\end{aligned} \tag{6.29}$$

Suppose that $\phi(1_i, 0_j, \mathbf{x}^{(ij)}) = 0$. It needs to be shown that $\phi(0_i, 1_j, \mathbf{x}^{(ij)}) = 0$, and thus $\phi(1_i, 0_j, \mathbf{x}^{(ij)}) \ge \phi(0_i, 1_j, \mathbf{x}^{(ij)})$ for all $\mathbf{x}^{(ij)}$. Assume that $\phi(1_i, 1_j, \mathbf{x}^{(ij)}) = 1$ (for otherwise, $\phi(0_i, 1_j, \mathbf{x}^{(ij)}) = 0$, and the statement holds). Now because $(1_i, 0_j, \mathbf{x}^{(ij)})$ is a cut and $(1_i, 1_j, \mathbf{x}^{(ij)})$ is not, there must be a minimal cut $C_k \in \overline{\mathscr{C}}_j^{(i)}$ and $C_k \subseteq N_0(1_i, 0_j, \mathbf{x}^{(ij)})$. Let $C_k = H \cup \{j\}$. Note that H is nonempty, for otherwise component j is a one-component cut; this implies that $I_{\mathrm{FV}^c}(j; \mathbf{p}) = q_j(1 - R(\mathbf{p}))^{-1}$, and consequently, part (ii) cannot hold when $p_i = p_j$. Let the binary vector $(\cdot_i, \cdot_j, \tilde{\mathbf{x}}^{(ij)})$ be defined by

$$\tilde{x}_k^{(ij)} = \begin{cases} 0 & \text{if } k \in H \\ 1 & \text{if } k \in N \backslash (H \cup \{i, j\}), \end{cases}$$

and let the probability vector $(\cdot_i, \cdot_j, \mathbf{p}^{(ij)})$ be such that for $k \in N \backslash \{i, j\}$,

$$p_k^{(ij)} = 1 - \Pr\{X_k = \tilde{x}_k^{(ij)}\} = \varepsilon, \quad 0 < \varepsilon < 1.$$

Then,

the right-hand side in Equation (6.29)

$$\begin{aligned}
&\ge \Pr\{(\cdot_j, \mathbf{X}) = (\cdot_j, \mathbf{0}^{C_k \backslash \{j\}}, \mathbf{X}^{N \backslash C_k}) : \exists C_k \in \overline{\mathscr{C}}_j^{(i)} \text{ s.t. } C_k \subseteq N_0(0_j, \mathbf{X}); X_i = 1\} \\
&\ge p_i \Pr\{\mathbf{X}^{(ij)} = \tilde{\mathbf{x}}^{(ij)}\},
\end{aligned}$$

and thus,

$$\text{the left-hand side in Equation (6.29)} > p_i(1-\varepsilon)^{n-2}. \qquad (6.30)$$

Letting ε approach 0, then $\Pr\{X^{(ij)} \neq \tilde{x}^{(ij)}\}$ approaches 0; thus, inequality (6.30) implies the existence of a minimal cut $C_s \in \overline{\mathscr{C}}_i^{(j)}$ and $C_s \setminus \{i\} \subseteq H$. Because $H = C_k \setminus \{j\} \subseteq N_0(1_i, 1_j, x^{(ij)})$ and $C_s \subseteq N_0(0_i, 1_j, x^{(ij)})$, then $\phi(0_i, 1_j, x^{(ij)}) = 0$. It has thus been shown that $\phi(1_i, 0_j, x^{(ij)}) \geq \phi(0_i, 1_j, x^{(ij)})$ for all $x^{(ij)}$. Note that strict inequality holds for some $x^{(ij)}$. Otherwise, from Theorem 6.5.8 and Equations (6.24) and (6.25), $i =_{\text{pe}} j$ implies that $I_{\text{FV}^c}(i; \mathbf{p}) = I_{\text{FV}^c}(j; \mathbf{p})$ when $p_i = p_j$, which contradicts the assumption. The proof is complete.

Corollary 6.13.6 is a special case of Theorem 6.13.5.

Corollary 6.13.6   *If $i >_{\text{pe}} j$, then $I_{\text{FV}^c}(i; \phi) > I_{\text{FV}^c}(j; \phi)$, and $I_{\text{FV}^p}(i; \phi) > I_{\text{FV}^p}(j; \phi)$.*

Theorem 6.13.7 $i =_{\text{pe}} j$ *if and only if either (i) or (ii) holds:*

(i) (if $\overline{\mathscr{C}}_i = \overline{\mathscr{C}}_j$) $I_{\text{FV}^c}(i; \mathbf{p}) = I_{\text{FV}^c}(j; \mathbf{p})$ for all $0 < \mathbf{p} < 1$;
(ii) (if $\overline{\mathscr{C}}_i \neq \overline{\mathscr{C}}_j$) $I_{\text{FV}^c}(i; \mathbf{p}) > I_{\text{FV}^c}(j; \mathbf{p})$ for $0 < \mathbf{p} < 1$ if and only if $p_i < p_j$.

Proof. ("Only if") The proof of part (i) is trivial. Supposing that $\overline{\mathscr{C}}_i \neq \overline{\mathscr{C}}_j$, Equation (6.26) becomes

$$\Pr\{(\cdot_j, \alpha_i, x^{(ij)}) \in B(j) \cup V_1(j)\} = \Pr\{(\cdot_i, \alpha_j, x^{(ij)}) \in B(i) \cup V_1(i)\}.$$

The result then follows from this and Equation (6.28).
   ("If") By using a similar proof to that of Theorem 6.13.5, the result is easily obtained and the proof is, therefore, omitted.

Examples 6.13.8 and 6.13.9 illustrate Theorem 6.13.7 as follows.

Example 6.13.8 (Example 6.8.3 continued)   Consider the system in Figure 2.4 in which $\phi(x_1, x_2, x_3) = 1 - (1 - x_1 x_2)(1 - x_1 x_3)$. Then, components $2 =_{\text{pe}} 3$. Components 2 and 3 satisfy condition (i) in Theorem 6.5.8; thus, $I_{\text{FV}^c}(2; \mathbf{p}) = I_{\text{FV}^c}(3; \mathbf{p}) = q_2 q_3/(1 - R(\mathbf{p}))$. The result of $I_{\text{FV}^c}(2; \mathbf{p}) = I_{\text{FV}^c}(3; \mathbf{p})$ can be directly obtained from condition (i) in Theorem 6.13.7.

Example 6.13.9 (Example 4.2.4 continued)   For 2-out-of-3 system, components $1 =_{\text{pe}} 2$. Furthermore, Example 4.2.4 shows that $I_{\text{FV}^c}(1; \mathbf{p}) = q_1(1 - p_2 p_3)/(1 - R(\mathbf{p}))$, and similarly $I_{\text{FV}^c}(2; \mathbf{p}) = q_2(1 - p_1 p_3)/(1 - R(\mathbf{p}))$. Hence, by condition (ii) in Theorem 6.13.7, $I_{\text{FV}^c}(1; \mathbf{p}) > I_{\text{FV}^c}(2; \mathbf{p})$ if and only if $p_1 < p_2$. Because in a k-out-of-n system $(n \geq k > 1)$ condition (ii) in Theorem 6.5.8 holds for all $i, j \in \{1, 2, \ldots, n\}$, and consequently, $i =_{\text{pe}} j$ for $i \neq j$, the least reliable component possesses the largest c-FV importance.

Analogous to Theorems 6.13.5 and 6.13.7, similar results for the p-FV reliability importance arise from the duality of the c-FV importance (Theorem 4.2.3) and of the permutation importance (Theorem 6.5.4).

Theorem 6.13.10 $i >_{\mathrm{pe}} j$ *if and only if either (i) or (ii) holds:*

(i) (if $\overline{\mathscr{P}}_j \subset \overline{\mathscr{P}}_i$) $I_{\mathrm{FV}^p}(i; \mathbf{p}) > I_{\mathrm{FV}^p}(j; \mathbf{p})$ for all $0 < \mathbf{p} < 1$;
(ii) (if $\overline{\mathscr{P}}_j \not\subset \overline{\mathscr{P}}_i$) $I_{\mathrm{FV}^p}(i; \mathbf{p}) > I_{\mathrm{FV}^p}(j; \mathbf{p})$ for all $0 < \mathbf{p} < 1$ satisfying $p_i \geq p_j$.

Theorem 6.13.11 $i =_{\mathrm{pe}} j$ *if and only if either (i) or (ii) holds:*

(i) (if $\overline{\mathscr{P}}_i = \overline{\mathscr{P}}_j$) $I_{\mathrm{FV}^p}(i; \mathbf{p}) = I_{\mathrm{FV}^p}(j; \mathbf{p})$ for all $0 < \mathbf{p} < 1$;
(ii) (if $\overline{\mathscr{P}}_i \neq \overline{\mathscr{P}}_j$) $I_{\mathrm{FV}^p}(i; \mathbf{p}) > I_{\mathrm{FV}^p}(j; \mathbf{p})$ for $0 < \mathbf{p} < 1$ if and only if $p_i > p_j$.

Combining Theorems 6.13.5, 6.13.7, 6.13.10, and 6.13.11, it is concluded that if $i \geq_{\mathrm{pe}} j$ and $0 < \mathbf{p} < 1$, then $I_{\mathrm{FV}^c}(i; \mathbf{p}) \geq I_{\mathrm{FV}^c}(j; \mathbf{p})$ for $p_i \leq p_j$, and $I_{\mathrm{FV}^p}(i; \mathbf{p}) \geq I_{\mathrm{FV}^p}(j; \mathbf{p})$ for $p_i \geq p_j$.

6.13.5   The Permutation Importance with the Cut-path Importance, Min-cut Importance, and Min-path Importance

The permutation importance is stronger than the cut-path importance, as shown in the following theorem.

Theorem 6.13.12   *(i)* $i >_{\mathrm{pe}} j$ *implies* $i >_{\mathrm{cp}} j$. *(ii)* $i =_{\mathrm{pe}} j$ *implies* $i =_{\mathrm{cp}} j$.

Proof. Let S be a subset containing neither component i nor j.

For part (i), by Theorem 6.5.7, if $i >_{\mathrm{pe}} j$, then for any S such that $\{j\} \cup S \in \mathscr{C}_j$, $\{i\} \cup S \in \mathscr{C}_i$; thus $|\mathscr{C}_i(d)| \geq |\mathscr{C}_j(d)|$ for all d. Furthermore, if $i >_{\mathrm{pe}} j$, then there exists a subset S_0 containing neither component i nor j such that $\{i\} \cup S_0 \in \mathscr{C}_i$ but $\{j\} \cup S_0 \notin \mathscr{C}_j$. Letting $d_0 = |S_0| + 1$, then $|\mathscr{C}_i(d_0)| > |\mathscr{C}_j(d_0)|$. This completes the proof of part (i).

For part (ii), by Theorem 6.5.8, if $i =_{\mathrm{pe}} j$, then $\{j\} \cup S \in \mathscr{C}_j$ if and only if $\{i\} \cup S \in \mathscr{C}_i$, and thus, $|\mathscr{C}_i(d)| = |\mathscr{C}_j(d)|$ for all d. That is, $i =_{\mathrm{cp}} j$.

However, being more permutation important does not necessarily imply being more min-cut important or more min-path important, as shown in the following example.

Example 6.13.13 (Example 6.8.3 continued)   Considering the system in Figure 2.4, clearly, $1 >_{\mathrm{pe}} 2$. Note that $|\overline{\mathscr{C}}_1(1)| = 1$, $|\overline{\mathscr{C}}_1(2)| = 0$, $|\overline{\mathscr{C}}_2(1)| = 0$, and $|\overline{\mathscr{C}}_2(2)| = 1$; thus, $1 \not\succ_{\mathrm{mc}} 2$. Actually, components 1 and 2 cannot be compared according to the min-cut importance. Similarly, for the system in Figure 2.5, $1 >_{\mathrm{pe}} 2$. Note that $|\overline{\mathscr{P}}_1(1)| = 1$, $|\overline{\mathscr{P}}_1(2)| = 0$, $|\overline{\mathscr{P}}_2(1)| = 0$, and $|\overline{\mathscr{P}}_2(2)| = 1$; thus, $1 \not\succ_{\mathrm{mp}} 2$. Actually, components 1 and 2 in this system cannot be compared according to the min-path importance.

For the system in Figure 2.4, although $1 \not\succ_{\mathrm{mc}} 2$, $1 >_{\mathrm{mp}} 2$ since $|\overline{\mathscr{P}}_1(1)| = |\overline{\mathscr{P}}_2(1)| = 0$, $|\overline{\mathscr{P}}_1(2)| = 2$, and $|\overline{\mathscr{P}}_2(2)| = 1$. For the system in Figure 2.5, although $1 \not\succ_{\mathrm{mp}} 2$, $1 >_{\mathrm{mc}} 2$ since $|\overline{\mathscr{C}}_1(1)| = |\overline{\mathscr{C}}_2(1)| = 0$, $|\overline{\mathscr{C}}_1(2)| = 2$, and $|\overline{\mathscr{C}}_2(2)| = 1$. On the basis of these observations, the following conjecture was made.

Conjecture 6.13.14 $i >_{\text{pe}} j$ *implies either* $i >_{\text{mc}} j$ *or* $i >_{\text{mp}} j$.

Furthermore, although $i >_{\text{pe}} j$ does not necessarily imply that $i >_{\text{mc}} j$ and $i >_{\text{mp}} j$, $i =_{\text{pe}} j$ does imply that $i =_{\text{mc}} j$ and $i =_{\text{mp}} j$, as shown in Theorem 6.13.15.

Theorem 6.13.15 $i =_{\text{pe}} j$ *implies* $i =_{\text{mc}} j$ *and* $i =_{\text{mp}} j$.

Proof. Let S be a subset containing neither component i nor j.

To prove that $i =_{\text{pe}} j$ implies $i =_{\text{mc}} j$, it first needs to be shown that if $i =_{\text{pe}} j$, then for any S, $\{i\} \cup S \in \overline{\mathscr{C}}_i$ if $\{j\} \cup S \in \overline{\mathscr{C}}_j$. By contradiction, suppose that $i =_{\text{pe}} j$, $\{j\} \cup S \in \overline{\mathscr{C}}_j$, but $\{i\} \cup S \notin \overline{\mathscr{C}}_i$. Due to Theorem 6.5.8, $\{i\} \cup S \in \mathscr{C}_i$. Since $\{i\} \cup S \notin \overline{\mathscr{C}}_i$, then there exists a subset $S' \subset S$ such that $\{i\} \cup S' \in \overline{\mathscr{C}}_i$. By Theorem 6.5.8, because $i =_{\text{pe}} j$ and $\{i\} \cup S' \in \overline{\mathscr{C}}_i \subset \mathscr{C}_i$, then $\{j\} \cup S' \in \mathscr{C}_j$. Therefore, $\{j\} \cup S \notin \overline{\mathscr{C}}_j$, which contradicts the assumption of $\{j\} \cup S \in \overline{\mathscr{C}}_j$. Hence, if $i =_{\text{pe}} j$ and $\{j\} \cup S \in \overline{\mathscr{C}}_j$, then $\{i\} \cup S \in \overline{\mathscr{C}}_i$.

Similarly, it can be shown that if $i =_{\text{pe}} j$, then for any S, $\{j\} \cup S \in \overline{\mathscr{C}}_j$ if $\{i\} \cup S \in \overline{\mathscr{C}}_i$. Hence, if $i =_{\text{pe}} j$, then for any S, $\{j\} \cup S \in \overline{\mathscr{C}}_j$ if and only if $\{i\} \cup S \in \overline{\mathscr{C}}_i$. Consequently, $|\overline{\mathscr{C}}_i(d)| = |\overline{\mathscr{C}}_j(d)|$ for all d, that is, $i =_{\text{mc}} j$.

A similar argument on the basis of paths can prove that $i =_{\text{pe}} j$ implies $i =_{\text{mp}} j$.

6.13.6   The Cut-path Importance with the Cut Importance and Path Importance

Theorem 6.13.17 (Hwang 2001) shows that the cut-path importance is stronger than the cut importance and path importance, whose proof needs Lemma 6.3.16. Recall from Definition 6.7.1 of the cut importance that $b_i(d)$ is used in defining the cut importance, and $b_i(d) = \sum_{\ell=1}^{u} (-1)^{\ell-1} a_i(\ell, d)$, where $a_i(\ell, d)$ for $1 \le \ell \le u$ and $1 \le d \le n$ is the number of collections of ℓ distinct minimal cuts such that the union of each collection contains exactly d components and includes component i. Similarly, let $a_{(i)}(\ell, d)$ be the number of collections of ℓ distinct minimal cuts such that the union of each collection contains exactly d components and *does not* include component i, $1 \le \ell \le u$, $1 \le d \le n$. Let $\tilde{a}(\ell, d) = a_i(\ell, d) + a_{(i)}(\ell, d)$ and $\tilde{b}(d) = \sum_{\ell=1}^{u} (-1)^{\ell-1} \tilde{a}(\ell, d)$.

Lemma 6.13.16 $\displaystyle |\mathscr{C}_i(d)| = \sum_{k=1}^{d} \left[b_i(k) \binom{n-k-1}{d-k} + \tilde{b}(k) \binom{n-k-1}{d-k-1} \right]$.

Proof. Each cut counted in $a_i(1, k)$ (i.e., minimal cut of size k) can become a cut in $\mathscr{C}_i(d)$ by adding some $d-k$ components from the remaining $n-k$ components, and there are $\binom{n-k}{d-k}$ such choices. Similarly, each cut counted in $a_{(i)}(1, k)$ can become a cut in $\mathscr{C}_i(d)$ by adding component i and some other $(d-k-1)$ components from the remaining $(n-k-1)$ components, and there are $\binom{n-k-1}{d-k-1}$ such choices. However, these additions could induce the same cut in $\mathscr{C}_i(d)$. $a_i(2, d)$ counts the number of pairs of minimal cuts whose additions overlap. But if three cuts overlap, then $a_i(2, d)$ overcorrects; thus, $a_i(3, d)$ is added back. By the inclusion-exclusion principle,

$$
|\mathscr{C}_i(d)| = \sum_{k=1}^{d} \sum_{\ell=1}^{u} (-1)^{\ell-1} \left[a_i(\ell, k) \binom{n-k}{d-k} + a_{(i)}(\ell, k) \binom{n-k-1}{d-k-1} \right]
$$

$$
= \sum_{k=1}^{d} \sum_{\ell=1}^{u} (-1)^{\ell-1} \left[a_i(\ell, k) \binom{n-k-1}{d-k} + (a_i(\ell, k) + a_{(i)}(\ell, k)) \binom{n-k-1}{d-k-1} \right]
$$

$$
= \sum_{k=1}^{d} \left[b_i(k) \binom{n-k-1}{d-k} + \tilde{b}(k) \binom{n-k-1}{d-k-1} \right].
$$

Theorem 6.13.17   *(i)* $i >_{cp} j$ *implies* $i >_c j$ *and* $i >_p j$. *(ii)* $i =_{cp} j$ *implies* $i =_c j$ *and* $i =_p j$.

Proof. Suppose that $i >_{cp} j$, that is, there exists a $d_0 \geq 1$ such that $|\mathscr{C}_i(d)| = |\mathscr{C}_j(d)|$ for $d < d_0$ and $|\mathscr{C}_i(d_0)| > |\mathscr{C}_j(d_0)|$. Then, by the induction for $d = 1, 2, \ldots, d_0$ and using Lemma 6.13.16, $b_i(d) = b_j(d)$ for $d < d_0$ and $b_i(d_0) > b_j(d_0)$. Hence, $i >_c j$.

Similar to Lemma 6.13.16, express $|\mathscr{P}_i(d)|$ in terms of $g_i(d)$, which is used to define the path importance in Definition 6.7.3. Then, note that $i >_{cp} j$ implies that $|\mathscr{P}_i(d)| \geq |\mathscr{P}_j(d)|$ for all d and $|\mathscr{P}_i(d)| > |\mathscr{P}_j(d)|$ for some d; thus, $i >_{cp} j$ implies $i >_p j$.

Part (ii) follows similarly.

6.13.7   The Cut-path Importance with the B-i.i.d. Importance

By Theorem 6.13.17, the cut-path importance is stronger than the cut importance and path importance, and by Theorem 6.7.4, the cut importance and path importance are equivalent to the B-i.i.d. importance as p approaches 1 and 0, respectively. Thus, intuitively, the cut-path importance should be stronger than the uniform B-i.i.d. importance. Theorem 6.13.18 (Hwang 2001) formally shows such a relation.

Theorem 6.13.18   *(i)* $i >_{cp} j$ *implies* $I_B(i; p) > I_B(j; p)$ *for all* $0 < p < 1$. *(ii)* $i =_{cp} j$ *implies* $I_B(i; p) = I_B(j; p)$ *for all* $0 < p < 1$. *(iii)* $i >_{cp} j$ *implies* $I_B(i; \phi) > I_B(j; \phi)$.

Proof. By Equation (6.3) and noting that $|\mathscr{C}_{(i)}(d)| = |\mathscr{C}(d)| - |\mathscr{C}_i(d)|$,

$$
I_B(i; p) = \sum_{d=1}^{n} |\mathscr{C}_i(d)| p^{n-d-1} q^{d-1} - \sum_{d=1}^{n} |\mathscr{C}(d)| p^{n-d-1} q^d.
$$

Hence, $I_B(i; p) > I_B(j; p)$ if and only if

$$
\sum_{d=1}^{n} \left(|\mathscr{C}_i(d)| - |\mathscr{C}_j(d)| \right) p^{n-d-1} q^{d-1} > 0. \tag{6.31}
$$

If $i >_{cp} j$, by definition, then

$$
|\mathscr{C}_i(d)| \geq |\mathscr{C}_j(d)| \text{ for all } d \text{ and } |\mathscr{C}_i(d)| > |\mathscr{C}_j(d)| \text{ for some } d, \tag{6.32}
$$

which clearly implies (6.31).

Part (ii) can be shown by changing inequalities (6.31) and (6.32) to equalities. Part (iii) follows by noting $I_B(i; \phi) = I_B(i; 1/2)$.

6.13.8 The B-i.i.d. Importance with the BP Importance

Theorem 6.13.19 *(i) If $I_B(i; p) > I_B(j; p)$ for all $0 < p < 1$, then $I_{BP}(i; \phi) > I_{BP}(j; \phi)$. (ii) If $I_B(i; p) = I_B(j; p)$ for all $0 < p < 1$, then $I_{BP}(i; \phi) = I_{BP}(j; \phi)$.*

Proof. The proof is straightforward by noting Equation (6.9) that $I_{BP}(i; \phi) = \int_0^1 I_B(i; p)\mathrm{d}p$.

6.14   Summary

Figure 6.3 presents a hierarchy of dominant relations among structure importance measures. Part of this figure was presented by Hwang (2005) with some errors, and was also presented by Zhu and Kuo (2008) and Kuo and Zhu (2012a) with corrections.

Note that Figure 6.3 is drawn according to the greater than or equal to relation. The arrow from A to B implies that if the importance of component i is greater than or equal to that of component j according to an A-type importance measure, then it is also true according to a B-type importance measure. Note that the relations of the cut importance to the first-term importance and the path importance to the rare-event importance in Figure 6.3 are drawn according to the discussion in Section 6.10. If only the greater than (i.e., strictly more important than) relation is considered, then there should be no arrow from the cut importance to the

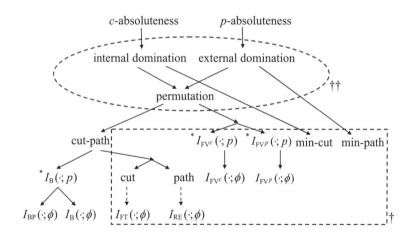

*: for all $0 < p < 1$.

†: Importance measures inside rectangle are defined on the basis of minimal cuts or minimal paths.

††: Importance measures inside oval can be defined on the basis of both cuts and paths as well as minimal cuts and minimal paths.

Figure 6.3   Relations among structure importance measures

first-term importance nor from the path importance to the rare-event importance. Because by the discussion in Section 6.10, if $i >_c j$, it may be either $I_{FT}(i; \phi) = I_{FT}(j; \phi)$ or $I_{FT}(i; \phi) > I_{FT}(j; \phi)$, and if $i >_p j$, it may be either $I_{RE}(i; \phi) = I_{RE}(j; \phi)$ or $I_{RE}(i; \phi) > I_{RE}(j; \phi)$.

As seen from Figure 6.3, the importance measures that can be defined on the basis of cuts or paths include the c-absoluteness importance and p-absoluteness importance, internal and external domination importance measures, permutation importance, cut-path importance, B-i.i.d. importance, and BP importance. The importance measures that can be defined on the basis of minimal cuts or minimal paths include the internal and external domination importance measures, permutation importance, cut importance and path importance, c-FV importance and p-FV importance, min-cut importance and min-path importance, and first-term importance and rare-event importance. Note that the domination importance measures and permutation importance can be defined based on both cuts and paths as well as minimal cuts and minimal paths.

By transitivity properties along the direction of the arrows in Figure 6.3, more relations among importance measures can be obtained. For example, it is easy to see that both c-absoluteness and p-absoluteness importance measures imply the permutation importance, which was proved by Hwang (2005) directly, and that the permutation importance is stronger than the B-structure importance, which was proved by Boland et al. (1989) directly. For another example, from Theorem 6.13.12 that the permutation importance implies the cut-path importance and from Theorem 6.13.18 that the cut-path importance implies the uniform B-i.i.d. importance, it follows that the permutation importance implies the uniform B-i.i.d. importance. Also, by Theorem 6.13.17, the cut-path importance implies the cut and path importance measures; thus, the permutation importance implies the cut and path importance measures. Meng (1994) proved some of these results directly, starting from the permutation importance. In particular, Freixas and Pons (2008b) showed the following lower bounds on $I_B(i, \phi) - I_B(j, \phi)$ given that $i >_{pe} j$.

Theorem 6.14.1 *If $i >_{pe} j$, then $I_B(i; \phi) \geq I_B(j; \phi) + 2^{-n+2}$, and $I_{BP}(i; \phi) \geq I_{BP}(j; \phi) + (s-1)!(n-s-1)!/(n-1)!$ for some s with $1 \leq s \leq n-1$.*

As in Theorem 6.5.2, components i and j are permutation equivalent if and only if they are structurally symmetric. The result can be generalized as in Corollary 6.14.2 using the dominant relations in Figure 6.3 and Theorem 6.13.15.

Corollary 6.14.2 *If components i and j are structurally symmetric, that is, $\phi(\mathbf{x})$ is permutation symmetric in x_i and x_j, then they are equally important according to every structure importance except the absoluteness and the domination importance measures.*

The following example shows that even if components i and j are structurally symmetric, they may not be compared according to the c-absoluteness and p-absoluteness importance measures.

Example 6.14.3 (Example 6.8.3 continued) Consider the system in Figure 2.4 in which components 2 and 3 are structurally symmetric. Then $\mathscr{C}_2 = \{\{1, 2\}, \{2, 3\}, \{1, 2, 3\}\}$, $\mathscr{C}_3 = \{\{1, 3\}, \{2, 3\}, \{1, 2, 3\}\}$, $\mathscr{P}_2 = \{\{1, 2\}, \{1, 2, 3\}\}$, and $\mathscr{P}_3 = \{\{1, 3\}, \{1, 2, 3\}\}$; thus, neither c-absoluteness nor p-absoluteness importance can be used to compare components 2 and 3.

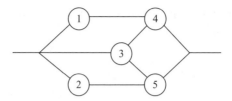

Figure 6.4 System of five components

Note that conclusions cannot be made along the reversed direction of the arrows in Figure 6.3. For example, the cut importance does not imply the permutation importance, as shown in Example 6.14.4 from Butler (1977) and Meng (1994), and the uniform B-i.i.d. importance does not imply the permutation importance, as shown in Example 6.14.5.

Example 6.14.4 (Example 6.7.8 continued) Example 6.7.8 shows that $2 >_c 1 >_c 5 >_c 3 =_c 6 >_c 4$ by directly comparing the vectors $\mathbf{b}_1, \mathbf{b}_2, \ldots, \mathbf{b}_6$. However, components 1 and 2 are not comparable by the permutation importance because $\phi(0, 1, 1, 1, 0, 1) < \phi(1, 0, 1, 1, 0, 1)$ and $\phi(1, 0, 0, 1, 1, 1) < \phi(0, 1, 0, 1, 1, 1)$.

On the other hand, it can be easily verified that $(1 >_{pe} 3, 4, 6)$, $(2 >_{pe} 4, 5)$, and $3 =_{pe} 6$. Hence, $(1 >_c 3, 4, 6)$, $(2 >_c 4, 5)$, and $3 =_c 6$ directly follow from the relations indicated in Figure 6.3. Thus, if two components in a system can be ranked by their permutation importance, then they can be ordered by the cut importance without calculating the vectors $\mathbf{b}_i$, which may otherwise be laborious for large systems.

Example 6.14.5 Considering a system of five components connected as shown in Figure 6.4, the reliability of the system is $R(\mathbf{p}) = p_3 p_4 + p_3 q_4 p_5 + q_3 (p_1 p_4 + p_2 p_5 - p_1 p_4 p_2 p_5)$. Using Equation (4.11), the B-i.i.d. importance values of components for $0 < p < 1$ are $I_B(1; p) = I_B(2; p) = pq - p^3 + p^4$; $I_B(3; p) = pq + p - 2p^2 + p^4$; and $I_B(4; p) = I_B(5; p) = pq + p - p^2 - p^3 + p^4$. Thus, $I_B(4; p) = I_B(5; p) > I_B(3; p) > I_B(1; p) = I_B(2; p)$ for all $0 < p < 1$.

Note that $\overline{\mathscr{P}}_1 = \{\{1, 4\}\} \subset \overline{\mathscr{P}}_4 = \{\{1, 4\}, \{3, 4\}\}$; thus, $4 >_{pe} 1$. Furthermore, $\overline{\mathscr{C}} = \{\{1, 2, 3\}, \{1, 3, 5\}, \{2, 3, 4\}, \{4, 5\}\}$. Thus, $\overline{\mathscr{C}}_1 \subset \overline{\mathscr{C}}_3$, and consequently, $3 >_{pe} 1$. However, neither $4 >_{pe} 3$ nor $3 >_{pe} 4$ holds because $\phi(0, 0, 0, 1, 1) = 0 < \phi(0, 0, 1, 0, 1) = 1$ and $\phi(1, 0, 1, 0, 0) = 0 < \phi(1, 0, 0, 1, 0) = 1$.

Example 6.14.6 illustrates the differences among the B-structure importance, BP structure importance, cut importance, path importance, and B-i.i.d. importance, which are related to each other as shown in the bottom-left corner of Figure 6.3. The B-structure importance is equivalent to the B-i.i.d. importance with $p = 1/2$, and the BP structure importance is equal to the integral of the B-i.i.d. importance over p in interval $[0, 1]$. However, sometimes the reliability values of components, although not precisely known, are high, for example, the components in electronic devices, nuclear reactors, oil refineries, spacecrafts, and aircrafts. In these situations, the cut importance seems preferable since it is the same as the B-i.i.d. importance for high values of p. Although it is a very rare case, if all component reliability values are rather low, the path importance could be used since it is the same as the B-i.i.d. importance for low values of p. Thus, the importance measure should be chosen on the basis of the range of component reliability.

Example 6.14.6 (Example 6.7.8 continued)   Assuming a same reliability p for all of the components in the system in Figure 6.2, the B-i.i.d. importance can be written in terms of $1 - p$:

$$I_B(1; p) = 2(1 - p) - 3(1 - p)^2 - (1 - p)^3 + 3(1 - p)^4 - (1 - p)^5$$
$$I_B(2; p) = 3(1 - p) - 4(1 - p)^2 - (1 - p)^3 + 3(1 - p)^4 - (1 - p)^5$$
$$I_B(3; p) = (1 - p) - (1 - p)^2 - 2(1 - p)^3 + 3(1 - p)^4 - (1 - p)^5$$
$$I_B(4; p) = 2(1 - p)^2 - 5(1 - p)^3 + 4(1 - p)^4 - (1 - p)^5$$
$$I_B(5; p) = (1 - p) + (1 - p)^2 - 5(1 - p)^3 + 4(1 - p)^4 - (1 - p)^5$$
$$I_B(6; p) = (1 - p) - (1 - p)^2 - 2(1 - p)^3 + 3(1 - p)^4 - (1 - p)^5.$$

Note that $\mathbf{b}_i$ is the vector of coefficients in the polynomial expression in terms of $1 - p$ for $I_B(i; p)$, as shown in Equation (6.22). Apparently, $3 =_c 6$ by comparing the expressions of $I_B(3; p)$ and $I_B(6; p)$. For high values of p, the lowest order terms in the polynomial dominate the rest. Thus, in this example, by looking only at the lowest order terms in the formulas for $I_B(i; p)$, it is apparent that for high values of p, $2 >_c 1$, $1 >_c 5$, $1 >_c 3$, $1 >_c 6$, $5 >_c 4$, $3 >_c 4$, and $6 >_c 4$. Furthermore, by examining the lowest and the second lowest order terms, $5 >_c 3$, and $5 >_c 6$. In summary, by analyzing $I_B(i; p)$ for high values of p, the cut importance ranking is $2 >_c 1 >_c 5 >_c 3 =_c 6 >_c 4$, consistent with the ranking obtained in Example 6.7.8 by analyzing the minimal cuts directly.

The B-i.i.d. importance can also be written in terms of p:

$$I_B(1; p) = 2p^2 - p^3 - 2p^4 + p^5 \qquad I_B(2; p) = p + p^2 - p^3 - 2p^4 + p^5$$
$$I_B(3; p) = p^2 - 2p^4 + p^5 \qquad I_B(4; p) = p^2 - p^3 - p^4 + p^5$$
$$I_B(5; p) = p - p^3 - p^4 + p^5 \qquad I_B(6; p) = p^2 - 2p^4 + p^5.$$

Then, according to Equation (6.23), $\mathbf{g}_1 = (0, 0, 2, -1, -2, 1)$, $\mathbf{g}_2 = (0, 1, 1, -1, -2, 1)$, $\mathbf{g}_3 = (0, 0, 1, 0, -2, 1)$, $\mathbf{g}_4 = (0, 0, 1, -1, -1, 1)$, $\mathbf{g}_5 = (0, 1, 0, -1, -1, 1)$, and $\mathbf{g}_6 = (0, 0, 1, 0, -2, 1)$. Therefore, $2 >_p 5 >_p 1 >_p 3 =_p 6 >_p 4$. These results can be verified using Definition 6.7.3 of the path importance.

Setting $p = 1/2$, then

$$I_B(2; \phi) > I_B(5; \phi) > I_B(1; \phi) > I_B(3; \phi) = I_B(6; \phi) > I_B(4; \phi),$$

and integrating $I_B(\cdot; p)$ over p in interval $[0, 1]$, then

$$I_{BP}(2; \phi) > I_{BP}(5; \phi) > I_{BP}(1; \phi) > I_{BP}(3; \phi) = I_{BP}(6; \phi) > I_{BP}(4; \phi).$$

In this example, the ranking of component importance according to the path importance is different from the cut importance but identical to the B-structure importance and the BP structure importance.

Note that for the B-i.i.d. importance,

$$I_B(2; p) > I_B(5; p), \; I_B(1; p) > I_B(3; p) = I_B(6; p) > I_B(4; p) \text{ for all } p.$$

For $p < (-1 + \sqrt{5})/2 \simeq 0.618$, $I_B(5; p) > I_B(1; p)$, and the B-i.i.d. importance ordering is identical to the rankings of the path importance, B-structure importance, and BP structure importance. But for $p > (-1 + \sqrt{5})/2$, $I_B(1; p) > I_B(5; p)$.

References

Barlow RE and Proschan F. 1975. Importance of system components and fault tree events. *Statistic Processes and Their Applications* **3**, 153–172.

Boland PJ, Proschan F, and Tong YL. 1989. Optimal arrangement of components via pairwise rearrangements. *Naval Research Logistics* **36**, 807–815.

Butler DA. 1977. An importance ranking for system components based upon cuts. *Operations Research* **25**, 874–879.

Butler DA. 1979. A complete importance ranking for components of binary coherent systems with extensions to multi-state systems. *Naval Research Logistics* **4**, 565–578.

Chang HW, Chen RJ, and Hwang FK. 2002. The structural Birnbaum importance of consecutive-k systems. *Journal of Combinatorial Optimization* **6**, 183–197.

Chang HW and Hwang FK. 2002. Rare-event component importance for the consecutive-k system. *Naval Research Logistics* **49**, 159–166.

Freixas J and Pons M. 2008a. Identifying optimal components in a reliability system. *IEEE Transactions on Reliability* **57**, 163–170.

Freixas J and Pons M. 2008b. The influence of the node criticality relation on some measures of component importance. *Operations Research Letters* **36**, 557–560.

Hwang FK. 2001. A new index of component importance. *Operations Research Letters* **28**, 75–79.

Hwang FK. 2005. A hierarchy of importance indices. *IEEE Transactions on Reliability* **54**, 169–172.

Koutras MV, Papadopoylos G, and Papastavridis SG. 1994. Note: Pairwise rearrangements in reliability structures. *Naval Research Logistics* **41**, 683–687.

Kuo W and Zhu X. 2012a. Relations and generalizations of importance measures in reliability. *IEEE Transactions on Reliability* **61**.

Kuo W and Zhu X. 2012b. Some recent advances on importance measures in reliability. *IEEE Transactions on Reliability* **61**.

Kuo W and Zuo MJ. 2003. *Optimal Reliability Modeling: Principles and Applications.* John Wiley & Sons, New York.

Lin FH and Kuo W. 2002. Reliability importance and invariant optimal allocation. *Journal of Heuristics* **8**, 155–171.

Meng FC. 1994. Comparing criticality of nodes via minimal cut (path) sets for coherent systems. *Probability in the Engineering and Informational Sciences* **8**, 79–87.

Meng FC. 1995. Some further results on ranking the importance of system components. *Reliability Engineering and System Safety* **47**, 97–101.

Meng FC. 1996. Comparing the importance of system components by some structural characteristics. *IEEE Transactions on Reliability* **45**, 59–65.

Meng FC. 2000. Relationships of Fussell-Vesely and Birnbaum importance to structural importance in coherent systems. *Reliability Engineering and System Safety* **67**, 55–60.

Mi J. 2003. A unified way of comparing the reliability of coherent systems. *IEEE Transactions on Reliability* **52**, 38–43.

Santha M and Zhang Y. 1987. Consecutive-2 systems on trees. *Probability in the Engineering and Informational Sciences* **1**, 441–456.

Shohat JA and Tamarkin JD. 1943. *The Problem of Moments.* American Mathematical Society, Providence, RI.

Zhu X and Kuo W. 2008. Comments on "A hierarchy of importance indices". *IEEE Transactions on Reliability* **57**, 529–531.

7

Importance Measures of Pairs and Groups of Components

The importance measures that are presented in Chapters 4–6 evaluate the strength of an individual component or of a special set of components (i.e., a modular set, minimal cut, and minimal path) with respect to system performance. Yet some decisions may affect groups of components, which raises the problem of importance measures for groups of components in Cheok et al. (1998b). This chapter discusses importance measures of any pair or group of components, considering the effects and higher order interactions of the components on system performance. They can provide additional information that the importance measures of individual components cannot. These importance measures of pairs and groups of components are commonly used in sensitivity analysis and risk/safety analysis, which are further addressed in Chapters 18 and 19.

This chapter focuses on several typical importance measures of pairs and groups of components, including the joint reliability importance (JRI), joint failure importance (JFI), differential importance measure (DIM), reliability achievement worth (RAW), and reliability reduction worth (RRW). Note that the total order importance (TOI), which is indeed an importance measure of individual components, is also presented in this chapter because it has close relationships with the JRI and DIM. In addition, Ryabinin (1994) introduced some reliability importance measures (namely, weight, contribution, and significance) of components in coherent systems based on a Boolean difference, reporting a study on the case of two components and claiming that the results can be generalized for more than two components. However, the importance measures in Ryabinin (1994) have limited applications; thus, they are not presented here.

Note that for all importance measures of pairs of components, the importance of components i and j is exactly the same as that of components j and i. That is, $I(i, j; \mathbf{p}) = I(j, i; \mathbf{p})$ where I denotes any importance measure of pairs of components. Similarly, the order of components in a group does not matter for any importance measure of the group.

Importance Measures in Reliability, Risk, and Optimization: Principles and Applications, First Edition.
Way Kuo and Xiaoyan Zhu. © 2012 John Wiley & Sons, Ltd. Published 2012 by John Wiley & Sons, Ltd.

7.1 The Joint Reliability Importance and Joint Failure Importance

Recall that the B-reliability importance evaluates the rate at which systems reliability improves as component reliability improves, as in Equation (4.10). Extending this concept to two independent components in a coherent system, Hong and Lie (1993) defined the JRI and the JFI in the context of an undirected graph where an edge corresponds to a component and the probability that the source and sink are connected by functioning edges corresponds to the systems reliability.

Definition 7.1.1 *The JRI of independent components i and j, $i \neq j$, denoted by $I_{\mathrm{JR}^{ll}}(i, j; \mathbf{p})$, is defined as*

$$I_{\mathrm{JR}^{ll}}(i, j; \mathbf{p}) = \frac{\partial^2 R(\mathbf{p})}{\partial p_i \partial p_j}, \tag{7.1}$$

and the JFI of components i and j, $i \neq j$, denoted by $I_{\mathrm{JF}^{ll}}(i, j; \mathbf{p})$, is defined as

$$I_{\mathrm{JF}^{ll}}(i, j; \mathbf{p}) = \frac{\partial^2 (1 - R(\mathbf{p}))}{\partial q_i \partial q_j} \tag{7.2}$$

$$= -\frac{\partial^2 R(\mathbf{p})}{\partial p_i \partial p_j} = -I_{\mathrm{JR}^{ll}}(i, j; \mathbf{p}).$$

Since $I_{\mathrm{JF}^{ll}}(i, j; \mathbf{p}) = -I_{\mathrm{JR}^{ll}}(i, j; \mathbf{p})$, the rest of this section focuses on the JRI only. The results for the JRI can be similarly derived for the JFI. The JRI provides useful insights on how two components interact with each other when their reliability changes, which the B-reliability importance cannot. *The sign and the value of the JRI represent the type and the degree of interactions between two components with respect to systems reliability.*

Example 7.1.2 (Example 2.4.3 continued) Consider a parallel–series system with i.i.d. components of a same reliability p. Supposing that the system has m series embedded systems with n_k (≥ 2) components in the kth embedded system for $k = 1, 2, \ldots, m$, then the reliability of the system is $R(p) = 1 - \prod_{k=1}^{m}(1 - p^{n_k})$. The JRI of any two components in the same embedded system is

$$I_{\mathrm{JR}^{ll}}(i, j; p) = p^{n_k-2} \prod_{r \neq k}(1 - p^{n_r}) \geq 0,$$

where components i and j are in the kth embedded system. The JRI of two components in different embedded systems is

$$I_{\mathrm{JR}^{ll}}(i, j; p) = -p^{n_k+n_{k'}-2} \prod_{r \neq k,k'}(1 - p^{n_r}) \leq 0,$$

where component i is in the kth embedded system and component j is in the k'th embedded system with $k' \neq k$.

Theorem 7.1.3 gives another expression of the JRI, which shows that $I_{\mathrm{JR}^{ll}}(i, j; \mathbf{p})$ is independent of the reliability of components i and j. Following this, Theorem 7.1.4 shows the relationships between the JRI and the B-reliability importance. Hong and Lie (1993) presented Theorems 7.1.3 and 7.1.4.

Theorem 7.1.3 *In a coherent system with independent components,*

$$I_{JR^{II}}(i, j; \mathbf{p}) = R(1_i, 1_j, \mathbf{p}^{(ij)}) - R(1_i, 0_j, \mathbf{p}^{(ij)}) - R(0_i, 1_j, \mathbf{p}^{(ij)}) + R(0_i, 0_j, \mathbf{p}^{(ij)}). \quad (7.3)$$

Proof. According to Equations (4.10) and (4.11),

$$I_{JR^{II}}(i, j; \mathbf{p}) = \frac{\partial^2 R(\mathbf{p})}{\partial p_i \partial p_j} = \frac{\partial [R(1_i, \mathbf{p}) - R(0_i, \mathbf{p})]}{\partial p_j} = \frac{\partial R(1_i, \mathbf{p})}{\partial p_j} - \frac{\partial R(0_i, \mathbf{p})}{\partial p_j} \quad (7.4)$$

$$= R(1_i, 1_j, \mathbf{p}^{(ij)}) - R(1_i, 0_j, \mathbf{p}^{(ij)}) - [R(0_i, 1_j, \mathbf{p}^{(ij)}) - R(0_i, 0_j, \mathbf{p}^{(ij)})].$$

Theorem 7.1.4 *In a coherent system with independent components,*

$$I_{JR^{II}}(i, j; \mathbf{p}) = I_B(j; 1_i, \mathbf{p}) - I_B(j; 0_i, \mathbf{p}) \quad (7.5)$$

$$= I_B(i; 1_j, \mathbf{p}) - I_B(i; 0_j, \mathbf{p}) \quad (7.6)$$

$$= \frac{1}{p_i}[I_B(j; \mathbf{p}) - I_B(j; 0_i, \mathbf{p})] \quad (7.7)$$

$$= \frac{1}{p_j}[I_B(i; \mathbf{p}) - I_B(i; 0_j, \mathbf{p})]. \quad (7.8)$$

Proof. From Equations (7.4) and (4.10), Equation (7.5) is immediately established. Furthermore, $R(\mathbf{p}) - R(0_i, \mathbf{p}) = p_i[R(1_i, \mathbf{p}) - R(0_i, \mathbf{p})]$; thus,

$$I_{JR^{II}}(i, j; \mathbf{p}) = \frac{1}{p_i} \frac{\partial [R(\mathbf{p}) - R(0_i, \mathbf{p})]}{\partial p_j},$$

and Equation (7.7) follows. Equations (7.6) and (7.8) can be similarly established.

Corollary 7.1.5 *For any pair of components i and j, $i \neq j$, $-1 \leq I_{JR^{II}}(i, j; \mathbf{p}) \leq 1$.*

Proof. The result follows from Corollary 4.1.3 and Equation (7.5).

Remark 7.1.6 Two components are called reliability *complements* (*substitutes*) if the sign of their JRI is nonnegative (nonpositive) (Hagstrom 1990). From Theorem 7.1.4, if $I_{JR^{II}}(i, j; \mathbf{p}) > 0$ (complements), then $I_B(i; 1_j, \mathbf{p}) > I_B(i; 0_j, \mathbf{p})$. This implies that component i is more important for systems reliability when component j functions than when it fails. Similarly, $I_{JR^{II}}(i, j; \mathbf{p}) < 0$ (substitutes) indicates that one component becomes more important when the other fails. These interpretations show that the JRI measures the interactions of two components on systems reliability (Hagstrom 1990; Hong and Lie 1993).

With respect to minimal cuts and minimal paths, there are three situations for the relations of two components i and j in the coherent system.

(i) There exists no minimal path containing both components i and j, that is, there exist some minimal cuts containing components i and j.
(ii) There exists no minimal cut containing both components i and j, that is, there exist some minimal paths containing components i and j.
(iii) There exist some minimal paths and some minimal cuts containing components i and j.

Note that it is impossible to have neither minimal cut nor minimal path containing both components i and j. Since the calculation of the JRI is generally difficult, it is desirable to know the sign of the JRI. Hong and Lie (1993) and Armstrong (1997) showed the sign of the JRI for situations (i) and (ii) as follows.

Theorem 7.1.7 $I_{\mathrm{JR}^{II}}(i, j; \mathbf{p}) \geq (\leq) \, 0$ *for all* $\mathbf{0} < \mathbf{p} < \mathbf{1}$ *if and only if there exists no minimal cut (path) containing both components i and j.*

Corollary 7.1.8 is a special case of Theorem 7.1.7 because no minimal cut contains two components that are connected in series and no minimal path contains two components that are connected in parallel. For series and parallel connected components, the sign of the JRI can also be easily determined from Equation (7.3).

Corollary 7.1.8   *If components i and j are connected in series (parallel), then* $I_{\mathrm{JR}^{II}}(i, j; \mathbf{p}) \geq (\leq) \, 0.$

Supposing that components i and j are connected in series, if component i functions, then component j is more important for system functioning than the situation where component i already fails, since the module consisting of components i and j fails when component i fails no matter whether component j fails or functions. Thus, in the series connection case, $I_{\mathrm{JR}^{II}}(i, j; \mathbf{p}) \geq 0$. Now, supposing that two components are connected in parallel, if one of these two components fails, then the other becomes more important for system functioning than the situation where both components functions since no or fewer backup components exist. Thus, in the parallel connection case, $I_{\mathrm{JR}^{II}}(i, j; \mathbf{p}) \leq 0$. These results are consistent with the analysis in Remark 7.1.6.

The following theorem gives the relationship between the JRI and the reliability function by means of its Schur-concavity and convexity (Hong et al. 2002).

Theorem 7.1.9   *If the reliability function $R(\mathbf{p})$ is Schur-concave (convex), then* $I_{\mathrm{JR}^{II}}(i, j; \mathbf{p}) \geq (\leq) \, 0.$

7.1.1   The Joint Reliability Importance of Dependent Components

This subsection assumes that the components in a coherent system are dependent. In this case, Armstrong (1995) studied the JRI as expressed in Equation (7.9); for the evaluation of the terms in Equation (7.9), see Hagstrom and Mak (1987).

Definition 7.1.10 *The JRI of dependent components i and j, $i \neq j$, is defined as*

$$I_{\mathrm{JR}^{II}}(i, j; \mathbf{p}) = \mathbb{E}\left(\phi(1_i, 1_j, \mathbf{X}^{(ij)}) - \phi(1_i, 0_j, \mathbf{X}^{(ij)}) - \phi(0_i, 1_j, \mathbf{X}^{(ij)}) + \phi(0_i, 0_j, \mathbf{X}^{(ij)})\right),$$
(7.9)

and the JFI of dependent components i and j, $i \neq j$, is defined as

$$I_{\mathrm{JF}^{II}}(i, j; \mathbf{p}) = -I_{\mathrm{JR}^{II}}(i, j; \mathbf{p}).$$

Recall from Section 4.1 that the B-reliability importance of dependent components is defined as

$$I_B(i; \mathbf{p}) = \mathbb{E}(\phi(1_i, \mathbf{X}) - \phi(0_i, \mathbf{X})).$$

Thus, Equations (7.5) and (7.6) and, consequently, Corollary 7.1.5 are also valid for the case of dependent components. However, Equations (7.7) and (7.8) are not valid for the dependent components. Armstrong (1995) showed that Theorem 7.1.7 and Corollary 7.1.8 are still valid for dependent components and, additionally, presented two cases, as follows, under which the JRI is always nonzero.

Corollary 7.1.11 *Suppose that $0 < p_i < 1$ for $i = 1, 2, \ldots, n$ and $\mathrm{Corr}(X_i, X_j) < 1$ (correlation coefficient) for all pairs of components i and j. (i) If there exists one minimal cut (path) containing both components i and j, then $I_{\mathrm{JR}^{II}}(i, j; \mathbf{p}) > (<) 0$. (ii) In a parallel–series system, $I_{\mathrm{JR}^{II}}(i, j; \mathbf{p}) \neq 0$ for all pairs of components i and j.*

In addition, Hong et al. (2002) considered the statistical dependence between any two components and showed that the error of the systems reliability caused by assuming statistical independence of two pairwise dependent components is bounded by their covariance and the JRI, as follows.

Theorem 7.1.12 *Let components i and j be pairwise dependent. Let R be the systems reliability and R^{ij} be the systems reliability in assuming statistical independence between components i and j. Then, the error is*

$$R - R^{ij} = \mathrm{Cov}(X_i, X_j) I_{\mathrm{JR}^{II}}(i, j; \mathbf{p}).$$

7.1.2 The Joint Reliability Importance of Two Gate Events

Hong et al. (2000) presented a formula for computing the JRI of two gate events in a fault tree. When a fault tree is used to describe system behavior, a component and a subsystem in the system, respectively, correspond to a basic event and a gate event in the corresponding fault tree, but not vice versa. In fact, some basic events and gate events stand for logical events.

Definition 7.1.13 *In a fault tree, the JRI of two distinct gate events, G_i and G_j, is defined as*

$$
\begin{aligned}
I_{\mathrm{JR}^{II}}(G_i, G_j; \mathbf{p}) &= \Pr\{G_0|G_i = 1, G_j = 1; \mathbf{p}\} - \Pr\{G_0|G_i = 1, G_j = 0; \mathbf{p}\} \\
&\quad - \Pr\{G_0|G_i = 0, G_j = 1; \mathbf{p}\} + \Pr\{G_0|G_i = 0, G_j = 0; \mathbf{p}\},
\end{aligned}
$$

where G_0 is the top event of the system.

Two gate events may contain topological relations (i.e., one gate is in the subtree of another) and/or may share some basic events in common; thus, statistical dependencies exist between gate events. The formula in Definition 7.1.13 removes the effect of these dependencies and can be used to calculate the JRI of any pair of gate events in a fault tree in which the basic events are mutually independent. Moreover, Hong et al. (2000) showed that if two gate events represent the minimal cuts (paths) of a fault tree, then the corresponding JRI are nonpositive (nonnegative).

7.1.3   The Joint Reliability Importance for k-out-of-n Systems

In the i.i.d. case, Hong et al. (2002) analyzed the JRI for k-out-of-n systems as in Theorem 7.1.14. Furthermore, Theorem 7.1.15 compares the JRI for k-out-of-n systems of different values for k, n, and p.

Theorem 7.1.14   *Let $n \geq 3$ and $2 \leq k \leq n$. In the k-out-of-n system with i.i.d. components of a same reliability p, the JRI of components i and j has the simple form:*

$$I_{\mathrm{JR}^{II}}(i,j;p) = p^{k-2}q^{n-k-1}\left[\binom{n-2}{k-2} - \binom{n-1}{k-1}p\right].$$

Theorem 7.1.15   *Suppose that the system has i.i.d. components of a same reliability p.*

(i) *For an arbitrary value of $n \geq 4$ of the 2-out-of-n system, the JRI decreases from 1 to some negative value as reliability p increases from 0 to some value less than 1; it then increases to approaching 0 as p continues increasing to 1.*
(ii) *Let JRI_n denote the JRI for the 2-out-of-n system. Then $JRI_n < JRI_{n+1}$, when $p > 1/2$.*
(iii) *Let JRI_k denote the JRI for the k-out-of-$2k$ system. Then $JRI_k < JRI_{k+1} < 0$, when $p > (5k-3)/(6k-3)$.*

From Theorem 7.1.15(iii), the k-out-of-$2k$ system is more robust for statistically dependent failures as k increases because the degree of interactions between two components decreases as k increases. Jan and Chang (2006) provided additional but similar discussion on the JRI for the k-out-of-n systems starting from Theorem 7.1.14.

In the case of the k-out-of-n system with independent but nonidentical components, the JRI cannot be presented in a simple closed form like that of the i.i.d. case in Theorem 7.1.14. Boland and Proschan (1983) showed that the reliability function of a k-out-of-n system is Schur-concave (convex) if $p_r \leq (\geq)(k-1)/(n-1)$ for all $r = 1, 2, \ldots, n$. Thus, according to Theorem 7.1.9, the sign of the JRI (i.e., the type of interactions) in the k-out-of-n system can be determined by the range of component reliability, as shown in Theorem 7.1.16.

Theorem 7.1.16   *In the k-out-of-n system with independent and nonidentical components ($n \geq 3$ and $2 \leq k \leq n$), $I_{\mathrm{JR}^{II}}(i,j;\mathbf{p}) \geq (\leq) 0$ if $p_r \leq (\geq)(k-1)/(n-1)$ for all $r = 1, 2, \ldots, n$.*

7.1.4 The Joint Reliability Importance of Order k

According to Equations (7.1) and (7.2), the JRI and JFI can be easily extended to groups of more than two components (Gao et al. 2007).

Definition 7.1.17 *The JRI of order k (JRI^k) of k distinct components $i_1, i_2, \ldots, i_k$, $k < n$, denoted by $I_{\mathrm{JR}^k}(i_1, i_2, \ldots, i_k; \mathbf{p})$, is defined as*

$$I_{\mathrm{JR}^k}(i_1, i_2, \ldots, i_k; \mathbf{p}) = \frac{\partial^k R(\mathbf{p})}{\partial p_{i_1} \partial p_{i_2}, \ldots, \partial p_{i_k}}.$$

Note that the first-order JRI becomes the B-reliability importance and that the second-order JRI is the aforementioned JRI and often simply referred to as the JRI. Example 4.1.6 provides a result relative to the B-reliability importance, which is similar to Theorem 7.1.16 for the (second-order) JRI. Formal properties of the (second-order) JRI hold also for the JRI of order k in coherent systems. However, the JRI of more than two components is ambiguous and too complicated to analyze and calculate.

7.2 The Differential Importance Measure

7.2.1 The First-order Differential Importance Measure

The introduction of the DIM is motivated by the need to evaluate the strength of simultaneous changes in the reliability of components. The first-order DIM of individual components in a coherent system, based on the work in Borgonovo and Apostolakis (2001), is defined as follows.

Definition 7.2.1 *The first-order DIM (DIM^I) of component i, denoted by $I_{\mathrm{DIM}^I}(i; \mathbf{p})$, is defined as*

$$I_{\mathrm{DIM}^I}(i; \mathbf{p}) = \frac{\frac{\partial R(\mathbf{p})}{\partial p_i}\mathrm{d}p_i}{\mathrm{d}R(\mathbf{p})} = \frac{\frac{\partial R(\mathbf{p})}{\partial p_i}\mathrm{d}p_i}{\sum_{j=1}^n \frac{\partial R(\mathbf{p})}{\partial p_j}\mathrm{d}p_j}. \tag{7.10}$$

The denominator in Equation (7.10) represents the total change in systems reliability due to a small change in the reliability of all components. The numerator represents the change of the systems reliability caused by the change of the reliability of component i. Thus, the DIM^I of component i is the fraction of the total change in systems reliability due to a change in the reliability of component i. Monte Carlo simulation methods and perturbation analysis can be used to calculate the DIM^I (Marseguerra and Zio 2004) (see Subsection 18.1.4).

Let a coherent system (N, ϕ) have a module (M, χ). Let R_χ and R_ϕ denote the reliability of module (M, χ) and system (N, ϕ), respectively. Then,

$$\mathrm{d}R_\chi = \sum_{j \in M} \frac{\partial R_\chi}{\partial p_j}\mathrm{d}p_j$$

$$\frac{\partial R_\phi}{\partial R_\chi}\mathrm{d}R_\chi = \frac{\partial R_\phi}{\partial R_\chi} \sum_{j \in M} \frac{\partial R_\chi}{\partial p_j}\mathrm{d}p_j = \sum_{j \in M} \frac{\partial R_\phi}{\partial R_\chi} \frac{\partial R_\chi}{\partial p_j}\mathrm{d}p_j = \sum_{j \in M} \frac{\partial R_\phi}{\partial p_j}\mathrm{d}p_j.$$

According to Definition 7.2.1, the DIM^I of module (M, χ) is

$$
\begin{aligned}
I_{DIM^I}^M (M; \mathbf{p}) &= \frac{\frac{\partial R_\phi}{\partial R_\chi} dR_\chi}{\sum_{j \in N \backslash M} \frac{\partial R_\phi}{\partial p_j} dp_j + \frac{\partial R_\phi}{\partial R_\chi} dR_\chi} \\
&= \frac{\sum_{j \in M} \frac{\partial R_\phi}{\partial p_j} dp_j}{\sum_{j \in N} \frac{\partial R_\phi}{\partial p_j} dp_j} \\
&= \sum_{j \in M} I_{DIM^I} (j; \mathbf{p}).
\end{aligned}
\tag{7.11}
$$

From Equation (7.11), the DIM^I is *additive*: the DIM^I of a modular set is the sum of the DIM^I of the components in the modular set. In fact, according to the definition of the DIM^I in Equation (7.10), the DIM^I of a group of components is the sum of the DIM^I of the individual components in the group (Borgonovo and Apostolakis 2001); the group is not necessarily a module set. This is particularly relevant in the presence of large reliability systems as it allows for obtaining component importance without further system evaluation. It is often the case in safety analysis and has applications as the evaluation of changes in maintenance policies and graded quality assurance programs, in which groups of components are simultaneously affected by changes (Borgonovo 2010). The DIM^I of all components is unity.

The DIM^I is defined on the basis of the change of systems reliability caused by the changes of component reliability. The different assumptions of the changes of component reliability result in different formats of DIM^I. Two assumptions are commonly used.

Assumption 1: All reliability values of components are changed by the same small amount δ, that is, $\Delta p_i = dp_i = \delta$ for $i = 1, 2, \ldots, n$.

Assumption 2: All reliability values of components are changed by the same small percentage $0 < \gamma < 1$, that is, $\Delta p_i / q_i = dp_i / q_i = \gamma$ for $i = 1, 2, \ldots, n$.

Under assumption 1,

$$
I_{DIM^I} (i; \mathbf{p}) = \frac{\frac{\partial R(\mathbf{p})}{\partial p_i}}{\sum_{j=1}^n \frac{\partial R(\mathbf{p})}{\partial p_j}} = \frac{I_B (i; \mathbf{p})}{\sum_{j=1}^n I_B (i; \mathbf{p})},
$$

that is, the DIM^I and B-reliability importance produce the same ranking of components. Under assumption 2,

$$
I_{DIM^I} (i; \mathbf{p}) = \frac{\frac{\partial R(\mathbf{p})}{\partial p_i} q_i}{\sum_{j=1}^n \frac{\partial R(\mathbf{p})}{\partial p_j} q_j} = \frac{I_{Bs} (i; \mathbf{p})}{\sum_{j=1}^n I_{Bs} (i; \mathbf{p})},
$$

that is, the DIM^I and improvement potential importance produce the same ranking of components. However, the B-reliability importance and the improvement potential importance do not have the additivity property of Equation (7.11) because assumptions 1 and 2 are applied for the reliability of a component and not for a module of a system.

7.2.2 The Second-order Differential Importance Measure

The JRI of a pair of components does not measure the overall importance of the group formed by the two components but instead measures their interaction. Conversely, the DIM^I measures the overall importance of the group as the sum of their individual effects. Zio and Podofillini (2006) presented a second-order DIM to take into account both the individual effects and the interactions of pairs of components when evaluating a change in systems reliability due to changes of component reliability. This extension supplements the DIM^I with the second-order information provided by the JRI.

Definition 7.2.2   *The second-order DIM (DIM^{II}) of components i and j, $i \neq j$, denoted by $I_{DIM^{II}}(i, j; \mathbf{p})$, is defined as*

$$I_{DIM^{II}}(i, j; \mathbf{p}) = \frac{I_B(i; \mathbf{p})\Delta p_i + I_B(j; \mathbf{p})\Delta p_j + I_{JR^{II}}(i, j; \mathbf{p})\Delta p_i \Delta p_j}{\sum_{i=1}^{n} I_B(i; \mathbf{p})\Delta p_i + \sum_{i=1}^{n}\sum_{j>i}^{n} I_{JR^{II}}(i, j; \mathbf{p})\Delta p_i \Delta p_j}. \tag{7.12}$$

Note that $I_B(i; \mathbf{p}) = \partial R(\mathbf{p})/\partial p_i$ and $I_{JR^{II}}(i, j; \mathbf{p}) = \partial^2 R(\mathbf{p})/(\partial p_i \partial p_j)$. Thus, the denominator in Equation (7.12) is the second-order (Taylor) approximation of a variation in the systems reliability in terms of the B-reliability importance and the JRI. The numerator is the variation of systems reliability due to the variations of ∂p_i and ∂p_j. Unlike the DIM^I, the DIM^{II} does not reserve the additivity property. However, the DIM^{II} of all pairs of components is unity.

The DIM^{II} in Equation (7.12) is useful in choosing pairs of components that have to be modified, for example, for reducing system operating costs with the goal of not causing the overall system performance to deteriorate excessively. In this case, the pair of components with the smallest DIM^{II} should be chosen. In a similar situation, where the goal is to get the greatest improvement in systems reliability with a limited budget by improving the reliability of pairs of components, the pair of components with the largest DIM^{II} should be chosen. In another situation that can occur in risk-informed decision-making, improving the availability of a component may degrade that of another. Again the use of the DIM^{II} is appropriate since it models not only the individual effects but also the interactions of the changes on two components. Zio and Podofillini (2006) demonstrated the applications of the DIM^I and DIM^{II} in a numerical example and discussed their applications in risk-informed decision-making.

7.2.3 The Differential Importance Measure of Order k

Straightforwardly, the DIM can be extended to order k, $k < n$, for evaluating the overall effects of reliability changes of k distinct components in a group. It counts for the effects of each individual component in the group and any order of interactions among the components in the group.

Definition 7.2.3   *The DIM of order k (DIM^k) of k distinct components $i_1, i_2, \ldots, i_k, k < n$, denoted by $I_{\mathrm{DIM}^k}(i_1, i_2, \ldots, i_k; \mathbf{p})$, is defined as*

$$
I_{\mathrm{DIM}^k}(i_1, i_2, \ldots, i_k; \mathbf{p})
$$

$$
= \frac{\sum_{s=1}^{k} I_{\mathrm{B}}(i_s; \mathbf{p}) \Delta p_{i_s} + \sum_{s=2}^{k} \sum_{\substack{r_1 < r_2 < \cdots < r_s \\ r_1, r_2, \ldots, r_s \in \{i_1, i_2, \ldots, i_k\}}} I_{\mathrm{JR}^s}(r_1, r_2, \ldots, r_s; \mathbf{p}) \prod_{t=1}^{s} \Delta p_{r_t}}{\sum_{i=1}^{n} I_{\mathrm{B}}(i; \mathbf{p}) \Delta p_i + \sum_{s=2}^{k} \sum_{r_1 < r_2 < \cdots < r_s} I_{\mathrm{JR}^s}(r_1, r_2, \ldots, r_s; \mathbf{p}) \prod_{t=1}^{s} \Delta p_{r_t}}.
$$

7.3   The Total Order Importance

Motivated by the DIM^I, Kuo and Zhu (2012) proposed a TOI of individual components, which is defined for various order k, $k < n$, and based on the multilinear expression of the systems reliability function and the kth order Taylor (Maclaurin) expansion (see Section 2.10).

Definition 7.3.1   *The TOI of order k (TOI^k) of component i, $k < n$, denoted by $I_{\mathrm{TOI}^k}(i; \mathbf{p})$, is defined as*

$$
I_{\mathrm{TOI}^k}(i; \mathbf{p}) = \frac{\sum_{s=1}^{k} \sum_{\substack{r_1 < r_2 < \cdots < r_s \\ i \in \{r_1, r_2, \ldots, r_s\}}} R^{(s)}(r_1, r_2, \ldots, r_s; \mathbf{p}) \prod_{t=1}^{s} \Delta p_{r_t}}{\sum_{s=1}^{k} \sum_{r_1 < r_2 < \cdots < r_s} R^{(s)}(r_1, r_2, \ldots, r_s; \mathbf{p}) \prod_{t=1}^{s} \Delta p_{r_t}} \tag{7.13}
$$

$$
= \frac{I_{\mathrm{B}}(i; \mathbf{p}) \Delta p_i + \sum_{s=2}^{k} \sum_{\substack{r_1 < r_2 < \cdots < r_s \\ i \in \{r_1, r_2, \ldots, r_s\}}} I_{\mathrm{JR}^s}(r_1, r_2, \ldots, r_s; \mathbf{p}) \prod_{t=1}^{s} \Delta p_{r_t}}{\sum_{i=1}^{n} I_{\mathrm{B}}(i; \mathbf{p}) \Delta p_i + \sum_{s=2}^{k} \sum_{r_1 < r_2 < \cdots < r_s} I_{\mathrm{JR}^s}(r_1, r_2, \ldots, r_s; \mathbf{p}) \prod_{t=1}^{s} \Delta p_{r_t}}, \tag{7.14}
$$

where $R^{(s)}$ denotes the sth order of partial derivative of R. The denominators in Equations (7.13) and (7.14) denote the kth order Taylor approximation of the change in systems reliability, and the numerators denote the fraction of the denominators associated with component i.

The TOI^k of component i considers the fraction of the kth order Taylor expansion of a systems reliability change associated with component i. Note that when $k = 1$, the TOI in Equation (7.14) becomes the DIM^I. When $k = 2$, it is

$$
I_{\mathrm{TOI}^{II}}(i; \mathbf{p}) = \frac{I_{\mathrm{B}}(i; \mathbf{p}) \Delta p_i + \sum_{j \neq i}^{n} I_{\mathrm{JR}^{II}}(i, j; \mathbf{p}) \Delta p_i \Delta p_j}{\sum_{i=1}^{n} I_{\mathrm{B}}(i; \mathbf{p}) \Delta p_i + \sum_{i=1}^{n} \sum_{j > i}^{n} I_{\mathrm{JR}^{II}}(i, j; \mathbf{p}) \Delta p_i \Delta p_j}.
$$

By this definition, the TOI^{II} captures the second-order interaction effects of component i with all other components in determining the importance of component i. Equation (7.14) defines the relationship between $I_{\mathrm{TOI}^k}(i; \mathbf{p})$, $I_{\mathrm{B}}(i; \mathbf{p})$ (i.e., the first-order JRI), and $I_{\mathrm{JR}^s}(r_1, r_2, \ldots, r_s; \mathbf{p})$, $s = 2, 3, \ldots, k$. The TOI^k does not have the additivity property.

By Theorems 2.10.2 and 2.10.3, the Taylor expansion of the systems reliability function R is exact at order $O = \min\{n, r_{\mathrm{largest}}\}$, where r_{largest} is the size of the largest product in the multilinear expression of R. Therefore, the bound on the Taylor series (i.e., the highest order of the Taylor expansion) in the denominators in Equations (7.13) and (7.14) is O, at which one needs to stop the Taylor expansion. Consequently, it is clear that O is the minimum k for which the TOI of order k can provide the overall importance ranking. Because of this discovery, Borgonovo (2010) introduced the TOI^O that is the exact fraction of the change in systems reliability caused by generic (finite or infinitesimal) changes in component reliability. Borgonovo (2010), indeed, named the TOI^O as the TOI measure. The TOI^O turns out to be a sensitivity measure that synthesizes the B-reliability importance and JRI of all orders in one unique indicator.

Note that the TOI is an importance measure of individual components but it measures the impact of a component as a result of its individual effects and of all of its possible interactions with other components. The B-reliability importance and JRI^k provide information on the sign and magnitude of the individual and interaction effects, respectively. Thus, the simultaneous utilization of the B-reliability importance, JRI^k, and TOI^k obtains a complete assessment of system performance and determines exactly how each component contributes to system performance.

The TOI^k differs from various orders significantly as interaction effects become relevant, that is, when changes are finite. However, when the changes become small, the TOI^k reduces to the DIM^I (i.e., the first-order TOI), as presented in Theorem 7.3.2 (Borgonovo 2010).

Theorem 7.3.2 *Let $1 < k < O$. As changes $\mathbf{\Delta p} = (\Delta p_1, \Delta p_2, \ldots, \Delta p_n)$ become small,*

$$\lim_{\mathbf{\Delta p} \to 0} I_{\text{TOI}^O}(i; \mathbf{p}) = \lim_{\mathbf{\Delta p} \to 0} I_{\text{TOI}^k}(i; \mathbf{p}) = I_{\text{DIM}^I}(i; \mathbf{p}).$$

Theorem 7.3.3 (Borgonovo 2010) enables the numerical estimation of the TOI^O by varying one component reliability at a time, making it suitable for the analysis of complex systems. Then the computational complexity of the TOI^O for n components is equal to $n+2$ systems reliability evaluations. Theorem 7.3.3 is induced from Theorem 2.10.3.

Theorem 7.3.3 *Let $\mathbf{p}$ and $\tilde{\mathbf{p}} = \mathbf{p} + \mathbf{\Delta p}$ denote the component reliability values before and after finite changes. Let $(p_i, \tilde{\mathbf{p}})$ denote the component reliability vector when the reliability values of all components except component i are changed. Then*

$$I_{\text{TOI}^O}(i; \mathbf{p}) = \frac{R(\tilde{\mathbf{p}}) - R(p_i, \tilde{\mathbf{p}})}{R(\tilde{\mathbf{p}}) - R(\mathbf{p})}.$$

Theorem 7.3.3 immediately implies the following (Borgonovo 2010):

$$R(\tilde{\mathbf{p}}) - R(\mathbf{p}) = \sum_{i=1}^{n} I_B(i; \mathbf{p})\Delta p_i + \sum_{s=2}^{O} \sum_{r_1 < r_2 < \ldots < r_s} I_{\text{JR}^s}(r_1, r_2, \ldots, r_s; \mathbf{p}) \prod_{t=1}^{s} \Delta p_{r_t}.$$

7.4 The Reliability Achievement Worth and Reliability Reduction Worth

The RAW, as reviewed by Vasseur and Llory (1999), quantifies the maximum percentage increase in systems reliability generated by a particular component. The RRW, presented by Levitin et al. (2003), evaluates the potential damage caused to a system by a particular component.

Definition 7.4.1 *The RAW of component i, denoted by $I_{\text{RAW}}(i; \mathbf{p})$, is defined as*

$$I_{\text{RAW}}(i; \mathbf{p}) = \frac{\Pr\{\phi(1_i, \mathbf{X}) = 1\}}{\Pr\{\phi(\mathbf{X}) = 1\}}. \tag{7.15}$$

The RRW of component i, denoted by $I_{\text{RRW}}(i; \mathbf{p})$, is defined as

$$I_{\text{RRW}}(i; \mathbf{p}) = \frac{\Pr\{\phi(\mathbf{X}) = 1\}}{\Pr\{\phi(0_i, \mathbf{X}) = 1\}}. \tag{7.16}$$

It is easy to verify that

$$I_{\text{RAW}}(i; \mathbf{p}) = 1 + \frac{q_i}{R(\mathbf{p})} I_{\text{B}}(i; \mathbf{p}),$$

and

$$I_{\text{RRW}}(i; \mathbf{p}) = \frac{1}{1 - \frac{p_i}{R(\mathbf{p})} I_{\text{B}}(i; \mathbf{p})}.$$

The RAW and the RRW can be extended to assess the importance of pairs of components (Zio and Podofillini 2006).

Definition 7.4.2 *The (second-order) RAW and the (second-order) RRW of components i and j, $i \neq j$, denoted by $I_{\text{RAW}^{II}}(i, j; \mathbf{p})$ and $I_{\text{RRW}^{II}}(i, j; \mathbf{p})$, respectively, are defined as*

$$I_{\text{RAW}^{II}}(i, j; \mathbf{p}) = \frac{\Pr\{\phi(1_i, 1_j, \mathbf{X}^{(ij)}) = 1\}}{\Pr\{\phi(\mathbf{X}) = 1\}},$$

$$I_{\text{RRW}^{II}}(i, j; \mathbf{p}) = \frac{\Pr\{\phi(\mathbf{X}) = 1\}}{\Pr\{\phi(0_i, 0_j, \mathbf{X}^{(ij)}) = 1\}}.$$

$I_{\text{RAW}^{II}}(i, j; \mathbf{p})$ is the ratio of systems reliability (or other terms of system performance) when components i and j always function to the current systems reliability. Thus, $I_{\text{RAW}^{II}}(i, j; \mathbf{p})$ represents the maximum potential to increase systems reliability by improving components i and j to perfect. Instead, $I_{\text{RRW}^{II}}(i, j; \mathbf{p})$ is the ratio of the current systems reliability to the systems reliability when components i and j always fail. Thus, it ranks pairs of components according to their potential to reduce the systems reliability.

The extensions of the RAW and RRW in ranking groups of components are not straightforward. It is not appropriate to extend the RAW by directly setting the reliability of each component in the group to one because this simple substitution can generate nonminimal paths (considering the case of components in parallel), and the RAW of the group would therefore be larger than expected, with the magnitude of the difference being dependent on the number of components appearing in parallel. To overcome this problem, Cheok et al. (1998b) suggested treating all components in the group as one component with its reliability to one, which may only be applied when the group of components forms a module. For extending the RRW, setting the reliability of each component in the group to zero may generate nonminimal cuts (considering the case of components in series) and, thus, is not appropriate either. Cheok et al. (1998a, 1998b) discussed extensions of the RAW and RRW in ranking groups of basic events in the context of fault trees.

References

Armstrong MJ. 1995. Joint reliability-importance of components. *IEEE Transactions on Reliability* **44**, 408–412.
Armstrong MJ. 1997. Reliability-importance and dual failure-mode components. *IEEE Transactions on Reliability* **46**, 212–221.
Boland PJ and Proschan F. 1983. The reliability of *k* out of *n* systems. *Annals of Probability* **11**, 760–764.
Borgonovo E. 2010. The reliability importance of components and prime implicants in coherent and non-coherent systems including total-order interactions. *European Journal of Operational Research* **204**, 485–495.
Borgonovo E and Apostolakis GE. 2001. A new importance measure for risk-informed decision making. *Reliability Engineering and System Safety* **72**, 193–212.
Cheok MC, Parry GW, and Sherry RR. 1998a. Response to "Supplemental viewpoints on the use of importance measures in risk-informed regulatory applications". *Reliability Engineering and System Safety* **60**, 261–261.
Cheok MC, Parry GW, and Sherry RR. 1998b. Use of importance measures in risk-informed regulatory applications. *Reliability Engineering and System Safety* **60**, 213–226.
Gao X, Cui L, and Li J. 2007. Analysis for joint importance of components in a coherent system. *European Journal of Operational Research* **182**, 282–299.
Hagstrom JN. 1990. Redundancy, substitutes and complements in system reliability. Technical report, College of Business Administration, University of Illinois, Chicago.
Hagstrom JN and Mak KT. 1987. System reliability analysis in the presence of dependent component failures. *Probability in the Engineering and Informational Sciences* **1**, 425–440.
Hong JS, Koo HY, and Lie CH. 2000. Computation of joint reliability importance of two gate events in a fault tree. *Reliability Engineering and System Safety* **68**, 1–5.
Hong JS, Koo HY, and Lie CH. 2002. Joint reliability-importance of *k*-out-of-*n* systems. *European Journal of Operational Research* **142**, 539–547.
Hong JS and Lie CH. 1993. Joint reliability-importance of two edges in an undirected network. *IEEE Transactions on Reliability* **42**, 17–33.
Jan S and Chang HW. 2006. Joint reliability importance of *k*-out-of-*n* systems and series-parallel systems. *Proceedings of PDPTA'06*, pp. 395–398.
Kuo W and Zhu X. 2012. Some recent advances on importance measures in reliability. *IEEE Transactions on Reliability* **61**.
Levitin G, Podofillini L, and Zio E. 2003. Generalised importance measures for multi-state elements based on performance level restrictions. *Reliability Engineering and System Safety* **82**, 287–298.
Marseguerra M and Zio E. 2004. Monte Carlo estimation of the differential importance measure: Application to the protection system of a nuclear reactor. *Reliability Engineering and System Safety* **86**, 11–24.
Ryabinin IA. 1994. A suggestion of a new measure of system components importance by means of a boolean difference. *Microelectronics and Reliability* **34**, 603–613.
Vasseur D and Llory M. 1999. International survey on PSA figures of merit. *Reliability Engineering and System Safety* **66**, 261–274.
Zio E and Podofillini L. 2006. Accounting for components interactions in the differential importance measure. *Reliability Engineering and System Safety* **91**, 1163–1174.

8

Importance Measures for Consecutive-*k*-out-of-*n* Systems

Con/k/n systems are of great applications in describing some complex reliability scenarios, as illustrated in Section 2.13. Recall that a *Con/k/n*:F (G) system fails (functions) if and only if at least *k*-consecutive components in the system fail (function). Furthermore, each type of *Con/k/n* system can be classified as either linear or circular, depending on whether the components are arranged in a line or a circle. If there is no specification on the arrangement for a *Con/k/n* system, it could be either linear or circular. In a circular case, assume that the components are labeled clockwise from 1 to n.

For *Con/k/n* systems, Section 8.1 gives the expressions of the B-reliability importance, B-i.i.d. importance, and B-structure importance; Section 8.2 presents the rankings of the importance of components according to these importance measures; and Section 8.3 discusses the structure importance measures.

8.1 Formulas for the B-importance

The following notation is used to simplify descriptions of the reliability of various *Con/k/n* systems and subsystems.

Notation

- $R_C(k, n)$: reliability of a *Cir/Con/k/n*:F system
- $R'_C(k, n)$: reliability of a *Cir/Con/k/n*:G system
- $R_L(k, n)$: reliability of a *Lin/Con/k/n*:F system
- $R'_L(k, n)$: reliability of a *Lin/Con/k/n*:G system
- $R_L(k, (i, j))$: reliability of a *Lin/Con/k/j − i + 1*:F subsystem consisting of components $i, i + 1, \ldots, j$, where $R_L(k, (i, j)) = 1$ if $j - i + 1 < k$. Note that $R_L(k, (1, n)) = R_L(k, n)$.
- $R'_L(k, (i, j))$: reliability of a *Lin/Con/k/j − i + 1*:G subsystem consisting of components $i, i + 1, \ldots, j$, where $R'_L(k, (i, j)) = 0$ if $j - i + 1 < k$. Note that $R'_L(k, (1, n)) = R'_L(k, n)$.

Importance Measures in Reliability, Risk, and Optimization: Principles and Applications, First Edition.
Way Kuo and Xiaoyan Zhu. © 2012 John Wiley & Sons, Ltd. Published 2012 by John Wiley & Sons, Ltd.

- $R_L(k, (1, i - 1), (i + 1, n))$: reliability of a $Lin/Con/k/n - 1{:}F$ subsystem consisting of components $1, 2, \ldots, i - 1, i + 1, \ldots, n$
- $R'_L(k, (1, i - 1), (i + 1, n))$: reliability of a $Lin/Con/k/n - 1{:}G$ subsystem consisting of components $1, 2, \ldots, i - 1, i + 1, \ldots, n$

8.1.1 The B-reliability Importance and B-i.i.d. Importance

Theorems 8.1.1–8.1.4 present the expressions of the B-reliability importance of components in $Lin/Con/k/n{:}F$, $Lin/Con/k/n{:}G$, $Cir/Con/k/n{:}F$, and $Cir/Con/k/n{:}G$ systems, respectively, in terms of the reliability of the whole system and subsystems. For $Lin/Con/k/n{:}F$ systems, Griffith and Govindarajulu (1985) and Papastavridis (1987) developed formula (8.1); the former restricted to the case of i.i.d. components and used the Markov chain approach. For linear and circular $Con/k/n{:}G$ systems, Kuo et al. (1990) developed formulas (8.3) and (8.7), respectively.

Theorem 8.1.1 *The B-reliability importance of component i in a $Lin/Con/k/n{:}F$ system is*

$$I_B(i; \mathbf{p}) = \frac{1}{q_i}[R_L(k, (1, i - 1))R_L(k, (i + 1, n)) - R_L(k, n)]. \tag{8.1}$$

When all components are i.i.d., $R_L(k, (i + 1, n)) = R_L(k, (1, n - i))$.

Proof. By pivotal decomposition (2.10) on component i,

$$R_L(0_i, \mathbf{p}) = \frac{1}{q_i}[R_L(\mathbf{p}) - p_i R_L(1_i, \mathbf{p})],$$

where $R_L(\mathbf{p})$, $R_L(1_i, \mathbf{p})$, and $R_L(0_i, \mathbf{p})$ represent the reliability of a $Lin/Con/k/n{:}F$ system when the reliability of component i is p_i, 1, and 0, respectively. Then, according to Equation (4.11), the B-reliability importance of component i is

$$I_B(i; \mathbf{p}) = R_L(1_i, \mathbf{p}) - \frac{1}{q_i}[R_L(\mathbf{p}) - p_i R_L(1_i, \mathbf{p})] = \frac{1}{q_i}[R_L(1_i, \mathbf{p}) - R_L(\mathbf{p})]. \tag{8.2}$$

By the definition of a $Lin/Con/k/n{:}F$ system,

$$R_L(1_i, \mathbf{p}) = R_L(k, (1, i - 1))R_L(k, (i + 1, n)).$$

Incorporating this into Equation (8.2) completes the proof.

Theorem 8.1.2 *The B-reliability importance of component i in a $Lin/Con/k/n{:}G$ system is*

$$I_B(i; \mathbf{p}) = \frac{1}{p_i}[R'_L(k, n) - R'_L(k, (1, i - 1)) - R'_L(k, (i + 1, n))$$
$$- R'_L(k, (1, i - 1))R'_L(k, (i + 1, n))]. \tag{8.3}$$

When all components are i.i.d., $R'_L(k, (i + 1, n)) = R'_L(k, (1, n - i))$.

Proof. Let $I_B(i; \phi_G, \mathbf{p})$ denote the B-reliability importance of component i in a $Lin/Con/k/n$:G system with component reliability $\mathbf{p} = (p_1, p_2, \ldots, p_n)$. Let $I_B(i; \phi_F, \mathbf{q})$ denote the B-reliability importance of component i in a $Lin/Con/k/n$:F system with component reliability $\mathbf{q} = (q_1, q_2, \ldots, q_n) = \mathbf{1} - \mathbf{p}$. By Theorems 2.13.2 and 4.1.10,

$$I_B(i; \phi_G, \mathbf{p}) = I_B(i; \phi_F, \mathbf{q}). \tag{8.4}$$

Using formula (8.1) and Corollary 2.13.3,

$$I_B(i; \phi_G, \mathbf{p}) = \frac{1}{p_i}[(1 - R'_L(k, (1, i-1)))(1 - R'_L(k, (i+1, n))) - (1 - R'_L(k, n))].$$

After simple manipulations, the expression is clear in the statement.

Theorem 8.1.3   *The B-reliability importance of component i in a $Cir/Con/k/n$:F system is*

$$I_B(i; \mathbf{p}) = \frac{1}{q_i}[R_L(k, (1, i-1), (i+1, n)) - R_C(k, n)]. \tag{8.5}$$

When all components are i.i.d., it reduces to

$$I_B(i; p) = \frac{1}{q}[R_L(k, n-1) - R_C(k, n)];$$

thus, the B-i.i.d. importance of all components is the same.

Proof. Similar to the proof of Theorem 8.1.1, the B-reliability importance of component i in a $Cir/Con/k/n$:F system can be expressed as

$$I_B(i; \mathbf{p}) = \frac{1}{q_i}[R_C(1_i, \mathbf{p}) - R_C(\mathbf{p})], \tag{8.6}$$

where $R_C(\mathbf{p})$ and $R_C(1_i, \mathbf{p})$ represent the reliability of a $Cin/Con/k/n$:F system when the reliability of component i is p_i and 1, respectively. By the definition of a $Cir/Con/k/n$:F system,

$$R_C(1_i, \mathbf{p}) = R_L(k, (1, i-1), (i+1, n)).$$

Incorporating this into Equation (8.6) completes the proof.

Theorem 8.1.4   *The B-reliability importance of component i in a $Cir/Con/k/n$:G system is*

$$I_B(i; \mathbf{p}) = \frac{1}{p_i}[R_C(k, n) - R'_L(k, (1, i-1), (i+1, n))]. \tag{8.7}$$

When all components are i.i.d., it reduces to

$$I_B(i; p) = \frac{1}{p}[R_C(k, n) - R'_L(k, n-1)];$$

thus, the B-i.i.d. importance of all components is the same.

Proof. On the basis of Theorems 2.13.2 and 4.1.10 and Corollary 2.13.3, the result can be proved similarly to the proof of Theorem 8.1.2.

Of course, formulas (8.1), (8.3), (8.5), and (8.7) can be used to compute importance measures that can be expressed in terms of the B-reliability importance. For example, by Equations (4.18) and (8.1), the improvement potential importance of component i in a $Lin/Con/k/n$:F system can be expressed as

$$I_{\text{Bs}}(i; \mathbf{p}) = R_L(k, (1, i-1))R_L(k, (i+1, n)) - R_L(k, n).$$

From formulas (8.1), (8.3), (8.5), and (8.7), the calculation of the B-reliability importance for a $Con/k/n$ system comes down to the reliability evaluation of the $Con/k/\ell$ systems for some values of $\ell \le n$. Thus, formulas that recursively calculate the reliability of the $Con/k/\ell$ systems for $\ell = k, k+1, \ldots, n$ are of good use here. Kuo and Zuo (2003) have given a comprehensive review on and detailed descriptions of the reliability evaluation of $Con/k/n$ systems.

In addition to the aforementioned formulas of the B-i.i.d. importance for $Con/k/n$ systems, the B-i.i.d. importance can be expressed for any coherent system in terms of the numbers of critical cuts and critical paths, respectively, as shown in Equations (6.1) and (6.2), and in terms of sets of cuts and paths, respectively, as shown in Equations (6.3) and (6.4). Furthermore, the B-i.i.d. importance is a polynomial function of component reliability p as in Equations (6.22) and (6.23), whose coefficients are related to the minimal cuts and minimal paths, respectively, and used in defining and analyzing the cut importance in Definition 6.7.1 and the path importance in Definition 6.7.3, respectively.

From formulas (8.1) and (8.3), the following theorem shows the symmetric property of the B-i.i.d. importance of components i and $(n-i+1)$ in a $Lin/Con/k/n$ system.

Theorem 8.1.5 (Symmetry) *For a $Lin/Con/k/n$:F or G system, for $0 < p < 1$,*

$$I_{\text{B}}(i; p) = I_{\text{B}}(n-i+1; p).$$

8.1.2 The B-structure Importance

Recall that the B-structure importance is the B-i.i.d. importance when all components have a same reliability of $1/2$. Lin et al. (1999) obtained a closed form of the B-structure importance for a component in a $Lin/Con/k/n$ system through the Fibonacci sequence with order k, $f_{k,n}$, which is defined as (Miles 1960):

$$f_{k,n} = \begin{cases} 0 & 1 \le n \le k-1 \\ 1 & n = k \\ \sum_{j=n-k}^{n-1} f_{k,j} & n \ge k+1. \end{cases}$$

Theorem 8.1.6 *For a $Lin/Con/k/n$ system with i.i.d. components of a same reliability $p = 1/2$,*

$$R_L(k, n) = 2^{-n} f_{k,n+k+1}, \tag{8.8}$$

$$R'_L(k, n) = 1 - 2^{-n} f_{k,n+k+1}, \tag{8.9}$$

$$I_{\text{B}}(i; \phi_F) = I_{\text{B}}(i; \phi_G) = 2^{-(n-1)}(2f_{k,i+k}f_{k,n-i+k+1} - f_{k,n+k+1}), \tag{8.10}$$

where $I_B(i; \phi_F)$ and $I_B(i; \phi_G)$ denote the B-structure importance of component i in $Lin/Con/k/n$:F and G systems, respectively.

Proof. Let k be arbitrary. For $1 \leq n \leq k - 1$, $R_L(k, n) = 1 = 2^{-n} f_{k,n+k+1}$.

For $n \geq k$, let j denote the last functioning component. Then, the $Lin/Con/k/n$:F system functions if and only if $n - k + 1 \leq j \leq n$ and the first $j-1$ components form a functioning $Lin/Con/k/j - 1$:F system. Hence, for $n = k$,

$$R_L(k, k) = \sum_{j=1}^{k} 2^{-(k-j+1)} R_L(k, j - 1) = \sum_{j=1}^{k} 2^{-(k-j+1)} 2^{-(j-1)} f_{k,j+k}$$

$$= 2^{-k} f_{k,2k+1} = 2^{-n} f_{k,n+k+1}.$$

For $n > k$, formula (8.8) can be obtained by induction.

Then, formula (8.9) can be obtained by duality since $R'_L(k, n) = 1 - R_L(k, n)$ when $p = q = 1/2$.

By formulas (8.1) and (8.8),

$$I_B(i; \phi_F) = 2[R_L(k, i - 1) R_L(k, n - i) - R_L(k, n)]$$

$$= 2[2^{-(i-1)} f_{k,i+k} 2^{-(n-i)} f_{k,n-i+k+1} - 2^{-n} f_{k,n+k+1}]$$

$$= 2^{-(n-1)} (2 f_{k,i+k} f_{k,n-i+k+1} - f_{k,n+k+1}).$$

By Corollary 2.13.3 and Theorem 6.12.1(i), $I_B(i; \phi_F) = I_B(i; \phi_G)$. This completes the proof. □

By formula (8.8), $f_{k,n+k+1}$ can be interpreted as the *number of path vectors in the $Lin/Con/k/n$:F system*; by formula (8.9), $f_{k,n+k+1}$ also represents the *number of cut vectors in the $Lin/Con/k/n$:G system*. This is correct because the $Lin/Con/k/n$:F and G systems are dual of each other. Formula (8.10) provides a physical meaning to the B-structure importance. For example, considering a $Lin/Con/k/n$:F system, let S_i denote the set of component state vectors where its first $i-1$ components form a functioning $Lin/Con/k/i - 1$:F system and its last $n-i$ components form a functioning $Lin/Con/k/n - i$:F system. Since component i can be either functioning or failed, $|S_i| = 2 f_{k,i+k} f_{k,n-i+k+1}$. Let H denote the set of path vectors. Then $|H| = f_{k,n+k+1}$. Note that $H \subseteq S_i$. Furthermore, any component state vector in S_i with component i functioning is also in H. Therefore, $S_i \setminus H$ is the set of failing component state vectors with component i failed such that if component i functions, then the system would also function.

8.2   Patterns of the B-importance for $Lin/Con/k/n$ Systems

There exist certain patterns of the component B-reliability importance (i.e., the relative ordering of the importance values of the individual components) for $Lin/Con/k/n$ systems. These patterns are good indicators for some reliability optimization problems, such as CAP in Chapter 12. The B-reliability importance patterns depend on the relative magnitudes of component reliability. When all components are assumed to be i.i.d. of a same reliability p, the B-i.i.d. importance and its patterns are conditioned on the value of p. Recall from Definition 6.1.4 that component i is said to be more *uniformly* B-i.i.d. important than component j if $I_B(i; p) > I_B(j; p)$ for all $0 < p < 1$, more *half-line* B-i.i.d. important if $I_B(i; p) > I_B(j; p)$

for all $1/2 \le p < 1$, and more B-structure important if $I_B(i; p) > I_B(j; p)$ for $p = 1/2$. This section uses $i >_{B^u} j$ ($i >_{B^h} j$) to denote that component i is more uniformly (half-line) B-i.i.d. important than component j. *Note that any uniform B-i.i.d. importance patterns hold for the half-line B-i.i.d. importance, and any half-line B-i.i.d. importance patterns hold for the B-structure importance.*

For $Lin/Con/k/n$ systems, Subsection 8.2.1 discusses the B-reliability importance patterns; Subsections 8.2.2 and 8.2.3 present the uniform and half-line B-i.i.d. importance patterns, respectively; Subsection 8.2.4 demonstrates the nature of the B-i.i.d. importance patterns; Subsections 8.2.5 and 8.2.6 present the changes of the B-i.i.d. importance patterns with respect to p and n, respectively; finally, Subsection 8.2.7 presents disproved and conjectured B-i.i.d. importance patterns. Throughout this section, *the B-i.i.d. importance $I_B(i; p)$ is discussed only for $i \le \lceil n/2 \rceil$*, unless otherwise specified, because the importance of components $\lceil n/2 \rceil + 1$, $\lceil n/2 \rceil + 2, \ldots, n$ can be compared using the symmetric property in Theorem 8.1.5 (i.e., $I_B(i; p) = I_B(n - i + 1; p)$). For convenience, the upper bound $\lceil n/2 \rceil$ is not explicitly stated each time. Consequently, the default range of n is assumed so that all components involved are the first half components (i.e., components 1 to $\lceil n/2 \rceil$). Moreover, the results in this section are mainly presented for $Lin/Con/k/n$:F systems, while they are also true for the G systems if component reliability vector is changed from $\mathbf{p}$ to $1 - \mathbf{p}$, as shown in Equation (8.4).

8.2.1 The B-reliability Importance

Chadjiconstantinidis and Koutras (1999) compared the B-reliability importance of components in a $Lin/Con/k/n$:F system using the Markov chain approach. The following theorem presents the principle results and extends them to a $Lin/Con/k/n$:G system according to Equation (8.4) of duality between $Lin/Con/k/n$:F and G systems.

Theorem 8.2.1 *The B-reliability importance of nonidentical components in a $Lin/Con/k/n$:F system has the following patterns:*

(i) when $n/2 < k < n$:	*(a)* $I_B(i; \mathbf{p}) < I_B(i + 1; \mathbf{p})$	*for* $1 \le i \le n - k$	*if* $p_i \le p_{i+1}$
	(b) $I_B(i; \mathbf{p}) < (=, >) I_B(i + 1; \mathbf{p})$	*for* $n - k < i < k$	*if* $p_i < (=, >) p_{i+1}$
	(c) $I_B(i; \mathbf{p}) > I_B(i + 1; \mathbf{p})$	*for* $k \le i < n$	*if* $p_i \ge p_{i+1}$
(ii) when $k \le n/2$:	*(a)* $I_B(i; \mathbf{p}) < I_B(i + 1; \mathbf{p})$	*for* $1 \le i < k$	*if* $p_i \le p_{i+1}$
	(b) $I_B(i; \mathbf{p}) > I_B(i + 1; \mathbf{p})$	*for* $n - k < i < n$	*if* $p_i \ge p_{i+1}$
(iii) when $k \le n - 2$:	$I_B(1; \mathbf{p}) < I_B(k + 1; \mathbf{p})$		*if* $p_1 \ge p_{k+1}$
(iv) when $k = n - 1$:	$I_B(1; \mathbf{p}) < (=, >) I_B(k + 1; \mathbf{p})$		*if* $p_1 < (=, >) p_{k+1}$,

where notation "$a < (=, >) b$ if $c < (=, >) d$" means that $a < b$ if $c < d$; $a = b$ if $c = d$; and $a > b$ if $c > d$. These patterns hold for a $Lin/Con/k/n$:G system under reversed conditions of component reliability (e.g., condition $p_i \le p_{i+1}$ should change to $p_i \ge p_{i+1}$, and so on).

8.2.2 The Uniform B-i.i.d. Importance

According to Equation (8.4), the B-i.i.d. importance of component i with a same component reliability p in a $Lin/Con/k/n$:G system is equal to that with a same reliability $1 - p$ in a $Lin/Con/k/n$:F system. Thus, as presented in Lemma 8.2.2, if component i is more uniformly

B-i.i.d. important than component j in a $Lin/Con/k/n$:F system, then it is also true in the $Lin/Con/k/n$:G system, and vice versa (Hwang et al. 2000; Zuo 1993).

Lemma 8.2.2 *If* $i \geq_{B^u} j$ *in a* $Lin/Con/k/n$:F *system, then* $i \geq_{B^u} j$ *in the* $Lin/Con/k/n$:G *system, and vice versa.*

By Lemma 8.2.2, the uniform B-i.i.d. importance patterns are the same for a $Lin/Con/k/n$:F and G system. Theorem 8.2.3 summarizes all existing uniform B-i.i.d. importance patterns for $Lin/Con/k/n$:F and G systems (Zhu et al. 2011).

Theorem 8.2.3 *The uniform B-i.i.d. importance of components (i.e., for all* $0 < p < 1$*) in a* $Lin/Con/k/n$:F *or G system has the following patterns:*

    (i) when $k = 1$ *or* n: $i =_{B^u} j$ *for* $1 \leq i < j \leq n$
    (ii) when $k = 2$:   *(a)* $(2t - 1) <_{B^u} 2t$
                              (b) $(2t + 2) <_{B^u} 2t$
                              (c) $(2t - 1) <_{B^u} (2t + 1)$
    (iii) when $n/2 < k < n$:   *(a)* $i <_{B^u} (i + 1)$ *for* $1 \leq i \leq n - k$
                              (b) $i =_{B^u} (i + 1)$ *for* $i > n - k$
    (iv) when $2 < k \leq n/2$:   *(a)* $i <_{B^u} (i + 1)$ *for* $1 \leq i < k$
                              (b) $1 <_{B^u} i$ *for* $i > 1$
                              (c) $k >_{B^u} i$ *for* $i \geq 1, i \neq k$
    (v) when $n \geq 2k + 3$: $(k + 1) <_{B^u} (k + 2)$
    (vi) when $n \geq 4k + 1$: $(2k + 1) <_{B^u} 2k$
    (vii) when $n = 4k - 1$: $(2k - 1) <_{B^u} 2k$
    (viii) when $n = 6k + 1$: $(3k + 1) <_{B^u} 3k.$

In part (ii), $t \geq 1$ *is an integer.*

Theorem 8.2.3 is accompanied for illustration by Figure 8.1, in which the horizontal axis stands for the component index i and the vertical axis stands for the corresponding B-i.i.d. importance. Figure 8.1 shows the B-structure importance (i.e., $p = 1/2$) rankings for $Lin/Con/k/14$:F (the same for G) systems with $k = 2, 3, 7$, and 12, representing the different relative magnitudes of k and n in Theorem 8.2.3. It also demonstrates the symmetric property of the B-i.i.d. importance around the middle (two) component(s) in Theorem 8.1.5. Note that

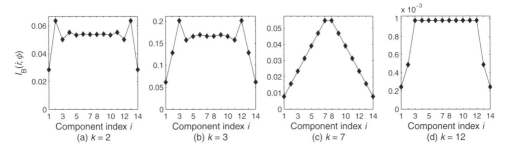

Figure 8.1 The B-structure importance for $Lin/Con/k/14$:F systems with $k = 2, 3, 7$, and 12

in all other figures in this section, the horizontal and vertical axes have the same meanings as in Figure 8.1 unless otherwise specified, and only components 1 to $\lceil n/2 \rceil$ are shown, to make the presentation clear and to save space.

Part (i) in Theorem 8.2.3 indicates that when $k = 1$ in an F system or $k = n$ in a G system (i.e., a series system) or when $k = n$ in an F system or $k = 1$ in a G system (i.e., a parallel system), all components are equally important according to the uniform B-i.i.d. importance. This is evident because, for the series system, the failure of any component is equally likely to cause system failure, and for the parallel system, the survival of any one of n components guarantees the functioning of the system.

Part (ii) completely identifies the uniform B-i.i.d. importance patterns for a $Lin/Con/2/n$ system, which were discovered by Zuo and Kuo (1988, 1990). For the $Lin/Con/2/n$ system (e.g., $Lin/Con/2/14$:F system in Figure 8.1(a)), the B-i.i.d. importance value alternatively rises and falls, and the relative difference of the B-i.i.d. importance values of two consecutive components decreases as the component index i increases from 1 to $\lceil n/2 \rceil$.

Part (iii) was presented by Zuo (1993), completely identifying the patterns of the uniform B-i.i.d. importance when $k > n/2$. When $k > n/2$, as shown in Figure 8.1(d), the B-i.i.d. importance value increases from components 1 to $(n-k+1)$ and decreases from components k to n because starting from component 1 until component $(n-k+1)$, each component gradually has more adjacent components, resulting in a heavier impact on systems reliability than its preceding components. Moreover, the components between $(n-k+1)$ and k maintain the same B-i.i.d. importance value:

$$I_B(i; p) = \frac{1}{q}[1 - R_L(k, n)] \quad \text{for F system and}$$

$$I_B(i; p) = \frac{1}{p}R'_L(k, n) \quad \text{for G system}$$

because the survival (failure) of any one of these components guarantees the functioning (failure) of the F (G) system. Furthermore, these components are more uniformly B-i.i.d. important than any others.

When n is even and $k = n/2$ (or n is odd and $k = (n + 1)/2$) (e.g., $Lin/Con/7/14$:F system in Figure 8.1(c)), the middle two components (or the middle component) have (has) the largest B-i.i.d. importance value, and the B-i.i.d. importance value increases from components 1 to k and decreases from components $k+1$ (or k) to n.

When $3 \leq k < n/2$, parts (iv)–(viii) identify the partial ordering of the uniform B-i.i.d. importance. Specifically, part (iv) indicates that the uniform B-i.i.d. importance monotonically increases from components 1 to k and that the B-i.i.d. importance values of components 1 and k, respectively, are the lower and upper bounds on the B-i.i.d. importance of all other components (e.g., $Lin/Con/3/14$:F system in Figure 8.1(b)). Part (iv)a was first presented by Zuo (1993) and was extended to parts (iv)b and (iv)c by Chang et al. (1999). Part (v) was proved by Chadjiconstantinidis and Koutras (1999) using the Markov chain approach. Parts (v)–(vii) were presented by Chang et al. (1999) and later corrected by Chang et al. (2000). Parts (vii) and (viii) were proved by Chang et al. (2002). Zakaria et al. (1992) showed that component k is more B-i.i.d. important than component $k + 1$ under the condition of $2k + 1 \leq n \leq 3k + 1$ or $n \geq 3k + 1$ and $(1 - q^k)^k \geq p$ (for an F system). However, Theorem 8.2.3(iv)c shows that component k is always more B-i.i.d. important than component $k + 1$.

To summarize Theorem 8.2.3, for the uniform B-i.i.d. importance, components k and $(n-k+1)$ have the largest B-i.i.d. importance, components 1 and n have the smallest B-i.i.d. importance, and the B-i.i.d. importance values of components 1 to k are nondecreasing. The patterns of uniform B-i.i.d. importance of components for a $Lin/Con/k/n$ system with $k = 1, 2$, or $k \geq n/2$ are completely identified in Theorem 8.2.3. However, the patterns of uniform B-i.i.d. importance of components in a $Lin/Con/k/n$ system with $3 \leq k < n/2$ have not yet been completely identified.

8.2.3 The Half-line B-i.i.d. Importance

Except for the uniform B-i.i.d. importance patterns presented in Theorem 8.2.3, Theorem 8.2.4 presents all additional patterns for the half-line B-i.i.d. importance for the $Lin/Con/k/n$:F system with $3 \leq k < n/2$. The patterns in Theorem 8.2.4 were proved by Chang et al. (2002). Some of them were first presented by Lin et al. (1999) for the B-structure importance and were then proved by Chang et al. (2002) for the half-line B-i.i.d. importance.

Theorem 8.2.4   *The half-line B-i.i.d. importance of components (i.e., for all $1/2 \leq p < 1$) in a $Lin/Con/k/n$:F system has the following patterns:*

(i) $(k-1) <_{B^h} (k+1)$
(ii) $i <_{B^h} (i+1)$ *for $k < i < 2k$*
(iii) $(k+1) <_{B^h} i$ *for $i > k+1$*
(iv) $2k >_{B^h} i$ *for $i > 2k$*
(v) $(2k+1) <_{B^h} (2k+2)$
(vi) $(3k+1) <_{B^h} 3k$.

These patterns also hold for a $Lin/Con/k/n$:G system when $0 < p \leq 1/2$ because of Equation (8.4).

With the restriction of $p \geq 1/2$, the patterns in Theorem 8.2.3(iv) are extended to Theorem 8.2.4(ii)–(iv). Note that Theorem 8.2.3(iv) specifies the uniform B-i.i.d. importance patterns of the first k components, and Theorem 8.2.4(ii)–(iv) specify the same half-line B-i.i.d. importance patterns of the second k components (i.e., components $k+1$ to $2k$), for example, in Figure 8.1(b). However, even with the addition of patterns in Theorem 8.2.4, the half-line B-i.i.d. importance ordering of the $Lin/Con/k/n$:F system with $3 \leq k < n/2$ is not completely determined. If further restricted to the B-structure importance of $p = 1/2$, no additional patterns are obtained. As shown in Subsection 8.2.7, the half-line B-i.i.d. importance patterns in parts (i), (ii), (iv), and (vi) in Theorem 8.2.4 cannot be generalized to the uniform case, and the ones in parts (iii) and (v) are conjectured for the uniform case.

8.2.4 The Nature of the B-i.i.d. Importance Patterns

As discussed in Subsections 8.2.2 and 8.2.3, for the $Lin/Con/k/n$:F system with $3 \leq k < n/2$, the uniform B-i.i.d. and half-line B-i.i.d. importance orderings cannot been completely identified. Research (Chang et al. 1999, 2000; Chang and Hwang 20002; Chang et al. 2002; Zuo 1993) initially done toward completing the orderings has instead resulted in some disproofs of the intuitively expected patterns by chance and has left some other patterns open to conjecture. Zhu et al. (2011) explored the nature of B-i.i.d. importance patterns theoretically and numerically. The rest of this section is based on their work.

The nonexistence of the fixed-length B-i.i.d. importance ordering

Zhu et al. (2011) showed that an ever-expected fixed-length B-i.i.d. importance ordering does not exist in general for the $Lin/Con/k/n$ system with $3 \leq k < n/2$, although it may hold for some special cases determined by the values of k, n, and p (e.g., $Lin/Con/3/14$:F system in Figure 8.1(b) with $p = 1/2$). This fixed-length B-i.i.d. importance ordering embodies the B-i.i.d. importance patterns in Theorems 8.2.3 and 8.2.4 and can be specified by the following four patterns (here $1 \leq t \leq \lceil n/(2k) \rceil$):

(i) $I_B(i; p) < I_B(i + 1; p)$ for $(t - 1)k < i < tk$
(ii) $I_B((t - 1)k + 1; p) < I_B(i; p)$ for $i > (t - 1)k + 1$
(iii) $I_B(tk; p) > I_B(i; p)$ for $i > tk$
(iv) $I_B(tk - 1; p) < I_B(tk + 1; p)$.

As its name implies, the fixed-length B-i.i.d. importance ordering is based on the k-interval division (see Definition 8.2.5), and the patterns are specified within a k-interval and among k-intervals. Mimicking Theorem 8.2.3(iv) and Theorem 8.2.4(ii)–(iv), patterns (i)–(iii) describe that the B-i.i.d. importance values of the components within a k-interval strictly increase and that the B-i.i.d. importance value of the first (last) component in a k-interval is the lower (upper) bound on the B-i.i.d. importance values of its succeeding components. Mimicking Theorem 8.2.4(i), pattern (iv) shows that the B-i.i.d. importance value of component $tk+1$ is between those of components $tk-1$ and tk, which describes the relations of the B-i.i.d. importance across two consecutive k-intervals. The patterns look like a water wave, and the wave form repeats itself at intervals of *fixed* length, k.

Definition 8.2.5 *For a $Lin/Con/k/n$ system with $n > 2k$, a k-interval (or simply, interval) is a set of k consecutive components starting from component $(t-1)k+1$ for each $t = 1, 2, \ldots, \lceil n/(2k) \rceil$, except that the last interval ($t = \lceil n/(2k) \rceil$) may include less than k components: $(t-1)k+1, (t-1)k+2, \ldots, \lceil n/2 \rceil$.*

The nonexistence of this fixed-length ordering can be illustrated by the example in Figure 8.2, which shows the B-i.i.d. importance of components 1 to 64 in $Lin/Con/8/128$:F system ($n > 2k$) with $p = 0.2$. In Figure 8.2, only relative orderings of the B-i.i.d. importance values are considered; the absolute values are omitted ($1 \times 10^{-4} < I_B(i; 0.2) < 10 \times 10^{-4}$ for $1 \leq i \leq 64$), and the B-i.i.d. importance values of the components in the eight k-intervals are drawn separately in order to further amplify the differences of the B-i.i.d. importance values. As shown in Figure 8.2, neither the monotonic increasing pattern (as in the first k-interval and as expected) nor the more general increasing-then-decreasing pattern (as in the second k-interval) universally exists. Zhu et al. (2011) observed this phenomenon for various values of $0 < p < 1$. Thus, when $3 \leq k < n/2$, the effort to find similar B-i.i.d. importance patterns within each of the k-intervals or among k-intervals is futile and cannot result in general B-i.i.d. importance patterns.

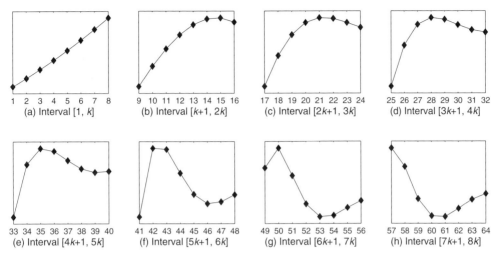

(a) Interval [1, k]   (b) Interval [k+1, 2k]   (c) Interval [2k+1, 3k]   (d) Interval [3k+1, 4k]

(e) Interval [4k+1, 5k]   (f) Interval [5k+1, 6k]   (g) Interval [6k+1, 7k]   (h) Interval [7k+1, 8k]

Figure 8.2 k-intervals for $Lin/Con/8/128$:F system with $p = 0.2$

Segments and their lengths and peaks

Since the fixed-length B-i.i.d. importance ordering based on the k-interval division does not generally hold, Zhu et al. (2011) proposed a method for dividing the first half components (i.e., components 1 to $\lceil n/2 \rceil$) of a $Lin/Con/k/n$ system with $n > 2k$ by defining a segment and its length and peak as follows.

Definition 8.2.6   *For a $Lin/Con/k/n$ system with $n > 2k$, a segment is a maximal set of consecutive first half components whose B-i.i.d. importance values first increase and then decrease, satisfying that the union of all segments covers the first half components and two consecutive segments contain exactly one component in common. The length of a segment is the number of components in the segment minus 1 since one common component is shared by two consecutive segments. The peak of a segment is the index of the component corresponding to the maximum B-i.i.d. importance value in the segment.*

Then, according to Theorem 8.2.3(iv) and (v), in any $Lin/Con/k/n$ system with $n > 2k$, the first segment always contains components 1 to k and $k+1$, and both its length and its peak are k. Figure 8.3 shows the nine segments for $Lin/Con/8/128$:F system, which is also associated with Figure 8.2, demonstrating that the lengths of segments can be different. Starting from k for the first segment, the length of segments may decrease in subsequent segments, increase again, and so on. These segments look like a water wave, but the wave form repeats at intervals of *different* lengths. Zhu et al. (2011) conjectured that the length of segments is never greater than k, the case for all of their computational tests. Meanwhile, the peaks can appear at different relative positions in the segments. In Figure 8.3, for the first two segments, the peaks appear at the back end of each segment; for the other segments except the last one, the peaks appear at the middle of each segment. Although no exact formulas can determine the length and peak of each segment (it may be difficult or impossible to determine), at least the increasing-then-decreasing

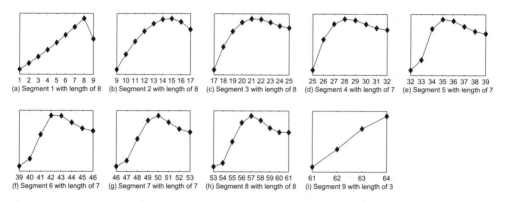

Figure 8.3   Segments for $Lin/Con/8/128$:F system with $p = 0.2$

B-i.i.d. importance pattern is consistent in all segments. Note that the last segment may be monotonically increasing because it is truncated at the middle component of the system.

8.2.5   Patterns with Respect to p

Theorem 8.2.7 (Zhu et al. 2011) gives the smallest possible value of the peak of the second segment for different p values, which implies the lower bound on the length of the second segment. Appendix A.1 provides the proof of Theorem 8.2.7.

Theorem 8.2.7   *For a $Lin/Con/k/n$:F system with $k > 2$, if $p \geq 1/(k - s)$, where $0 \leq s \leq k - 3$, then*

(i) $I_B(k + 1; p) > I_B(s + 1; p)$ *for $n \geq 2k + 1$ and*
(ii) $I_B(i; p) < I_B(i + 1; p)$ *for $k + 1 \leq i \leq k + s + 1$ and $n \geq 4k + s + 2$.*

As shown in Theorem 8.2.7(ii), $I_B(k + 1; p) < I_B(k + 2; p) < \cdots < I_B(k + s + 1; p)$ when $p \geq 1/(k - s)$ with $0 \leq s \leq k - 3$. Thus, the peak of the second segment is no less than $(k + s + 1)$, and, consequently, the length of the second segment is no less than $s + 1$. Thus, as $p \geq 1/(k - s)$ increases, that is, s increases, the lower bounds on the peak and the length increase.

Zhu et al. (2011) conducted the numerical studies on the relations of the length and the peak of segments to p. As an example, Figure 8.4 presents the B-i.i.d. importance for $Lin/Con/5/29$:F system for $p = 0.1, 0.3, 0.5, 0.7$, and 0.9, in which the B-i.i.d. importance of the first k components are not presented since the patterns of these components are known and fixed. The following main results were obtained.

Remark 8.2.8

(i) Given a $Lin/Con/k/n$:F system, as p becomes smaller, it is more likely to have short length segments and the length of the segments decreases more quickly. With a low p, the length of the segments may become less than k starting from the third segment, but with a high p, the length of the segments may maintain k for the first few segments.

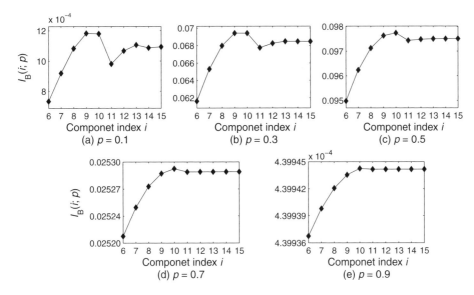

Figure 8.4   The B-i.i.d. importance for $Lin/Con/5/29$:F systems with $p = 0.1, 0.3, 0.5, 0.7$, and 0.9

(ii) As p increases, the patterns gradually change. With a low p (e.g., 0.1 in Figure 8.4(a)), each segment except the first segment has both a clear increasing and decreasing part, and its peak usually appears before the second component from the end of each segment. With a high p, the increasing part is much longer than the decreasing part in each segment, and the peak usually appears at the second component from the end of each segment. When p is very high (e.g., 0.9 in Figure 8.4(e)), the B-i.i.d. importance values are almost the same starting from the second segment, which is proved in Theorem 8.2.9.

(iii) When $p \geq 1/2$, the length of the second segment is always k, and the peak of the second segment is $2k$ (Theorem 8.2.4(ii), (iv), and (v)). When $p < 1/2$, it is observed from computational tests (Zhu et al. 2011) that the B-i.i.d. importance value in the second segment first increases and then decreases until component $2k+1$. Additionally, the peak of the second segment can appear earlier than $2k$, but the length of the second segment is always k. This leads to conjecture (8.31), that is, $(2k + 1) <_{\mathrm{B}^u} (2k + 2)$ (see Subsection 8.2.7).

The aforementioned discoveries can explain why some half-line B-i.i.d. importance patterns in Theorem 8.2.4 hold for $p \geq 1/2$, but cannot be extended to $p < 1/2$. Specifically, the B-i.i.d. importance patterns, in particular the peaks and lengths of segments, may vary for different p values. For example, if $p \geq 1/2$ (e.g., Figure 8.4(c)–(e)), the peak of the second segment is $2k = 10$, obeying Theorem 8.2.4(ii); however, when p becomes small (e.g., Figure 8.4(a)), the peak of the second segment is 9, less than $2k$, and the decreasing part in the second segment starts from components 9 until 11, which violates Theorem 8.2.4(ii).

8.2.6   Patterns with Respect to *n*

For the $Lin/Con/k/n$:F system with $3 \leq k < n/2$, Theorem 8.2.3 presents the uniform B-i.i.d. importance ordering for the first $k+2$ components, and Theorem 8.2.4 presents the half-line

B-i.i.d. importance ordering for the first $2k+2$ components. Because the lengths of the segments are different, the B-i.i.d. importance patterns that exist for the first or the first few k-intervals cannot generally be extended further. Fortunately, as n increases, only the B-i.i.d. importance values of components in the first few segments make a difference; the components in the middle of the n-component line have almost the same B-i.i.d. importance values. This statement is supported by the following three theorems, whose proofs can be found in Zhu et al. (2011). To investigate the effect of n on the B-i.i.d. importance patterns, this subsection uses $I_n(i; p)$ to denote the B-i.i.d. importance of component i with a same reliability p in a $Lin/Con/k/n$:F system.

Theorem 8.2.9   *For a $Lin/Con/k/n$:F system with $n > 2k$, the difference of $I_n(k; p) - I_n(k + 1; p)$ decreases as n increases and approaches 0 as p approaches 1.*

According to Theorem 8.2.4(iii) and Theorem 8.2.3(iv)c, in the half-line case, $I_n(k + 1; p)$ and $I_n(k; p)$ are the lower and upper bounds on the B-i.i.d. importance of components $k + 1$, $k + 2, \ldots, \lceil n/2 \rceil$, respectively (note that $I_n(2k; p)$ is a tighter upper bound). In addition, if conjecture (8.29) can be proved, then these lower and upper bounds can be extended to the uniform case. The significance of Theorem 8.2.9 is that, for the half-line case, when n is large enough or p is close to 1, there is no significant difference among the B-i.i.d. importance values of components $k + 2, k + 3, \ldots, n-k-1$, and thus, the B-i.i.d. importance patterns of these components may not deserve further investigation. This conclusion is consistent with the numerical observation in Remark 8.2.8(ii) that the B-i.i.d. importance values are almost the same starting from the second segment when p is very high.

For the half-line B-i.i.d. importance, similar results for other special pairs of components are presented in Theorem 8.2.10. Theorem 8.2.10 shows that when n is large, $I_n(i; p) \simeq I_n(i - k; p)$ for $p \geq 1/2$ and $i = k+1, k+2, \ldots, 2k+1$, and $3k$.

Theorem 8.2.10   *For a $Lin/Con/k/n$:F system with $k > 2$ and $1/2 \leq p < 1$, the following statements hold.*

  (i) *The difference of $I_n(i; p) - I_n(i - k; p)$ decreases as n increases for $i = k + 1$, $k + 2, \ldots, 2k - 1$ and $n \geq 4k - 1$.*
  (ii) *The difference of $I_n(k; p) - I_n(2k; p)$ decreases as n increases for $n \geq 4k$.*
  (iii) *The difference of $I_n(2k + 1; p) - I_n(k + 1; p)$ decreases as n increases for $n \geq 4k + 2$.*
  (iv) *The difference of $I_n(2k; p) - I_n(3k; p)$ decreases as n increases for $n \geq 6k$.*

Considering a $Lin/Con/k/n$:F system and a $Lin/Con/k/n + 1$:F system, define

$$\Delta I_n(i; p) = I_n(i; p) - I_{n+1}(i; p)$$

for $i = 1, 2, \ldots, n$ and $0 < p < 1$. Theorem 8.2.11 presents the main results related to $\Delta I_n(i; p)$, which demonstrate how the B-i.i.d. importance of a single component changes as n increases.

Theorem 8.2.11   *For a $Lin/Con/k/n$:F system and a $Lin/Con/k/n + 1$:F system with $n > 2k$ and $0 < p < 1$, the following statements hold.*

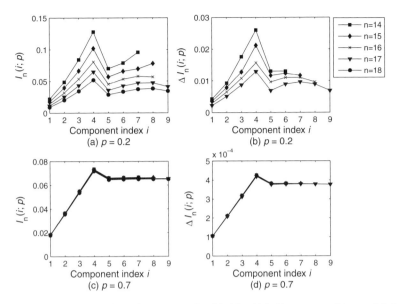

Figure 8.5   The changes of B-i.i.d. importance for $Lin/Con/4/n$:F systems with $n = 14, 15, 16, 17$, and 18

(i) $\Delta I_n(i; p) > 0$ (i.e., $I_n(i; p)$ decreases as n increases), and $\Delta I_n(i; p)$ approaches 0 (i.e., $I_{n+1}(i; p) \simeq I_n(i; p)$) as p approaches 1, for $i = 1, 2, \ldots, n - k$.

(ii) $\Delta I_n(i; p)$ decreases (i.e., the decreasing rate of $I_n(i; p)$ slows down) as n increases for $i = 1, 2, \ldots, n - 2k$.

To illustrate Theorem 8.2.11, Figure 8.5 demonstrates the changes of the B-i.i.d. importance for the $Lin/Con/4/n$:F systems for $n = 14, 15, 16, 17$, and 18, in which (a) and (b) are for the case of $p = 0.2$ and (c) and (d) are for $p = 0.7$. Figure 8.5(a) clearly verifies that $I_n(i; p)$ decreases as n increases for $i = 1, 2, \ldots, n - k$, as shown in Theorem 8.2.11(i). Furthermore, Figure 8.5(b) indicates that $\Delta I_n(i; p)$ decreases as n increases for $i = 1, 2, \ldots, n - 2k$, that is, the decreasing rate of $I_n(i; p)$ slows down as n increases, as shown in Theorem 8.2.11(ii). When $p = 0.7$, a relatively high component reliability, Figure 8.5(c) shows that $I_{n+1}(i; p) \simeq I_n(i; p)$, and Figure 8.5(d) shows that $\Delta I_n(i; p)$ is almost zero ($< 5 \times 10^{-4}$). In other words, when p is high, there is no significant decrease of the B-i.i.d. importance of a given component in the $Lin/Con/4/n$:F systems as n changes from 14 to 18, as proved in Theorem 8.2.11(i).

8.2.7 Disproved Patterns and Conjectures

On the basis of computation of $I_B(i; \phi)$ and $I_B(i; p)$ for the $Lin/Con/k/n$:F systems for $3 \leq k \leq 8$, $1 \leq n \leq 50$, and also $n = \{60, 70, \ldots, 200\}$, Chang et al. (2002) disproved some patterns by counterexamples and also made some conjectures. Along this line, Zhu et al. (2011) conducted more comprehensive computational tests so that more patterns could be disproved or conjectured. Their tests range in size of $3 \leq k \leq 20$, $2k + 1 \leq n \leq 10k + 1$, and

$p = \{0.01, 0.02, \ldots, 0.99\}$ and are computed using symbolic operations in Matlab; thus, they are sufficiently accurate even when p is small. All patterns that are disproved or conjectured are listed in the following text. Note that the disproving of a B-structure importance pattern implies that neither the corresponding half-line B-i.i.d. nor uniform B-i.i.d. importance pattern holds.

Disproved patterns

For a $Lin/Con/k/n$:F or G system, the following patterns are disproved by their counterexamples (Zhu et al. 2011).

$$(n - 2k) <_{B^u} (k + 1) \qquad \text{for } 2k + 2 \le n < 3k \tag{8.11}$$

$$I_B(tk + 1; \phi) < I_B(tk; \phi) \qquad \text{for } t \ge 4 \tag{8.12}$$

$$(tk + 1) <_{B^u} (tk + 2) \qquad \text{for } t \ge 4 \tag{8.13}$$

$$I_B(tk - 1; \phi) < I_B(tk; \phi) \qquad \text{for } n = 2tk - 1 \text{ and } t \ge 3 \tag{8.14}$$

$$I_B((t + 1)k; \phi) < I_B(tk; \phi) \qquad \text{for } t \ge 3 \tag{8.15}$$

$$I_B(i; \phi) < I_B(i + 1; \phi) \qquad \text{for } 3k < i < 4k - 1 \tag{8.16}$$

$$i <_{B^u} (i + 1) \qquad \text{for } 2k < i < 3k - 1 \tag{8.17}$$

$$(tk + 1) <_{B^u} i \qquad \text{for } i > tk + 1 \text{ and } t \ge 3 \tag{8.18}$$

$$(k - 1) <_{B^u} (k + 1) \tag{8.19}$$

$$i <_{B^u} (i + 1) \qquad \text{for } k < i < 2k \tag{8.20}$$

$$(3k + 1) <_{B^u} 3k \tag{8.21}$$

$$i <_{B^u} 2k \qquad \text{for } i > 2k \tag{8.22}$$

$$3k <_{B^u} 2k \tag{8.23}$$

$$(tk + 1) <_{B^u} tk \qquad \text{for } n = 2tk + 1 \text{ and } t \ge 4. \tag{8.24}$$

Chadjiconstantinidis and Koutras (1999) claimed that they proved (8.11). However, their proof is incorrect and (8.11) was disproved by the following counterexample in Zhu et al. (2011).

Example 8.2.12 For $k = 4$, $n = 11$, and $p = 0.01, 0.02, \ldots, 0.34$, $I_B(n - 2k; p) > I_B(k + 1; p)$. For example, when $p = 0.01$, $I_B(n - 2k; p) = 0.0008673900 > I_B(k + 1; p) = 0.0006810248$.

Zuo (1993) and Chang et al. (1999) claimed, respectively, that they proved (8.12). Chang et al. (1999) claimed that they proved (8.13) and (8.14). However, Chang et al. (2000) pointed out that (8.12) is true only when $t = 1$ and 2, (8.13) is true only when $t = 1$, and (8.14) is true only when $t = 1$ and 2. Theorem 8.2.4(vi) shows that (8.12) is also true for $t = 3$. Chang et al. (2002) and Zhu et al. (2011) disproved (8.12), (8.13), and (8.14) using counterexamples

8.2.13, 8.2.14, and 8.2.15, respectively. Note that (8.13) is disproved only for $t \geq 4$ and is conjectured for $t = 2$ and 3, as later shown in conjectures (8.31) and (8.32).

Example 8.2.13   For $k = 3$, $n = 50$, and $t = 8$, $I_B(8k; \phi) = 0.00817316129022 < I_B(8k + 1; \phi) = 0.00817316129208$. For $k = 5$, $n = 70$, and $t = 6$, $I_B(6k; \phi) = 0.04822129731578 < I_B(6k + 1; \phi) = 0.04822129731696$.

Example 8.2.14   (a) For $k = 6$, $n = 100$, $p = 0.2$, and $t = 7$, $I_B(7k + 1; p) = 0.000042881154 > I_B(7k + 2; p) = 0.000042881153$.
   (b) For $k = 18$, $n = 145$, $p = 0.06, 0.07, 0.08$, and $t = 4$, $I_B(4k + 1; p) > I_B(4k + 2; p)$. For example, when $p = 0.06$, $I_B(4k + 1; p) = 0.0017098709 > I_B(4k + 2; p) = 0.0017098693$.

Example 8.2.15   For $k = 3$, $n = 35$, and $t = 6$, $I_B(6k - 1; \phi) = 0.02871431695530 > I_B(6k; \phi) = 0.02871431043604$. For $k = 4$, $n = 31$, and $t = 4$, $I_B(4k - 1; \phi) = 0.09231528453529 > I_B(4k; \phi) = 0.09231519699097$.

Patterns (8.15)–(8.18) were disproved by the counterexamples presented by Chang et al. (2002). Pattern (8.15) is valid for $t = 1$ and 2, but it does not hold generally for $t \geq 3$ as shown in Example 8.2.16. Pattern (8.16) was disproved by counterexample 8.2.17. Note that conjecture (8.26) is made for the B-structure importance but the corresponding pattern (8.17) for the uniform B-i.i.d. importance does not hold as shown in Example 8.2.18. Pattern (8.18) was disproved by counterexample 8.2.19 and is conjectured for $t = 1$ and 2, as later shown in conjectures (8.29) and (8.28).

Example 8.2.16   For $k = 3$, $n = 47$, and $t = 7$, $I_B(7k; \phi) = 0.01050828216397 < I_B(8k; \phi) = 0.01050828216627$.   For $k = 5$, $n = 60$,   and $t = 5$, $I_B(5k; \phi) = 0.05725561176026 < I_B(6k; \phi) = 0.05725561183744$.

Example 8.2.17   For $k = 6$ and $n = 48$, $I_B(3k + 4; \phi) = 0.06068431501809 > I_B(3k + 5; \phi) = 0.06068431266068$.

Example 8.2.18   For $k = 5$, $n = 29$, and $p = 0.3$, $I_B(2k + 3; p) = 0.06847519 > I_B(2k + 4; p) = 0.06846597$.

Example 8.2.19   For $k = 6$, $n = 100$, and $p = 0.2$, $I_B(6k + 1; p) = 0.000042881164 > I_B(7k + 1; p) = 0.000042881154 > I_B(8k + 1; p) = 0.000042881149$.

Although the half-line B-i.i.d. importance patterns in Theorem 8.2.4(i), (ii), and (vi) have been proved, the corresponding uniform B-i.i.d. importance patterns (8.19), (8.20), and (8.21) do not hold. Counterexamples 8.2.20 and 8.2.21 in Chang et al. (2002) disprove (8.19) and (8.20), respectively. Counterexample 8.2.22 in Zhu et al. (2011) disproves (8.21).

Example 8.2.20   For $k = 3$, $n = 7$, and $p = 0.3$, $I_B(k - 1; p) = 0.24357900 > I_B(k + 1; p) = 0.21564900$.

Example 8.2.21 For $k = 4$, $n = 16$, and $p = 0.3$, $I_B(2k; p) = 0.12337090 < I_B(2k - 1; p) = 0.12376224$.

Example 8.2.22 For $k = 14$, $n = 92$, and $p = 0.03, 0.04, \ldots, 0.1$, $I_B(3k + 1; p) > I_B(3k; p)$. For example, when $p = 0.03$, $I_B(3k + 1; p) = 0.0000493310 > I_B(3k; p) = 0.0000492933$.

Patterns (8.22) and (8.23) were conjectured by Chang et al. (2002). Chang and Hwang (2002) disproved (8.22) using the rare-event importance as later shown in Example 8.3.6. Example 8.2.23 is a direct counterexample to (8.22). In fact, (8.23) is a special case of (8.22) and can be disproved by counterexample 8.2.24 in Zhu et al. (2011).

Example 8.2.23 For $k = 6$, $n = 30$, and $p = 0.01$, $I_B(2k; p) = 0.0000006067 < I_B(15; p) = 0.0000006341$.

Example 8.2.24 For $k = 15$, $n = 96$, and $p = 0.04, 0.05, 0.06, 0.07$, $I_B(2k; p) < I_B(3k; p)$. For example, when $p = 0.04$, $I_B(2k; p) = 0.0004558502 < I_B(3k; p) = 0.0004559212$.

Motivated by Theorem 8.2.3(iv)c, (vi), and (viii), where $(tk + 1) <_{B^u} tk$ for $t = 1, 2$, and 3, respectively, Chang et al. (2002) conjectured (8.24) for all $t \geq 4$. However, the following counterexample in Zhu et al. (2011) disproves it.

Example 8.2.25 For $k = 4$, $n = 57$, $p = 0.08, 0.09, \ldots, 0.46$, and $t = 7$, $I_B(7k; p) < I_B(7k + 1; p)$. For example, when $p = 0.30$, $I_B(7k; p) = 0.0006396002 < I_B(7k + 1; p) = 0.0006396003$.

The findings in Subsection 8.2.4 and knowledge of the relationship of the B-i.i.d. importance patterns to the values of p and n in Subsections 8.2.5 and 8.2.6 can facilitate the process of seeking counterexamples. Take the disproved patterns (8.21), (8.22), and (8.23) as examples, and note that the length of segments may decrease, especially when component reliability p is small. If the length of the third segment decreases to $k-1$ or smaller, then both components $3k$ and $3k+1$ are in the increasing part of the fourth segment, resulting in $I_B(3k + 1; p) > I_B(3k; p)$. Following this analysis, a counterexample to (8.21) is found as in Example 8.2.22 with a small value of p. As for (8.22) and (8.23), if p is small enough, the peak of the second segment may appear earlier than component $2k$, and then the B-i.i.d. importance of component $2k$ could be less than the B-i.i.d. importance of the peak and its surrounding components in the third segment. Thus, the counterexamples to (8.22) and (8.23) are found by using small p values as in Examples 8.2.23 and 8.2.24, respectively.

Conjectures

Chang et al. (2002) made conjectures (8.25)–(8.30), and Zhu et al. (2011) made conjectures (8.31) and (8.32). To the best of our knowledge, one neither proves nor finds counterexamples to disprove them.

Conjecture 8.2.26 *For a $Lin/Con/k/n{:}F$ or G system, the following conjectures are made.*

$$I_B(2k+1; \phi) < I_B(2k-1; \phi) \qquad \text{for } k \geq 4 \tag{8.25}$$

$$I_B(i; \phi) < I_B(i+1; \phi) \qquad \text{for } 2k < i < 3k-1 \tag{8.26}$$

$$I_B(tk+1; \phi) < I_B(i; \phi) \qquad \text{for } i > tk+1 \text{ and } t \geq 2 \tag{8.27}$$

$$(2k+1) <_{B^u} i \qquad \text{for } i > 2k+1 \tag{8.28}$$

$$(k+1) <_{B^u} i \qquad \text{for } i > k+1 \tag{8.29}$$

$$(k+1) <_{B^u} (2k+1) \tag{8.30}$$

$$(2k+1) <_{B^u} (2k+2) \tag{8.31}$$

$$(3k+1) <_{B^u} (3k+2). \tag{8.32}$$

Chang et al. (2002) made conjecture (8.25) for $k \geq 4$ because no counterexample was found for $4 \leq k \leq 8$ in their tests. Meanwhile, they showed that when $k = 3$, all relative magnitudes of $I_B(2k-1; \phi)$ and $I_B(2k+1; \phi)$ are possible. Pattern (8.26) was first presented by Lin et al. (1999). Chang et al. (2002) pointed out the insufficiency in the proof of (8.26) but were not able to prove or disprove (8.26). Conjecture (8.26) can neither be extended to the next k consecutive components, as in disproved pattern (8.16), nor to the uniform case, as in disproved pattern (8.17). Note that conjecture (8.27) cannot be extended to the uniform case as in disproved pattern (8.18). Conjecture (8.28) is an extension of (8.27) with $t = 2$. Patterns (8.29) and (8.31) are conjectured because the corresponding half-line B-i.i.d. patterns in Theorem 8.2.4(iii) and (v) have been proved. Although (8.30) is a special case of (8.29), (8.30) has not been proved. It is worthwhile presenting them separately, as it may be relatively easier in the future to prove special cases rather than general ones. Pattern (8.32) is an extension of (8.31) and is conjectured because no counterexample has yet been found. But the related pattern (8.13) has been disproved.

The nature and rules of the B-i.i.d. importance patterns in Subsections 8.2.4–8.2.6 give a better understanding of these conjectures. For example, conjecture (8.31) is consistent with the observation that the length of the second segment is always k as in Remark 8.2.8(iii). Note that all conjectures except (8.27) are made for the first few segments rather than for all segments. Because the length of segments varies and complex increasing-then-decreasing patterns exist for the general cases, it becomes very difficult to make conjectures that hold in general. Furthermore, conjecture (8.27) is made for the case of $p = 1/2$ only, which may simplify the change of the length of segments and the complexity of the increasing-then-decreasing patterns, thus possibly making (8.27) hold for all segments under the special condition of $p = 1/2$.

8.3 Structure Importance Measures

Knowing the structure importance of positions for a $Lin/Con/k/n$ system helps determine which positions are more significant to systems reliability, especially when the reliability of components is unknown or of no concern. According to structure importance measures, all positions in a $Cir/Con/k/n$ system are either equally important (e.g., the B-i.i.d. importance, cut-path importance, BP structure importance, first-term importance, and rare-event importance) or not comparable (e.g., the permutation importance by Theorem 6.5.6). Thus, this section focuses on $Lin/Con/k/n$ systems.

8.3.1 The Permutation Importance

Theorem 8.3.1 *For a $Lin/Con/k/n$:F or G system with $n < 2k$, $(i+1) >_{\text{pe}} i$ for $1 \leq i \leq n-k$, $i >_{\text{pe}} (i+1)$ for $k \leq i < n$, and $i =_{\text{pe}} (i+1)$ for $n-k+1 \leq i \leq k-1$. For a $Lin/Con/k/n$:F or G system with $n \geq 2k$, $(i+1) >_{\text{pe}} i$ for $1 \leq i \leq k-1$, $i >_{\text{pe}} (i+1)$ for $n-k+1 \leq i < n$, and the intermediate components $k, k+1, \ldots, n-k+1$ are not comparable according to the permutation importance.*

Proof. The minimal cuts for a $Lin/Con/k/n$:F system (the minimal paths for a $Lin/Con/k/n$:G system) are $\{1, 2, \ldots, k\}, \{2, 3, \ldots, k+1\}, \ldots, \{n-k+1, \ldots, n\}$. Thus, when $n < 2k$, $\overline{\mathscr{C}}_i \supseteq \overline{\mathscr{C}}_j$ $(\mathscr{P}_i \supseteq \mathscr{P}_j)$ holds for all $1 \leq j < i \leq k$ and $n-k+1 \leq i < j \leq n$. By Theorems 6.5.6 and 6.5.8, $i >_{\text{pe}} j$ for all i, j satisfying these inequalities unless i and j are permutation equivalent.

Similarly, when $n \geq 2k$, $k >_{\text{pe}} k-1 >_{\text{pe}} \ldots >_{\text{pe}} 1$ and $n-k+1 >_{\text{pe}} n-k+2 >_{\text{pe}} \ldots >_{\text{pe}} n$. A direct application of Theorem 6.5.6 shows that the intermediate components $k, k+1, \ldots, n-k+1$ are not comparable according to the permutation importance.

According to the relations of structure importance measures in Figure 6.3, some of the uniform B-i.i.d. importance patterns presented in Theorem 8.2.3 can be directly obtained from the patterns of the permutation importance in Theorem 8.3.1.

8.3.2 The Cut-path Importance

Theorem 8.3.2 shows that components k and $(n-k+1)$ are the most important in a $Lin/Con/k/n$ system with $n \geq 2k$ according to the cut-path importance. It is extended from the result in Chang et al. (1999) that for a $Lin/Con/k/n$:F system with $n \geq 2k$, $k \geq_{\text{cp}} i$ for $i = 1, 2, \ldots, n$.

Theorem 8.3.2 *For a $Lin/Con/k/n$:F or G system with $n \geq 2k$, $n-k+1 =_{\text{cp}} k \geq_{\text{cp}} i$ for $i = 1, 2, \ldots, n$.*

8.3.3 The BP Structure Importance

The following corollary shows that components k and $(n-k+1)$ are the most important in a $Lin/Con/k/n$ system with $n \geq 2k$ according to the BP structure importance. This result is obtained from Theorem 8.2.3(iv)c relative to the uniform B-i.i.d. importance and Equation (6.9) of the BP structure importance definition, which specifies that $I_{\text{BP}}(i; \phi) = \int_0^1 I_{\text{B}}(i; p)\mathrm{d}p$.

Corollary 8.3.3 *For a $Lin/Con/k/n$:F or G system with $n \geq 2k$, $I_{\text{BP}}(n-k+1; \phi) = I_{\text{BP}}(k; \phi) > I_{\text{BP}}(i; \phi)$ for $i \neq k, n-k+1$.*

More results can be similarly induced from Theorem 8.2.3 and Equation (6.9) of the BP structure importance definition. Under certain conditions, Zakaria et al. (1992) specified some relations of the BP structure importance values of two components in a *Lin/Con/k/n*:G system; these relations are covered by Corollary 8.3.3.

8.3.4 The First-term Importance and Rare-event Importance

Let $I_{\mathrm{FT}}(i; \phi_F)$ and $I_{\mathrm{FT}}(i; \phi_G)$ ($I_{\mathrm{RE}}(i; \phi_F)$ and $I_{\mathrm{RE}}(i; \phi_G)$) denote the first-term (rare-event) importance of component i in a *Lin/Con/k/n*:F and G system, respectively. According to Definition 6.10.1 of the first-term importance and Definition 6.10.2 of the rare-event importance, $I_{\mathrm{FT}}(i; \phi_F)$ and $I_{\mathrm{RE}}(i; \phi_G)$ equal the number of distinct sets of k consecutive components containing component i, that is,

$$I_{\mathrm{FT}}(i; \phi_F) = I_{\mathrm{RE}}(i; \phi_G) = \begin{cases} i & \text{for } i < k - 1 \\ k & \text{for } k - 1 \leq i \leq n - k \\ n - i + 1 & \text{for } i > n - k. \end{cases}$$

Let $ps_\ell(k, n)$ ($ps_{i,\ell}(k, n)$) denote the number of paths of size ℓ (containing component i) in a *Lin/Con/k/n*:F system. By Theorem 2.5.2(ii) and duality of *Lin/Con/k/n*:F and G systems in Theorem 2.13.2, a cut in one type of *Lin/Con/k/n* system is a path in another type of *Lin/Con/k/n* system, and vice versa. Then $ps_\ell(k, n)$ ($ps_{i,\ell}(k, n)$) is also the number of cuts of size ℓ (containing component i) in a *Lin/Con/k/n*:G system. Therefore,

$$I_{\mathrm{RE}}(i; \phi_F) = I_{\mathrm{FT}}(i; \phi_G) = ps_{i,\ell}(k, n). \tag{8.33}$$

For a *Lin/Con/k/n* system, represent n as $n = \ell k + r$ where $0 \leq r < k$ and i as $i = tk + b$ where $1 \leq b \leq k$. Theorem 8.3.4 (Chang and Hwang, 2002) gives the expressions of $ps_\ell(k, n)$ and $ps_{i,\ell}(k, n)$ in terms of $k, \ell, r, t,$ and b. Define $\binom{x}{y} = 0$ for $y < 0$ as usual.

Theorem 8.3.4   *For a* Lin/Con/k/n:F *or* G *system with* $n = \ell k + r$ *and* $0 \leq r < k$,

$$ps_\ell(k, n) = \binom{\ell + k - r - 1}{k - r - 1},$$

and considering component i *where* $i = tk + b$ *and* $1 \leq b \leq k$,

$$ps_{i,\ell}(k, n) = \binom{t + k - b}{k - b}\binom{\ell - t + b - r - 2}{b - r - 1}. \tag{8.34}$$

Furthermore, for each $t = 0, 1, \ldots, \ell - 1$,

$$\sum_{i=tk+1}^{tk+k} ps_{i,\ell}(k, n) = ps_\ell(k, n).$$

On the basis of Equation (8.34), Theorem 8.3.5 presents some patterns of the rare-event importance and first-term importance. Chang and Hwang (2002) demonstrated similar results.

Theorem 8.3.5 *For a Lin/Con/k/n:F system,*

(i) $I_{RE}(tk + k; \phi_F)$ *is nonincreasing in t. Furthermore, it is strictly decreasing except for* $r = k - 1$;

(ii) $I_{RE}(tk + b; \phi_F) \le I_{RE}((t + 1)k + b; \phi_F)$ *for* $1 \le b \le (k+1)/2$ *and* $(t+1)k + b \le \lceil n/2 \rceil$;

(iii) $I_{RE}(tk + 1; \phi_F) < I_{RE}(tk; \phi_F)$ *for all t satisfying* $tk + 1 \le \lceil n/2 \rceil$; *and*

(iv) $I_{RE}(tk + 1; \phi_F) \le I_{RE}(j; \phi_F)$ *for* $j = tk + 2, tk + 3, \ldots, \lceil n/2 \rceil$.

These results also hold for the first-term importance for a Lin/Con/k/n:G system by Equation (8.33).

Although in Theorem 8.3.5(ii)–(iv), only components $1, 2, \ldots, \lceil n/2 \rceil$ are dealt with, components $\lceil n/2 \rceil + 1, \lceil n/2 \rceil + 2, \ldots, n$ can be compared by noting $I_{RE}(i; \phi_F) = I_{RE}(n + 1 - i; \phi_F)$ and $I_{FT}(i; \phi_G) = I_{FT}(n + 1 - i; \phi_G)$ because of the structural symmetry.

As discussed in Section 6.10, showing the nonexistence of a pattern $I_{RE}(i; \phi) > I_{RE}(j; \phi)$ is an alternative way to disprove the uniform B-i.i.d. importance of $I_B(i; p) > I_B(j; p)$ for all $0 < p < 1$. For example, in Theorem 8.2.4(iv), $I_B(2k; p) > I_B(i; p)$ holds for $2k < i \le \lceil n/2 \rceil$ and $1/2 \le p < 1$. But this pattern as in (8.22) does not hold for all $0 < p < 1$ because $I_{RE}(2k; \phi_F) < I_{RE}(i; \phi_F)$ for some k and $2k < i \le \lceil n/2 \rceil$, as in the following example.

Example 8.3.6 For $Lin/Con/5/25$:F system, $I_B(10; p) < I_B(13; p)$ when $p = 0.024$ (a small value), which is consistent with the rare-event importance result that $ps_{10,5}(5, 25) = 35 < ps_{13,5}(5, 25) = 36$.

However, a pattern that holds true for the rare-event importance does not necessarily hold for the uniform B-i.i.d. importance or even for the B-structure importance. For example, in Theorem 8.3.5, part (iii) does not hold true for the uniform B-i.i.d. importance and part (iv) holds for the B-structure importance only when $t = 1$.

References

Chadjiconstantinidis S and Koutras MV. 1999. Measures of component importance for Markov chain imbeddable reliability structures. *Naval Research Logistics* **46**, 613–639.

Chang GJ, Cui L, and Hwang FK. 1999. New comparisons in Birnbaum importance for the consecutive-k-out-of-n system. *Probability in the Engineering and Informational Sciences* **13**, 187–192.

Chang GJ, Hwang FK, and Cui L. 2000. Corrigenda on "New comparisons in Birnbaum importance for the consecutive-k-out-of-n system". *Probability in the Engineering and Informational Sciences* **14**, 405.

Chang HW, Chen RJ, and Hwang FK. 2002. The structural Birnbaum importance of consecutive-k systems. *Journal of Combinatorial Optimization* **6**, 183–197.

Chang HW and Hwang FK. 2002. Rare-event component importance for the consecutive-k system. *Naval Research Logistics* **49**, 159–166.

Griffith WS and Govindarajulu Z. 1985. Consecutive k-out-of-n failure systems: reliability, availability, component importance, and multistate extensions. *American Journal of Mathematical and Management Sciences* **5**, 125–160.

Hwang FK, Cui LR, Chang JC, and Lin WD. 2000. Comments on "Reliability and component importance of a consecutive-k-out-of-n systems" by Zuo. *Microelectronics and Reliability* **40**, 1061–1063.

Kuo W, Zhang W, and Zuo MJ. 1990. A consecutive k-out-of-n:G: The mirror image of a consecutive k-out-of-n:F system. *IEEE Transactions on Reliability* **R-39**, 244–253.

Kuo W and Zuo MJ. 2003. *Optimal Reliability Modeling: Principles and Applications*. John Wiley & Sons, New York.

Lin FH, Kuo W, and Hwang F. 1999. Structure importance of consecutive-k-out-of-n systems. *Operations Research Letters* **25**, 101–107.

Miles Jr. EP. 1960. Generalized Fibonacci numbers and associated metrices. *The American Mathematical Monthly* **67**, 745–752.

Papastavridis S. 1987. The most important component in a consecutive-k-out-of-n:F system. *IEEE Transactions on Reliability* **R-36**, 266–268.

Zakaria RS, David HA, and Kuo W. 1992. A counter-intuitive aspect of component importance measures in linear consecutive-k-out-of-n systems. *IIE Transactions* **24**, 147–154.

Zhu X, Yao Q, and Kuo W. 2011. Patterns of Birnbaum importance for linear consecutive-k-out-of-n systems. *IIE Transactions* **44**, 277–290.

Zuo MJ. 1993. Reliability and component importance of a consecutive-k-out-of-n system. *Microelectronics and Reliability* **33**, 243–258.

Zuo MJ and Kuo W. 1988. Reliability design for high system utilization. In *Advances in Reliability and Quality Control* (ed. Hamza MH). ACTA Press, Zurich, pp. 53–56.

Zuo MJ and Kuo W. 1990. Design and performance analysis of consecutive k-out-of-n structure. *Naval Research Logistics* **37**, 203–230.

Part Three

Importance Measures for Reliability Design

Introduction

In applications to reliability design, importance measures, such as the permutation importance and B-reliability importance, are expected to give a good guide for optimal solutions. By examining the importance measures of a given system, systems analysts can gain better insight into the system design without resorting to mathematical programming methods, as described by Kuo et al. (2006).

Chapters 4–8 have discussed many types of importance measures that can be used to solve various problems in reliability design, including component redundancy allocation, system upgrading with respect to certain system performance, and CAP, which is to assign a set of available components to different positions in a system with the objective of maximizing the systems reliability. Apparently, different situations call for different importance measures. In general, the design of importance measures for a given situation highly depends on the type of problem under consideration, the system structure, and the range of component reliability. The previously introduced importance measures may be appropriate in some applications, whereas variations or even new importance measures need to be derived for a better fit for other applications.

This part contains five chapters. Chapters 9 and 10 discuss the importance measures in redundancy allocation and system upgrading, respectively. Chapters 11–13 specifically investigate CAP and discuss their relevance to various importance measures. Chapter 11 discusses CAP in general coherent systems and their applications, Chapter 12 focuses on the CAP in $Con/k/n$ systems and some variations of $Con/k/n$ systems, and Chapter 13 presents heuristics that use the B-reliability importance to solve a CAP. Part Five presents other broad applications. Zhu and Kuo (2012) provided a review of applications of importance measures in reliability and mathematical programming, some of which serve as an important reference for this part.

References

Kuo W, Prasad VR, Tillman FA, and Hwang CL. 2006. *Optimal Reliability Design: Fundamentals and Applications, 2nd edn.* Cambridge University Press, Cambridge, UK.
Zhu X and Kuo W. 2012. Importance measures in reliability and mathematical programming. *Annals of Operations Research*, Forthcoming.

9

Redundancy Allocation

This chapter studies a typical reliability design problem, redundancy allocation, focusing on the applications of importance measures and the importance-measure-based methods. Let (N, ϕ) be a coherent system of independent components where $\mathbf{p} = (p_1, p_2, \ldots, p_n)$ is the vector of reliability of all components. Consider a type of redundancy allocation problem in which for each component i, a spare component i^* of reliability p_i^* is available for redundancy. Let $N^* = \{1^*, 2^*, \ldots, n^*\}$ be the set of available spares with respective reliability $\mathbf{p}^* = (p_1^*, p_2^*, \ldots, p_n^*)$. The general redundancy and redundancy allocation can be found in Kuo et al. (2006).

The redundant component i^* can function simultaneously with component i (active redundancy) or function if and only if component i fails (standby redundancy). In some situations of active redundancy, it is desirable to use parallel redundancy in order to increase systems reliability. In other instances, series redundancy is implemented in order to decrease systems reliability. Relative to the set of available spares, N^*, the redundancy allocation problem is to allocate one or more redundant components to the system so that systems reliability is maximally increased (in the case of parallel redundancy) or decreased (in the case of series redundancy). Boland et al. (1988, 1991) gave many examples considering redundancy in k-out-of-n systems. The following examples are typical.

Example 9.0.1 (Parallel redundancy)   Weather instruments are located in n areas, and signals are periodically sent to a central weather station. An accurate picture of the current weather may be obtained if the instruments in at least k locations are accurately relaying information. $r\ (< n)$ new devices are obtained and are to be distributed to r of n locations to improve local weather detection. The allocation is to be made in order to maximize the probability that the central station obtains a true picture of the weather.

Example 9.0.2 (Series redundancy) n patrol boats are policing a coastal area to stop drug trafficking. The probability that a drug trafficker can pass undetected by the patrol boats is then the reliability of a 1-out-of-n (parallel) system. r extra detection devices are to be distributed to r of n boats so as to minimize the chance of a drug trafficker slipping through.

Importance Measures in Reliability, Risk, and Optimization: Principles and Applications, First Edition.
Way Kuo and Xiaoyan Zhu. © 2012 John Wiley & Sons, Ltd. Published 2012 by John Wiley & Sons, Ltd.

A well-known principle among design engineers states that parallel (series) active redundancy at the component level is always better than at the system level in terms of increasing (decreasing) systems reliability (Meng 1996). However, by cold standby redundancy, the system lifetime improved at the component level is stochastically greater (less) than that improved at the system level if and only if the system is series (parallel) (Meng 1996; Shen and Xie 1991). This chapter studies redundancy at the component level.

This chapter focuses mainly on the active redundancy allocation. Section 9.1 presents parallel and series redundancy importance measures. Sections 9.2 and 9.3 discuss allocating a single active spare in several special structure systems using the redundancy importance measures as well as the permutation importance, cut importance, and path importance. When the reliability values of components or even their orderings are unknown, the later three structure importance measures can be used to guide the redundancy allocation. Sections 9.4 and 9.5 study the allocation of more than one active spare. Finally, Section 9.6 investigates the cold standby redundancy allocation.

9.1   Redundancy Importance Measures

For active redundancy, it is meaningful to investigate both the parallel and series redundancy allocations although a series redundancy problem for system ϕ could be transformed to a parallel redundancy problem for its dual system ϕ^D. Adding spare i^* in parallel with component i (in position i) would increase the reliability of position i to $p_i + q_i p_i^*$ and the systems reliability to $R(p_1, p_2, \ldots, p_{i-1}, p_i + q_i p_i^*, p_{i+1}, \ldots, p_n)$. Adding spare i^* in series with component i would decrease the reliability of position i to $p_i p_i^*$ and the systems reliability to $R(p_1, p_2, \ldots, p_{i-1}, p_i p_i^*, p_{i+1}, \ldots, p_n)$. These led Boland et al. (1988, 1991, 1992), Shen and Xie (1990), and Xie and Shen (1989) to define and investigate parallel and series redundancy importance measures.

Definition 9.1.1   *The parallel (series) redundancy importance of component i, denoted by $I_{PR}(i; \mathbf{p})$ ($I_{SR}(i; \mathbf{p})$), is defined as the increment (reduction) in systems reliability which is achieved by adding spare i^* of reliability p_i^* in parallel (series) active redundancy with component i. Mathematically,*

$$I_{PR}(i; \mathbf{p}) = R(p_1, p_2, \ldots, p_{i-1}, p_i + q_i p_i^*, p_{i+1}, \ldots, p_n) - R(\mathbf{p}) = p_i^* q_i I_B(i; \mathbf{p}) \quad (9.1)$$

$$I_{SR}(i; \mathbf{p}) = R(\mathbf{p}) - R(p_1, p_2, \ldots, p_{i-1}, p_i p_i^*, p_{i+1}, \ldots, p_n) = q_i^* p_i I_B(i; \mathbf{p}). \quad (9.2)$$

By Theorem 3.2.1, the TDL type of redundancy importance measures can be defined similarly as Equations (9.1) and (9.2) if p_i is replaced with $\overline{F}_i(t)$ and p_i^* with $\overline{F}_i^*(t)$ for $i = 1, 2, \ldots, n$. For simplification, the discussion in this chapter is in terms of reliability (i.e., Equations (9.1) and (9.2)), but all of the results can be straightforwardly extended to the case of the reliability functions of components and spares and the TDL type of redundancy importance measures.

According to Equations (9.1) and (9.2), the parallel and series redundancy importance measures of component i depend on its reliability p_i, its B-reliability importance $I_B(i; \mathbf{p})$, and the reliability of the available spare, p_i^*. By these definitions, maximizing the parallel (series) redundancy importance is equivalent to maximizing (minimizing) the increased (decreased) systems reliability if only one active redundant component is allowed to be added to the system.

Example 9.1.2   If component i is in parallel with the rest of the system, then the increment in systems reliability by adding a parallel redundant component of reliability p_i^* with component i is equal to

$$I_{PR}(i; \mathbf{p}) = p_i^*(1 - R(\mathbf{p})). \tag{9.3}$$

If component i is in series with the rest of the system, then the reduction of systems reliability by adding a series redundant component of reliability p_i^* with component i is equal to

$$I_{SR}(i; \mathbf{p}) = q_i^* R(\mathbf{p}). \tag{9.4}$$

Recall Theorem 4.1.13 that for a modular decomposition $\{(M_k, \chi_k)\}_{k=1}^m$ of a system (N, ϕ) with an organizing structure ψ, $I_B(i; \phi, \mathbf{p}) = I_B^M(M_k; \psi, \mathbf{p})I_B(i; \chi_k, \mathbf{p}^{M_k})$ for $i \in M_k$; thus,

$$I_{PR}(i; \phi, \mathbf{p}) = p_i^* q_i I_B(M_k; \psi, \mathbf{p})I_B(i; \chi_k, \mathbf{p}^{M_k}) = I_B(M_k; \psi, \mathbf{p})I_{PR}(i; \chi_k, \mathbf{p}^{M_k}), \tag{9.5}$$

$$I_{SR}(i; \phi, \mathbf{p}) = q_i^* p_i I_B(M_k; \psi, \mathbf{p})I_B(i; \chi_k, \mathbf{p}^{M_k}) = I_B(M_k; \psi, \mathbf{p})I_{SR}(i; \chi_k, \mathbf{p}^{M_k}), \tag{9.6}$$

where $I_{PR}(i; \chi_k, \mathbf{p}^{M_k})$ and $I_{SR}(i; \chi_k, \mathbf{p}^{M_k})$ are the parallel and series redundancy importance measures of component i in embedded system (M_k, χ_k), respectively.

Example 9.1.3 (Example 2.4.3 continued)   A series–parallel system (N, ϕ), as shown in Figure 2.6, has a modular structure $\{(M_k, \chi_k)\}_{k=1}^m$ with each module (M_k, χ_k) being a parallel system and the organizing structure ψ being a series structure. Then, by Equations (9.5) and (9.6), for $i \in M_k$,

$$I_{PR}(i; \phi, \mathbf{p}) = p_i^* q_i I_B(M_k; \psi, \mathbf{p})I_B(i; \chi_k, \mathbf{p}^{M_k}) = p_i^* \frac{1 - R_{\chi_k}(\mathbf{p}^{M_k})}{R_{\chi_k}(\mathbf{p}^{M_k})} R_\phi(\mathbf{p}), \tag{9.7}$$

$$I_{SR}(i; \phi, \mathbf{p}) = q_i^* p_i I_B(M_k; \psi, \mathbf{p})I_B(i; \chi_k, \mathbf{p}^{M_k}) = q_i^* \frac{p_i}{q_i} \frac{1 - R_{\chi_k}(\mathbf{p}^{M_k})}{R_{\chi_k}(\mathbf{p}^{M_k})} R_\phi(\mathbf{p}), \tag{9.8}$$

where $R_{\chi_k}(\mathbf{p}^{M_k}) = 1 - \prod_{j \in M_k} q_j$. Hence, finding the component with the maximum parallel redundancy importance is equivalent to maximizing $p_i^*(1 - R_{\chi_k}(\mathbf{p}^{M_k}))/R_{\chi_k}(\mathbf{p}^{M_k})$, and finding the component with the maximum series redundancy importance is equivalent to maximizing $q_i^* p_i q_i^{-1}(1 - R_{\chi_k}(\mathbf{p}^{M_k}))/R_{\chi_k}(\mathbf{p}^{M_k})$.

Similarly, for the parallel–series system (N, ϕ) as shown in Figure 2.7, there are m series embedded systems connected in parallel. Then, by Equations (9.5) and (9.6), for $i \in M_k$,

$$I_{\text{PR}}(i; \phi, \mathbf{p}) = p_i^* q_i I_{\text{B}}(M_k; \psi, \mathbf{p}) I_{\text{B}}(i; \chi_k, \mathbf{p}^{M_k})$$

$$= p_i^* q_i \left[\prod_{\substack{\ell=1 \\ \ell \neq k}}^{m} \left(1 - \prod_{j \in M_\ell} p_j \right) \right] \prod_{\substack{j \in M_k \\ j \neq i}} p_j = p_i^* \frac{q_i}{p_i} \frac{R_{\chi_k}(\mathbf{p}^{M_k})}{1 - R_{\chi_k}(\mathbf{p}^{M_k})} [1 - R_\phi(\mathbf{p})],$$

$$\tag{9.9}$$

$$I_{\text{SR}}(i; \phi, \mathbf{p}) = q_i^* \frac{R_{\chi_k}(\mathbf{p}^{M_k})}{1 - R_{\chi_k}(\mathbf{p}^{M_k})} [1 - R_\phi(\mathbf{p})], \tag{9.10}$$

where $R_{\chi_k}(\mathbf{p}^{M_k}) = \prod_{j \in M_k} p_j$.

9.2   A Common Spare

In this section, suppose that a common spare c^* of reliability p^* is available for active redundancy with any of n components in a system. This could happen, for example, in a k-out-of-n system where, at least structurally, the components are all equivalent. In such a case,

$$p_i^* = p^*, \quad i = 1, 2, \ldots, n.$$

9.2.1   The Redundancy Importance Measures

According to Equation (9.1) (Equation (9.2)), in order to maximally increase (decrease) the systems reliability, the common spare should be allocated in parallel (series) redundancy to the component, which maximizes $q_i I_{\text{B}}(i; \mathbf{p})$ ($p_i I_{\text{B}}(i; \mathbf{p})$) for $i = 1, 2, \ldots, n$. Note that the decision as to where the single spare c^* should be allocated in active redundancy is made based only on the current system, no matter what the reliability p^* of spare c^*.

Parallel systems and series systems

Referring to Example 9.1.2, for parallel redundancy in a parallel system, spare c^* can be added with any component, and the effect is the same as adding it with the parallel system. The result of adding a parallel active spare in a 1-out-of-n system (parallel system) is a 1-out-of-$(n + 1)$ system, no matter where the spare is added. For series redundancy in a series system, spare c^* can be added with any component, and the effect is the same as adding it with the series system. That is, adding a series active spare in an n-out-of-n system (series system) results in an $(n + 1)$-out-of-$(n + 1)$ system, no matter where the spare is added. These results can be obtained from Equations (9.3) and (9.4), noting that $p_i^* = p^*$ for $i = 1, 2, \ldots, n$ in the case of a common spare.

Series–parallel systems

In a series–parallel system, according to Equation (9.7), spare c^* should be allocated in parallel redundancy to any component in the least reliable parallel embedded system. According to Equation (9.8), for series redundancy, spare c^* should be allocated to the component that maximizes $p_i q_i^{-1}(1 - R_{\chi_k}(\mathbf{p}^{M_k}))/R_{\chi_k}(\mathbf{p}^{M_k})$.

If all components in (N, ϕ) are equally reliable, then clearly spare c^* would be allocated in parallel or series redundancy to any component within the embedded system of the fewest components. If all embedded systems are of the same size n/m and the components in embedded system (M_k, χ_k) have the same reliability $\bar{p}_k$ but $\bar{p}_1, \bar{p}_2, \ldots, \bar{p}_m$ may be different, then the parallel redundancy importance in Equation (9.7) becomes $p^* \bar{q}_k^{n/m} R_\phi(\mathbf{p})/(1 - \bar{q}_k^{n/m})$, and the series redundancy importance in Equation (9.8) becomes $q^* \bar{q}_k^{n/m-1} R_\phi(\mathbf{p})/(\bar{q}_k^{n/m-1} + \ldots + \bar{q}_k + 1)$. Thus, for both parallel and series redundancy, spare c^* should be allocated to any component within the embedded system with a minimum $\bar{p}_k$, which maximizes the parallel and series redundancy importance measures.

Parallel–series systems

In a parallel–series system, according to Equation (9.9), maximizing the parallel redundancy importance is equivalent to maximizing $q_i p_i^{-1} R_{\chi_k}(\mathbf{p}^{M_k})/(1 - R_{\chi_k}(\mathbf{p}^{M_k}))$. According to Equation (9.10), maximizing the series redundancy importance is equivalent to maximizing $R_{\chi_k}(\mathbf{p}^{M_k})/(1 - R_{\chi_k}(\mathbf{p}^{M_k}))$, and therefore spare c^* should be allocated in series redundancy to any component in the most reliable series embedded system.

If all components are equally reliable, then clearly spare c^* would be allocated in parallel or series redundancy to any component within the embedded system of the fewest components. If all embedded systems are of the same size n/m and the components in embedded system (M_k, χ_k) have the same reliability $\bar{p}_k$ but $\bar{p}_1, \bar{p}_2, \ldots, \bar{p}_m$ may be different, then according to Equations (9.9) and (9.10), for both parallel and series redundancy, spare c^* should be allocated to any component within the embedded system with a maximum $\bar{p}_k$, which maximizes the parallel redundancy importance (equal to $p^* \bar{p}_k^{n/m-1}(1 - R_\phi(\mathbf{p}))/(\bar{p}_k^{n/m-1} + \ldots + \bar{p}_k + 1)$) and the series redundancy importance (equal to $q^* \bar{p}_k^{n/m}(1 - R_\phi(\mathbf{p}))/(1 - \bar{p}_k^{n/m}))$.

k-out-of-n systems

Theorem 9.2.1 *In a k-out-of-n system, assuming that $p_1 \leq p_2 \leq \cdots \leq p_n$, then $q_1 I_B(1; \mathbf{p}) \geq q_2 I_B(2; \mathbf{p}) \geq \cdots \geq q_n I_B(n; \mathbf{p})$ and $p_1 I_B(1; \mathbf{p}) \leq p_2 I_B(2; \mathbf{p}) \leq \cdots \leq p_n I_B(n; \mathbf{p})$.*

Proof. Let $R(1_i, \mathbf{p})$ be the reliability of a k-out-of-n system whose n components have reliability $p_1, p_2, \ldots, p_{i-1}, p_i = 1, p_{i+1}, \ldots, p_n$. Then, the k-out-of-n system associated with $R(1_{i+1}, \mathbf{p})$ differs from the system associated with $R(1_i, \mathbf{p})$ by only one component that has reliability p_i in the former and p_{i+1} in the latter. Because $p_i \leq p_{i+1}$, $R(1_i, \mathbf{p}) \geq R(1_{i+1}, \mathbf{p})$. Furthermore, $R(1_i, \mathbf{p}) - R(\mathbf{p}) = q_i I_B(i; \mathbf{p})$; thus, $q_i I_B(i; \mathbf{p}) \geq q_{i+1} I_B(i + 1; \mathbf{p})$ for $i = 1, 2, \ldots, n - 1$.

Noting that the dual of a k-out-of-n structure is an $(n - k + 1)$-out-of-n structure and that $I_{\mathrm{B}}(i; \phi, \mathbf{p}) = I_{\mathrm{B}}(i; \phi^D, \mathbf{q})$ in Theorem 4.1.10, then duality shows that $p_1 I_{\mathrm{B}}(1; \mathbf{p}) \leq p_2 I_{\mathrm{B}}(2; \mathbf{p}) \leq \cdots \leq p_n I_{\mathrm{B}}(n; \mathbf{p})$ based on the fact that $q_1 I_{\mathrm{B}}(1; \mathbf{p}) \geq q_2 I_{\mathrm{B}}(2; \mathbf{p}) \geq \cdots \geq q_n I_{\mathrm{B}}(n; \mathbf{p})$. The proof is complete.

Boland et al. (1988, 1991) presented Theorem 9.2.1, which, in the case of a common spare, implies that $I_{\mathrm{PR}}(1; \mathbf{p}) \geq I_{\mathrm{PR}}(2; \mathbf{p}) \geq \cdots \geq I_{\mathrm{PR}}(n; \mathbf{p})$ and $I_{\mathrm{SR}}(1; \mathbf{p}) \leq I_{\mathrm{SR}}(2; \mathbf{p}) \leq \cdots \leq I_{\mathrm{SR}}(n; \mathbf{p})$. Note that using Theorem 6.5.15, Mi (2003) generalized the results in Theorem 9.2.1 to the coherent system where $1 \geq_{\mathrm{pe}} 2 \geq_{\mathrm{pe}} \cdots \geq_{\mathrm{pe}} n$, of which a k-out-of-n system is a special case with $1 =_{\mathrm{pe}} 2 =_{\mathrm{pe}} \cdots =_{\mathrm{pe}} n$.

By Theorem 9.2.1, for parallel redundancy in a k-out-of-n system, spare c^* should always be allocated to the least reliable component to maximize the systems reliability. The result can be extended to parallel redundancy for a k-out-of-m-parallel system, which is a special k-out-of-m system with modular structure $\{(M_\ell, \chi_\ell)\}_{\ell=1}^m$ where each module is a parallel embedded system. Then, spare c^* should be allocated to any component in the least reliable module (i.e., the one with the smallest value of $1 - \prod_{i \in M_\ell} q_i$). When $k = m$, it is a series–parallel system, for which the results are given in the aforementioned discussion.

In the case of series redundancy in a k-out-of-n system, spare c^* should be allocated to the most reliable component to minimize the systems reliability. Similarly, the result can be extended to a k-out-of-m-series system in which each module is a series embedded system and the modules form a k-out-of-m structure. The greatest reduction in the systems reliability would be made by allocating spare c^* to any component in the most reliable embedded system (i.e., the one with the largest value of $\prod_{i \in M_\ell} p_i$). When $k = 1$, it is a parallel–series system, for which the results are given in the aforementioned discussion.

9.2.2 The Permutation Importance

The following theorem (Meng 1996) shows that under some conditions, the optimal allocation can be determined according to the permutation importance.

Theorem 9.2.2 *Suppose that a spare is available for parallel active redundancy. If $i >_{\mathrm{pe}} j$ and $p_i = p_j$ or $i =_{\mathrm{pe}} j$ and $p_i \leq p_j$, then allocating the spare to component i produces a larger increment in systems reliability than allocating it to component j.*

In a k-out-of-n system, all components are permutation equivalent. Thus, Theorem 9.2.2 directly implies that in a k-out-of-n system, spare c^* should be allocated in parallel redundancy to the least reliable component in order to stochastically maximize systems reliability. Boland et al. (1992) also showed the same result. Meng (1996) interpreted this result based on the minimal cuts as follows. Suppose that components i and j are permutation equivalent. Considering a given minimal cut C such that $i \in C$, if $j \notin C$, then, by replacing i with j, the set $C \cup \{j\} \setminus \{i\}$ is also a minimal cut. Clearly, if components i and j appear together in a minimal cut, then it makes no difference where to add the parallel redundant component in this minimal cut (because the components in a minimal cut are connected in parallel). Now each pair of the minimal cuts $\{i\} \cup (C \setminus \{i\})$ and $\{j\} \cup (C \setminus \{i\})$ may be merged into one cut $\{i \wedge j\} \cup (C \setminus \{i\})$, where $i \wedge j$ means components i and j are connected in series. Furthermore, because now components i and j are connected in series in the cut $\{i \wedge j\} \cup (C \setminus \{i\})$, the spare should be allocated to the less reliable component to maximize the probability that both components i and j function, hence minimizing the probability that the cut $\{i \wedge j\} \cup (C \setminus \{i\})$ causes system failure.

9.2.3   The Cut Importance and Path Importance

The cut importance and path importance are related to the redundancy importance measures as shown in Corollary 9.2.3. According to it, when the reliability values of components are equal and high (low), then component i should be selected for redundancy rather than component j if $i >_c j$ $(i >_p j)$.

Corollary 9.2.3   *Suppose that a spare of reliability p^* is available for active redundancy and that all components in the system have the same reliability p. If p is sufficiently close to 1 (0), then $i >_c j$ $(i >_p j)$ implies $I_{PR}(i; p) \geq I_{PR}(j; p)$ and $I_{SR}(i; p) \geq I_{SR}(j; p)$.*

In Corollary 9.2.3, $I_{PR}(i; p) = p^* q I_B(i; p)$ and $I_{SR}(i; p) = q^* p I_B(i; p)$; thus, Corollary 9.2.3 is a direct consequence of Theorems 6.7.4 and 6.7.6 that the cut importance and path importance are equivalent to the B-i.i.d. importance as p approaches 1 and 0, respectively. The relationship between the cut importance and the parallel redundancy importance was first discovered by Meng (1993).

9.3   Spare Identical to the Respective Component

Another scenario for redundancy might be where a set of spares $N^* = \{1^*, 2^*, \ldots, n^*\}$ is available with respective reliability values identical to those already in the system, that is,

$$p_i^* = p_i, \quad i = 1, 2, \ldots, n,$$

and only one spare is allowed to be added. This might happen, for example, if all components are different, but time or resources allow for only one spare with a component identical to that already in position.

9.3.1   The Redundancy Importance Measures

According to Equations (9.1) and (9.2), the most important component for both parallel and series redundancy is the one that maximizes

$$I_{PR}(i; \mathbf{p}) = I_{SR}(i; \mathbf{p}) = p_i q_i I_B(i; \mathbf{p}). \tag{9.11}$$

Additionally, $I_{PR}(i; \mathbf{p}) = I_{SR}(i; \mathbf{p}) = \mathrm{Cov}(X_i, \phi(\mathbf{X}))$ (see the proof of Theorem 4.1.2).

By Lemma 5.1.2, an upper bound on $I_{PR}(\cdot; \mathbf{p})$ and $I_{SR}(\cdot; \mathbf{p})$ is obtained as follows.

Corollary 9.3.1 *When $p_i^* = p_i$ for $i = 1, 2, \ldots, n$, the increment (reduction) in systems reliability by adding the respective parallel (series) active spare with component i is bounded from above as*

$$I_{PR}(i; \mathbf{p}) = I_{SR}(i; \mathbf{p}) \leq \min\{q_i R(\mathbf{p}), p_i(1 - R(\mathbf{p}))\}.$$

Parallel systems and series systems

Referring to Example 9.1.2, when $p_i^* = p_i$ for $i = 1, 2, \ldots, n$, for a parallel system,

$$I_{PR}(i; \mathbf{p}) = I_{SR}(i; \mathbf{p}) = p_i(1 - R(\mathbf{p}));$$

thus, the spare should be allocated in parallel or series redundancy to the most reliable component to maximally increase or decrease the systems reliability. For a series system,

$$I_{PR}(i; \mathbf{p}) = I_{SR}(i; \mathbf{p}) = q_i R(\mathbf{p});$$

thus, the least reliable component is the most important for both parallel and series redundancy.

Series–parallel systems

According to Equations (9.7) and (9.8), in a series–parallel system, both the parallel and series redundancy allocation problems reduce to allocate the spare to component $i \in M_k$, which maximizes $p_i(1 - R_{\chi_k}(\mathbf{p}^{M_k}))/R_{\chi_k}(\mathbf{p}^{M_k})$ for $k = 1, 2, \ldots, m$. If all components in (N, ϕ) are equally reliable, then the spare should be allocated to any component within the embedded system of the fewest components. If all embedded systems are of the same size n/m and components in embedded system (M_k, χ_k) have the same reliability $\bar{p}_k$ but $\bar{p}_1, \bar{p}_2, \ldots, \bar{p}_m$ may be different, then maximizing the redundancy importance is equivalent to maximizing $\bar{q}_k^{-n/m}/(\bar{q}_k^{-n/m-1} + \ldots + \bar{q}_k + 1)$; therefore, the spare should be allocated to any component within the embedded system with a minimum $\bar{p}_k$.

Parallel–series systems

In a parallel–series system, both the parallel and series redundancy allocations reduce to maximizing $q_i R_{\chi_k}(\mathbf{p}^{M_k})/(1 - R_{\chi_k}(\mathbf{p}^{M_k}))$ for $i \in M_k$. Similar results as for a series–parallel system can be obtained for a parallel–series system.

***k*-out-of-*n* systems**

Although for a k-out-of-n system the sequences $\{q_i I_B(i; \mathbf{p})\}_{i=1}^n$ and $\{p_i I_B(i; \mathbf{p})\}_{i=1}^n$ are both monotone in i whenever $p_1 \leq p_2 \leq \cdots \leq p_n$ (Theorem 9.2.1), the same cannot be concluded in general about $\{p_i q_i I_B(i; \mathbf{p})\}_{i=1}^n$. Thus, for the k-out-of-n system with $1 < k < n$ there is no direct correspondence between the component reliability ranking and $I_{PR}(i; \mathbf{p})$ and $I_{SR}(i; \mathbf{p})$

when $p_i^* = p_i$ for $i = 1, 2, \ldots, n$. However, Theorem 9.3.2 (Shen and Xie 1989) shows that more reliable components have a tendency to be less important.

Theorem 9.3.2 *In a k-out-of-n system, assuming that $p_i^* = p_i$ for $i = 1, 2, \ldots, n$, $I_{PR}(i; \mathbf{p})$ − $I_{PR}(j; \mathbf{p})$ and $I_{SR}(i; \mathbf{p}) - I_{SR}(j; \mathbf{p})$ approach a nonnegative constant as p_j approaches 1.*

Proof. First, note that $I_{PR}(i; \mathbf{p}) = I_{SR}(i; \mathbf{p})$ for $i = 1, 2, \ldots, n$. Then, for $I_{PR}(i; \mathbf{p})$ and $I_{PR}(j; \mathbf{p})$,

$$I_{PR}(i; \mathbf{p}) = p_i q_i (R(1_i, \mathbf{p}) - R(0_i, \mathbf{p})), \tag{9.12}$$

$$I_{PR}(j; \mathbf{p}) = p_j q_j (R(1_j, \mathbf{p}) - R(0_j, \mathbf{p})). \tag{9.13}$$

Note that $R(1_i, 1_j, \mathbf{p}^{(ij)}) = R_{k,n-2}(\mathbf{p}^{(ij)})$, $R(0_i, 0_j, \mathbf{p}^{(ij)}) = R_{k-2,n-2}(\mathbf{p}^{(ij)})$, and $R(1_i, 0_j, \mathbf{p}^{(ij)}) = R(0_i, 1_j, \mathbf{p}^{(ij)}) = R_{k-1,n-2}(\mathbf{p}^{(ij)})$, where $R_{k,n-2}$ denotes the reliability of k-out-of-$(n-2)$ system (excluding components i and j); $R_{k-1,n-2}$ and $R_{k-2,n-2}$ are defined similarly. By using these relations and pivoting on component j (component i) in Equation (9.12) (Equation (9.13)), it follows that

$$I_{PR}(i; \mathbf{p}) - I_{PR}(j; \mathbf{p}) = I_{SR}(i; \mathbf{p}) - I_{SR}(j; \mathbf{p}) = (p_j - p_i)$$
$$\times \left[q_i q_j (R_{k,n-2}(\mathbf{p}^{(ij)}) - R_{k-1,n-2}(\mathbf{p}^{(ij)})) + p_i p_j (R_{k-2,n-2}(\mathbf{p}^{(ij)}) - R_{k-1,n-2}(\mathbf{p}^{(ij)})) \right]. \tag{9.14}$$

Finally, note that $R_{k-2,n-2}(\mathbf{p}^{(ij)}) \geq R_{k-1,n-2}(\mathbf{p}^{(ij)})$ because a $(k-2)$-out-of-$(n-2)$ system is more reliable than a $(k-1)$-out-of-$(n-2)$ system. As p_j approaches 1, $q_i q_j$ approaches 0, and then by Equation (9.14), $I_{PR}(i; \mathbf{p}) - I_{PR}(j; \mathbf{p}) = I_{SR}(i; \mathbf{p}) - I_{SR}(j; \mathbf{p})$ approaches $p_i q_i (R_{k-2,n-2}(\mathbf{p}^{(ij)}) - R_{k-1,n-2}(\mathbf{p}^{(ij)})) \geq 0$.

9.3.2 The Permutation Importance

Note that Theorem 9.3.3 (Meng 1993) and its proof are similar to Theorem 14.2.1, which pertains to the permutation importance and B-reliability importance and is more essential.

Theorem 9.3.3 *Suppose that spare i^* identical to component i (i.e., $p_i^* = p_i$) is available for active redundancy, $i = 1, 2, \ldots, n$. Assume that $\mathbf{p} : \mathbb{R} \mapsto \mathbb{R}^n$ satisfies $0 < \mathbf{p}(\varepsilon) < 1$ for all $\varepsilon \in \mathbb{R}$ and that $\lim_{\varepsilon \to 0} p_i(\varepsilon) = \lim_{\varepsilon \to 0} p_j(\varepsilon) = p$ for some $0 < p < 1$, $i \neq j$. Then there exists a ε_0 such that for all $\varepsilon < \varepsilon_0$, $i >_{pe} j$ implies $I_{PR}(i; \mathbf{p}(\varepsilon)) > I_{PR}(j; \mathbf{p}(\varepsilon))$ and $I_{SR}(i; \mathbf{p}(\varepsilon)) > I_{SR}(j; \mathbf{p}(\varepsilon))$.*

According to Theorem 9.3.3, supposing that the reliability values of components (known or unknown) are quite close to one another, then component i should be selected for redundancy rather than component j if $i >_{pe} j$.

9.4   Several Spares in a k-out-of-n System

Note that if only one spare is allowed to be added, then maximizing the redundancy importance measures of components in Equations (9.1) and (9.2) is equivalent to maximizing (in the parallel active redundancy case) or minimizing (in the series active redundancy case) the reliability of the resulting system, since $R(\mathbf{p})$ in the definition of redundancy importance measures is a constant. However, calculating the redundancy importance measures is no easier than directly calculating the reliability of the resulting system unless the B-reliability importance values of the components are easily known or unless the special cases as discussed in Sections 9.1–9.3 are involved.

Suppose that r ($r \leq n$) spares are available for active redundancy with any r distinct components in a k-out-of-n system. Boland et al. (1988, 1991) addressed this problem using the arrangement decreasing functions (Marshall and Olkin 1979) and concluded that if parallel allocation is contemplated, the greatest increment in systems reliability results from allocating the most reliable spare to the least reliable component, the second most reliable spare to the second least reliable component, and so on. On the other hand, if series redundancy is contemplated, the greatest reduction in systems reliability is obtained by allocating the least reliable spare to the most reliable component, the second least reliable spare to the second most reliable component, and so on. Mi (1999) studied a similar active redundancy allocation problem in a k-out-of-n system, and Mi (2003) generalized it to the coherent system where $1 \geq_{\text{pe}} 2 \geq_{\text{pe}} \cdots \geq_{\text{pe}} n$ using Theorem 6.5.15.

Extending these results straightforwardly, suppose that r spares are available for parallel redundancy in the k-out-of-m-parallel system, $r \leq m$, each spare to a different parallel embedded system. Then, the optimal allocation is to allocate the most reliable spare to the least reliable parallel embedded system, the second most reliable spare to the second least reliable embedded system, and so on. If r spares are available for series redundancy in the k-out-of-m-series system, $r \leq m$, each spare to a different series embedded system, then the optimal allocation is to allocate the least reliable spare to the most reliable series embedded system, the second least reliable spare to the second most reliable embedded system, and so on.

Using the arrangement decreasing functions and Schur functions (Kuo and Zuo 2003), Boland et al. (1988, 1991) also investigated other redundancy allocation problems in a k-out-of-n system, and Boland et al. (1991) extended them to multistate systems where the system and its components have more than two states, from complete failure up to perfect functioning (see Sections 15.2 – 15.5). But these studies do not involve the importance measures and are thus outside the scope of this book.

9.5   Several Spares in an Arbitrary Coherent System

Suppose that spare i^* of reliability p_i^* is available for redundancy with component i for $i = 1, 2, \ldots, n$ and that in total r ($1 < r < n$) spares are allowed to be added, *one for each component*. We propose two greedy methods in Tables 9.1 and 9.2 using the parallel redundancy importance for determining the parallel redundancy allocation. These two methods can also solve the series redundancy allocation by using $I_{\text{SR}}(i; \mathbf{p})$ in Equation (9.2) instead of $I_{\text{PR}}(i; \mathbf{p})$ in the algorithms and using $p_{i_0} = p_{i_0} p_{i_0}^*$ in step 4 of greedy method 2 in Table 9.2.

These two greedy methods cannot guarantee an optimal solution. The following example demonstrates their performance on a bridge system. The optimal solution can be obtained by an enumeration method for small instances, which is to enumerate all possible choices of r out of n available spares and to select the one with the largest (smallest) resulting systems reliability as the optimal solution for parallel (series) redundancy. The number of enumerations is $\binom{n}{r}$.

Example 9.5.1 Considering the parallel active redundancy allocation in the bridge system in Figure 2.1, let $(p_1, p_2, p_3, p_4, p_5) = (0.5, 0.5, 0.5, 0.5, 0.5)$. Then, $R(\mathbf{p}) = 0.5$. Furthermore, let $r = 2$, and $(p_1^*, p_2^*, p_3^*, p_4^*, p_5^*) = (0.95, 0.99, 0.5, 0.95, 0.1)$.

Greedy method 1: The corresponding spares of reliability 0.99 and 0.95 are chosen to be added with components 2 and 1, respectively, and the resulting systems reliability is 0.7464.

Greedy method 2: The corresponding spares of reliability 0.99 and 0.95 are chosen to be added with components 2 and 4, respectively, and the resulting systems reliability is 0.8638.

Using the enumeration method, the optimal solution is to add the corresponding spares of reliability 0.95 and 0.95 with components 1 and 4, respectively, and the optimal improved systems reliability is 0.9691.

For the case of the series active redundancy allocation, the corresponding spares of reliability 0.1 and 0.5 should be optimally added with components 5 and 3, respectively, and the resulting systems reliability is 0.3. Both of these two greedy methods obtain the optimal solution by luck.

Consider another redundancy allocation problem, in which for each component i, there are $r_i \geq 1$ identical spares available and more than one spare can be added for active redundancy. The spares for distinct components may be different, and in total r spares are allowed to be added. We propose a greedy method in Table 9.3 using $I_{PR}(i; \mathbf{p})$ for determing the parallel

Table 9.1 Greedy method 1

1. Calculate $I_{PR}(i; \mathbf{p})$ for $i = 1, 2, \ldots, n$ according to Equation (9.1).
2. Add r spares with the respective r components that correspond to the first r largest values of $I_{PR}(i; \mathbf{p})$.

Table 9.2 Greedy method 2

1. Set $S = \{1, 2, \ldots, n\}$ (S is a set of current available spares).
2. Calculate $I_{PR}(i; \mathbf{p})$ for $i \in S$ according to Equation (9.1).
3. Choose the spare i_0^* to be added with component $i_0 \in S$ that has the largest value of $I_{PR}(i; \mathbf{p})$ among $i \in S$.
4. Update $S = S \setminus \{i_0\}$ and $p_{i_0} = p_{i_0} + q_{i_0} p_{i_0}^*$.
5. If r spares have already been added, stop; otherwise, go to step 2.

Table 9.3 Greedy method 3

1. Set $S = \{1, 2, \ldots, n\}$ (S is a set of currently available spares).
2. Calculate $I_{PR}(i; \mathbf{p})$ for $i \in S$ according to Equation (9.1).
3. Choose the spare i_0^* to be added with component $i_0 \in S$ that has the largest value of $I_{PR}(i; \mathbf{p})$ among $i \in S$.
4. Decrease r_{i_0} by 1. If $r_{i_0} = 0$, then update $S = S \setminus \{i_0\}$.
5. Set $p_{i_0} = p_{i_0} + q_{i_0} p_{i_0}^*$.
6. If r spares have already been added, stop; otherwise, go to step 2.

redundancy allocation. This method can solve the corresponding series redundancy allocation by using $I_{SR}(i; \mathbf{p})$ in Equation (9.2) instead of $I_{PR}(i; \mathbf{p})$ and $p_{i_0} = p_{i_0} p_{i_0}^*$ in step 5. The difference between greedy method 2 and method 3 is that method 3 adjusts set S in step 4 based on the current value of r_i so that more than one spare can be added with the same component.

Misra and Agnihotri (1979) studied a similar problem in a bridge system subject to a single cost constraint. For each component, more than one spare can be added, and they all have the same reliability as the component. They concluded that the interconnecting component (i.e., component 3 in Figure 2.1) is the least important, and no spare tends to be added with it even if the cost limitation is loose. Their conclusion is consistent with the results in Example 9.5.1.

9.6   Cold Standby Redundancy

This section briefly presents cold standby redundancy allocation problems. Cold standby redundancy provides higher reliability than parallel active redundancy because a spare does not fail in cold standby, but it is hard to implement because of the difficulty of detecting failure (Kuo and Wan 2007). However, an actual optimal design may include active redundancy, cold standby redundancy, or both.

Assume that in a coherent system, the component lifetime distributions satisfy $F_1(t) \leq F_2(t) \leq \cdots \leq F_n(t)$ for all $t \geq 0$ and their lifetime densities $f_1(t), f_2(t), \ldots, f_n(t)$ exist. Furthermore, assume that $f_i \in \mathscr{F}_\theta$ for each i. Here, $\mathscr{F}_\theta$ denotes a one-parameter family of lifetime distributions whose density functions $\{f_\theta(t) : \theta \in \Theta\}$ or mass functions $\{p_\theta(t) : \theta \in \Theta\}$ possess the Reverse Rule of order 2 property (RR_2) in θ and $t \geq 0$, which means that if $\theta_1 > \theta_2$ and $t_1 < t_2$, then $f_{\theta_1}(t_1)f_{\theta_2}(t_2) \geq f_{\theta_1}(t_2)f_{\theta_2}(t_1)$.

Under these assumptions and further assuming that an identical cold standby spare is available for each component and only one standby spare is allowed to be added, Boland et al. (1992) showed that in a parallel system, the cold standby of the less reliable component should be made to stochastically maximize systems reliability. In another situation where only one common cold standby is available, to maximally increase systems reliability, the cold standby should be allocated to the most reliable component in a series system; it should be allocated to the least reliable component in a parallel system; however, in a k-out-of-n system, allocation depends on the formats and parameters of component lifetime distributions. These results have been further generalized by Shaked and Shanthikumar (1992) and El-Neweihi and Sethuraman (1993) in more complicated settings.

The permutation importance can be used when selecting the cold standby, as shown in the following theorem (Meng 1996), which is easily obtained from Theorem 6.5.13.

Theorem 9.6.1   *Suppose that a spare is available for cold standby redundancy. If $i >_{pe} j$ and $F_i(t) = F_j(t)$ for all $t \geq 0$, then allocating the spare to component i stochastically produces a larger increment in systems reliability than allocating it to component j.*

References

Boland PJ, El-Neweihi E, and Proschan F. 1988. Active redundancy allocation in coherent systems. *Probability in the Engineering and Informational Sciences* **2**, 343–353.

Boland PJ, El-Neweihi E, and Proschan F. 1991. Redundancy importance and allocation of spares in coherent systems. *Journal of Statistical Planning and Inference* **29**, 55–65.

Boland PJ, El-Neweihi E, and Proschan F. 1992. Stochastic order for redundancy allocations in series and parallel systems. *Advances in Applied Probability* **24**, 161–171.

El-Neweihi E and Sethuraman J. 1993. Optimal allocation under partial ordering of lifetimes of components. *Advances in Applied Probability* **25**, 914–925.

Kuo W, Prasad VR, Tillman FA, and Hwang CL. 2006. *Optimal Reliability Design: Fundamentals and Applications, 2nd edn.* Cambridge University Press, Cambridge, UK.

Kuo W and Wan R. 2007. Recent advances in optimal reliability allocation. *IEEE Transactions on Systems, Man, and Cybernetics, Series A* **37**, 143–156.

Kuo W and Zuo MJ. 2003. *Optimal Reliability Modeling: Principles and Applications.* John Wiley & Sons, New York.

Marshall A and Olkin I. 1979. *Inequalities: Theory of Majorization and Its Applications.* Academic Press, New York.

Meng FC. 1993. On selecting components for redundancy in coherent systems. *Reliability Engineering and System Safety* **41**, 121–126.

Meng FC. 1996. More on optimal allocation of components in coherent systems. *Journal of Applied Probability* **33**, 548–556.

Mi J. 1999. Optimal active redundancy in k-out-of-n system. *Journal of Applied Probability* **36**, 927–933.

Mi J. 2003. A unified way of comparing the reliability of coherent systems. *IEEE Transactions on Reliability* **52**, 38–43.

Misra RB and Agnihotri G. 1979. Peculiarities in optimal redundancy for a bridge network. *IEEE Transactions on Reliability* **R-28**, 70–72.

Shaked M and Shanthikumar JG. 1992. Optimal allocation of resources to nodes of parallel and series systems. *Advances in Applied Probability* **24**, 894–914.

Shen K and Xie M. 1989. The increase of reliability of k-out-of-n systems through improving a component. *Reliability Engineering and System Safety* **26**, 189–195.

Shen K and Xie M. 1990. On the increase of system reliability by parallel redundancy. *IEEE Transactions on Reliability* **R-39**, 607–611.

Shen K and Xie M. 1991. The effectiveness of adding standby redundancy at system and component levels. *IEEE Transactions on Reliability* **40**, 53–55.

Xie M and Shen K. 1989. On ranking of system components with respect to different improvement actions. *Microelectronics and Reliability* **29**, 159–164.

10

Upgrading System Performance

In the design process of a system, a reliability analyst shall present the reliability structure of the system and evaluate the system performance. Beyond that, it is important to point out the critical components or subsystems so that the system performance can be improved when additional resources are available (Bergman, 1985b). As in Example 1.0.5, when there is a fixed budget for system upgrading, the more important components should get attention first, taking into account other information if necessary (e.g., the associated marginal cost for improving the components).

This chapter discusses the system upgrading associated with various component improvement actions, such as increasing component reliability, parallel active redundancy, standby redundancy, repair operations, burn-in, reducing working stress level, and so on. The problem is to decide which component(s) should be further improved to gain the greatest improvement in system performance at a limited cost or effort. Helpful tools for identifying the critical components include various types of importance measures that have to be closely connected with component improvement actions and system performance measures.

Suppose that a certain improvement action on component i increases the reliability of component i from p_i to g_i or upgrades its reliability function from $\overline{F}_i(t)$ to $\overline{G}_i(t)$. The importance measures should reflect the impact of the improved reliability of g_i or $\overline{G}_i(t)$ on the system performance of interest. The rule of thumb for using the importance measures to evaluate the strength of components is that the comparisons must be made *on the basis of the same improvement to each of the components*.

The commonly considered system performance measures include

(i) systems reliability at a fixed time t_0, $\Pr\{T_\phi \geq t_0\}$,
(ii) expected system lifetime, $\mathbb{E}(T_\phi)$, and
(iii) expected yield, $\mathbb{E}(Y(T_\phi))$, assuming $Y(\cdot)$ to be an increasing random process—the accumulated yield during the system lifetime.

Obviously, performance measures (i) and (ii) are special cases of the general performance measure (iii), which is a natural measure in many situations. For different system performance measures, different importance measures of components have been developed. If the reliability

Importance Measures in Reliability, Risk, and Optimization: Principles and Applications, First Edition.
Way Kuo and Xiaoyan Zhu. © 2012 John Wiley & Sons, Ltd. Published 2012 by John Wiley & Sons, Ltd.

of a system with a fixed mission time is of concern as with performance measure (i), then the reliability importance measures may be adequate. If multiple points in time are of concern, then the TDL importance measures may be needed. If the long lifetime of the system is expected as with performance measures (ii) and (iii), the TIL importance measures may be necessary. Because structure importance measures only consider the structural aspect of a system, they may not be adequate for analyzing a system upgrade where the system performance is related to current and new reliability values of components.

Other system performance measures for upgrading systems have been studied. Singh and Misra (1994) considered supplying redundancy to the components in a parallel or series system. The criterion is to prefer system A to B if $\Pr\{T_{\phi,A} > T_{\phi,B}\} \geq \Pr\{T_{\phi,B} > T_{\phi,A}\}$, where $T_{\phi,A}$ and $T_{\phi,B}$ are the lifetimes of resulting systems A and B, respectively. Mi (1998) used the redundancy, burn-in, and reducing working stress level to bolster components in order to stochastically maximize the lifetime of a parallel, series, or k-out-of-n system. Another possible performance measure is the expected restricted system lifetime, $\mathbb{E}(\min\{T_{\phi}, t_0\})$. However, few studies on importance measures are related to these performance measures, possibly because of mathematical inconvenience. Therefore, this chapter focuses on the system performance measures (i), (ii), and (iii).

With respect to a particular system performance measure and a particular improvement action on components, an importance measure of a component can be defined to evaluate the gain to the system performance by implementing that improvement to the component. Sections 10.1, 10.2, and 10.3 investigate the importance measures that are appropriate in application to improving systems reliability, expected system lifetime, and expected system yield, respectively. Finally, Section 10.4 gives a further discussion on practical implementation.

10.1 Improving Systems Reliability

A commonly considered objective is to improve systems reliability or the systems reliability function of time. Then, the problem may arise of deciding on which components additional research and development should be done to improve their reliability so that the greatest gain is achieved in the systems reliability.

When the reliability of component i increases from p_i to g_i, the improvement in the systems reliability, ΔR_i, is

$$\Delta R_i = R(g_i, \mathbf{p}) - R(\mathbf{p}) = (g_i - p_i)I_\mathrm{B}(i; \mathbf{p}), \tag{10.1}$$

or, in terms of lifetime distributions,

$$\Delta R_i(t) = R(\overline{G}_i(t), \overline{\mathbf{F}}(t)) - R(\overline{\mathbf{F}}(t)) = (\overline{G}_i(t) - \overline{F}_i(t))I_\mathrm{B}(i; \overline{\mathbf{F}}(t)). \tag{10.2}$$

ΔR_i and $\Delta R_i(t)$ are proportional to the B-importance and are related to g_i and $\overline{G}_i(t)$, respectively. Because $I_\mathrm{B}(i; \mathbf{p})$ and $I_\mathrm{B}(i; \overline{\mathbf{F}}(t))$ are always less than 1 according to Corollary 4.1.3, the increment in systems reliability due to the increase in reliability of component i is always less than $g_i - p_i$ or $\overline{G}_i(t) - \overline{F}_i(t)$. The different component improvement actions correspond to the different g_i and $\overline{G}_i(t)$ and thus result in the different formats of ΔR_i and $\Delta R_i(t)$. The importance measures can be defined in terms of ΔR_i or $\Delta R_i(t)$. On the basis of this whole picture, the remainder of this section discusses the effects of several improvement actions on increasing systems reliability. Note that all of the importance measures in this section are proportional to the B-importance, thus demonstrating the essential role of the B-importance.

10.1.1   Same Amount of Improvement in Component Reliability

When each component receives the same amount of increment in reliability (see also the assumption 1 for the DIM^I in Subsection 7.2.1), denoted by δ, that is,

$$g_i = p_i + \delta,$$

using Equation (10.1),

$$\Delta R_i = \delta I_B(i; \mathbf{p}).$$

Thus, the component with the largest B-reliability importance should be improved to achieve the highest improvement in systems reliability. This result is consistent with the monotone property of the B-reliability importance in Equation (4.14), that is,

$$R(\mathbf{p} + \Delta\mathbf{p}) - R(\mathbf{p}) \simeq \sum_{i=1}^{n} I_B(i; \mathbf{p}) \Delta p_i,$$

where $\Delta\mathbf{p} = (\Delta p_1, \Delta p_2, \ldots, \Delta p_n)$ and $\Delta p_i = g_i - p_i$. The B-reliability importance compares improvement in systems reliability when each component receives the same small amount of increment in reliability.

Note that further improvement of components of very high reliability may be very costly. However, the B-reliability importance of a component is independent of its reliability value, thus not reflecting this concern. To attack this shortcoming, cost constraints and other factors may be involved to limit the optimal action of a system upgrade. In Birnbaum (1969), the marginal cost of improving the component reliability is considered and assumed to be an increasing function of p_i, $\lambda_i(p_i)$, such that $\lambda_i(0) = 0$ and $\lim_{p \to 1} \lambda_i(p) = \infty$. The total cost of improving all components is $\sum_{i=1}^{n} \lambda_i(p_i) \Delta p_i$, and the gain in systems reliability per unit cost is

$$\frac{R(\mathbf{p} + \Delta\mathbf{p}) - R(\mathbf{p})}{\sum_{i=1}^{n} \lambda_i(p_i) \Delta p_i}. \tag{10.3}$$

The direction of steepest ascent of this gain is achieved in the following sense. Let

$$\Delta p_i = \alpha_i t, \quad i = 1, 2, \ldots, n$$

with

$$\sum_{i=1}^{n} \alpha_i^2 = 1. \tag{10.4}$$

It turns out to determine the vector of direction cosines $(\alpha_1, \alpha_2, \ldots, \alpha_n) = \boldsymbol{\alpha}$ so that, for all Δp_i small, expression (10.3) is maximized. Because expression (10.3) now is

$$\frac{R(\mathbf{p} + \alpha t) - R(\mathbf{p})}{t \sum_{i=1}^{n} \lambda_i(p_i)\alpha_i} \longrightarrow \frac{\left.\frac{dR(\mathbf{p}+\alpha t)}{dt}\right|_{t=0}}{\sum_{i=1}^{n} \lambda_i(p_i)\alpha_i}$$

as t approaches 0, the problem becomes maximizing

$$\frac{\sum_{i=1}^{n} \frac{\partial R(\mathbf{p})}{\partial p_i} \alpha_i}{\sum_{i=1}^{n} \lambda_i(p_i)\alpha_i} = \frac{\sum_{i=1}^{n} I_B(i;\mathbf{p})\alpha_i}{\sum_{i=1}^{n} \lambda_i(p_i)\alpha_i} = Q(\boldsymbol{\alpha})$$

under the restriction (10.4). The maximum of $Q(\boldsymbol{\alpha})$ is attained by selecting component i_0 for which the importance-to-cost ratio $I_B(i;\mathbf{p})/\lambda_i(p_i)$ is maximum and by setting $\alpha_{i_0} = 1$ and $\alpha_i = 0$ for $i \neq i_0$, because

$$Q(\boldsymbol{\alpha}) = \frac{I_B(i_0;\mathbf{p})\alpha_{i_0} + \sum_{i=1,i\neq i_0}^{n} I_B(i;\mathbf{p})\alpha_i}{\sum_{i=1}^{n} \lambda_i(p_i)\alpha_i}$$

$$< \frac{I_B(i_0;\mathbf{p})\alpha_{i_0} + \sum_{i=1,i\neq i_0}^{n} \frac{I_B(i_0;\mathbf{p})}{\lambda_{i_0}(p_{i_0})}\lambda_i(p_i)\alpha_i}{\sum_{i=1}^{n} \lambda_i(p_i)\alpha_i} = \frac{I_B(i_0;\mathbf{p})}{\lambda_{i_0}(p_{i_0})} = Q(1_{i_0}, \mathbf{0}).$$

For a reliable system, the component reliability typically varies between 0.9 and $1 - 10^{-8}$; thus, adding given increment δ in each component reliability is not a good test for a system upgrade because of the largeness and variability of component reliability. To overcome this disadvantage, some different improvement actions were considered, as shown in the next three subsections.

10.1.2 A Fractional Change in Component Reliability

Lambert (1975) assumed that each component lifetime distribution is improved by the same fractional change (see also the assumption 2 for the DIM^I in Subsection 7.2.1), that is,

$$g_i = p_i + \gamma q_i, \tag{10.5}$$

where γ is a given constant and could be any value between 0 and 1. Then, according to Equation (10.1), the ratio of the increment of the systems reliability to the constant γ is

$$\frac{\Delta R_i}{\gamma} = \frac{\gamma q_i I_B(i;\mathbf{p})}{\gamma} = q_i I_B(i;\mathbf{p}) = I_{Bs}(i;\mathbf{p}). \tag{10.6}$$

Thus, the component that has the largest value of the improvement potential importance $I_{Bs}(i;\mathbf{p})$ (see Subsection 4.1.1) should be improved under assumption (10.5) to gain the greatest possible increase in the systems reliability.

A perfect component

Considering a special case of $\gamma = 1$, then according to Equation (10.5),

$$g_i = 1.$$

That is, the original component is replaced with a perfect component whose reliability, approximately, is one. The replacement should be made to the components according to the ordering of their improvement potential importance values.

Note that the maximum improvement on a component is to replace it with a perfect one. In such a case, ΔR_i, equal to the improvement potential importance in Equation (10.6), is the maximum possible improvement in systems reliability that can be obtained by improving the reliability of component i. Thus, the systems reliability improvement gained by increasing an individual component cannot exceed

$$\max_{i=1}^{n}\{I_{Bs}(i; \mathbf{p})\}.$$

10.1.3   Cold Standby Redundancy

An improvement action is to have cold standby redundancy. Adding with component i one identical cold standby is equivalent to conducting exactly one instant perfect repair for component i. Then, the component is repaired to have the same distribution of remaining lifetime as originally, that is,

$$\overline{G}_i(t) = \overline{F}_i(t) + \int_0^t f_i(t - u)\overline{F}_i(u)du. \tag{10.7}$$

Then, using Equation (10.2),

$$\Delta R_i(t) = \left(\int_0^t f_i(t - u)\overline{F}_i(u)du\right) I_B(i; \overline{\mathbf{F}}_i(t)).$$

After the mission time t_0 is specified, the importance of components can be ranked according to $(\int_0^{t_0} f_i(t_0 - u)\overline{F}_i(u)du)I_B(i; \overline{\mathbf{F}}_i(t_0)))$.

10.1.4   Parallel Redundancy

One widely used improvement action is to add spares in parallel active redundancy with components in a system; this is thoroughly discussed in Chapter 9 where the parallel redundancy importance is defined in Equation (9.1) and is used to order the importance of components.

10.1.5   Example and Discussion

The following example shows that for different improvement actions, the ranking of the importance of components may be different. That is, the extent of importance of components to a system upgrade depends on the improvement actions. Thus, different importance measures should be used for different improvement actions.

Example 10.1.1   Considering the bridge structure in Figure 2.1 and supposing that the component reliability vector $\mathbf{p} = (0.8, 0.3, 0.5, 0.6, 0.7)$, then the current systems reliability $R(\mathbf{p}) = 0.673$.

If the improvement is to increase component reliability by the same small amount and the marginal cost for all components is assumed to be the same, then according to the B-reliability importance, where

$$(I_B(1; \mathbf{p}), I_B(2; \mathbf{p}), I_B(3; \mathbf{p}), I_B(4; \mathbf{p}), I_B(5; \mathbf{p})) = (0.545, 0.270, 0.168, 0.445, 0.250),$$

component 1 is the most important.

If the improvement is to make the same fractional changes in component reliability or to replace component i with a perfect one ($\gamma = 1$), then according to the improvement potential importance, where

$$(I_{Bs}(1; \mathbf{p}), I_{Bs}(2; \mathbf{p}), I_{Bs}(3; \mathbf{p}), I_{Bs}(4; \mathbf{p}), I_{Bs}(5; \mathbf{p})) = (0.109, 0.189, 0.084, 0.178, 0.075),$$

component 2 is the most important. In the case of updating with a perfect component ($\gamma = 1$), the maximal improved systems reliability is $R(\mathbf{p}) + I_{Bs}(2; \mathbf{p}) = 0.862$.

If the improvement action is to add the parallel active spare identical to the respective component in a system, then the increment in the systems reliability is calculated according to the parallel redundancy importance (9.11) where $p_i^* = p_i$ for $i = 1, 2, \ldots, 5$, as

$$(I_{PR}(1; \mathbf{p}), I_{PR}(2; \mathbf{p}), I_{PR}(3; \mathbf{p}), I_{PR}(4; \mathbf{p}), I_{PR}(5; \mathbf{p})) = (0.087, 0.057, 0.042, 0.107, 0.053).$$

Hence, component 4 is the most important, and the corresponding improved systems reliability is $R(\mathbf{p}) + I_{PR}(4; \mathbf{p}) = 0.780$.

10.2 Improving Expected System Lifetime

Some systems do not have a fixed mission time and are expected to function as long as possible. For example, microwave ovens are commonly used until they break down. In this situation, people may want to determine which components should receive the most urgent attention in order to increase the expected system lifetime.

Recall from Subsection 2.7.2 that random variable T_ϕ represents the lifetime of system ϕ and that $\overline{F}_\phi(t)$ is its reliability function. Then,

$$\mathbb{E}(T_\phi) = \int_0^\infty \overline{F}_\phi(t)dt$$

$$= \int_0^\infty \left[\overline{F}_i(t)(R(1_i, \overline{\mathbf{F}}(t)) - R(0_i, \overline{\mathbf{F}}(t))) + R(0_i, \overline{\mathbf{F}}(t)) \right] dt$$

$$= \int_0^\infty \overline{F}_i(t)I_B(i; \overline{\mathbf{F}}(t))dt + \int_0^\infty R(0_i, \overline{\mathbf{F}}(t))dt.$$

Let $T_{\phi,i}$ be the lifetime corresponding to the system in which the reliability function of component i is upgraded from $\overline{F}_i(t)$ to $\overline{G}_i(t)$. Then, the difference in expected system lifetime is

$$\Delta E_i = \mathbb{E}(T_{\phi,i}) - \mathbb{E}(T_\phi) = \int_0^\infty \left(\overline{G}_i(t) - \overline{F}_i(t) \right) I_B(i; \overline{\mathbf{F}}(t))dt \qquad (10.8)$$

and a normalized importance measure can be defined as

$$\frac{\Delta E_i}{\sum_{j=1}^n \Delta E_j}. \qquad (10.9)$$

With respect to a particular improvement action, $\overline{G}_i(t)$ can be specified and the corresponding importance measure in the format of (10.9) follows. Note that the importance measures defined by (10.8) and (10.9) are all weighted averages of the B-TDL importance with certain weight functions, again demonstrating the essential role of the B-importance. This section discusses several formats of improvement in component lifetime distributions. Subsections 10.2.1–10.2.8 investigate the different component improvements and the related importance measures. Subsection 10.2.9 further analyzes and compares all these TIL importance measures, showing that the BP importance and L_1 importance are most reasonable and acceptable.

10.2.1   A Shift in Component Lifetime Distributions

One improvement would be to upgrade $\overline{F}_i(t)$ by

$$\overline{G}_i(t) = \overline{F}_i(t - \varepsilon), \tag{10.10}$$

where ε is a small positive shift (transposition), the same for all components. In this case, it is natural to assume ε to be infinitesimal, and using Equation (10.8) then

$$\lim_{\varepsilon \to 0} \frac{\Delta E_i}{\varepsilon} = \lim_{\varepsilon \to 0} \int_0^\infty \frac{(F_i(t) - F_i(t - \varepsilon))}{\varepsilon} I_B(i; \overline{\mathbf{F}}(t)) dt$$

$$= \int_0^\infty I_B(i; \overline{\mathbf{F}}(t)) dF_i(t)$$

$$= I_{BP}(i; \overline{\mathbf{F}}).$$

This gives another interpretation of the BP TIL importance (see Section 5.3) in that it evaluates the contributions of the transposition improvements of component lifetimes to the expected system lifetime. Because $\sum_{i=1}^n I_{BP}(i; \overline{\mathbf{F}}) = 1$ by Theorem 5.3.3, there is no need for a normalization.

10.2.2   Exactly One Minimal Repair

Considering the component improvement action as exactly one minimal repair (i.e., the component will be given a minimal repair for its first failure but will not receive any repair for its second and later failures), then, $\overline{F}_i(t)$ is upgraded to (Norros 1986b)

$$\overline{G}_i(t) = \overline{F}_i(t) + \int_0^t f_i(t - u) \frac{\overline{F}_i(t)}{\overline{F}_i(t - u)} du = \overline{F}_i(t) \left[1 - \ln \overline{F}_i(t) \right]. \tag{10.11}$$

Under this situation, for $i = 1, 2, \ldots, n$, define a random variable

$$Z_i = T_{\phi,i} - T_\phi.$$

Then, Z_i and $\mathbb{E}(Z_i)$ represent, respectively, the increase in the system lifetime and the increment in the expected system lifetime that are caused by exactly one minimal repair of component i. Z_i can also be interpreted as

$$Z_i = Y_i^1 - Y_i^0,$$

where random variables $Y_i^0 =$ remaining system lifetime right after the failure of component i, and $Y_i^1 =$ remaining system lifetime right after the failure of component i, which, however, immediately undergoes a minimal repair, that is, it is repaired to have the same distribution of remaining lifetime as it had just before failing. Y_i^1 can also be interpreted as the remaining system lifetime just before the (first) failure of component i. *Thus, Z_i and $\mathbb{E}(Z_i)$ represent the reduction and the* expected *reduction in remaining system lifetime due to the failure of component i*. Note that $\mathbb{E}(Z_i) = \mathbb{E}(Y_i^1) - \mathbb{E}(Y_i^0)$; thus, $\mathbb{E}(Z_i)$ can also be interpreted as the reduction in *expected* remaining system lifetime due to the failure of component i.

According to Natvig (1979) and Natvig (1985), respectively, Lemma 10.2.1 and Theorem 10.2.2 present two expressions of $\mathbb{E}(Z_i)$.

Lemma 10.2.1 *Let*

$$c_{i,t}^1(u) = \frac{\overline{F}_i(t+u)}{\overline{F}_i(t)}, \; c_{i,t}^0(u) = 0, \; and \; \mathbf{c}_t^{\mathbf{x}}(u) = \left(c_{1,t}^{x_1}(u), c_{2,t}^{x_2}(u), \ldots, c_{n,t}^{x_n}(u)\right).$$

Then

$$\mathbb{E}(Z_i) = \int_0^\infty \sum_{(\cdot_i,\mathbf{x})} \left[\prod_{j \neq i} F_j(t)^{1-x_j} \overline{F}_j(t)^{x_j} \int_0^\infty \left[R(\mathbf{c}_t^{(1_i,\mathbf{x})}(u)) - R(\mathbf{c}_t^{(0_i,\mathbf{x})}(u)) \right] du \right] f_i(t) dt.$$

(10.12)

Proof. First note that the vector $\mathbf{c}_t^{\mathbf{x}}(u)$ gives the conditional reliability of the components at time $t+u$, given the component state vector, $\mathbf{x}$, at time t. Introduce the random variable $Y_t^{\mathbf{x}} =$ remaining lifetime for the system given the component state vector, $\mathbf{x}$, at time t. Then

$$\Pr\{Y_t^{\mathbf{x}} > u\} = \Pr\{\phi(\mathbf{X}(t+u)) = 1 | \mathbf{X}(t) = \mathbf{x}\} = R(\mathbf{c}_t^{\mathbf{x}}(u)),$$

and

$$\mathbb{E}(Y_t^{\mathbf{x}}) = \int_0^\infty \Pr\{Y_t^{\mathbf{x}} > u\} du = \int_0^\infty R(\mathbf{c}_t^{\mathbf{x}}(u)) du.$$

Hence,

$$\int_0^\infty \left[R(\mathbf{c}_t^{(1_i,\mathbf{x})}(u)) - R(\mathbf{c}_t^{(0_i,\mathbf{x})}(u)) \right] du$$

equals the conditional expected reduction in remaining system lifetime, given that component i fails at time t and that the component state vector just before time t is $(1_i, \mathbf{x})$. Now the expression of $\mathbb{E}(Z_i)$ follows by an ordinary conditional expectation argument.

Theorem 10.2.2

$$\mathbb{E}(Z_i) = \int_0^\infty \overline{F}_i(t) \left[-\ln \overline{F}_i(t) \right] I_B(i; \overline{\mathbf{F}}(t)) dt.$$

(10.13)

Proof. From Equation (10.12),

$$\mathbb{E}(Z_i) = \int_0^\infty \int_0^\infty \left[R\left(\frac{\overline{F}_i(t+u)}{\overline{F}_i(t)}, \overline{\mathbf{F}}(t+u) \right)_i - R(0_i, \overline{\mathbf{F}}(t+u)) \right] du f_i(t) dt$$

$$= \int_0^\infty \int_0^\infty \left[R\left(1_i, \overline{\mathbf{F}}(t+u)\right) - R(0_i, \overline{\mathbf{F}}(t+u)) \right] \frac{\overline{F}_i(t+u)}{\overline{F}_i(t)} du f_i(t) dt$$

$$= \int_0^\infty \overline{F}_i(v) \left[-\ln \overline{F}_i(v) \right] I_B(i; \overline{\mathbf{F}}(v)) dv.$$

Actually, Equation (10.13) follows directly from Equations (10.8) and (10.11).

Compared to Equation (10.12), the idea in Equation (10.13) is that conditioning on the states of the components excluding component i (i.e., $(\cdot_i, \mathbf{x})$) at the time when component i fails is unnecessary.

According to Equations (10.8) and (10.9), Natvig (1979) proposed the L_1 TIL importance using $\mathbb{E}(Z_i)$ as follows.

Definition 10.2.3   *The L_1 TIL importance of component i, denoted by $I_{L_1}(i; \overline{\mathbf{F}})$, is defined as*

$$I_{L_1}(i; \overline{\mathbf{F}}) = \frac{\mathbb{E}(Z_i)}{\sum_{j=1}^{n} \mathbb{E}(Z_j)} \tag{10.14}$$

where $\mathbb{E}(Z_i)$ is expressed in Equation (10.13) and is tacitly assumed to be finite, $i = 1, 2, \ldots, n$.

Norros (1986b) and Bueno (2000) extended the L_1 TIL importance to dependent components using the martingale notion and methods. This subsection investigates only the case of independent components. Definition 10.2.3 immediately implies the following theorem.

Theorem 10.2.4   *(i) $0 \leq I_{L_1}(i; \overline{\mathbf{F}}) \leq 1$. (ii) $\sum_{i=1}^{n} I_{L_1}(i; \overline{\mathbf{F}}) = 1$.*

Associated with the two aforementioned interpretations of Z_i and $\mathbb{E}(Z_i)$, the L_1 TIL importance, from a positive point of view, takes into account the improvements of component lifetimes due to exactly one minimal repair and then treats as most important those components that strongly improve the expected system lifetime. The negative viewpoint would be to consider the effect of component failure in reducing the remaining system lifetime and treat as most important those components whose failure strongly reduces the remaining system lifetime.

As in Corollary 10.2.5 (Natvig 1979), the expressions of $I_{L_1}(i; \overline{\mathbf{F}})$ for a series and a parallel system can be obtained from the expressions of $I_B(i; \overline{\mathbf{F}}(t))$ straightforwardly according to Equation (10.13). Note the similarity and difference between expressions (10.15) and (10.16).

Corollary 10.2.5   *The L_1 TIL importance of component i for a series system is*

$$I_{L_1}(i; \overline{\mathbf{F}}) = \frac{\int_0^\infty \overline{F}_i(t) \ln(\overline{F}_i(t)) \prod_{j \neq i} \overline{F}_j(t) dt}{\sum_{k=1}^{n} \int_0^\infty \overline{F}_k(t) \ln(\overline{F}_k(t)) \prod_{j \neq k} \overline{F}_j(t) dt}, \tag{10.15}$$

whereas for a parallel system it is

$$I_{L_1}(i; \overline{\mathbf{F}}) = \frac{\int_0^\infty \overline{F}_i(t) \ln(\overline{F}_i(t)) \prod_{j \neq i} F_j(t) dt}{\sum_{k=1}^{n} \int_0^\infty \overline{F}_k(t) \ln(\overline{F}_k(t)) \prod_{j \neq k} F_j(t) dt}. \tag{10.16}$$

Natvig (1982) developed the entire distribution of Z_i, which consists of a probability mass in $Z_i = 0$ and an absolutely continuous distribution for $Z_i > 0$. He also specified the distributions of Z_i for series and parallel systems. Since the expressions of the distributions are so complicated and do not contribute much to the L_1 importance, they are not presented here. Interested readers can refer to Natvig (1982) for detailed expressions and discussion.

Properties

The following Lemma 10.2.6 and Theorem 10.2.7 for the L_1 TIL importance correspond to Lemma 5.3.8 and Theorem 5.3.9 for the BP TIL importance. Natvig (1985) showed similar results to Lemma 10.2.6 but considered the special case that component i is in series (parallel) with the rest of the system. In contrast, Lemma 10.2.6 is for the general case that component i is in series (parallel) with a module of the system. Xie (1988) showed Theorem 10.2.7, whose proof does not benefit from Lemma 10.2.6.

Lemma 10.2.6 *Let component i be in series (parallel) with a module containing component j.*

(i) If $\overline{F}_j(t) > 0$ for $t \geq 0$, then $\mathbb{E}(Z_i)$ is strictly increasing in $\overline{F}_i(t)(-\ln \overline{F}_i(t))$.
(ii) If $0 < \overline{F}_i(t) < 1$ for $t > 0$, then $\mathbb{E}(Z_i)$ is strictly increasing (decreasing) in $\overline{F}_j(t)$, $0 < \overline{F}_j(t) < 1$.
(iii) If $F_i(t) = F_j(t)$ for $t \geq 0$, then $I_{L_1}(i; \overline{\mathbf{F}}) \geq I_{L_1}(j; \overline{\mathbf{F}})$.

Proof. Assuming that component i is in series with a module containing component j, let (M, χ) be a module consisting of component i and the module containing component j. The following proves the results for the series connection case and the proof is similar for the parallel connection case.

Parts (i) and (ii): When component i fails, the reliability of module (M, χ) is zero, that is, $R_\chi(0_i, \overline{\mathbf{F}}(t)^M) = 0$. From Equation (10.13), by the chain rule of the B-importance in Theorem 4.1.13,

$$\mathbb{E}(Z_i) = \int_0^\infty \overline{F}_i(t) \left[-\ln \overline{F}_i(t) \right] I_B^M(M; \overline{\mathbf{F}}(t)) R_\chi(1_i, \overline{\mathbf{F}}(t)^M) dt. \qquad (10.17)$$

Hence, $\mathbb{E}(Z_i)$ is strictly increasing in $\overline{F}_i(t)(-\ln \overline{F}_i(t))$ because $I_B^M(M; \overline{\mathbf{F}}(t)) R_\chi(1_i; \overline{\mathbf{F}}(t)) > 0$ for $t \geq 0$. Furthermore, $R_\chi(1_i; \overline{\mathbf{F}}(t)^M)$ is strictly increasing in $\overline{F}_j(t), 0 < \overline{F}_j(t) < 1$. The same is true for $\mathbb{E}(Z_i)$ since $\overline{F}_i(t)(-\ln \overline{F}_i(t)) > 0$ for $t > 0$.

Part (iii): Supposing that $F_i(t) = F_j(t) = F(t)$ for $t \geq 0$, then from Equation (10.17),

$$\mathbb{E}(Z_i) = \int_0^\infty \overline{F}(t) \left(-\ln \overline{F}(t) \right) I_B^M(M; \overline{\mathbf{F}}(t)) \left[\overline{F}(t) R_\chi(1_i, 1_j, \overline{\mathbf{F}}(t)^{M\backslash\{i,j\}}) \right.$$
$$\left. + F(t) R_\chi(1_i, 0_j, \overline{\mathbf{F}}(t)^{M\backslash\{i,j\}}) \right] dt$$

$$\mathbb{E}(Z_j) = \int_0^\infty \overline{F}(t) \left(-\ln \overline{F}(t) \right) I_B^M(M; \overline{\mathbf{F}}(t)) [R_\chi(1_i, 1_j, \overline{\mathbf{F}}(t)^{M\backslash\{i,j\}}) - R_\chi(1_i, 0_j, \overline{\mathbf{F}}(t)^{M\backslash\{i,j\}})] dt$$

$$= \mathbb{E}(Z_i) - \int_0^\infty \overline{F}(t) \left(-\ln \overline{F}(t) \right) I_B^M(M; \overline{\mathbf{F}}(t)) R_\chi(1_i, 0_j, \overline{\mathbf{F}}(t)^{M\backslash\{i,j\}}) dt$$

$$\leq \mathbb{E}(Z_i),$$

where $(1_i, 1_j, \overline{\mathbf{F}}(t)^{M\backslash\{i,j\}})$ represents a vector with elements $\overline{F}_k(t)$, $k \in M \backslash \{i, j\}$ and 1 in positions of $\overline{F}_i(t)$ and $\overline{F}_j(t)$ for $i, j \in M$; vector $(1_i, 0_j, \overline{\mathbf{F}}(t)^{M\backslash\{i,j\}})$ is defined similarly. Through normalization, inequality is established.

Note that Lemma 10.2.6(i) and (ii) are not true when replacing $\mathbb{E}(Z_i)$ with $I_{L_1}(i; \overline{\mathbf{F}})$ because of the normalization of $I_{L_1}(i; \overline{\mathbf{F}})$.

Theorem 10.2.7   *Let component i be in series (parallel) with a module containing component j. If $F_i(t) \geq F_j(t)$ $(\overline{F}_i(t) \geq \overline{F}_j(t))$ for $t \geq 0$, then $I_{L_1}(i; \overline{\mathbf{F}}) \geq I_{L_1}(j; \overline{\mathbf{F}})$.*

Proof. In both cases, it needs to be shown that $\mathbb{E}(Z_i) \geq \mathbb{E}(Z_j)$. Let $G(t)$ be the lifetime distribution of the module and $\overline{G}(t) = 1 - G(t)$. In the series connection case, let $\overline{F}_M(t) = \overline{F}_i(t)\overline{G}(t)$. Then,

$$\mathbb{E}(Z_i) - \mathbb{E}(Z_j) = \int_0^\infty \left(\overline{F}_i(t)(-\ln \overline{F}_i(t))\frac{\partial \overline{F}_\phi}{\partial \overline{F}_M}\frac{\partial \overline{F}_M}{\partial \overline{F}_i} - \overline{F}_j(t)(-\ln \overline{F}_j(t))\frac{\partial \overline{F}_\phi}{\partial \overline{F}_M}\frac{\partial \overline{F}_M}{\partial \overline{F}_j} \right) dt$$

$$= \int_0^\infty \frac{\partial \overline{F}_\phi}{\partial \overline{F}_M} \left(\overline{F}_i(t)(-\ln \overline{F}_i(t))\overline{G}(t) - \overline{F}_j(t)(-\ln \overline{F}_j(t))\overline{F}_i(t)\frac{\partial \overline{G}}{\partial \overline{F}_j} \right) dt.$$

This is nonnegative because by Equation (5.6), $\partial \overline{G}/\partial \overline{F}_j \leq \overline{G}(t)/\overline{F}_j(t)$, and then

$$(-\ln \overline{F}_i(t))\overline{G}(t) - \overline{F}_j(t)(-\ln \overline{F}_j(t))\frac{\partial \overline{G}}{\partial \overline{F}_j} \geq (-\ln \overline{F}_i(t))\overline{G}(t) - (-\ln \overline{F}_j(t))\overline{G}(t) \geq 0.$$

Hence, the statement of the series connection case is proved.

To prove the parallel connection case, let $F_M(t) = F_i(t)G(t)$. Then,

$$\mathbb{E}(Z_i) - \mathbb{E}(Z_j) = \int_0^\infty \frac{\partial \overline{F}_\phi}{\partial \overline{F}_M} \left(\overline{F}_i(t)(-\ln \overline{F}_i(t))G(t) - \overline{F}_j(t)(-\ln \overline{F}_j(t))F_i(t)\frac{\partial G}{\partial F_j} \right) dt.$$

By Equation (5.6), $\partial G/\partial F_j \leq G(t)/F_j(t)$; thus,

$$\overline{F}_i(t)(-\ln \overline{F}_i(t))G(t) - \overline{F}_j(t)(-\ln \overline{F}_j(t))F_i(t)\frac{\partial G}{\partial F_j}$$

$$\geq F_i(t)G(t)\left(\frac{\overline{F}_i(t)(-\ln \overline{F}_i(t))}{F_i(t)} - \frac{\overline{F}_j(t)(-\ln \overline{F}_j(t))}{F_j(t)} \right) \geq 0.$$

The last inequality holds because $\overline{F}_i(t) \geq \overline{F}_j(t)$, and $\overline{F}_i(t)(-\ln \overline{F}_i(t))/F_i(t)$ is a decreasing function of $F_i(t)$ that can be shown by straightforward algebra. This completes the proof.

Corollary 10.2.8   *Let components i and j be in series (parallel) with the rest of the system, $i \neq j$. If $F_i(t) \geq F_j(t)$ $(\overline{F}_i(t) \geq \overline{F}_j(t))$ for $t \geq 0$, then $I_{L_1}(i; \overline{\mathbf{F}}) \geq I_{L_1}(j; \overline{\mathbf{F}})$.*

Corollary 10.2.9   *Consider components i and j in a series (parallel) system, $i \neq j$. If $F_i(t) \geq F_j(t)$ $(\overline{F}_i(t) \geq \overline{F}_j(t))$ for $t \geq 0$, then $I_{L_1}(i; \overline{\mathbf{F}}) \geq I_{L_1}(j; \overline{\mathbf{F}})$.*

Note that Corollaries 10.2.8 and 10.2.9 are special cases of Theorem 10.2.7. They have been proved by Natvig (1985) and Natvig (1979), respectively. Similar to Theorem 5.3.11, Natvig and Gåsemyr (2009) extended Corollary 10.2.8 under its same condition by giving lower bounds on $\mathbb{E}(Z_i) - \mathbb{E}(Z_j)$ as

$$\mathbb{E}(Z_i) - \mathbb{E}(Z_j) \geq \int_0^\infty (-\ln \overline{F}_j(t)) R(0_j, \overline{\mathbf{F}}(t)) dt$$

when component i is in series with the rest of the system and as

$$\mathbb{E}(Z_i) - \mathbb{E}(Z_j) \geq \int_0^\infty (-\ln \overline{F}_j(t)) \frac{\overline{F}_j(t)}{F_j(t)} (1 - R(1_j, \overline{\mathbf{F}}(t))) dt$$

when component i is in parallel with the rest of the system. The right-hand-side terms of the inequalities are zero when component j is, respectively, in series and in parallel with the rest of the system. Further results are obtained for series–parallel systems, as shown in Theorem 10.2.10 (Natvig 1979). Appendix A.2 provides the proof of Theorem 10.2.10.

Theorem 10.2.10 *Assume that the lifetime distributions of components have proportional hazards as in Equation (2.16), that is, $\overline{F}_i(t) = e^{-\lambda_i h(t)}$, $\lambda_i > 0, t \geq 0, i = 1, 2, \ldots, n$, where*

$$\int_0^\infty h(t) \exp\left(\left(-\sum_{i=1}^n \lambda_i\right) h(t)\right) dt < \infty. \tag{10.18}$$

(i) Then for a series system,

$$I_{L_1}(i; \overline{\mathbf{F}}) = \frac{\lambda_i}{\sum_{j=1}^n \lambda_j}. \tag{10.19}$$

(ii) Assume, furthermore, that $h(t) = t^\alpha$, $t \geq 0$, $\alpha > 0$, that is, the lifetimes of components are Weibull distributed with the same shape parameter α as in Equation (2.17). Then, for a parallel system,

$$I_{L_1}(i; \overline{\mathbf{F}}) = \frac{\lambda_i \left[\lambda_i^{-\beta} - \sum_{j \neq i} (\lambda_i + \lambda_j)^{-\beta} + \sum_{\substack{j<k \\ j,k \neq i}} (\lambda_i + \lambda_j + \lambda_k)^{-\beta} - \cdots + (-1)^{n-1}(\lambda_1 + \lambda_2 + \cdots + \lambda_n)^{-\beta} \right]}{\sum_{\ell=1}^n \lambda_\ell \left[\lambda_\ell^{-\beta} - \sum_{j \neq \ell} (\lambda_\ell + \lambda_j)^{-\beta} + \sum_{\substack{j<k \\ j,k \neq \ell}} (\lambda_\ell + \lambda_j + \lambda_k)^{-\beta} - \cdots + (-1)^{n-1}(\lambda_1 + \lambda_2 + \cdots + \lambda_n)^{-\beta} \right]}, \tag{10.20}$$

where $\beta = (1 + 1/\alpha)$.

(iii) Make the same assumptions as in parts (i) and (ii). For a series (parallel) system of two components, $I_{L_1}(i; \overline{\mathbf{F}})$ is increasing (decreasing) in $F_i(t)$ and in $\overline{F}_j(t)$, $j \neq i$; furthermore,

$$I_{L_1}(i; \overline{\mathbf{F}}) \geq I_{L_1}(j; \overline{\mathbf{F}}) \text{ if and only if } F_i(t) \geq F_j(t) \ (\overline{F}_i(t) \geq \overline{F}_j(t)) \text{ for } t \geq 0. \tag{10.21}$$

Parts (i) and (ii) in Theorem 10.2.10 are analogous to Theorem 5.3.4 that pertains to the BP importance. For a series system, the results are similar except for assumption (10.18), and for a parallel system, the L_1 importance is identical to the BP importance if α approaches infinity, that is, if all components cease functioning at $t = 1$. The first result in Theorem 10.2.10(iii) corresponds to Lemma 5.3.8(i) for the BP importance, which, however, does not require the assumptions made in Theorem 10.2.10 and is not limited to a simple two-component system. Obviously, under the assumptions in Theorem 10.2.10, the rankings of components according to the BP importance and the L_1 importance are the same. However, this is not always true for a two-component series or parallel system when these assumptions do not hold, as shown in the following example.

Example 10.2.11 (Example 2.2.1 continued)   Consider a series system of two components, where

$$\overline{F}_i(t) = \exp(-\lambda_i t^{\alpha_i}), \quad \lambda_i > 0, \alpha_1 = 2, \alpha_2 = 1, t \geq 0.$$

Then,

$$I_{BP}(2; \overline{\mathbf{F}}) = \int_0^\infty \lambda_2 \exp(-(\lambda_1 t^2 + \lambda_2 t)) dt,$$

$$\mathbb{E}(Z_1) = \int_0^\infty \lambda_1 t^2 \exp(-(\lambda_1 t^2 + \lambda_2 t)) dt,$$

$$\mathbb{E}(Z_2) = \int_0^\infty \lambda_2 t \exp(-(\lambda_1 t^2 + \lambda_2 t)) dt.$$

Let $\gamma = \lambda_2/\sqrt{2\lambda_1}$, $G(t)$ be the distribution of the standard normal distribution, and $\bar{G}(t) = 1 - G(t)$. Then, by substituting $u = \sqrt{2\lambda_1}(t + \lambda_2/2\lambda_1)$,

$$I_{BP}(2; \overline{\mathbf{F}}) = \gamma \exp\left(\frac{\gamma^2}{2}\right) \sqrt{2\pi} \bar{G}(\gamma),$$

$$\mathbb{E}(Z_1) = \left(\frac{\gamma}{2\lambda_2}\right) \exp\left(\frac{\gamma^2}{2}\right) \left[(1 + \gamma^2)\sqrt{2\pi}\bar{G}(\gamma) - \gamma \exp\left(-\frac{\gamma^2}{2}\right)\right],$$

$$\mathbb{E}(Z_2) = \left(\frac{\gamma}{2\lambda_2}\right) \exp\left(\frac{\gamma^2}{2}\right) \left[2\gamma \exp\left(-\frac{\gamma^2}{2}\right) - 2\gamma^2\sqrt{2\pi}\bar{G}(\gamma)\right].$$

The last two expressions are obtained after some integration, and, from them, it follows that

$$I_{L_1}(2; \overline{\mathbf{F}}) = \frac{2\gamma \exp\left(-\frac{\gamma^2}{2}\right)(1 - I_{BP}(2; \overline{\mathbf{F}}))}{(1 - \gamma^2)\sqrt{2\pi}\,\bar{G}(\gamma) + \gamma \exp\left(-\frac{\gamma^2}{2}\right)}.$$

For $\gamma = 0.6$, $I_{BP}(2; \overline{\mathbf{F}}) = 0.494 < I_{BP}(1; \overline{\mathbf{F}}) = 0.506$, and $I_{L_1}(2; \overline{\mathbf{F}}) = 0.539 > I_{L_1}(1; \overline{\mathbf{F}}) = 0.461$.

The L_1 TIL importance of a modular set

Let a coherent system (N, ϕ) have a modular decomposition $\{(M_k, \chi_k)\}_{k=1}^m$.

Definition 10.2.12   *For $k = 1, 2, \ldots, m$, defining a random variable*

$$Z_{M_k} = reduction\ in\ remaining\ system\ lifetime\ due\ to\ the\ failure\ of\ modular\ set\ M_k,$$

then the L_1 TIL importance of modular set M_k, denoted by $I_{L_1}^M(M_k; \overline{\mathbf{F}})$, is defined as

$$I_{L_1}^M(M_k; \overline{\mathbf{F}}) = \frac{\mathbb{E}(Z_{M_k})}{\sum_{j=1}^m \mathbb{E}(Z_{M_j})},$$

tacitly assuming $\mathbb{E}(Z_{M_k}) < \infty, k = 1, 2, \ldots, m$.

Using the same notation as in Lemma 10.2.1, two expressions of $\mathbb{E}(Z_{M_k})$ as in Equations (10.22) and (10.23) (Natvig 1979) can be obtained almost the same as Equations (10.12) and (10.13), respectively, noting that the component whose failure coincides with the failure of the module must be critical for the failure of the module.

Theorem 10.2.13

$$\mathbb{E}(Z_{M_k}) = \sum_{i \in M_k} \int_0^\infty \sum_{(\cdot_i, \mathbf{x})} [\chi_k(1_i, \mathbf{x}^{M_k}) - \chi_k(0_i, \mathbf{x}^{M_k})] \left[\prod_{j \neq i} F_j(t)^{1-x_j} \overline{F}_j(t)^{x_j} \right.$$

$$\left. \times \int_0^\infty [R(\mathbf{c}_t^{(1_i, \mathbf{x})}(u)) - R(\mathbf{c}_t^{(0_i, \mathbf{x})}(u))]du \right] f_i(t)dt \qquad (10.22)$$

$$= \sum_{i \in M_k} \int_0^\infty \sum_{(\cdot_i, \mathbf{x}^{M_k})} [\chi_k(1_i, \mathbf{x}^{M_k}) - \chi_k(0_i, \mathbf{x}^{M_k})] \prod_{j \in M_k \setminus \{i\}} F_j(t)^{1-x_j} \overline{F}_j(t)^{x_j}$$

$$\times \int_0^\infty \left[R(\mathbf{c}_t^{(1_i, \mathbf{x}^{M_k \setminus \{i\}})}(u), (\overline{\mathbf{F}}(t+u))^{N \setminus M_k}) \right.$$

$$\left. - R(\mathbf{c}_t^{(0_i, \mathbf{x}^{M_k \setminus \{i\}})}(u), (\overline{\mathbf{F}}(t+u))^{N \setminus M_k}) \right] du f_i(t)dt, \qquad (10.23)$$

where $(\mathbf{c}_t^{(1_i, \mathbf{x}^{M_k \setminus \{i\}})}(u), (\overline{\mathbf{F}}(t+u))^{N \setminus M_k})$ *is a vector with* $\overline{F}_i(t+u)/\overline{F}_i(t)$ *in position* i, $c_{j,t}^{x_j}(u)$ *for* $j \in M_k \setminus \{i\}$, *and* $\overline{F}_j(t+u)$ *for all* $j \notin M_k$; *vector* $(\mathbf{c}_t^{(0_i, \mathbf{x}^{M_k \setminus \{i\}})}(u), (\overline{\mathbf{F}}(t+u))^{N \setminus M_k})$ *is defined similarly.*

Compared to Equation (10.22), the idea behind Equation (10.23) is that conditioning on the states of components outside the module at the time of the module failure is unnecessary. Additionally, Natvig (1982) developed the entire distribution of Z_{M_k}. Interested readers can refer to Natvig (1982) for detailed expressions.

Different from the BP importance in which $I_{BP}^M(M_k; \overline{\mathbf{F}}) = \sum_{i \in M_k} I_{BP}(i; \overline{\mathbf{F}})$ as shown in Theorem 5.3.7(ii), for the L_1 importance $\mathbb{E}(Z_{M_k}) \leq \sum_{i \in M_k} \mathbb{E}(Z_i)$. Note also that the L_1 importance of a modular set depends on the whole modular decomposition. That means a module of a system may have different values of the L_1 importance relative to the different modular decompositions of the system.

The L_1 TIL importance of a minimal cut

Let u be the number of minimal cuts in the system.

Definition 10.2.14   *For* $k = 1, 2, \ldots, u$, *defining a random variable*

$$Z_{C_k} = \text{reduction in remaining system lifetime due to the failure of minimal cut } C_k,$$

then the L_1 TIL importance of minimal cut C_k, *denoted by* $I_{L_1}^C(C_k; \overline{\mathbf{F}})$, *is defined as*

$$I_{L_1}^C(C_k; \overline{\mathbf{F}}) = \frac{\mathbb{E}(Z_{C_k})}{\sum_{j=1}^u \mathbb{E}(Z_{C_j})},$$

tacitly assuming $\mathbb{E}(Z_{C_k}) < \infty$, $k = 1, 2, \ldots, u$.

Z_{C_k} can be interpreted as the remaining system lifetime right after the failure of C_k, where, however, the last component to fail within this minimal cut immediately undergoes a minimal repair. Using the same notation as in Lemma 10.2.1 and Theorem 10.2.13, Natvig (1979, 1985) presented the distribution and expectation of random variable Z_{C_k} as in Theorem 10.2.15, whose proof is similar to those of Lemma 10.2.1 and Theorem 10.2.2.

Theorem 10.2.15

$$\Pr\{Z_{C_k} \le z\} = 1 - \sum_{i \in C_k} \int_0^\infty \prod_{j \in C_k \setminus \{i\}} F_j(t) R\left(\frac{\overline{F}_i(t+z)}{\overline{F}_i(t)}, \mathbf{0}^{C_k \setminus \{i\}}, (\overline{\mathbf{F}}(t+z))^{N \setminus C_k}\right)_i f_i(t)\,dt, \quad (10.24)$$

$$\mathbb{E}(Z_{C_k}) = \sum_{i \in C_k} \int_0^\infty \sum_{(\cdot_{C_k}, \mathbf{x})} \left[\prod_{j \notin C_k} F_j(t)^{1-x_j} \overline{F}_j(t)^{x_j} \prod_{j \in C_k \setminus \{i\}} F_j(t)\right.$$

$$\left. \times \int_0^\infty R(\mathbf{c}_t^{(1_i, \mathbf{0}^{C_k \setminus \{i\}}, \mathbf{x})}(u))\,du\right] f_i(t)\,dt \qquad (10.25)$$

$$= \sum_{i \in C_k} \int_0^\infty \left[\prod_{j \in C_k \setminus \{i\}} F_j(t) \int_0^\infty R\left(\frac{\overline{F}_i(t+u)}{\overline{F}_i(t)}, \mathbf{0}^{C_k \setminus \{i\}}, (\overline{\mathbf{F}}(t+u))^{N \setminus C_k}\right)_i du\right] f_i(t)\,dt.$$

$$(10.26)$$

Note that the component whose failure coincides with the failure of the minimal cut must be the last one to fail within the cut. Hence, Z_{C_k} cannot be interpreted as an increase in the system lifetime by improving the lifetime distributions of some components, and $\mathbb{E}(Z_{C_k})$ cannot be interpreted as a reduction in the expected remaining system lifetime due to the failure of the minimal cut.

Equation (10.26) is a simplified version of Equation (10.25), in which conditioning on the states of the components outside the minimal cut at the time of the minimal cut failing is unnecessary. In addition, Natvig (1982) gave another expression of distribution of Z_{C_k}, which is complicated and can be simplified to Equation (10.24).

10.2.3 Reduction in the Proportional Hazards

Assume that the lifetimes of components have proportional hazards as in Equation (2.16). For this special case, Natvig (1982) considered the component improvement of making the same, infinitesimal reduction in the proportional hazard rates and speculated on a measure as

$$\frac{-\frac{\partial \mathbb{E}(T_\phi)}{\partial \lambda_i}}{\sum_{j=1}^n -\frac{\partial \mathbb{E}(T_\phi)}{\partial \lambda_j}}, \qquad (10.27)$$

tacitly assuming $-\partial \mathbb{E}(T_\phi)/\partial \lambda_i < \infty$, $i = 1, 2, \ldots, n$. Recall that T_ϕ is the system lifetime and that

$$\mathbb{E}(T_\phi) = \int_0^\infty R(\overline{\mathbf{F}}(t))\,dt.$$

Assuming that the order of differentiation and integration is allowed to reverse, then

$$\frac{\partial \mathbb{E}(T_\phi)}{\partial \lambda_i} = \int_0^\infty \frac{\partial R(\overline{\mathbf{F}}(t))}{\partial \lambda_i} dt = \int_0^\infty \frac{\partial R(\overline{\mathbf{F}}(t))}{\partial \overline{F}_i(t)} \frac{\partial \overline{F}_i(t)}{\partial \lambda_i} dt$$

$$= \int_0^\infty -h(t) \exp(-\lambda_i h(t)) I_B(i; \overline{\mathbf{F}}(t)) dt \qquad (10.28)$$

$$= -\lambda_i^{-1} \mathbb{E}(Z_i).$$

The last equation follows from Equation (10.13). On the basis of Equation (10.28), the most reliable component (having the smallest λ_i) in a parallel system is the most important in terms of the measure (10.27). However, all components in a series system are equally important irrespective of the λ_i values. Natvig (1982) abandoned the measure (10.27) due to this observation.

Motivated by the fact that λ_i^{-1} is the expected lifetime of component i, Natvig (1982) proposed the L_2 TIL importance, which is proportional to the derivative of the expected system lifetime with respect to the inverse of the component's proportional hazard rate. Assuming that the order of differentiation and integration is allowed to reverse, then

$$\frac{\partial \mathbb{E}(T_\phi)}{\partial \lambda_i^{-1}} = \int_0^\infty \frac{\partial R(\overline{\mathbf{F}}(t))}{\partial \lambda_i^{-1}} dt = \int_0^\infty \frac{\partial R(\overline{\mathbf{F}}(t))}{\partial \overline{F}_i(t)} \frac{\partial \overline{F}_i(t)}{\partial \lambda_i^{-1}} dt$$

$$= \int_0^\infty \lambda_i^2 h(t) \exp(-\lambda_i h(t)) I_B(i; \overline{\mathbf{F}}(t)) dt \qquad (10.29)$$

$$= \lambda_i \mathbb{E}(Z_i).$$

Definition 10.2.16   *Assume that the lifetimes of components have proportional hazards as in Equation (2.16). The L_2 TIL importance of component i, denoted by $I_{L_2}(i; \overline{\mathbf{F}})$, is defined as*

$$I_{L_2}(i; \overline{\mathbf{F}}) = \frac{\dfrac{\partial \mathbb{E}(T_\phi)}{\partial \lambda_i^{-1}}}{\sum_{j=1}^n \dfrac{\partial \mathbb{E}(T_\phi)}{\partial \lambda_j^{-1}}} = \frac{\lambda_i \mathbb{E}(Z_i)}{\sum_{j=1}^n \lambda_j \mathbb{E}(Z_j)},$$

where $\mathbb{E}(Z_i)$ is expressed in Equation (10.13) and $\partial \mathbb{E}(T_\phi)/\partial \lambda_i^{-1}$ is tacitly assumed to be finite, $i = 1, 2, \ldots, n$.

Natvig (1985) presented Theorem 10.2.17 for the L_2 importance, which corresponds to Corollary 10.2.8 and Theorem 10.2.10(iii) for the L_1 importance. Appendix A.3 provides the proof of Theorem 10.2.17.

Theorem 10.2.17   *Let components i and j be in series (parallel) with the rest of the system, $i \neq j$. If $F_i(t) \geq F_j(t)$ for $t \geq 0$ (for the parallel connection case, $\overline{F}_i(t) \geq \overline{F}_j(t)$, and $h(t)/h'(t)$ is increasing for $t \geq 0$ where $h'(t)$ is the derivative of $h(t)$), then $I_{L_2}(i; \overline{\mathbf{F}}) \geq I_{L_2}(j; \overline{\mathbf{F}})$. In addition, for a two-component parallel system with $\overline{F}_2(t) \geq \overline{F}_1(t)$ for $t \geq 0$, if $h(t)/h'(t)$ is strictly increasing (constant, strictly decreasing), then $I_{L_2}(2; \overline{\mathbf{F}}) < (=, >) I_{L_2}(1; \overline{\mathbf{F}})$.*

As shown in Theorem 10.2.17, for a series system, the least reliable component is the most important according to the L_2 importance; however, for a two-component parallel system, the most reliable component could be the most or the least important, depending on the monotonicity of $h(t)/h'(t)$. For example, $h(t)/h'(t)$ is strictly increasing for $h(t) = t^\alpha, \alpha > 0$ (i.e., Weibull distributions with the same shape parameter α in Equation (2.17)), constant for $h(t) = e^t$, and strictly decreasing for $h(t) = e^{t^2}$. Under the knowledge of $\overline{F}_2(t) \geq \overline{F}_1(t)$ for $t \geq 0$ in a two-component parallel system, all three possible relations between $I_{L_2}(2; \overline{\mathbf{F}})$ and $I_{L_2}(1; \overline{\mathbf{F}})$ may occur (e.g., $I_{L_2}(2; \overline{\mathbf{F}}) < I_{L_2}(1; \overline{\mathbf{F}})$ when $h(t) = t^\alpha, \alpha > 0$; $I_{L_2}(2; \overline{\mathbf{F}}) = I_{L_2}(1; \overline{\mathbf{F}})$ when $h(t) = e^t$; $I_{L_2}(2; \overline{\mathbf{F}}) > I_{L_2}(1; \overline{\mathbf{F}})$ when $h(t) = e^{t^2}$). Compared to other importance measures, this property is unique and results in the doubt of the L_2 importance.

10.2.4 Cold Standby Redundancy

Compared to the situation of exactly one minimal repair in Subsection 10.2.2, if instead exactly one perfect repair of component i is undergone immediately after the failure of component i, then the repaired component has the same distribution of remaining lifetime as originally, that is, $\overline{F}_i(t)$ is upgraded to $\overline{G}_i(t) = \overline{F}_i(t) + \int_0^t f_i(t-u)\overline{F}_i(u)du$ as in Equation (10.7). The reliability operation is to replace component i with a new component of the same type if component i fails for the first time, for example, having an identical cold standby available for component i.

For $i = 1, 2, \ldots, n$, define a random variable

U_i = increase in system lifetime due to a perfect repair of component i.

Then, the expected increase in system lifetime according to Equation (10.8) is given by

$$\mathbb{E}(U_i) = \int_0^\infty \int_0^t f_i(t-u)\overline{F}_i(u)du\, I_\mathrm{B}(i; \overline{\mathbf{F}}(t))dt. \tag{10.30}$$

Note that for exponentially distributed lifetimes, a minimal repair and a perfect repair are the same and $\mathbb{E}(Z_i) = \mathbb{E}(U_i)$. Using $\mathbb{E}(U_i)$, the L_3 TIL importance is defined as follows (Natvig 1985).

Definition 10.2.18 *The L_3 TIL importance of component i, denoted by $I_{L_3}(i; \overline{\mathbf{F}})$, is defined as*

$$I_{L_3}(i; \overline{\mathbf{F}}) = \frac{\mathbb{E}(U_i)}{\sum_{j=1}^n \mathbb{E}(U_j)},$$

where $\mathbb{E}(U_i)$ is expressed in Equation (10.30) and is tacitly assumed to be finite, $i = 1, 2, \ldots, n$.

For a special case in which the lifetimes of the investigated components are Gamma distributions, Natvig (1985) proved Theorem 10.2.19 for the L_3 importance, which is similar to Corollary 10.2.8 for the L_1 importance.

Theorem 10.2.19 *Let components i and j be in series with the rest of the system, $i \neq j$. (Consider a parallel system.) Assume that the lifetimes of components i and j (all components)* *are Gamma distributed as in Equation* (2.19), *that is,*

$$\overline{F}_r(t) = \sum_{k=0}^{\alpha-1} \frac{(\lambda_r t)^k}{k!} \exp(-\lambda_r t), \quad \lambda_r > 0, t \geq 0, r = i, j \ (r = 1, 2, \ldots, n),$$

where $\alpha \geq 1$ *is an integer. If* $F_i(t) \geq F_j(t)$ ($\overline{F}_i(t) \geq \overline{F}_j(t)$) *for* $t \geq 0$, *then* $I_{L_3}(i; \overline{\mathbf{F}}) \geq I_{L_3}(j; \overline{\mathbf{F}})$.

Proof. After some algebra or from a direct argument, it follows that

$$\int_0^t f_i(t-u)\overline{F}_i(u)du = \sum_{k=\alpha}^{2\alpha-1} \frac{(\lambda_i t)^k}{k!} \exp(-\lambda_i t).$$

Note that $F_i(t) \geq F_j(t)$ ($\overline{F}_i(t) \geq \overline{F}_j(t)$) for $t \geq 0$ is equivalent to $\lambda_i \geq \lambda_j$ ($\lambda_i \leq \lambda_j$). Assume without loss of generality that $i = 1$ and $j = 2$. Consider first the series connection case, and thus $\lambda_1 \geq \lambda_2$. Then from Equation (10.30),

$$\mathbb{E}(U_1) - \mathbb{E}(U_2) = \sum_{j_1=\alpha}^{2\alpha-1}\sum_{j_2=0}^{\alpha-1}\int_0^{\infty} R(1_i, 1_j, \overline{\mathbf{F}}(t)^{(ij)})t^{j_1+j_2}\exp(-(\lambda_1+\lambda_2)t)dt$$

$$\times \frac{(\lambda_1\lambda_2)^{j_2}}{j_1!j_2!}(\lambda_1^{j_1-j_2} - \lambda_2^{j_1-j_2}) \geq 0,$$

and this part of the proof is complete.

Now considering the parallel system and thus $\lambda_1 \leq \lambda_2$, then

$$\mathbb{E}(U_1) = \sum_{j_1=\alpha}^{2\alpha-1}\sum_{j_2=\alpha}^{\infty}\cdots\sum_{j_n=\alpha}^{\infty}\frac{(j_1+j_2+\cdots+j_n)!}{(\lambda_1+\lambda_2+\cdots+\lambda_n)^{j_1+j_2+\cdots+j_n+1}}\prod_{r=1}^{n}\frac{(\lambda_r)^{j_r}}{j_r!}$$

$$= \sum_{j_1=\alpha}^{2\alpha-1}\frac{1}{\lambda_1}\sum_{j_2=\alpha}^{\infty}\cdots\sum_{j_n=\alpha}^{\infty}\frac{(j_1+j_2+\cdots+j_n)!}{j_1!j_2!\ldots j_n!}p_1^{j_1+1}p_2^{j_2}\ldots p_n^{j_n},$$

where $p_i = \lambda_i/(\lambda_1 + \lambda_2 + \cdots + \lambda_n)$, $i = 1, 2, \ldots, n$, can be interpreted as probabilities for multinomial events $A_1, A_2, \ldots, A_n$, respectively. Then

$$\sum_{j_2=\alpha}^{\infty}\sum_{j_3=\alpha}^{\infty}\cdots\sum_{j_n=\alpha}^{\infty}\frac{(j_1+j_2+\cdots+j_n)!}{j_1!j_2!\ldots j_n!}p_1^{j_1+1}p_2^{j_2}\ldots p_n^{j_n}$$

is the probability of A_r happening at least α times for all $r = 2, 3, \ldots, n$ before A_1 happens $j_1 + 1$ times. $\mathbb{E}(U_2)$ is obtained from $\mathbb{E}(U_1)$ by exchanging λ_1 and λ_2. The associated multinomial probability is now the one discussed earlier with A_1 and A_2 exchanged. Because $p_1 \leq p_2$, this new probability is smaller. Furthermore, $\lambda_1^{-1} \geq \lambda_2^{-1}$. Hence, $\mathbb{E}(U_1) \geq \mathbb{E}(U_2)$, and the proof is complete.

It is not easy to establish a similar result to Theorem 10.2.19 for general component lifetime distributions. In conclusion, the L_3 importance is mathematically rather inconvenient.

10.2.5 A Perfect Component

Different from the L_1 and L_3 TIL importance measures, Natvig (1985) considered the expected increase in system lifetime by replacing component i with a perfect one when component i fails. In reliability terms, replacing component i at failure with a perfect component is equivalent to having a perfect component in that position from the beginning. That is,

$$\overline{G}_i(t) = 1. \tag{10.31}$$

For $i = 1, 2, \ldots, n$, define a random variable

V_i = increase in system lifetime by replacing component i with a perfect one.

Then, the corresponding expected increase in system lifetime according to Equation (10.8) is given by

$$\mathbb{E}(V_i) = \int_0^\infty F_i(t) I_B(i; \overline{\mathbf{F}}(t)) dt. \tag{10.32}$$

Consequently, the L_4 TIL importance is defined as follows, which was also suggested by Bergman (1985a).

Definition 10.2.20 *The L_4 TIL importance of component i, denoted by $I_{L_4}(i; \overline{\mathbf{F}})$, is defined as*

$$I_{L_4}(i; \overline{\mathbf{F}}) = \frac{\mathbb{E}(V_i)}{\sum_{j=1}^n \mathbb{E}(V_j)},$$

where $\mathbb{E}(V_i)$ is expressed in Equation (10.32) and is tacitly assumed to be finite, $i = 1, 2, \ldots, n$.

When component i is in parallel with the rest of the system, $\mathbb{E}(V_i) = \int_0^\infty R(1_i, \overline{\mathbf{F}}(t)) dt - \int_0^\infty R(\overline{\mathbf{F}}(t)) dt = \int_0^\infty 1 dt - \mathbb{E}(T_\phi) = \infty - \mathbb{E}(T_\phi)$. This is a degenerate case. If this relation is formally applied, then $I_{L_4}(i; \overline{\mathbf{F}}) = I_{L_4}(j; \overline{\mathbf{F}}), i \neq j$, when components i and j are both in parallel with the rest of the system, irrespective of their lifetime distributions. Thus, the L_4 importance is not well defined. Basically, the motivation of replacing component i with a perfect one is suspect.

10.2.6 An Imperfect Repair

Norros (1986a) considered a "β-improvement" of component i, which is to upgrade $\overline{F}_i(t)$ by

$$\overline{G}_i(t) = \overline{F}_i(t)^\beta, \quad 0 < \beta < 1. \tag{10.33}$$

Then, according to Equations (10.8) and (10.9), the difference in expected system lifetime is

$$\Delta E_i = \int_0^\infty \left(\overline{F}_i(t)^\beta - \overline{F}_i(t) \right) I_B(i; \overline{\mathbf{F}}(t)) dt,$$

and the corresponding importance measure can be defined as follows.

Definition 10.2.21 *The L_5 TIL importance of component i, denoted by $I_{L_5}(i; \overline{\mathbf{F}})$, is defined as*

$$I_{L_5}(i; \overline{\mathbf{F}}) = \frac{\int_0^\infty \left(\overline{F}_i(t)^\beta - \overline{F}_i(t)\right) I_{\mathrm{B}}(i; \overline{\mathbf{F}}(t))\mathrm{d}t}{\sum_{j=1}^n \int_0^\infty \left(\overline{F}_j(t)^\beta - \overline{F}_j(t)\right) I_{\mathrm{B}}(j; \overline{\mathbf{F}}(t))\mathrm{d}t}.$$

This transformation can model an imperfect repair (Brown and Proschan 1983) because $\overline{F}_i(0)^\beta \geq \overline{G}_i(t) = \overline{F}_i(t)^\beta \geq \overline{F}_i(t)$ for $t > 0$. If the distribution $\overline{F}_i(t)$ is specified, the effects of the imperfect repair can be further addressed. The situations for a minimal repair and perfect repair have been discussed in Subsections 10.2.2 and 10.2.4, respectively. When $\beta = 0$ and thus $\overline{G}_i(t) = 1$, this improvement represents a replacement with a perfect component as in Subsection 10.2.5.

10.2.7 A Scale Change in Component Lifetime Distributions

An improvement of a scale change of the component lifetime distribution is to upgrade $\overline{F}_i(t)$ by

$$\overline{G}_i(t) = \overline{F}_i\left(\frac{t}{\omega}\right), \quad \omega > 1.$$

This improvement is motivated by the fact that the most natural type of reliability improvement is the decrease in relative working stress of the component, which may be effectuated either by protecting the component or increasing its strength. Almost all published models relating lifetime distributions on several stress levels to each other are of this type.

It is natural to make only small changes, meaning that ω should be close to 1. Using Equation (10.8) and making the scale changes infinitesimal, Bergman (1985b) considered

$$\lim_{\omega \to 1} \frac{\Delta E_i}{\omega - 1} = \lim_{\omega \to 1} \int_0^\infty \frac{1}{\omega - 1} \left(\overline{F}_i\left(\frac{t}{\omega}\right) - \overline{F}_i(t)\right) I_{\mathrm{B}}(i; \overline{\mathbf{F}}(t))\mathrm{d}t = \int_0^\infty t I_{\mathrm{B}}(i; \overline{\mathbf{F}}(t))\mathrm{d}F_i(t).$$

After normalization, this importance measure is referred to as the L_6 TIL importance.

Definition 10.2.22 *The L_6 TIL importance of component i, denoted by $I_{L_6}(i; \overline{\mathbf{F}})$, is defined as*

$$I_{L_6}(i; \overline{\mathbf{F}}) = \frac{\int_0^\infty t I_{\mathrm{B}}(i; \overline{\mathbf{F}}(t))\mathrm{d}F_i(t)}{\sum_{j=1}^n \int_0^\infty t I_{\mathrm{B}}(j; \overline{\mathbf{F}}(t))\mathrm{d}F_j(t)}.$$

In the special case that all components have Weibull lifetime distributions with the same shape parameter as in Equation (2.17), it is easy to show that $I_{L_6}(i; \overline{\mathbf{F}})$ coincides with $I_{L_1}(i; \overline{\mathbf{F}})$ (see Subsection 10.2.2); thus, the results about $I_{L_1}(i; \overline{\mathbf{F}})$ hold for $I_{L_6}(i; \overline{\mathbf{F}})$ in this special case.

10.2.8 Parallel Redundancy

Boland and El-Neweihi (1995) studied a case where component i is improved by adding an independent parallel redundant component of lifetime distribution $F_i^*(t)$. Then, the corresponding

$$\overline{G}_i(t) = 1 - F_i(t)F_i^*(t). \tag{10.34}$$

According to Equation (10.8), a TIL type of parallel redundancy importance is obtained as follows.

Definition 10.2.23 *The parallel redundancy TIL importance of component i, denoted by $I_{PR}(i; \overline{\mathbf{F}})$, is defined as*

$$I_{PR}(i; \overline{\mathbf{F}}) = \Delta E_i = \int_0^\infty \overline{F}_i^*(t) F_i(t) I_B(i; \overline{\mathbf{F}}(t)) dt = \int_0^\infty I_{PR}(i; \overline{\mathbf{F}}(t)) dt, \qquad (10.35)$$

where

$$I_{PR}(i; \overline{\mathbf{F}}(t)) = \overline{F}_i^*(t) F_i(t) I_B(i; \overline{\mathbf{F}}(t)) \qquad (10.36)$$

is the TDL type of parallel redundancy importance. The reliability type of parallel redundancy importance is defined in Equation (9.1) as $I_{PR}(i; \mathbf{p}) = p_i^* q_i I_B(i; \mathbf{p})$.

Chapter 9 discusses the effect of parallel redundancy on systems reliability. The parallel redundancy TIL importance here evaluates the effect on the expected system lifetime, representing the expected increase in system lifetime. As in Equation (10.35), $I_{PR}(i; \overline{\mathbf{F}})$ is an integral of $I_{PR}(i; \overline{\mathbf{F}}(t))$. Thus, some results for $I_{PR}(i; \overline{\mathbf{F}}(t))$ and $I_{PR}(i; \mathbf{p})$ can be extended to $I_{PR}(i; \overline{\mathbf{F}})$. For example, consider a case where one spare is available for parallel active redundancy anywhere in a k-out-of-n system. By Theorem 9.2.1, if component i is stochastically weaker than component j (i.e., $F_i(t) \geq F_j(t)$ for $t \geq 0$), then $I_{PR}(i; \overline{\mathbf{F}}) \geq I_{PR}(j; \overline{\mathbf{F}})$, and thus, the spare would be the most useful when added in parallel redundancy with the least reliable component in the system. Note that for a parallel system, the benefit is the same no matter where the spare is added. Räde (1989) proposed a similar TIL importance measure for only the special case of parallel active redundancy where the spare is identical to the respective component in the system.

10.2.9 Comparisons and Numerical Evaluation

Subsections 10.2.1–10.2.8 discuss eight TIL importance measures: $I_{BP}(\cdot; \overline{\mathbf{F}})$, $I_{L_k}(\cdot; \overline{\mathbf{F}})$ for $k = 1, 2, \ldots, 6$, and $I_{PR}(\cdot; \overline{\mathbf{F}})$. These importance measures judge the importance of components by the contributions of the different improvements of component lifetimes to expected system lifetime. The BP TIL importance assumes that the lifetimes of all components are improved by the same infinitesimal shift. The L_1, L_3, and L_4 TIL importance measures assume that the failed component undergoes a minimal repair, a perfect repair (i.e., switching to an identical cold standby), and a replacement with a perfect component, respectively. The L_2 importance is for the special case of components having proportional hazards and assumes that the expected lifetimes of components are improved by the same amount. The L_5 and L_6 TIL importance measures assume that the "β-improvement" and the scale change of component lifetimes, respectively, modeling the imperfect repair and the decrease in relative working stress of the component. The parallel redundancy TIL importance is for the case of adding a parallel redundant component.

All eight of these TIL importance measures have a common feature; they all are weighted averages of the B-TDL importance but with different weight functions. Define the weight function $w_{L_k}(i, t)$ for the L_k importance measures for $k = 1, 2, \ldots, 6$ such that

$I_{L_k}(i; \overline{\mathbf{F}}) = \int_0^\infty w_{L_k}(i, t) I_B(i; \overline{\mathbf{F}}(t)) dt / \text{const.}$, where const. are the denominators in their definitions. Similar definitions apply to the BP importance and parallel redundancy importance.

For the BP importance, the weight function $w_{BP}(i, t) = f_i(t)$. It gives most weight to the time point, say $t_{BP}(i)$, when a component is most likely to fail.

For the L_1 importance, the weight function $w_{L_1}(i, t) = \overline{F}_i(t)\left(-\ln \overline{F}_i(t)\right)$ according to Equations (10.14) and (10.13). It starts from 0 at $t = 0$, achieves a single maximum at the time point, say $t_{L_1}(i)$, such that $F_i(t_{L_1}(i)) = 1 - e^{-1}$, and asymptotically approaches 0 as t approaches infinity. In the case of proportional hazards, $h(t_{L_1}(i)) = \lambda_i^{-1}$.

For the L_2 importance, the weight function $w_{L_2}(i, t) = \lambda_i^2 h(t) \exp(-\lambda_i h(t))$ according to Equation (10.29). It achieves a maximum value $\lambda_i e^{-1}$ at the time $t_{L_2}(i)$ such that $h(t_{L_2}(i)) = 1/\lambda_i$, and its other properties depend on the common function $h(t)$. The L_2 importance seems unreasonable because it finds self-conflicted situations in a two-component parallel system, as shown in Theorem 10.2.17.

For the L_3 importance, the weight function $w_{L_3}(i, t) = \int_0^t f_i(t - u)\overline{F}_i(u)du = F_i(t) - \int_0^t f_i(u)F_i(t - u)du$. It starts from 0 at $t = 0$ and approaches 0 as t approaches infinity. It achieves a maximum point at the time $t_{L_3}(i)$ such that $f_i(t_{L_3}(i)) = \int_0^{t_{L_3}(i)} f_i(u)f_i(t_{L_3}(i) - u)du$.

For the L_4 importance, the weight function $w_{L_4}(i, t) = F_i(t)$, which always gives most weight to large t and has a degenerate case; thus, it is unreasonable.

For the L_5 importance, the weight function $w_{L_5}(i, t) = \overline{F}_i(t)^\beta - \overline{F}_i(t)$ for $0 < \beta < 1$. It starts from 0 at $t = 0$ and approaches 0 as t approaches infinity. It achieves a maximum point at the time $t_{L_5}(i)$ such that $F_i(t_{L_5}(i)) = 1 - \beta^{1/(1-\beta)}$. When β approaches 0, $1 - \beta^{1/(1-\beta)}$ approaches 1, and thus, $t_{L_5}(i)$ approaches infinity, which is the case for the L_4 importance. When $\beta = 0$, the imperfect repair associated with the L_5 importance becomes the replacement with a perfect component associated with the L_4 importance. When β approaches 1, $1 - \beta^{1/(1-\beta)}$ approaches $1 - e^{-1}$ and then $t_{L_5}(i)$ approaches $t_{L_1}(i)$.

For the L_6 importance, the weight function $w_{L_6}(i, t) = t f_i(t)$.

For the parallel redundancy importance, the weight function $w_{PR}(i, t) = \overline{F}_i^*(t)F_i(t)$. It starts from 0 at $t = 0$ and approaches 0 as t approaches infinity. Assuming the redundant component is identical to component i, then the weight function $w_{PR}(i, t)$ achieves the maximum point at $t_{PR}(i)$ such that $F_i(t_{PR}(i)) = 1/2$, that is, at the moment the component has the half chance to fail.

For further comparisons of these TIL importance measures (excluding the L_2 and L_4 importance measures), consider Gamma distributed component lifetimes as in Equation (2.18), that is, $f_i(t) = \lambda_i^\alpha t^{\alpha-1} e^{-\lambda_i t}/\Gamma(\alpha)$. For simplification, let $\lambda_i = 1$ for $i = 1, 2, \ldots, n$, and then $f_1(t) = f_2(t) = \ldots = f_n(t)$. Table 10.1 presents the expressions of expectation of Gamma distribution $\mathbb{E}(T_i)$, $t_{BP}(i)$, $t_{L_1}(i)$, $t_{L_3}(i)$, $t_{L_5}(i)$, $t_{L_6}(i)$, and $t_{PR}(i)$ in the first row and the corresponding values with respect to different values of α in the second to sixth rows. Note that the L_5 importance depends on the parameter of β in Equation (10.33), and in Table 10.1, $\beta = 0.5$.

As shown in Table 10.1, $t_{BP}(i) = \max\{0, (\alpha - 1)/\lambda_i\} < \alpha/\lambda_i$, that is, $t_{BP}(i)$ is less than the expected component lifetime, and so does $t_{PR}(i)$. $t_{L_6}(i)$ is exactly equal to the expected component lifetime. For $\alpha > 1$, $t_{L_3}(i) > (\alpha^\alpha)^{1/\alpha}/\lambda_i = \alpha/\lambda_i$, implying that $t_{L_3}(i)$ is larger than the expected component lifetime. Similarly, $t_{L_1}(i)$ and $t_{L_5}(i)$ are always larger than but close to the expected component lifetime. This analysis demonstrates the different behaviors of the weight functions and, thus, the importance measures. Essentially, these are due to the different assumptions on the component improvements associated with various importance measures.

Table 10.1   Comparisons of the TIL importance measures

	$\mathbb{E}(T_i)$	$t_{BP}(i)$	$t_{L_1}(i)$	$t_{L_3}(i)$	$t_{L_5}(i)$	$t_{L_6}(i)$	$t_{PR}(i)$
α	$\frac{\alpha}{\lambda_i}$	$\max\{0, \frac{\alpha-1}{\lambda_i}\}$	$\frac{1}{2\lambda_i}\chi^{2\,a}_{2\alpha,1-e^{-1}}$	$\left(\frac{\Gamma(2\alpha)}{\Gamma(\alpha)}\right)^{\frac{1}{\alpha}}\lambda_i^{-1}$	$\frac{1}{2\lambda_i}\chi^{2\,a}_{2\alpha,1-\beta\frac{1}{1-\beta}}$	$\frac{\alpha}{\lambda_i}$	$\frac{1}{2\lambda_i}\chi^{2\,a}_{2\alpha,0.5}$
0.5	0.5	0	0.41	0.32	0.66	0.5	0.23
1	1	0	1	1	1.39	1	0.69
2	2	1	2.15	2.45	2.69	2	1.68
4	4	3	4.35	5.38	5.11	4	3.67
10	10	9	10.75	14.21	11.91	10	9.67

[a] $\chi^2_{2\alpha,\gamma}$ is the lower γ point in the χ^2 distribution with 2α degrees of freedom.

Numerical comparisons of the BP TIL importance and L_1 TIL importance

To show the possibly different rankings induced from various TIL importance measures, the BP TIL importance and L_1 TIL importance are tested as an example for the two three-component systems in Figures 2.4 and 2.5. As in Section 5.5, assume that the lifetimes of components have proportional hazards as in Equation (2.16), that is, $\overline{F}_i(t) = e^{-\lambda_i h(t)}$ for $i = 1, 2, 3$, where $h(t) = t$ is the common hazard. Furthermore, assume that $\lambda_2 = \lambda_3$ so that $\overline{F}_2(t) = \overline{F}_3(t)$. In each of the systems in Figures 2.4 and 2.5, components 2 and 3 have the same value for any TIL importance measure because these two positions are structurally symmetric and $\overline{F}_2(t) = \overline{F}_3(t)$. Thus, only the importance measures of components 1 and 2 are compared.

Table 10.2 lists the results for different values of λ_1, λ_2, and λ_3. The BP TIL importance and L_1 TIL importance may or may not produce the same ranking of components, depending on the values of parameters. For a given importance measure, the ranking of the importance of components in a system depends on the component lifetime distributions.

10.3   Improving Expected System Yield

The system yield, $Y(t)$, $t \geq 0$, is assumed to be some nondecreasing, real-valued, function of the system lifetime, and its derivative, $Y'(t)$, is assumed to exist. A typical yield function

Table 10.2   Comparisons of the BP TIL importance and L_1 TIL importance for the systems in Figures 2.4 and 2.5

	System in Figure 2.4		System in Figure 2.5	
$(\lambda_1, \lambda_2, \lambda_3)$	BP	L_1	BP	L_1
(0.1, 1, 1)	2(0.433), 1(0.134)	2(0.447), 1(0.106)	1(0.952), 2(0.024)	1(0.995), 2(0.002)
(0.39, 1, 1)	1(0.398), 2(0.301)	2(0.336), 1(0.329)	1(0.837), 2(0.082)	1(0.943), 2(0.028)
(0.4, 1, 1)	1(0.405), 2(0.298)	1(0.335), 2(0.333)	1(0.833), 2(0.083)	1(0.941), 2(0.030)
(3.9, 1, 1)	1(0.931), 2(0.035)	1(0.892), 2(0.054)	1(0.339), 2(0.331)	2(0.377), 1(0.246)
(4, 1, 1)	1(0.933), 2(0.033)	1(0.895), 2(0.052)	1(0.333), 2(0.333)	2(0.381), 1(0.238)
(5, 1, 1)	1(0.952), 2(0.024)	1(0.923), 2(0.039)	2(0.357), 1(0.286)	2(0.412), 1(0.176)

is $Y(t) = t$, by which the system yield is the system lifetime. This section does not specify the system yield (i.e., the format of $Y(t)$); instead, it presents a general class of importance measures that are defined on an arbitrary yield function. For situations where a yield function is given, the importance measures with respect to the particular yield function could be further specified.

With the objective of increasing expected system yield, $\mathbb{E}(Y(T_\phi))$, one could improve the system at the component level. Assuming that the reliability function of component i is upgraded from $\overline{F}_i(t)$ to $\overline{G}_i(t)$, recall that $T_{\phi,i}$ represents the lifetime of the upgraded system. Then the expected system yield is changed from $\mathbb{E}(Y(T_\phi))$ to $\mathbb{E}(Y(T_{\phi,i}))$. Assuming that these expectations exist and $Y(0) = 0$, then

$$
\mathbb{E}(Y(T_\phi)) = \int_0^\infty Y(t) \mathrm{d}F_\phi(t)
$$

$$
= \int_0^\infty \overline{F}_\phi(t) \mathrm{d}Y(t) = \int_0^\infty \mathbb{E}(\phi(\mathbf{X}(t))) \mathrm{d}Y(t)
$$

$$
= \int_0^\infty \overline{F}_i(t) I_\mathrm{B}(i; \overline{\mathbf{F}}(t)) \mathrm{d}Y(t) + \int_0^\infty \mathbb{E}(\phi(0_i, \mathbf{X}(t))) \mathrm{d}Y(t).
$$

The increase in expected system yield, denoted by $\Delta E_{Y,i}$, is thus

$$
\Delta E_{Y,i} = \mathbb{E}(Y(T_{\phi,i})) - \mathbb{E}(Y(T_\phi)) = \int_0^\infty (\overline{G}_i(t) - \overline{F}_i(t)) I_\mathrm{B}(i; \overline{\mathbf{F}}(t)) \mathrm{d}Y(t), \qquad (10.37)
$$

and a class of normalized importance measures can be defined as

$$
\frac{\Delta E_{Y,i}}{\sum_{j=1}^n \Delta E_{Y,j}}.
$$

The most important component would be the one that maximizes $\Delta E_{Y,i}$ since the improvement action on such a component incurs the greatest increase in the expected system yield.

Similar to Sections 10.1 and 10.2, the importance measures related to improving expected system yield are derived using different formats of $\overline{G}_i(t)$ in expression (10.37), which correspond to various improvement actions at the component level. Subsection 10.3.1 focuses on the improvement that results in a shift in the lifetime distributions of components. Subsection 10.3.2 presents and compares $\Delta E_{Y,i}$ associated with four types of component improvement actions, including minimal repair operations, perfect repair operations, replacements with perfect components, and parallel redundancy.

10.3.1   A Shift in Component Lifetime Distributions

This subsection considers the improvement of the same infinitesimal shift (transposition) for the lifetime distributions of all components, that is, $\overline{G}_i(t) = \overline{F}_i(t - \varepsilon)$ for $\varepsilon > 0$ as in

Equation (10.10). Using Equation (10.37) and letting the shift ε approach 0, Xie (1987) defined a yield TIL importance, $I_Y(i; \overline{\mathbf{F}})$, which is proportional to

$$
\begin{aligned}
\lim_{\varepsilon \to 0} \frac{\Delta E_{Y,i}}{\varepsilon} &= \lim_{\varepsilon \to 0} \frac{1}{\varepsilon} \int_0^\infty (\overline{G}_i(t) - \overline{F}_i(t)) I_B(i; \overline{\mathbf{F}}(t)) \mathrm{d}Y(t) \\
&= \int_0^\infty \lim_{\varepsilon \to 0} \frac{1}{\varepsilon} (F_i(t) - F_i(t - \varepsilon)) I_B(i; \overline{\mathbf{F}}(t)) \mathrm{d}Y(t) \\
&= \int_0^\infty f_i(t) I_B(i; \overline{\mathbf{F}}(t)) \mathrm{d}Y(t) \\
&= \int_0^\infty Y'(t) I_B(i; \overline{\mathbf{F}}(t)) \mathrm{d}F_i(t).
\end{aligned}
$$

Following Lemma 10.3.1, Definition 10.3.2 gives the complete definition of $I_Y(i; \overline{\mathbf{F}})$.

Lemma 10.3.1 *Let EY' denote the expectation of $Y'(T_\phi)$, that is,*

$$
EY' = \mathbb{E}(Y'(T_\phi)) = \int_0^\infty Y'(t) \mathrm{d}F_\phi(t) = \int_0^\infty f_\phi(t) \mathrm{d}Y(t). \tag{10.38}
$$

The last equality holds whenever $F_\phi(t)$ has density $f_\phi(t)$. Then,

$$
\sum_{i=1}^n \int_0^\infty Y'(t) I_B(i; \overline{\mathbf{F}}(t)) \mathrm{d}F_i(t) = EY'. \tag{10.39}
$$

Proof. It suffices to show that $\sum_{i=1}^n I_B(i; \overline{\mathbf{F}}(t)) \mathrm{d}F_i(t) = \mathrm{d}F_\phi(t)$ for $t \geq 0$, and this follows from Equation (5.7).

Definition 10.3.2 *For any yield function $Y(t)$, the yield TIL importance of component i denoted by $I_Y(i; \overline{\mathbf{F}})$, is defined as*

$$
I_Y(i; \overline{\mathbf{F}}) = \frac{1}{EY'} \int_0^\infty Y'(t) I_B(i; \overline{\mathbf{F}}(t)) \mathrm{d}F_i(t), \tag{10.40}
$$

where EY' is referred to as proportionality constant and is defined in Equation (10.38), assuming that EY' exists and is finite.

The definition of the yield importance in (10.40) can be generalized to dependent components by using the B-importance of dependent components as in Equation (4.8). It is easily seen that $0 \leq I_Y(i; \overline{\mathbf{F}}) \leq 1$ since the yield function $Y(t)$ is nondecreasing and consequently $Y'(t) \geq 0$. Furthermore, the proportionality constant EY' in Equation (10.39) ensures that the sum of the yield importance over all components is unity.

Theorem 10.3.3   *(i)* $0 \leq I_Y(i; \overline{\mathbf{F}}) \leq 1$. *(ii)* $\sum_{i=1}^{n} I_Y(i; \overline{\mathbf{F}}) = 1$.

As in Definition 10.3.2, the yield TIL importance is a weighted B-TDL importance with weight function $Y'(t)f_i(t)/EY'$, which is related to the yield function $Y(t)$. Using Equations (10.38) and (10.40), Xie (1987) gave an expression of the yield TIL importance associated with a special yield function $Y(t)$ as in Corollary 10.3.4.

Corollary 10.3.4   *If $Y(t) = ct^{\alpha+1}$ for some constants c and $\alpha \geq 0$, then*

$$I_Y(i; \overline{\mathbf{F}}) = \frac{1}{\mu_\alpha} \int_0^\infty t^\alpha I_B(i; \overline{\mathbf{F}}(t)) \mathrm{d}F_i(t),$$

where $\mu_\alpha = \int_0^\infty t^\alpha \mathrm{d}F_\phi(t)$ is the αth moment of $F_\phi(t)$.

The function $ct^{\alpha+1}$ may be used as an approximation of the true yield function due to its simplicity. When $\alpha = 0$ and $c = 1$, then $Y(t) = t$, the system yield is the system lifetime, and

$$I_Y(i; \overline{\mathbf{F}}) = \int_0^\infty I_B(i; \overline{\mathbf{F}}(t)) \mathrm{d}F_i(t) = I_{BP}(i; \overline{\mathbf{F}}).$$

Thus, $I_{BP}(i; \overline{\mathbf{F}})$ is the simplest case of $I_Y(i; \overline{\mathbf{F}})$ since neither the proportionality constant ($EY' = 1$) nor the weight function ($Y'(t)f_i(t)/EY' = f_i(t)$) depends on the system structure. When $\alpha = 1$ and $c = 1$, then $Y(t) = t^2$ and

$$I_Y(i; \overline{\mathbf{F}}) = I_{L_6}(i; \overline{\mathbf{F}}).$$

Thus, the results for the yield importance are also applicable for the BP importance and L_6 importance since they are special cases of the yield importance.

The yield TIL importance of a modular set

Let a coherent system (N, ϕ) have a module (M, χ) and the lifetime distribution of module (M, χ) be $F_M(t)$. Treating a module as a whole, a yield TIL importance of a modular set can be defined as follows.

Definition 10.3.5   *The yield TIL importance of modular set M, denoted by $I_Y^M(M; \overline{\mathbf{F}})$, is defined as*

$$I_Y^M(M; \overline{\mathbf{F}}) = \frac{1}{EY'} \int_0^\infty Y'(t) I_B^M(M; \overline{\mathbf{F}}(t)) \mathrm{d}F_M(t),$$

where $I_B^M(M; \overline{\mathbf{F}}(t))$ is the B-TDL importance of modular set M.

This definition is consistent with the BP TIL importance of modular sets (Definition 5.3.6) in which $Y'(t) = 1$ and $EY' = 1$. Theorem 10.3.6 (Xie 1987) shows that the yield importance of a modular set coincides with the sum of the yield importance of all individual components in the modular set and that the sum of the yield importance over all disjoint modular sets is unity. It is a generalization of Theorem 5.3.7 that pertains to the BP importance. By Theorem 10.3.6, if $I_Y(i; \overline{\mathbf{F}}) > I_Y^M(M; \overline{\mathbf{F}})$, $i \notin M$ (i.e., component i is more yield important than a module

$(M, \chi))$, then $I_Y(i; \overline{\mathbf{F}}) > I_Y(j; \overline{\mathbf{F}})$ for all $j \in M$ (i.e., component i is more yield important than each component in modular set M).

Theorem 10.3.6

(i) $I_Y^M(M; \overline{\mathbf{F}}) = \sum_{i \in M} I_Y(i; \overline{\mathbf{F}})$.

(ii) Let a system (N, ϕ) have a modular decomposition $\{(M_k, \chi_k)\}_{k=1}^m$, then $\sum_{k=1}^m I_Y^M(M_k; \overline{\mathbf{F}})$

$= 1$.

Proof. Using Theorem 4.1.13,

$$I_B^M(M; \overline{\mathbf{F}}(t)) \frac{dF_M}{dt} = \sum_{i \in M} I_B^M(M; \overline{\mathbf{F}}(t)) \frac{\partial \overline{F}_M}{\partial \overline{F}_i} \frac{dF_i}{dt} = \sum_{i \in M} I_B(i; \overline{\mathbf{F}}(t)) \frac{dF_i}{dt}.$$

Thus, the result follows from the definitions of $I_Y^M(M; \overline{\mathbf{F}})$ and $I_Y(i; \overline{\mathbf{F}})$.

Furthermore, according to Theorem 10.3.3, the proportionality constant EY' also ensures that the sum of the yield importance over all disjoint modular sets is unity.

The importance of a component in an embedded system is defined by treating the embedded system as an isolated system. Generally, the ordering of the importance measures of two components in an embedded system cannot be presumed to be the same as the ordering in the system that contains the embedded system. For the yield importance, Xie and Bergman (1991) showed a simple condition to ensure this relationship, as in the following theorem.

Theorem 10.3.7 *If for two distinct components i and j in a module (M, χ),*

$$\frac{\partial \overline{F}_M}{\partial \overline{F}_i} f_i(t) \geq \frac{\partial \overline{F}_M}{\partial \overline{F}_j} f_j(t), \tag{10.41}$$

then component i is more yield important than component j in all systems containing this module.

Proof. Assuming that a system containing module (M, χ) has a lifetime distribution $F_\phi(t)$, then

$$I_B(i; \overline{\mathbf{F}}(t)) f_i(t) - I_B(j; \overline{\mathbf{F}}(t)) f_j(t)$$

$$= \frac{\partial \overline{F}_\phi}{\partial \overline{F}_M} \frac{\partial \overline{F}_M}{\partial \overline{F}_i} f_i(t) - \frac{\partial \overline{F}_\phi}{\partial \overline{F}_M} \frac{\partial \overline{F}_M}{\partial \overline{F}_j} f_j(t) = \frac{\partial \overline{F}_\phi}{\partial \overline{F}_M} \left(\frac{\partial \overline{F}_M}{\partial \overline{F}_i} f_i(t) - \frac{\partial \overline{F}_M}{\partial \overline{F}_j} f_j(t) \right) \geq 0.$$

Thus, by Theorem 10.3.11(i) later, component i is more yield important than component j in all systems containing module (M, χ).

According to the chain rule in Theorem 4.1.13, the B-TDL importance measures of two components in an embedded system always have the same ordering as in any system that contains the embedded system. According to Theorem 10.2.2, for the L_1 TIL importance, the conclusion in Theorem 10.3.7 holds when condition (10.41) is changed to

$$\frac{\partial \overline{F}_M}{\partial \overline{F}_i} \overline{F}_i(t)(-\ln \overline{F}_i(t)) \geq \frac{\partial \overline{F}_M}{\partial \overline{F}_j} \overline{F}_j(t)(-\ln \overline{F}_j(t)).$$

The results for the L_1 importance and the yield importance (including the BP importance and L_6 importance) benefit from the chain rule of the B-importance.

Properties

When component i is in series or in parallel with the rest of the system, expressions of $I_Y(i; \overline{\mathbf{F}})$ can be simplified as in Theorem 10.3.8 (Xie 1987).

Theorem 10.3.8   *If component i is in series with the rest of the system, then*

$$I_Y(i; \overline{\mathbf{F}}) = \frac{1}{EY'} \int_0^\infty Y'(t) \overline{F}_\phi(t) r_i(t) dt, \tag{10.42}$$

where $r_i(t)$ is the failure rate of component i, and if component i is in parallel with the rest of the system, then

$$I_Y(i; \overline{\mathbf{F}}) = \frac{1}{EY'} \int_0^\infty Y'(t) F_\phi(t) \frac{f_i(t)}{F_i(t)} dt. \tag{10.43}$$

Proof. Equations (10.42) and (10.43) follow, respectively, from Equations (5.8) and (5.9) in Theorem 5.1.3, which pertain to $I_B(i; \overline{\mathbf{F}}(t))$.

Like Theorem 5.1.3 relative to the B-importance, Theorem 10.3.8 implies that when component i is in series or in parallel with the rest of the system, $I_Y(i; \overline{\mathbf{F}})$ can be computed with only the knowledge of the lifetime distributions of component i and the whole system. No information about other components is needed; the components may even be dependent. This result may be useful when both component i and the system are tested and accurate empirical distributions are available. Generally, when only knowing $F_i(t)$ and $F_\phi(t)$, $I_B(i; \overline{\mathbf{F}}(t))$ and $I_Y(i; \overline{\mathbf{F}})$ cannot be computed. They also relate to the structure of the system and the lifetime distributions of other components; thus, the computation of $I_Y(i; \overline{\mathbf{F}})$ is a much more difficult task.

However, in many practical situations (e.g., identifying the most important portion of a system), only the ranking of component importance is of concern. Xie and Bergman (1991) obtained Theorem 10.3.9, Corollary 10.3.10, and Theorem 10.3.11, which are useful in comparing the yield importance of two components.

Theorem 10.3.9   *Supposing that component i is in series with component j and for some constant $\alpha > 0$, $r_i(t) = \alpha r_j(t)$, then $I_Y(i; \overline{\mathbf{F}}) = \alpha I_Y(j; \overline{\mathbf{F}})$.*

Proof. Considering a modular set $M = \{i, j\}$, the reliability function of the modular set is $\overline{F}_M(t) = \overline{F}_i(t) \overline{F}_j(t)$, and

$$I_B(i; \overline{\mathbf{F}}(t)) = \frac{\partial \overline{F}_\phi}{\partial \overline{F}_i} = \frac{\partial \overline{F}_\phi}{\partial \overline{F}_M} \frac{\partial \overline{F}_M}{\partial \overline{F}_i} = \frac{\partial \overline{F}_\phi}{\partial \overline{F}_M} \frac{\overline{F}_M(t)}{\overline{F}_i(t)}.$$

Hence, by definition,

$$I_Y(i; \overline{\mathbf{F}}) = \frac{1}{EY'} \int_0^\infty Y'(t) \frac{\partial \overline{F}_\phi}{\partial \overline{F}_M} \frac{\overline{F}_M(t)}{\overline{F}_i(t)} f_i(t) dt = \frac{1}{EY'} \int_0^\infty Y'(t) \frac{\partial \overline{F}_\phi}{\partial \overline{F}_M} \overline{F}_M(t) r_i(t) dt.$$

A similar equality holds for $I_B(j; \overline{\mathbf{F}}(t))$ with i replaced with j in the aforementioned equation, and the result follows if $r_i(t) = \alpha r_j(t)$ for some α.

Corollary 10.3.10   *For a series system, if the lifetimes of components have proportional hazards as in Equation (2.16), that is, $\overline{F}_i(t) = \exp(-\lambda_i h(t))$, $\lambda_i > 0$, $t \geq 0$, $i = 1, 2, \ldots, n$, then*

$$I_Y(i; \overline{\mathbf{F}}) = \frac{\lambda_i}{\sum_{j=1}^n \lambda_j}.$$

Proof. In this case, $r_i(t) = -\lambda_i h'(t)$ for $i = 1, 2, \ldots, n$. Thus, by Theorems 10.3.3 and 10.3.9, the result follows.

For a series system in which the lifetimes of components have proportional hazards, the yield importance of components can be ordered by their failure rates as shown in Corollary 5. The same result has been proved directly for the BP importance (Theorem 5.3.4) and for the L_1 importance under an additional assumption (Theorem 10.2.10), whereas Corollary 10.3.10 generally holds for any yield function.

Theorem 10.3.11

(i) *For two distinct components i and j in a system, if $I_B(i; \overline{\mathbf{F}}(t)) f_i(t) \geq I_B(j; \overline{\mathbf{F}}(t)) f_j(t)$ for $t \geq 0$, then $I_Y(i; \overline{\mathbf{F}}) \geq I_Y(j; \overline{\mathbf{F}})$.*

(ii) *If components i and j are in parallel with the rest of the system and $f_i(t) F_j(t) \geq f_j(t) F_i(t)$ for $t \geq 0$, then $I_Y(i; \overline{\mathbf{F}}) \geq I_Y(j; \overline{\mathbf{F}})$.*

(iii) *If components i and j are in series with the rest of the system and $r_i(t) \geq r_j(t)$ for $t \geq 0$, then $I_Y(i; \overline{\mathbf{F}}) \geq I_Y(j; \overline{\mathbf{F}})$.*

(iv) *If component i is in series with a module containing component j and $r_i(t) \geq r_j(t)$ for $t \geq 0$, then $I_Y(i; \overline{\mathbf{F}}) \geq I_Y(j; \overline{\mathbf{F}})$.*

Appendix A.4 provides the proof of Theorem 10.3.11. In Theorem 10.3.11, part (iv) is a generalization of part (iii). Part (iv) is valid for the BP TIL importance, since it is the yield TIL importance with yield function $Y(t) = t$. Note that, by using Equation (2.15), if $r_i(t) \geq r_j(t)$ for $t \geq 0$, then $F_i(t) \geq F_j(t)$ for $t \geq 0$; however, the reverse is not true. Therefore, $r_i(t) \geq r_j(t)$ for $t \geq 0$ is a stronger condition than $F_i(t) \geq F_j(t)$ for $t \geq 0$. Considering the BP TIL importance, the series connection part in Theorem 5.3.9, which assumes $F_i(t) \geq F_j(t)$ for $t \geq 0$, is thus more general than Theorem 10.3.11(iv). Note that for Weibull distributions with the same shape parameter as in Equation (2.17), these two conditions are equivalent. The next example (Xie and Bergman 1991) shows the case of Weibull distributions.

Example 10.3.12   Assume component i is in series with a module containing component j. Let $F_i(t)$ and $F_j(t)$ be Weibull distributions as in (2.17) with parameters (α_i, λ_i) and

(α_j, λ_j), respectively. Let $Y'(t) = 0$ for $t \geq t_0$. If $\alpha_i \neq \alpha_j$ and $t_0 \leq \left(\alpha_i \lambda_j^{\alpha_j} / \alpha_j \lambda_i^{\alpha_i}\right)^{1/(\alpha_j - \alpha_i)}$, then $I_Y(i; \overline{\mathbf{F}}) \geq I_Y(j; \overline{\mathbf{F}})$ because $r_i(t) \geq r_j(t)$ for $t \leq t_0$. If $\alpha_i = \alpha_j$ (i.e., the shape parameters are the same), then the condition $r_i(t) \geq r_j(t)$ for $t \geq 0$ is the same as $\lambda_i \leq \lambda_j$.

It should be noted that in many cases as demonstrated in the preceding text, the ranking of component importance is independent of the yield function. The simplest one, the BP importance, which corresponds to $Y(t) = t$, is thus sufficient in these cases.

Bounds on the yield TIL importance

Theorem 10.3.13 (Xie and Bergman 1991) includes the main results about the bounds on the yield TIL importance. Since the proportionality constant EY' may be computed whenever the system lifetime distribution is known, these results mainly concern the bounds on the numerator in Equation (10.40). Following Theorem 10.3.13, Example 10.3.14 (Xie and Bergman 1991) shows its use in a simple case.

Theorem 10.3.13

(i) *For component i in a system,*

$$I_Y(i; \overline{\mathbf{F}}) \leq \min\left\{ \frac{1}{EY'} \int_0^\infty Y'(t) \overline{F}_\phi(t) r_i(t) dt, \; \frac{1}{EY'} \int_0^\infty Y'(t) F_\phi(t) r_i(t) d \ln F_i(t) \right\}.$$

(ii) *If component i is in series with the rest of the system, then*

$$\inf_{t \geq 0} r_i(t) \cdot \frac{EY}{EY'} \leq I_Y(i; \overline{\mathbf{F}}) \leq \sup_{t \geq 0} r_i(t) \cdot \frac{EY}{EY'}.$$

(iii) *If (M, χ) is a module in series with the rest of the system containing component i, then*

$$I_Y(i; \overline{\mathbf{F}}) \leq I_Y^M(M; \overline{\mathbf{F}}) \leq \sup_{t \geq 0} r_M(t) \cdot \frac{EY}{EY'},$$

where $r_M(t) = f_M(t)/\overline{F}_M(t)$ is the failure rate function of the module.

Proof. Part (i): The results follow from the bounds on the B-TDL importance given in Lemma 5.1.2.

Part (ii): Note that

$$\int_0^\infty Y'(t) \overline{F}_\phi(t) dt = \int_0^\infty \overline{F}_\phi(t) dY(t) = \int_0^\infty Y(t) dF_\phi(t) = EY,$$

where the second equality follows from the fact that $Y(0) = 0$ and $\overline{F}_\phi(\infty) = 0$. Then, the results follow from Equation (10.42).

Part (iii): By Theorem 10.3.6(i), an obvious upper bound on $I_Y(i; \overline{\mathbf{F}})$ for component i in a module can be obtained from part (ii).

Example 10.3.14 Suppose that component i is in series with the rest of the system. If the failure rate of component i, $r_i(t)$, is increasing, and $Y(0) = 0$ and $Y'(t) = 0$ for $t \geq t_0$, then by Theorem 10.3.13(ii),

$$r_i(0)\frac{EY}{EY'} \leq I_Y(i; \overline{\mathbf{F}}) \leq r_i(t_0)\frac{EY}{EY'}.$$

These inequalities are reversed if the failure rate $r_i(t)$ is decreasing.

Consider a module that consists of a number of i.i.d. components in parallel. The yield importance of each component in the module decreases as the number of components in the module increases. For a special case, a bound on the reduction of $I_Y(i; \overline{\mathbf{F}})$ can be obtained as follows, whose proof is in Appendix A.5.

Theorem 10.3.15   *Let component i be in series with the rest of the system and assume that it is replaced with a module consisting of m i.i.d. components in parallel. If $Y'(t)$ is an increasing (decreasing) function of t, then the reduction of the yield TIL importance of each component in the module is bounded from above (below) by*

$$\frac{1}{E_\phi Y'} \int_0^\infty Y'(t) \frac{F_i(t)(1 - F_i^{m-1}(t))}{\overline{F}_i(t)} \overline{F}_\phi(t) r_i(t) dt, \tag{10.44}$$

where $E_\phi Y'$ is defined as in Equation (10.38) for the original system ϕ. Furthermore, expression (10.44) is the exact reduction of the BP TIL importance because then $E_\psi Y' = E_\phi Y' = 1$. Again, the BP importance is the most fundamental and the simplest of the yield importance class.

10.3.2   *Exactly One Minimal Repair, Cold Standby Redundancy, a Perfect Component, and Parallel Redundancy*

The improvement of the lifetime distribution of component i corresponding to exactly one minimal repair on component i can be expressed as $\overline{G}_i(t) = \overline{F}_i(t)[1 - \ln \overline{F}_i(t)]$ in Equation (10.11). Thus, according to Equation (10.37), the corresponding increase in expected system yield is

$$\Delta E_{Y,i}^{(1)} = \int_0^\infty \overline{F}_i(t)[-\ln \overline{F}_i(t)]I_B(i; \overline{\mathbf{F}}(t)) dY(t).$$

Restoring a failed component by exactly one perfect repair is equivalent to replacing it with a new one or to adding an identical cold standby with it. In this case, $\overline{G}_i(t) = \overline{F}_i(t) + \int_0^t f_i(t - u)\overline{F}_i(u)du$ as in Equation (10.7), and

$$\Delta E_{Y,i}^{(2)} = \int_0^\infty \int_0^t \overline{F}_i(t - u)dF_i(u)I_B(i; \overline{\mathbf{F}}(t)) dY(t).$$

The maximum improvement on a component is to replace it with a very reliable one, approximately a perfect component, $\overline{G}_i(t) = 1$ as in Equation (10.31). Then,

$$\Delta E_{Y,i}^{(3)} = \int_0^\infty F_i(t)I_B(i; \overline{\mathbf{F}}(t)) dY(t).$$

$\Delta E_{Y,i}^{(3)}$ can be infinite when component i is in parallel with the rest of the system (see also Subsection 10.2.5 for the case of $Y(t) = t$), although in other cases, it is usually finite. This is because placing a perfect component in parallel with the rest of the system increases the systems reliability to one.

Assuming that an identical parallel redundant component is added with component i, then $\overline{G}_i(t) = 1 - F_i(t)F_i(t)$ as in Equation (10.34). Thus,

$$\Delta E_{Y,i}^{(4)} = \int_0^\infty \overline{F}_i(t)F_i(t)I_B(i; \overline{\mathbf{F}}(t))\mathrm{d}Y(t) = \int_0^\infty I_{PR}(i; \overline{\mathbf{F}}(t))\mathrm{d}Y(t),$$

where $I_{PR}(i; \overline{\mathbf{F}}(t))$ is the parallel redundancy importance of component i as in Equation (10.36).

Theorem 10.3.16 (Xie and Shen 1990) states the relations among $\Delta E_{Y,i}^{(k)}$ for $k = 1, 2, 3, 4$. Of course, these relations hold true for the corresponding ΔE_i, which are defined in Section 10.2 for the case of $Y(t) = t$.

Theorem 10.3.16   *For $\Delta E_{Y,i}^{(k)}$, $k = 1, 2, 3, 4$, defined above,*

$$\Delta E_{Y,i}^{(4)} \leq \Delta E_{Y,i}^{(1)}, \Delta E_{Y,i}^{(2)} \leq \Delta E_{Y,i}^{(3)}.$$

Furthermore, if $\overline{F}_i(t + t_0) \leq (\geq) \overline{F}_i(t)\overline{F}_i(t_0)$ for $t \geq 0$ and $t_0 \geq 0$, then

$$\Delta E_{Y,i}^{(1)} \leq (\geq) \Delta E_{Y,i}^{(2)}.$$

Proof. $\Delta E_{Y,i}^{(4)} \leq \Delta E_{Y,i}^{(2)}$ follows because $\int_0^t \overline{F}_i(t - u)\mathrm{d}F_i(u) \geq \int_0^t \overline{F}_i(t)\mathrm{d}F_i(u) = \overline{F}_i(t)F_i(t)$. $\Delta E_{Y,i}^{(2)} \leq \Delta E_{Y,i}^{(3)}$ follows because $\int_0^t \overline{F}_i(t - u)\mathrm{d}F_i(u) \leq \int_0^t 1\mathrm{d}F_i(u) = F_i(t)$.

To show $\Delta E_{Y,i}^{(1)} \leq \Delta E_{Y,i}^{(3)}$ is equivalent to showing

$$-\ln \overline{F}_i(t) = -\ln(1 - F_i(t)) = \sum_{s=1}^\infty \frac{F_i^s(t)}{s} \leq \sum_{s=1}^\infty F_i^s(t) = \frac{F_i(t)}{\overline{F}_i(t)},$$

where the second equality follows from the Taylor expression. Similarly, $\Delta E_{Y,i}^{(4)} \leq \Delta E_{Y,i}^{(1)}$ follows because $-\ln \overline{F}_i(t) = -\ln(1 - F_i(t)) = \sum_{s=1}^\infty F_i^s(t)/s \geq F_i(t)$.

Now, supposing that $\overline{F}_i(t) \leq \overline{F}_i(t - u)\overline{F}_i(u)$ for $t \geq 0$ and $u \geq 0$, then

$$\int_0^t F_i(t - u)\mathrm{d}F_i(u) \geq \int_0^t \frac{\overline{F}_i(t)}{\overline{F}_i(u)}\mathrm{d}F_i(u) = \overline{F}_i(t)[-\ln \overline{F}_i(u)]\Big|_0^t = -\overline{F}_i(t)\ln \overline{F}_i(t).$$

Thus, $\Delta E_{Y,i}^{(2)} \geq \Delta E_{Y,i}^{(1)}$. The proof is similar for the case of $\overline{F}_i(t + t_0) \geq \overline{F}_i(t)\overline{F}_i(t_0)$.

The comparison is useful for choosing between adding a spare and using a repair facility. Replacing a component with a perfect one brings in an upper bound ($\Delta E_{Y,i}^{(3)}$) on the expected system yield improvement that can be achieved by an improvement action on the component. Because $\Delta E_{Y,i}^{(1)} \geq \Delta E_{Y,i}^{(4)}$ and $\Delta E_{Y,i}^{(2)} \geq \Delta E_{Y,i}^{(4)}$, conducting a minimal repair and adding an identical cold standby are always better than adding an identical parallel active spare for improving the expected system yield. But repair operations take time, and it is hard to implement the cold standby redundancy because of the difficulty of failure detection and switching (Kuo and Wan 2007).

Intuitively, a perfect repair should increase the expected system yield more than a minimal repair (i.e., $\Delta E_{Y,i}^{(2)} \geq \Delta E_{Y,i}^{(1)}$), but this is true only when the lifetime distribution of the component is new better than used (NBU) . A lifetime distribution $F(t)$ is said to be NBU (NWU – new worse than used) if $\overline{F}(t + t_0) \leq (\geq) \overline{F}(t)\overline{F}(t_0)$ for $t \geq 0$ and $t_0 \geq 0$. NBU (NWU) implies that the conditional survival probability $\overline{F}(t + t_0)/\overline{F}(t)$ of a component of age t is less (greater) than the corresponding survival probability $\overline{F}(t_0)$ of a new component.

10.4   Discussion

It would be tempting to take more information into consideration, for example, cost figures, before deciding on a specific reliability improvement. But cost information usually is rather difficult (and costly) to find, and therefore, it is unnecessary to gather such information for all components. Only for those components with large values of the importance measures of interest does it seem worthwhile to look for costs of improvement balanced against the improvement in system performance in order to make final decisions on how to allocate resources for reliability improvement.

The practical implementation is that work on improving component reliability only needs to be done on one component at a time, but that at each step of this process, the component has to be chosen according to the updated ranking of the importance measure or the importance-to-cost ratio of interest that is calculated on the basis of the improvements that have already been made.

References

Bergman B. 1985a. On reliability theory and its applications. *Scandinavian Journal of Statistics* **12**, 1–41.

Bergman B. 1985b. On some new reliability importance measures. *Proceedings of IFAC SAFECOMP'85* (ed. Quirk WJ), pp. 61–64.

Birnbaum ZW. 1969. On the importance of different components in a multicomponent system. In *Multivariate Analysis, Vol. 2* (ed. Krishnaiah PR). Academic Press, New York, pp. 581–592.

Boland PJ and El-Neweihi E. 1995. Measures of component importance in reliability theory. *Computers and Operations Research* **22**, 455–463.

Brown M and Proschan F. 1983. Imperfect repair. *Journal of Applied Probability* **20**, 851–859.

Bueno VC. 2000. Component importance in a random environment. *Statistics and Probability Letters* **48**, 173–179.

Kuo W and Wan R. 2007. Recent advances in optimal reliability allocation. *IEEE Transactions on Systems, Man, and Cybernetics, Series A* **37**, 143–156.

Lambert HE. 1975. Measure of importance of events and cut sets in fault trees. In *Reliability and Fault Tree Analysis* (eds. Barlow RE, Fussell JB, and Singpurwalla ND). Society for Industrial and Applied Mathematics, Philadelphia, pp. 77–100.

Mi J. 1998. Bolstering components for maximizing system life. *Naval Research Logistics* **45**, 497–509.

Natvig B. 1979. A suggestion of a new measure of importance of system component. *Stochastic Processes and Their Applications* **9**, 319–330.

Natvig B. 1982. On the reduction in remaining system lifetime due to the failure of a specific component. *Journal of Applied Probability* **19**, 642–652.

Natvig B. 1985. New light on measures of importance of system components. *Scandinavian Journal of Statistics* **12**, 43–54.

Natvig B and Gåsemyr J. 2009. New results on the Barlow-Proschan and Natvig measures of component importance in nonrepairable and repairable systems. *Methodology and Computing in Applied Probability* **11**, 603–620.

Norros I. 1986a. A compensator representation of multivariate life length distributions, with applications. *Scandinavian Journal of Statistics* **13**, 99–112.

Norros I. 1986b. Notes on Natvig's measure of importance of system components. *Journal of Applied Probability* **23**, 736–747.

Räde L. 1989. Expected time to failure of reliability systems. *Mathematical Scientist* **14**, 24–37.

Singh H and Misra N. 1994. On redundancy allocations in systems. *Journal of Applied Probability* **31**, 1004–1014.

Xie M. 1987. On some importance measures of system components. *Stochastic Processes and Their Applications* **25**, 273–280.

Xie M. 1988. A note on the Natvig measure. *Scandinavian Journal of Statistics* **15**, 211–214.

Xie M and Bergman B. 1991. On a general measure of component importance. *Journal of Statistical Planning and Inference* **29**, 211–220.

Xie M and Shen K. 1990. On the increase of system reliability due to some improvement at component level. *Reliability Engineering and System Safety* **28**, 111–120.

11

Component Assignment in Coherent Systems

The CAP have applications in cases where a given set of components are functionally inter-changeable and are available for being assigned to different positions in a system. By optimally assigning these components, the resulting systems reliability is maximized. For instance, Example 1.0.5 describes an oil pipeline pumping system, in which there are n pump stations along an oil pipeline, and each pump station is used to pump oil to the next k pump stations. Such a system is a $Lin/Con/k/n$:F system (Zuo and Shen 1992). The pumps in this pumping system are functionally interchangeable; thus, given n pumps of different reliability values, the reliability of this pumping system is determined by the assignment of n pumps to n pump stations. Intuitively, the more reliable pumps should be assigned to the more important positions in the system. Other examples in two-terminal networks include wireless and hard-wired telecommunication networks, computer networks, and electric power distribution networks (Gebre and Ramirez-Marques 2007; Kontoleon 1979). In a two-terminal network, links are considered as components and have different reliability values due to the different supporting resources behind the links. Given a fixed set of supporting resources, optimally arranging the components (i.e., allocating the different supporting resources) can maximize network reliability.

In addition, since components degrade differently due to their positions and uses in the system, rearranging them has the potential to improve the systems reliability and extend the system lifetime. The corresponding CAP can provide a new assignment of these components based on their current reliability values (i.e., the extents of degradation) so that the systems reliability is improved to a possible maximum level.

The following notation is used in this and the next two chapters about the CAP.

Notation

- Π: set of all $n!$ permutations of integers $\{1, 2, \ldots, n\}$
- π_i: index of the component assigned to position i for $i = 1, 2, \ldots, n$
- $\pi = (\pi_1, \pi_2, \ldots, \pi_n)$: a permutation of n available components that assigns component π_i to position i for $i = 1, 2, \ldots, n$, $\pi \in \Pi$

Importance Measures in Reliability, Risk, and Optimization: Principles and Applications, First Edition.
Way Kuo and Xiaoyan Zhu. © 2012 John Wiley & Sons, Ltd. Published 2012 by John Wiley & Sons, Ltd.

- $\pi(i, j)$: a permutation formed from permutation π by exchanging the integers in positions i and j, $i, j \in \{1, 2, \ldots, n\}$
- $\tilde{p}_i$: reliability of available component i for $i = 1, 2, \ldots, n$, without loss of generality, assuming $\tilde{p}_1 \le \tilde{p}_2 \le \cdots \le \tilde{p}_n$. If not, it is easy to sort and reorder them in this manner.
- $\mathbf{p} = (p_1, p_2, \ldots, p_n)$: vector of reliability of positions $1, 2, \ldots, n$ in a system
- $\mathbf{p}_\pi = (\tilde{p}_{\pi_1}, \tilde{p}_{\pi_2}, \ldots, \tilde{p}_{\pi_n})$: vector of reliability (of positions) specified by permutation π. Note that for $i = 1, 2, \ldots, n$,

$$p_i = \tilde{p}_{\pi_i}.$$

- $R(\mathbf{p}_\pi)$: reliability of the system corresponding to a permutation π.

Section 11.1 presents a mathematical model and detailed description of the CAP. Section 11.2 illustrates an enumeration method to find an optimal solution over the entire solution space and an induced randomization method for an approximate solution. Section 11.3 presents a method based on the permutation importance. The permutation importance can be used to reduce the solution space by means of an elimination process and achieves the optimal solution by enumerations over the reduced search space. Section 11.4 introduces an important concept of invariant arrangements of components. Section 11.5 investigates the CAP, especially invariant arrangements, in parallel–series and series–parallel systems. Section 11.6 addresses the relationships of the B-i.i.d. importance to invariant arrangements. Section 11.7 illustrates designs based on the B-reliability importance. Finally, Section 11.8 studies, as an extension of the CAP, an optimal assembly problem.

11.1   Description of Component Assignment Problems

Broadly, consider a coherent system of n positions and $\tilde{n}$ components of respective reliability $\tilde{p}_1, \tilde{p}_2, \ldots, \tilde{p}_{\tilde{n}}$. Assume that these components are functionally interchangeable and therefore can be assigned to any positions. A CAP is to find an optimal assignment of components to positions to maximize systems reliability or to enhance some other system performance. If the numbers of positions and components are the same (i.e., $n = \tilde{n}$) and each position is assigned exactly one component, then an assignment policy is also an arrangement policy (Boland et al. 1989). In this case, the CAP is also known as the component arrangement problem. Note that if the components are of different types and different types of components cannot be interchanged, then the corresponding problem is an optimal assembly problem, which is discussed in Section 11.8.

Here, the CAP is specified as a case where exact n components are available to be assigned to n positions, and each position must be assigned one and exactly one component. The optimal solution is an arrangement of n given components of respective reliability values to n system positions such that the reliability of the resulting system is maximized. In the literature, such a problem is referred to as a CAP, a component arrangement problem, an optimal design problem, or an optimal component allocation problem. Correspondingly, the optimal solution can be called the optimal assignment, optimal arrangement, optimal design, or optimal allocation of components. A CAP can also be understood as a reliability allocation problem, in which a system consists of n components, and n distinct reliability values are to be assigned to the components.

The CAP is combinatorial optimization in nature and generally NP-hard. Let u_{ij} be binary variables for $i, j \in \{1, 2, \ldots, n\}$. $u_{ij} = 1$ if component j is assigned to position i; 0 otherwise. Let p_i denote the reliability of position i in the system for $i = 1, 2, \ldots, n$, and let R denote the systems reliability function. Let $\tilde{p}_j$ denote the reliability of component j for $j = 1, 2, \ldots, n$. Then, given a set of components of reliability $\tilde{p}_1, \tilde{p}_2, \ldots, \tilde{p}_n$, the CAP can be modeled as a mixed integer nonlinear programming (MINLP) with a maximization nonlinear objective function (i.e., systems reliability function) and a set of linear constraints (Prasad et al. 1991b):

$$\max \quad R(p_1, p_2, \ldots, p_n) \tag{11.1}$$

$$\text{subject to} \quad p_i = \sum_{j=1}^{n} u_{ij} \tilde{p}_j \qquad i = 1, 2, \ldots, n \tag{11.2}$$

$$\sum_{j=1}^{n} u_{ij} = 1 \qquad i = 1, 2, \ldots, n \tag{11.3}$$

$$\sum_{i=1}^{n} u_{ij} = 1 \qquad j = 1, 2, \ldots, n \tag{11.4}$$

$$u_{ij} \in \{0, 1\} \qquad i, j \in \{1, 2, \ldots, n\} \tag{11.5}$$

$$p_i \geq 0 \qquad i = 1, 2, \ldots, n. \tag{11.6}$$

The objective function (11.1) is the systems reliability function with respect to the reliability of positions in the system. Equation (11.2) ensures that the reliability of position i equals the reliability of the component assigned to position i. Thus, the objective function (11.1) is indeed a function of the arrangements represented by u_{ij}, $i, j \in \{1, 2, \ldots, n\}$. Constraints (11.3) and (11.4) ensure that one and only one component is assigned to each position. The binary constraint (11.5) and the nonnegative constraint (11.6) are applied to decision variables u_{ij} and p_i, respectively.

A solution of the CAP is represented using a *permutation* $\pi = (\pi_1, \pi_2, \ldots, \pi_n)$, which means that component π_i is assigned to position i for $i = 1, 2, \ldots, n$. A permutation represents an arrangement of n components; thus, *the terms "arrangement" and "permutation" are interchangeably used.* $\pi(i, j)$ represents an arrangement that is obtained from arrangement π by exchanging the components assigned to positions i and j. For example, $\pi_3 = 4$ means that component 4 is assigned to position 3. If $\pi = (3, 1, 4, 5, 2)$, then $\pi(2, 4) = (3, 5, 4, 1, 2)$. Note that $p_i = \tilde{p}_{\pi_i}$ and $\pi_i = j$ implies $p_i = \tilde{p}_j$. For example, if the reliability of component 4 is $\tilde{p}_4 = 0.8$ and $\pi_3 = 4$, then the reliability of position 3 now is $p_3 = 0.8$. For a given permutation π, the corresponding reliability function of the system is $R(\mathbf{p}_\pi)$.

Let $\pi^* = (\pi_1^*, \pi_2^*, \ldots, \pi_n^*)$ denote the optimal permutation for which $R(\mathbf{p}_{\pi^*}) = \max_{\pi \in \Pi} R(\mathbf{p}_\pi)$. Conversely, the worst arrangement minimizes systems reliability. Let $\pi^0 = (\pi_1^0, \pi_2^0, \ldots, \pi_n^0)$ denote the worst permutation for which $R(\mathbf{p}_{\pi^0}) = \min_{\pi \in \Pi} R(\mathbf{p}_\pi)$. One may be interested in the worst arrangement since it gives the lower bound on systems reliability, and, in the field of risk analysis, identifying the worst arrangement is more important than identifying the optimal arrangement.

The following corollary indicates the relationship between the design of a system and its dual system, for example, $Con/k/n$:F and G systems. Its proof is trivial and similar to the one for Theorem 4.1.10.

Corollary 11.1.1 *The optimal arrangement for a system with respect to a set of component reliability $\tilde{p}_i$, for $i = 1, 2, \ldots, n$, is the worst arrangement for its dual system with respect to a set of component reliability $1 - \tilde{p}_i$, for $i = 1, 2, \ldots, n$, and vice versa.*

11.2 Enumeration and Randomization Methods

To the best of our knowledge, there exists no study on the exact optimal method for solving the CAP other than an enumeration method (i.e., the exhaustive search). The enumeration method generates all possible component arrangements, evaluates systems reliability under each component arrangement, and finds the arrangement with maximum (minimum) systems reliability as the optimal (worst) arrangement. There exist $n!$ possible arrangements for a system of n components; therefore, the complexity for an enumeration method is $n!$. When the number of components is relatively small, the enumeration method can find the optimal and the worst arrangements with respect to a set of component reliability values. However, the enumeration method is time-consuming and may even be impossible to solve a CAP in a system composed of a large number of components. For example, considering a CAP in a system with $n = 20$ positions and using a server with two Intel Dual-Core Xeon 5160 3.0 GHz processors and 8GB RAM, the enumeration method takes several days without in the end yielding the optimal arrangement (Yao et al. 2011).

The randomization method is to calculate the systems reliability corresponding to a limited number of, instead of all, arrangements and choose the best and/or worst one among them. The tested arrangements are generated randomly. As the number of tested arrangements increases, the generated solution is expected to improve, but the computation time increases as well.

The necessary conditions that optimal arrangements must satisfy can reduce the solution space, consequently reducing the computational effort of searching for the optimal solution. Only the arrangements that satisfy all of the necessary conditions need to be evaluated. The permutation importance can be used to derive such necessary conditions. Section 11.3 introduces an elimination procedure that reduces the solution space by means of the permutation importance and pairwise exchange.

11.3 Optimal Design Based on the Permutation Importance and Pairwise Exchange

Section 6.5 comprehensively presents the permutation importance. This section, based on the work in Boland et al. (1989), describes an elimination procedure, which uses the permutation importance and pairwise exchange to find the optimal permutation for the CAP in a coherent system. Assume knowledge only of the orderings of component reliability as $0 < \tilde{p}_1 \leq \tilde{p}_2 \leq \cdots \leq \tilde{p}_n < 1$ and not of their actual values, and for obvious reasons, avoid the trivial case

$\tilde{p}_i = 0$ or $\tilde{p}_i = 1$. In real applications, the true values of the $\tilde{p}_i$ may or may not be known, and in most cases, they are unknown.

Before presenting the elimination procedure, Definition 11.3.1 gives two concepts used in the elimination procedure.

Definition 11.3.1 *A permutation $\pi \in \Pi$ is said to be inadmissible if there exists a $\pi' \in \Pi$ such that $R(\mathbf{p}_\pi) \leq R(\mathbf{p}_{\pi'})$ for all sets of $\tilde{p}_1, \tilde{p}_2, \ldots, \tilde{p}_n$ satisfying $0 < \tilde{p}_1 \leq \tilde{p}_2 \leq \cdots \leq \tilde{p}_n < 1$ and strict inequality holds for some sets of $\tilde{p}_1, \tilde{p}_2, \ldots, \tilde{p}_n$. Permutations π and $\pi(i, j)$ are said to be equivalent, if $R(\mathbf{p}_{\pi(i,j)}) = R(\mathbf{p}_\pi)$ for all sets of $\tilde{p}_1, \tilde{p}_2, \ldots, \tilde{p}_n$.*

The following corollary shows that using the permutation importance can find the inadmissible permutations.

Corollary 11.3.2 *Let $\pi \in \Pi$ be a permutation such that $\pi_i < \pi_j$. If $i >_{\text{pe}} j$, then π is inadmissible.*

As in Corollary 11.3.2, if $\pi_i < \pi_j$, then $\tilde{p}_{\pi_i} \leq \tilde{p}_{\pi_j}$ because of the assumption $\tilde{p}_1 \leq \tilde{p}_2 \leq \cdots \leq \tilde{p}_n$. Thus, under the present permutation π, a less reliable component is assigned to position i. If the components assigned to positions i and j are exchanged and the components assigned to the other positions remain the same (i.e., *pairwise exchange*), then the reliability function changes from $R(\mathbf{p}_\pi)$ to $R(\mathbf{p}_{\pi(i,j)})$, and under the new permutation $\pi(i, j)$, a more reliable component is assigned to position i. If $i >_{\text{pe}} j$, by Theorem 6.5.13, $R(\mathbf{p}_{\pi(i,j)}) \geq R(\mathbf{p}_\pi)$, and thus, Corollary 11.3.2 follows. *As a consequence, π is inadmissible and should be eliminated from further consideration for the optimal permutation because after pairwise exchange, $\pi(i, j)$ is a uniformly better permutation.*

Similarly, Theorem 6.5.14 shows that using the permutation importance can find the equivalent permutations. Using Corollary 11.3.2 in step 1 and Theorem 6.5.14 in step 2, the elimination procedure in Table 11.1 can guarantee obtaining an optimal permutation from the subset of permutations that are not yet eliminated by means of pairwise exchange.

Note that the permutation importance depends only on the structure function of a system and not on the reliability values of components. It highlights the advantage of the use of importance measures. In the process of elimination, the permutation importance ranking of two components is compared to their reliability ordering even if exact reliability values are not known. This narrows the subset of permutations containing the optimal permutation. Accurate reliability values of components are only needed in the last step to test the optimal permutation among the narrowed subset of permutations. The following examples in Boland et al. (1989) illustrate this procedure.

Table 11.1 Elimination procedure

---
1. Eliminate all inadmissable permutations by means of the pairwise exchange principle.
2. Delete all but one of the equivalent permutations.
3. Let $\Pi_0 \subset \Pi$ denote the subsets of permutations that are not yet eliminated. Find a permutation π^* satisfying $R(\mathbf{p}_{\pi^*}) = \max_{\pi \in \Pi_0} R(\mathbf{p}_\pi)$, either analytically or (when necessary) from numerical calculations, where the $\tilde{p}_i$ values are known. Then, π^* is an optimal permutation.

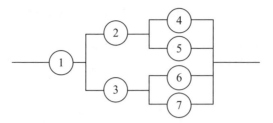

Figure 11.1   System of seven positions

Example 11.3.3 (Example 2.2.4 continued)   Suppose that oil is to be pumped through the system in Figure 2.4. Pumps with respective reliability $\tilde{p}_1 \leq \tilde{p}_2 \leq \tilde{p}_3$ are to be assigned to positions 1, 2, and 3. It is easy to verify that $1 >_{\text{pe}} 2$, $1 >_{\text{pe}} 3$, and $2 =_{\text{pe}} 3$. Thus, among the six permutations in Π, $(1, 2, 3)$, $(1, 3, 2)$, $(2, 1, 3)$, and $(2, 3, 1)$ are inadmissible, and both $(3, 1, 2)$ and $(3, 2, 1)$ are optimal.

Example 11.3.4 (Examples 4.1.7 and 6.1.8 continued)   Considering the system in Figure 4.1, it is easy to verify from the structure function of the system that (i) positions i and j are permutation equivalent for $1 \leq i < j \leq k$ and $k + 1 \leq i < j \leq n$, and (ii) $i >_{\text{pe}} j$ for all $i \leq k$ and $j > k$. Thus, under the optimal permutation, the k most reliable components are assigned to positions $1, 2, \ldots, k$ in any order.

Example 11.3.5   Consider a system consisting of seven positions connected as in Figure 11.1. Here, $1 >_{\text{pe}} j$ for all $j > 1$, $2 >_{\text{pe}} 4$, $2 >_{\text{pe}} 5$, $3 >_{\text{pe}} 6$, and $3 >_{\text{pe}} 7$. Furthermore, modular sets $\{2, 4, 5\}$ and $\{3, 6, 7\}$ are permutation equivalent. Thus, any permutation $\pi \in \Pi_0$ must satisfy $\pi_1 = 7$, $\pi_2 > \max\{\pi_4, \pi_5\}$, and $\pi_3 > \max\{\pi_6, \pi_7\}$. It follows that there are only $\binom{6}{3}/2 = 10$ permutations left in Π_0 for consideration. For example, $(7, 3, 6, 1, 2, 4, 5)$ and $(7, 4, 6, 1, 2, 3, 5)$ are in Π_0, but $(7, 3, 4, 1, 2, 6, 5)$ is inadmissible and is not in Π_0. Without this elimination process, all $7!/(2!)^3 = 630$ permutations need to be evaluated. If the $\tilde{p}_i$ values are known, then the values of $R(\mathbf{p}_\pi)$ can be computed for selecting the best permutation. The use of the elimination procedure reduces the computation time by a factor greater than 50.

After fixing component 7 to position 1 and treating modular sets $\{4, 5\}$ and $\{6, 7\}$ as supercomponents, Theorem 11.5.1, to be introduced later, can be used to reduce the number of permutations to be considered even further to eight.

11.4   Invariant Optimal and Invariant Worst Arrangements

In the CAP, the optimal arrangement of components to system positions generally depends on the reliability of the available components. The exact component reliability values may be hard to obtain, but it is relatively easy to obtain their orderings. For example, the reliability of components can be ordered according to their ages if they have the same failure rate, while their lifetime distributions might be unknown.

For some systems, the optimal arrangement is to assign components to system positions according to the orderings of component reliability. When this is the case, the optimal arrangement is called the invariant optimal arrangement, which is independent of the magnitudes of component reliability and depends only on the orderings of component reliability. Lin and

Kuo (2002) gave a formal definition of an invariant optimal arrangement, as shown in Definition 11.4.2. Similarly, the invariant worst arrangement is the worst arrangement that depends only on the orderings rather than the magnitudes of component reliability. Definition 11.4.2 gives a formal definition of an invariant worst arrangement.

Definition 11.4.1 *Let $\tilde{p}_1, \tilde{p}_2, \ldots, \tilde{p}_n$ be the reliability of components $1, 2, \ldots, n$ that are to be allocated to a system. The system admits an invariant optimal arrangement π^*, if $R(\mathbf{p}_{\pi^*}) \geq R(\mathbf{p}_\pi)$ for any permutation $\pi \in \Pi$ and for any set of $\tilde{p}_1, \tilde{p}_2, \ldots, \tilde{p}_n$ satisfying $\tilde{p}_1 \leq \tilde{p}_2 \leq \cdots \leq \tilde{p}_n$.*

Definition 11.4.2 *Let $\tilde{p}_1, \tilde{p}_2, \ldots, \tilde{p}_n$ be the reliability of components $1, 2, \ldots, n$ that are to be allocated to a system. The system admits an invariant worst arrangement π^0, if $R(\mathbf{p}_{\pi^0}) \leq R(\mathbf{p}_\pi)$ for any permutation $\pi \in \Pi$ and for any set of $\tilde{p}_1, \tilde{p}_2, \ldots, \tilde{p}_n$ satisfying $\tilde{p}_1 \leq \tilde{p}_2 \leq \cdots \leq \tilde{p}_n$.*

An attractive characteristic of an invariant optimal (worst) arrangement is that knowing only the ordering of component reliability, it is possible to determine the assignment that maximizes (minimizes) systems reliability. Importantly, whether a system admits an invariant arrangement depends only upon its structure.

The following corollary indicates the relationship between the invariant optimal and invariant worst arrangements for a system and its dual system (Lin and Kuo 2002). The proof derives straightforwardly from Corollary 11.1.1.

Corollary 11.4.3 *If a system admits an invariant optimal arrangement, its dual system admits an invariant worst arrangement, and vice versa.*

For a k-out-of-n system (including a parallel and a series system), any arrangement is optimal. Hwang (1989) investigated the invariant optimal arrangement for a two-stage k-out-of-n system. A two-stage k-out-of-n system is an ℓ-out-of-m system (the first stage) where each supercomponent is itself a k_i-out-of-n_i embedded system (the second stage) for $i = 1, 2, \ldots, m$, and $n = \sum_{i=1}^{m} n_i$. Without loss of generality, assume that $n_1 \leq n_2 \leq \cdots \leq n_m$. For some special two-stage k-out-of-n systems, Hwang (1989) obtained the following results.

(i) When each embedded system (i.e., the second stage) is a series system, an invariant optimal arrangement exists and assigns the n_1 most reliable components to embedded system 1, the next n_2 most reliable components to embedded system 2 and so on, and the n_m least reliable components to embedded system m.

(ii) When the first stage is a parallel system and the second stage is $(n_i - k + 1)$-out-of-n_i embedded systems ($1 \leq k \leq n_1$), then the aforementioned arrangement is the invariant optimal arrangement for this case, too.

(iii) When the first stage is a parallel system and the second stage is k-out-of-n_i systems (i.e., $k_i = k$ for any embedded system i), an invariant optimal arrangement exists and assigns the n_1 least reliable components to embedded system 1, the next n_2 least reliable components to embedded system 2 and so on, and the n_m most reliable components to embedded system m.

The next section presents the invariant arrangements for parallel–series and series–parallel systems. The investigation on invariant arrangements for $Con/k/n$ systems and their related systems is presented in Chapter 12.

11.5   Invariant Arrangements for Parallel–series and Series–parallel Systems

For the parallel–series system in Figure 2.7 that is a special two-stage k-out-of-n system, El-Neweihi et al. (1986) proved the existence of an invariant optimal arrangement as in Theorem 11.5.1, which can be directly obtained from points (i) and (ii) in Section 11.4. Corollary 11.5.2 gives a direct consequence of Theorem 11.5.1 in constructing a set of series systems.

Theorem 11.5.1   *For a parallel–series system that has m embedded series systems $1, 2, \ldots, m$ of sizes $n_1, n_2, \ldots, n_m$ satisfying $n_1 \leq n_2 \leq \cdots \leq n_m$, an invariant optimal arrangement assigns the n_1 most reliable components to embedded system 1, the next n_2 most reliable components to embedded system 2 and so on, and the n_m least reliable components to embedded system m.*

Corollary 11.5.2   *Suppose that $n\,(= n_1 + n_2 + \cdots + n_m)$ components of respective reliability values are to be allocated to m series systems of sizes $n_1, n_2, \ldots, n_m$ satisfying $n_1 \leq n_2 \leq \cdots \leq n_m$. Then the arrangement in Theorem 11.5.1 maximizes the expected number of functioning systems.*

For the series–parallel system in Figure 2.6, which is also a special two-stage k-out-of-n system as defined in Section 11.4, no invariant optimal arrangement exists in general (Baxter and Harche 1992a; El-Neweihi et al. 1986) and the corresponding CAP is NP-complete for general values of m and $n_1, n_2, \ldots, n_m$ (Prasad and Raghavachari 1998). An exception exists when each embedded system has only two components, as shown in Theorem 11.5.3 (Prasad et al. 1991b).

Theorem 11.5.3   *For a series–parallel system that has m embedded parallel systems $1, 2, \ldots, m$ of sizes $n_1 = n_2 = \cdots = n_m = 2$, suppose that $n = 2m$ components of reliability $\tilde{p}_1 \leq \tilde{p}_2 \leq \cdots \leq \tilde{p}_n$ are to be allocated among these embedded parallel systems. Then, an invariant optimal arrangement allocates components k and $(n-k+1)$ to embedded system k for $k = 1, 2, \ldots, m$.*

El-Neweihi et al. (1986) showed that there exists an invariant worst arrangement for series–parallel systems, as presented in Corollary 11.5.4, which can be obtained from the invariant optimal arrangement for parallel–series systems in Theorem 11.5.1 and the relationship between invariant worst and invariant optimal arrangements for dual systems in Corollary 11.4.3. Note that the series–parallel and parallel–series systems are dual of each other. Meanwhile, because the series–parallel system generally does not admit an invariant optimal arrangement, the parallel–series system generally does not admit an invariant worst arrangement.

Corollary 11.5.4 *For a series–parallel system that has m embedded parallel systems* $1, 2, \ldots, m$ *of sizes* $n_1, n_2, \ldots, n_m$ *satisfying* $n_1 \leq n_2 \leq \cdots \leq n_m$, *an invariant worst arrangement assigns the* n_1 *least reliable components to embedded system 1, the next* n_2 *least reliable components to embedded system 2 and so on, and the* n_m *most reliable components to embedded system m.*

It is observed that the reliability of a series–parallel system increases with the homogeneity among the embedded systems. On the basis of that, Prasad and Raghavachari (1998) developed an approximate mixed integer linear programming model in which the objective is to minimize the sum of absolute deviations from the average. They also provided an algorithm to improve further the resulting arrangement. For a special series–parallel system in which each embedded system contains exactly k components in parallel, Baxter and Harche (1992b) provided a simple top-down heuristic with an analytic error term. The top-down heuristic tries to allocate the components such that the resulting reliability of the embedded systems is as equal as possible.

Prasad et al. (1991b) considered a CAP in which each position i has a probability of being shock free, r_i, and the reliability of position i is $r_i \tilde{p}_j$ when component j of reliability $\tilde{p}_j$ is assigned to position i. They provided algorithms to find the optimal solution of such a problem in parallel–series systems. A more general case is considered by Prasad et al. (1991a) where the reliability of a component may be different when it is assigned to a different position in the system. Then $\tilde{p}_{ij}$ denotes the reliability of component j when it is assigned to position i. They proposed a heuristic to attack the problem in parallel–series systems. Given more than n available components, the heuristic chooses exact n components to allocate to n positions. Except for the aforementioned work, research of the CAP in parallel–series and series–parallel systems is mostly based on Schur functions and majorization, which are referred to Marshall and Olkin (1979) and Kuo and Zuo (2003).

11.6   Consistent B-i.i.d. Importance Ordering and Invariant Arrangements

In the CAP, the B-i.i.d. importance has a close relationship with the invariant arrangements that is established by means of a concept of consistent B-i.i.d. importance ordering, which is defined by Lin and Kuo (2002) as follows.

Definition 11.6.1 *For any pair of components i and j in a system, if* $I_B(i; p) \geq I_B(j; p)$ *for all* $0 < p < 1$, *(i.e., component i is more uniformly B-i.i.d. important than component j), the system is said to have consistent B-i.i.d. importance ordering; otherwise, the system is said to have inconsistent B-i.i.d. importance ordering.*

Corollary 11.6.2 is straightforward from Theorem 4.1.10 for dual systems. Theorem 11.6.3 and Corollary 11.6.4 establish a relation of the invariant optimal and invariant worst arrangements, respectively, to the consistent B-i.i.d. importance ordering. Corollary 11.6.4 is obtained from Theorem 11.6.3 and Corollaries 11.4.3 and 11.6.2. These results were derived by Lin and Kuo (2002).

Corollary 11.6.2 *A system has consistent B-i.i.d. importance ordering if and only if its dual system has the same consistent B-i.i.d. importance ordering.*

Theorem 11.6.3 *If a system admits an invariant optimal arrangement, the system must have consistent B-i.i.d. importance ordering. In addition, the invariant optimal arrangement matches the consistent B-i.i.d. importance ordering, that is, the invariant optimal arrangement assigns the least reliable component to the least uniform B-i.i.d. important position and the most reliable component to the most uniform B-i.i.d. important position.*

Proof. From Figure 6.3, the permutation importance implies the uniform B-i.i.d. importance. By Theorem 6.5.13, the results are straightforward.

Corollary 11.6.4 *If a system admits an invariant worst arrangement, the system must have consistent B-i.i.d. importance ordering. In addition, the invariant worst arrangement matches oppositely the consistent B-i.i.d. importance ordering, that is, the invariant worst arrangement assigns the most reliable component to the least uniform B-i.i.d. important position and the least reliable component to the most uniform B-i.i.d. important position.*

Figure 11.2 summarizes these relationships in a coherent system and its dual system, where symbol "⇒" means "implying," "⇔" means "if and only if," and the dashed line with an "×" means "not necessary implying." Note that by Theorem 11.6.3 and Corollary 11.6.4, a necessary condition for a system to admit an invariant optimal or invariant worst arrangement is that the system has a consistent B-i.i.d. importance ordering. However, this condition is not sufficient. That is, a system having a consistent B-i.i.d. importance ordering may or may not admit an invariant optimal or invariant worst arrangement. Table 11.2 presents some examples that admit both, one, or neither of the invariant optimal and invariant worst arrangements, respectively.

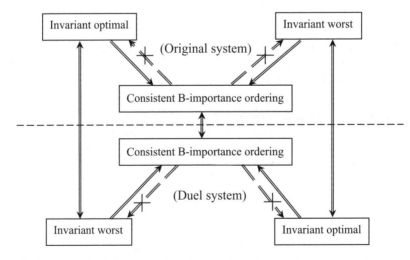

Figure 11.2   Relationships between consistent B-i.i.d. importance ordering and invariant arrangements

Table 11.2 Invariant optimal and invariant worst arrangements

Consistent B-i.i.d. importance	Invariant optimal arrangement	Invariant worst arrangement	Example system
Yes	Yes	No	Parallel–series
Yes	No	Yes	Series–parallel
Yes	Yes	Yes	Bridge system in Example 11.6.5
Yes	No	No	$Cir/Con/k/n$ systems with $3 \leq k < (n-1)/2$

As listed in Table 11.2, a general parallel–series system admits an invariant optimal arrangement but not an invariant worst arrangement, while its dual system—a series–parallel system—admits an invariant worst arrangement but not an invariant optimal arrangement; the parallel–series and series–parallel systems have the same consistent B-i.i.d. importance orderings. This also demonstrates that a system admitting an invariant optimal arrangement may not admit an invariant worst arrangement, and vice versa. The following example illustrates the case that a system may admit both invariant optimal and invariant worst arrangements, as listed in Table 11.2. For $Cir/Con/k/n$ systems, the B-i.i.d. importance of all positions is the same, as shown in Theorems 8.1.3 and 8.1.4. When $3 \leq k < (n-1)/2$, a $Cir/Con/k/n$ system admits neither invariant optimal arrangement nor invariant worst arrangement. Subsection 12.1.3 gives further analysis on this aspect for $Con/k/n$ systems.

Example 11.6.5   For the bridge system in Figure 2.1, $j >_{\text{pe}} 3$ for all $j = 1, 2, 4, 5$. Thus, the least (most) reliable component should be fixed to position 3 for maximizing (minimizing) the systems reliability. By pivotal decomposition (2.1) on position 3, the bridge system can be decomposed into systems (a) and (b) in Figure 11.3, respectively, corresponding to the functioning and failure of component in position 3. These two systems are dual systems of each other and admit the same invariant optimal and the same invariant worst arrangements according to Theorems 11.5.1 and 11.5.3. Assuming that $\tilde{p}_1 \leq \tilde{p}_2 \leq \cdots \leq \tilde{p}_5$, the arrangement $\pi_1^* = 5$, $\pi_2^* = 2$, $\pi_4^* = 4$, and $\pi_5^* = 3$ is invariant optimal for these two systems (after fixing component 1 to position 3); similarly, the arrangement $\pi_1^0 = 1$, $\pi_2^0 = 2$, $\pi_4^0 = 4$, and $\pi_5^0 = 3$ is invariant worst for these two systems (after fixing component 5 to position 3). Hence, $(5, 2, 1, 4, 3)$ is the invariant optimal arrangement, and $(1, 2, 5, 4, 3)$ is the invariant worst arrangement for the bridge system.

Note that the dual of the bridge system in Figure 2.1 is another bridge system as in Figure 11.4. The difference between the original and dual bridge systems lies in the exchange

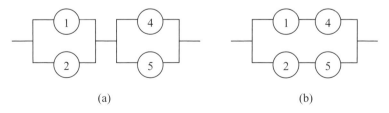

(a)   (b)

Figure 11.3   Decomposition of a bridge system

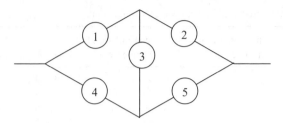

Figure 11.4 Dual of the bridge system in Figure 2.1

of labels between positions 2 and 4. Note that a system admits both the invariant optimal and invariant worst arrangements if and only if its dual system also admits both.

By Theorem 11.6.3 and Corollary 11.6.4, if a system does not have a consistent B-i.i.d. importance ordering, then it admits neither the invariant optimal nor the invariant worst arrangement. However, when a system has a consistent B-i.i.d. importance ordering, it is not easy to show whether or not the system admits the invariant optimal or invariant worst arrangement. Similar to Example 11.6.5, a possible method could be to first identify the permutation importance (maybe a partial ranking) and fix the components in some positions based on the permutation importance as in Section 11.3. After that, if any embedded system is identified not to admit the invariant optimal (worst) arrangement, then the overall system does not admit the invariant optimal (worst) arrangement. On the other hand, it is possible to prove that a system admits the invariant optimal or invariant worst arrangement by applying some techniques (e.g., pivotal decomposition (2.1) as in Example 11.6.5) on the simplified system. Unfortunately, except for these preliminary results, there is no general criterion to judge whether a system admits an invariant arrangement when it has consistent B-i.i.d. importance ordering.

11.7   Optimal Design Based on the B-reliability Importance

On the basis of the B-reliability importance, Kuo et al. (1990) suggested the following three general guidelines for optimal system design.

(i) Given a desired systems reliability with to-be-determined reliability of components, devote minimum effort to allocating components of higher reliability to more B-reliability important positions and components of lower reliability to less B-reliability important positions.
(ii) Given the reliability of n components, the sequence of assigning components to the system should follow the B-reliability importance patterns by allocating high reliable components to great B-reliability important positions.
(iii) The aforementioned arrangements may not be optimal if different costs or reliability are incurred when components are allocated at different positions.

When the relative differences of component reliability values are sufficiently small, the B-i.i.d. importance ordering can lead to an optimal solution for the CAP, as shown in Theorem

11.7.1 (Lin and Kuo 2002). The requisite degree of closeness of the component reliability values depends on the structure of the system.

Theorem 11.7.1 *Whenever the reliability values of available components are relatively close to each other, say $\tilde{p}_1 \leq \tilde{p}_2 \leq \cdots \leq \tilde{p}_n$ and $\tilde{p}_1 \simeq \tilde{p}_2 \simeq \cdots \simeq \tilde{p}_n$, the optimal arrangement matches the B-i.i.d. importance ordering associated with $\tilde{p}_1$.*

Note that as long as the component reliability values are sufficiently close to each other, the optimal arrangement matches the B-i.i.d. importance ordering associated with the minimum available component reliability no matter whether the system admits the invariant optimal arrangement or not. Example 11.7.2 shows this point.

Example 11.7.2 (Example 6.14.5 continued)   Considering the system in Figure 6.4 and five available components of reliability values $(0.70, 0.73, 0.76, 0.79, 0.82)$, the optimal arrangement by the enumeration method is $(2, 1, 3, 5, 4)$, which matches the consistent B-i.i.d. importance ordering $I_B(2; p) = I_B(1; p) < I_B(3; p) < I_B(5; p) = I_B(4; p)$ for all p, as shown in Example 6.14.5.

For another set of component reliability values $(0.1, 0.2, 0.3, 0.4, 0.5)$, the optimal arrangement is $(2, 1, 5, 4, 3)$, which is different from the consistent B-i.i.d. importance ordering; thus, according to Theorem 11.6.3, this system does not admit an invariant optimal arrangement. However, since the reliability values of available components are very close in the first case, the optimal arrangement matches the B-i.i.d. importance ordering, although the system does not admit an invariant optimal arrangement.

In fact, the optimal arrangement is not always consistent with the B-reliability importance ordering, when the system does not admit the invariant optimal arrangement. That is, when $p_i \leq p_j$, it is possible that $I_B(i; \mathbf{p}) > I_B(j; \mathbf{p})$ under the optimal arrangement, as shown in Example 11.7.3.

Example 11.7.3   Considering $Lin/Con/3/8$:F system and eight available components of reliability values $(0.1, 0.2, 0.3, 0.4, 0.5, 0.6, 0.7, 0.8)$, the optimal arrangement by the enumeration method is $\pi^* = (1, 4, 8, 3, 6, 7, 5, 2)$. Under this optimal arrangement, the increasing ordering of the B-reliability importance of positions is $1, 8, 4, 2, 5, 7, 6, 3$. Particularly, $I_B(7; \mathbf{p}_{\pi^*}) = 0.1593 > I_B(5; \mathbf{p}_{\pi^*}) = 0.1581$ but $p_7 = \tilde{p}_5 = 0.5 < p_5 = \tilde{p}_6 = 0.6$.

When data on component reliability are unavailable, for example, in the early stages of system design, structure importance measures can be used as a guideline for allocating components to different positions. As shown in Theorems 6.7.5 and 6.7.6, if $\tilde{p}_1$ approaches 1 ($\tilde{p}_n$ approaches 0), the cut importance (path importance) is equivalent to the B-i.i.d. importance; thus, by Theorem 11.7.1, the fact that position i is more cut important (path important) than position j dictates that a more reliable component should go to position i rather than position j. Similarly, the first-term and rare-event importance measures could apply in the two cases of $\tilde{p}_1$ approaches 1 and $\tilde{p}_n$ approaches 0, respectively.

For the general reliability values of components, the B-reliability importance can be used to solve the CAP approximately, if not optimally. Chapter 13 introduces heuristics that use the B-reliability importance.

11.8   Optimal Assembly Problems

This section considers the optimal assembly of n systems or modules (i.e., embedded systems) from r types of components in which components of different types cannot be interchanged. Such an assembly problem is known as *cannibalization* (Baxter 1988), which is different from the CAP in the preceding sections of this chapter in that the CAP deals with only a single type of components. For some special systems, this optimal assembly problem admits an invariant optimal solution, that is, the optimal solution depends only on the orderings of component reliability. This section briefly presents the results of invariant optimal designs for this optimal assembly problem.

Derman et al. (1972) studied the assembly of n series systems from r types of components, where each type has n components and each series system consists of r components of distinct types. The objective is to maximize the expected number of such assembled systems that function satisfactorily. For such a problem, an invariant optimal assignment exists and assigns the least reliable components of each type to construct the first system, the next least reliable components of each type to construct the second system and so on, and the most reliable components of each type to construct the nth system. Furthermore, if such assembled n systems are used as n embedded systems in a k-out-of-n system, then this invariant optimal solution also maximizes the reliability of the k-out-of-n system. Derman et al. (1974) considered the same situation but to assemble n parallel systems and concluded that no invariant optimal assignment exists.

El-Neweihi et al. (1987) extended the problem of assembling series systems that each may contain more than one component of each type. Assume that the ith series system consists of m_{ji} components of type j for $i = 1, 2, \ldots, n$ and $m_{j1} \geq m_{j2} \geq \cdots \geq m_{jn}$ for $j = 1, 2, \ldots, r$. The invariant optimal assignment exists for maximizing the expected number of functioning systems. It allocates, for each type j, the m_{j1} least reliable components to the first system, the next m_{j2} least reliable components to the second system and so on, and the m_{jn} most reliable components to the nth system. If such assembled n systems are used as n embedded systems in a k-out-of-n system, then this invariant optimal solution also maximizes the reliability of the k-out-of-n system.

Boland and Proschan (1984) investigated the assembly of two types of components, representing "generator" and "machines," into n systems. The ith system consists of one generator and $m_i \geq 1$ machines that are connected to the generator. Assume that $m_1 \geq m_2 \geq \cdots \geq m_n$. The machine functions only if both it and its connected generator function. If the objective is to maximize the expected number of functioning machines, then an invariant optimal solution exists. It allocates the most reliable generator and the m_1 most reliable machines into the first system, the second most reliable generator and the next m_2 most reliable machines into the second one, and so on.

Malon (1990) investigated the assembly of more than one type of components in a two-stage coherent system. A *two-stage coherent system* is built with identically structured modules, each module consisting of a number of different components. The problem is to assemble the modules out of a collection of available components and then to install the modules in the system in a way that maximizes systems reliability. A greedy assembly rule builds one module out of the available most reliable components, another out of the most reliable components remaining, and so on. Malon (1990) showed that the greedy assembly is invariant optimal whenever the modules have a series structure (no matter what the structure of the system in which the modules are used might be), provided that the modules, once assembled, are installed in the system in an optimal way. They also demonstrated that only series modules possess this property. In this context, suppose that modules do not each necessarily require the same number of components of each type but instead that the ith module uses m_{ji} components of type j for $j = 1, 2, \ldots, r$. Assuming that $m_{j1} \leq m_{j2} \leq \cdots \leq m_{jn}$ for all j, El-Neweihi et al. (1987) and Malon (1990) proved that the greedy assembly rule (which now assigns the m_{j1} most reliable components to the first module, the next m_{j2} most reliable components to the second module, and so on) remains optimal.

References

Baxter LA. 1988. On the theory of cannibalization. *Journal of Mathematical Analysis and Applications* **136**, 290–297.

Baxter LA and Harche F. 1992a. Note: On the greedy algorithm for optimal assembly. *Naval Research Logistics* **39**, 833–837.

Baxter LA and Harche F. 1992b. On the optimal assembly of series-parallel systems. *Operations Research Letters* **11**, 153–157.

Boland PJ and Proschan F. 1984. Optimal arrangement of systems. *Naval Research Logistics* **31**, 399–407.

Boland PJ, Proschan F, and Tong YL. 1989. Optimal arrangement of components via pairwise rearrangements. *Naval Research Logistics* **36**, 807–815.

Derman C, Lieberman GJ, and Ross SM. 1974. Assembly of systems having maximum reliability. *Naval Research Logistics* **21**, 1–12.

Derman C, Lieberman GJ, and Ross SM. 1972. On optimal assembly of system. *Naval Research Logistics* **19**, 569–574.

El-Neweihi E, Proschan F, and Sethuraman J. 1986. Optimal allocation of components in parallel-series and series-parallel systems. *Journal of Applied Probability* **23**, 770–777.

El-Neweihi E, Proschan F, and Sethuraman J. 1987. Optimal assembly of systems using Schur functions and majorization. *Naval Research logistics* **34**, 705–712.

Gebre BA and Ramirez-Marques JE. 2007. Element substitution algorithm for general two-terminal network reliability analyses. *IIE Transactions* **39**, 265–275.

Hwang FK. 1989. Optimal assignment of components to a two-stage k-out-of-n system. *Mathematics of Operations Research* **14**, 376–382.

Kontoleon JM. 1979. Optimal link allocation of fixed topology networks. *IEEE Transactions on Reliability* **28**, 145–147.

Kuo W, Zhang W, and Zuo MJ. 1990. A consecutive k-out-cf-n:G: The mirror image of a consecutive k-out-of-n:F system. *IEEE Transactions on Reliability* **R-39**, 244–253.

Kuo W and Zuo MJ. 2003. *Optimal Reliability Modeling: Principles and Applications*. John Wiley & Sons, New York.

Lin FH and Kuo W. 2002. Reliability importance and invariant optimal allocation. *Journal of Heuristics* **8**, 155–171.

Malon DM. 1990. When is greedy module assembly optimal?. *Naval Research Logistics* **37**, 847–854.

Marshall A and Olkin I. 1979. *Inequalities: Theory of Majorization and Its Applications*. Academic Press, New York.

Prasad VR, Aneja YP, and Nair KPK. 1991a. A heuristic approach to optimal assignment of components to parallel-series network. *IEEE Transactions on Reliability* **40**, 555–558.

Prasad VR, Nair KPK, and Aneja YP. 1991b. Optimal assignment of components to parallel-series and series-parallel systems. *Operations Research* **39**, 407–414.

Prasad VR and Raghavachari M. 1998. Optimal allocation of interchangeable components in a series-parallel system. *IEEE Transactions on Reliability* **R-47**, 255–260.

Yao Q, Zhu X, and Kuo W. 2011. Heuristics for component assignment problems based on the Birnbaum importance. *IIE Transactions* **43**, 1–14.

Zuo MJ and Shen J. 1992. System reliability enhancement through heuristic design. *1992 OMAE Volume II, Safety and Reliability, ASME*, pp. 301–304.

12

Component Assignment in Consecutive-k-out-of-n and Its Variant Systems

This chapter investigates the CAP in $Con/k/n$ and its variant systems based on importance measures. The methods and results descried in Chapter 11 are used for tackling the CAP in general systems and, of course, can be applied to $Con/k/n$ related systems. However, additional pertinent results have been derived for the CAP in these $Con/k/n$ related systems because they have special structures. Note that the reversed sequence of an arrangement of components for a $Con/k/n$ system is the same as the original one because of the symmetry of a $Con/k/n$ system.

   Section 12.1 shows that invariant arrangements exist for $Con/k/n$ systems with some special k and n values and that the B-reliability importance has a close relationship with the optimal arrangements of the CAP in $Con/k/n$ systems. Section 12.2 presents necessary conditions for optimal solutions for $Con/k/n$ systems; these conditions can reduce the solution space. Section 12.3 investigates the sequential CAP in the $Con/2/n$:F systems and presents the greedy algorithms that can produce the invariant optimal arrangement for the $Lin/Con/2/n$:F system as well as all possible optimal arrangements for the $Cir/Con/2/n$:F system. Sections 12.4–12.8 investigate the CAP in some variations of $Con/k/n$ systems, including consecutive-2 failure systems (Section 12.4), series $Con/k/n$ systems (Section 12.5), consecutive-k-out-of-r-from-n systems (Section 12.6), multidimensional and redundant $Con/k/n$ systems (Section 12.7), and others (Section 12.8). Some of these variations were introduced by Kuo and Zuo (2003). For the CAP in these systems, this chapter focuses on invariant and variant optimal arrangements and their relationships with the patterns of the permutation importance and the B-i.i.d. importance. This chapter continues to use the notation from Chapter 11.

Importance Measures in Reliability, Risk, and Optimization: Principles and Applications, First Edition.
Way Kuo and Xiaoyan Zhu. © 2012 John Wiley & Sons, Ltd. Published 2012 by John Wiley & Sons, Ltd.

12.1 Invariant Arrangements for $Con/k/n$ Systems

Recall from Section 11.4 that an optimal arrangement of the CAP is invariant if it depends solely on the ordering of component reliability, that is, this arrangement maximizes systems reliability no matter what the actual reliability values of components are as long as the same ordering of component reliability is maintained. For $Con/k/n$ systems with various values of k and n, this section presents invariant optimal arrangements and their relationships with the consistent B-i.i.d. importance ordering.

Note that linear and circular $Con/1/n$:F and $Con/n/n$:G systems are series systems and linear and circular $Con/n/n$:F and $Con/1/n$:G systems are parallel systems; thus, any arrangement for these systems is equivalent and optimal.

12.1.1 Invariant Optimal Arrangements for $Lin/Con/k/n$ Systems

Derman et al. (1982) first raised the CAP in a $Lin/Con/2/n$:F system and conjectured the invariant optimal arrangement to be $(1, n, 3, n-2, \ldots, n-3, 4, n-1, 2)$, given that $\tilde{p}_1 < \tilde{p}_2 < \ldots < \tilde{p}_n$. Wei et al. (1983) partially proved this conjecture, and Malon (1984) and Du and Hwang (1986), respectively, proved this conjecture.

Malon (1985) characterized all values of k and n for which the $Lin/Con/k/n$:F system admits an invariant optimal arrangement, as shown in Theorem 12.1.1. For $3 \leq k \leq n-3$, Malon (1985) proved that an invariant optimal arrangement does not exist.

Theorem 12.1.1 *The $Lin/Con/k/n$:F system admits an invariant optimal arrangement if and only if $k \in \{1, 2, n-2, n-1, n\}$. The invariant optimal arrangements are as tabulated in the second column in Table 12.1, given component reliability satisfying $\tilde{p}_1 \leq \tilde{p}_2 \leq \ldots \leq \tilde{p}_n$.*

Note that a $Lin/Con/k/2k$:F system does not admit an invariant optimal arrangement, but a $Lin/Con/k/2k$:G system does admit an invariant optimal arrangement, as shown in Theorem 12.1.2 (Kuo et al. 1990).

Table 12.1 Complete invariant optimal arrangements for $Lin/Con/k/n$ systems

k	F system	G system
$k = 1$	(any arrangement)	(any arrangement)
$k = 2$	$(1, n, 3, n-2, \ldots, n-3, 4, n-1, 2)$ Derman et al. (1982); Malon (1985) Zuo and Kuo (1990)	Does not exist
$2 < k < n/2$	Does not exist Malon (1985); Zuo and Kuo (1990)	Does not exist
$n/2 \leq k < n-2$	Does not exist Malon (1985)	$(1, 3, 5, \ldots, 2(n-k)-1,$
$k = n-2$	$(1, 4, (\text{any arrangement}), 3, 2)$ Malon (1985)	(any arrangement), $2(n-k), \ldots, 6, 4, 2)$
$k = n-1$	$(1, (\text{any arrangement}), 2)$ Malon (1985)	Kuo et al. (1990)
$k = n$	(any arrangement)	(any arrangement)

Theorem 12.1.2   *For a $Lin/Con/k/2k$:G system, the necessary conditions for the optimal arrangement of the CAP are*

$$
\begin{array}{ll}
p_i < p_j & \text{for } 1 \le i < j \le k \\
p_i > p_j & \text{for } k+1 \le i < j \le n \\
(p_i - p_j)(p_{i-1} - p_{j+1}) > 0 & \text{for } j = n - i + 1, i = 2, 3, \ldots, k \\
(p_i - p_j)(p_i - p_{j+1}) < 0 & \text{for } j = n - i + 1, i = 2, 3, \ldots, k,
\end{array}
$$

and the only permutation satisfying these necessary conditions is

$$(1, 3, 5, \ldots, 2k - 1, 2k, \ldots, 6, 4, 2),$$

given that $\tilde{p}_1 < \tilde{p}_2 < \ldots < \tilde{p}_n$. Thus, this permutation is the invariant optimal arrangement for a $Lin/Con/k/2k$:G system.

Theorem 12.1.2 assumes that $\tilde{p}_1, \tilde{p}_2, \ldots, \tilde{p}_n$ are all distinct since other cases can be viewed as limits of this case. Without this assumption, some strict inequalities in this theorem become nonstrict, and consequently, the optimal arrangement presented is unique up to the components of the same reliability.

Theorem 12.1.3 presents complete invariant optimal arrangements for $Lin/Con/k/n$:G systems, in which Kuo et al. (1990) identified the invariant optimal arrangement for the $Lin/Con/k/n$:G systems with $k \ge n/2$ and Jalali et al. (2005) and Cui and Hawkes (2008) gave other proofs for this case. Zuo and Kuo (1990) showed that no invariant optimal arrangement for the $Lin/Con/k/n$:G systems with $2 \le k < n/2$ exists.

Theorem 12.1.3   *The $Lin/Con/k/n$:G system admits an invariant optimal arrangement if and only if $k = 1$ or $n/2 \le k \le n$. The invariant optimal arrangements are as tabulated in the third column in Table 12.1, given component reliability satisfying $\tilde{p}_1 \le \tilde{p}_2 \le \ldots \le \tilde{p}_n$.*

With Theorems 12.1.1–12.1.3, Table 12.1 summarizes the invariant optimal arrangements for $Lin/Con/k/n$:F and G systems and related references, assuming that component reliability satisfies $\tilde{p}_1 \le \tilde{p}_2 \le \ldots \le \tilde{p}_n$. These results were presented by Zuo and Kuo (1990) and Kuo and Zuo (2003).

12.1.2   *Invariant Optimal Arrangements for $Cir/Con/k/n$ Systems*

Considering the $Cir/Con/k/n$:F systems, when $k = 2$, Hwang (1982) conjectured that arrangement $(1, n - 1, 3, n - 3, \ldots, n - 2, 2, n, 1)$ is invariant optimal, which was later proved by Du and Hwang (1986). This invariant optimal arrangement for $Cir/Con/2/n$:F systems and the one for $Lin/Con/2/n$:F systems essentially interlace the more reliable components with the less reliable components. When $k = n - 1$, Tong (1986) and Hwang (1989) discovered that the arrangement of components does not affect the reliability of a $Cir/Con/k/n$:F system, and, consequently, any arrangement is optimal. When $k = n - 2$, Hwang (1989) and Zuo and Kuo (1990) found the invariant optimal arrangement for $Cir/Con/k/n$:F systems; they also showed the nonexistence of the invariant optimal arrangement when $3 \le k \le n - 3$. In fact, a $Lin/Con/k/n$:F system can be considered to be a special case of the $Cir/Con/k/n + 1$:F system with $\tilde{p}_{n+1} = 1$, as first indicated by Hwang (1982). Thus, the nonexistence of the invariant optimal arrangement of a $Lin/Con/k/n$:F system with $3 \le k \le n - 3$ (Theorem 12.1.1) implies

Table 12.2 Complete invariant optimal arrangements for $Cir/Con/k/n$ systems

k	F system	G system
$k = 1$	(any arrangement)	(any arrangement)
$k = 2$	$(1, n-1, 3, n-3, \ldots, n-2, 2, n, 1)$ Du and Hwang (1986); Hwang (1982) Zuo and Kuo (1990)	Does not exist
$2 < k < (n-1)/2$	Does not exist Hwang (1989); Zuo and Kuo (1990)	Does not exist
$(n-1)/2 \le k < n-2$	Does not exist Hwang (1989)	$(1, 3, 5, 7, \ldots, n, \ldots, 6, 4, 2, 1)$ Zuo and Kuo (1990)
$k = n-2$	$(1, n-1, 3, n-3, \ldots, n-2, 2, n, 1)$ Hwang (1989)	
$k = n-1$	(any arrangement) Hwang (1989); Kuo et al. (1990); Tong (1986)	(any arrangement)
$k = n$	(any arrangement)	(any arrangement)

the same for a $Cir/Con/k/n$:F system with $3 \le k \le n-4$. Theorem 12.1.4 summarizes these results.

Theorem 12.1.4 *The $Cir/Con/k/n$:F system admits an invariant optimal arrangement if and only if $k \in \{1, 2, n-2, n-1, n\}$. The invariant optimal arrangements are as tabulated in the second column in Table 12.2, given component reliability satisfying $\tilde{p}_1 \le \tilde{p}_2 \le \ldots \le \tilde{p}_n$.*

Considering the $Cir/Con/k/n$:G system with $(n-1)/2 \le k < n-1$ as shown in Theorem 12.1.5, Zuo and Kuo (1990) derived a necessary condition for the optimal arrangement, which is stronger than the necessary condition derived by Kuo et al. (1990) for the case of $k = n-2$; they also found the invariant optimal arrangement satisfying this necessary condition. Later, Jalali et al. (2005) pointed out their proof as incomplete, and Du et al. (2001, 2002) completed the proofs for the invariant optimal arrangements for the $Cir/Con/k/2k+1$:G system and the $Cir/Con/k/n$:G system with $n \le 2k+1$, respectively. Similar to Theorem 12.1.2, in Theorem 12.1.5, if $\tilde{p}_1, \tilde{p}_2, \ldots, \tilde{p}_n$ are not all distinct, then the strict inequality becomes nonstrict and consequently the optimal arrangement presented is unique up to the components of the same reliability.

Theorem 12.1.5 *For a $Cir/Con/k/n$:G system with $(n-1)/2 \le k < n-1$, a necessary condition for the optimal arrangement of the CAP is*

$$(p_i - p_j)(p_{i-1} - p_{j+1}) > 0 \quad for \quad j > i, \ i = 2, 3, \ldots, n,$$

where $p_j = p_{j-n}$ if $j > n$, and the only permutation satisfying this necessary condition is

$$(1, 3, 5, 7, \ldots, n, \ldots, 6, 4, 2, 1),$$

given that $\tilde{p}_1 < \tilde{p}_2 < \ldots < \tilde{p}_n$. Thus, this permutation is the invariant optimal arrangement for a $Cir/Con/k/n$:G system with $(n-1)/2 \le k < n-1$.

Furthermore, Zuo and Kuo (1990) showed that there does not exist any invariant optimal arrangement for a $Cir/Con/k/n$:G system when $2 \le k < (n-1)/2$. Kuo et al. (1990) stated that all arrangements of components for $Cir/Con/k/n$:F and G systems with $k = n - 1$ result in the same systems reliability. Theorem 12.1.6 completes the invariant optimal arrangement for $Cir/Con/k/n$:G systems.

Theorem 12.1.6 *The $Cir/Con/k/n$:G system admits an invariant optimal arrangement if and only if $k = 1$ or $(n-1)/2 \le k \le n$. The invariant optimal arrangements are as tabulated in the third column in Table 12.2, given component reliability satisfying $\tilde{p}_1 \le \tilde{p}_2 \le \ldots \le \tilde{p}_n$.*

With Theorems 12.1.4–12.1.6, Table 12.2 summarizes the invariant optimal arrangements for $Cir/Con/k/n$:F and G systems and related references, assuming that component reliability satisfies $\tilde{p}_1 \le \tilde{p}_2 \le \ldots \le \tilde{p}_n$. These results were presented by Zuo and Kuo (1990) and Kuo and Zuo (2003).

12.1.3   *Consistent B-i.i.d. Importance Ordering and Invariant Arrangements*

Continuing the studies in Section 11.6, which are on the relationships between consistent B-i.i.d. importance ordering and the invariant arrangements for general coherent systems, this subsection specifies the relationships for $Con/k/n$ systems. For $Cir/Con/k/n$ systems, the B-i.i.d. importance values of all component positions are equal; thus, the following is for $Lin/Con/k/n$ systems.

For $Lin/Con/k/n$ systems, Table 12.3 summarizes the relationships of consistent B-i.i.d. importance ordering to invariant optimal and invariant worst arrangements, assuming that $\tilde{p}_1 \le \tilde{p}_2 \le \ldots \le \tilde{p}_n$. In Table 12.3, the second column lists the consistent B-i.i.d. importance orderings, which are stated in Theorem 8.2.3 and are the same as the permutation importance rankings. The third column lists the invariant optimal arrangements, which are also presented in Table 12.1. These invariant optimal arrangements match the consistent B-i.i.d. importance orderings in the second column, as proved in Theorem 11.6.3. The last column presents the invariant worst arrangements, which are obtained according to the invariant optimal arrangements of the dual systems that are listed in the third column. By Theorem 2.13.2, $Con/k/n$:F and G systems are dual of each other. Corollary 11.4.3 determines the existences of invariant worst arrangements, and Corollary 11.6.4 specifies the invariant worst arrangements. As shown in Table 12.3, when a $Lin/Con/k/n$ system has consistent B-i.i.d. importance ordering, then it admits either the invariant optimal arrangement, the invariant worst arrangement, or both.

12.2   Necessary Conditions for Component Assignment in $Con/k/n$ Systems

According to the permutation importance patterns in Theorem 8.3.1 and the optimal permutation rule in Corollary 11.3.2, the necessary conditions in the following theorem can be easily

Table 12.3 Relationships of consistent B-i.i.d. importance ordering to invariant arrangements for $Lin/Con/k/n$ systems

k	System	Consistent B-i.i.d. importance ordering	Invariant optimal arrangement	Invariant worst arrangement
$k = 1$	F and G	$I_1 = I_2 = \ldots = I_n$	(any arrangement)	(any arrangement)
$k = 2$	F	$I_1 = I_n < I_3 = I_{n-2} < \cdots$ $< I_{n-3} = I_4 < I_{n-1} = I_2$	$(1, n, 3, n-2, \ldots, n-3, 4, n-1, 2)$	Does not exist
	G		Does not exist	$(n, 1, n-2, 3, \ldots, 4, n-3,$ $2, n-1)$
$2 < k < n/2$	F and G	Does not exist	Does not exist	Does not exist
$k = n/2$	F	$I_1 = I_n < I_2 = I_{n-1} < \cdots$	Does not exist	π_2
	G	$< I_{n-k-1} = I_{k+2} < I_{n-k} = I_{k+1}$	π_1	Does not exist
$n/2 < k < n-2$	F		Does not exist	π_2
	G	$I_1 = I_n < I_2 = I_{n-1} < \cdots$	π_1	Does not exist
$k = n-2$	F	$< I_{n-k-1} = I_{k+2} < I_{n-k} = I_{k+1}$	$(1, 4, (\text{any arrangement}), 3, 2)$	π_2
	G	$< I_{n-k+1} = I_{n-k+2} = \cdots$ $= I_{k-1} = I_k$	π_1	$(n, n-3, (\text{any arrangement}),$ $n-2, n-1)$
$k = n-1$	F	$I_1 = I_n < I_2 = I_{n-1} < \cdots$	$(1, (\text{any arrangement}), 2)$	π_2
	G	$< I_{n-k-1} = I_{k+2} < I_{n-k} = I_{k+1}$ $< I_{n-k+1} = I_{n-k+2} = \cdots$ $= I_{k-1} = I_k$	π_1	$(n, (\text{any arrangement}), n-1)$
$k = n$	F and G	$I_1 = I_2 = \ldots = I_n$	(any arrangement)	(any arrangement)

The simplified notation I_j refers to $I_B(j; p)$ since the B-i.i.d. importance ordering is consistent for all p.

$\pi_1 = (1, 3, 5, \ldots, 2(n-k) - 1, (\text{any arrangement}), 2(n-k), \ldots, 6, 4, 2)$.

$\pi_2 = (n, n-2, n-4, \ldots, 2k - n + 2, (\text{any arrangement}), 2k - n + 1, \ldots, n-5, n-3, n-1)$.

obtained. Similar results were also presented by Kuo et al. (1990), Malon (1985), Papastavridis and Hadzichristos (1988), Tong (1985), and Zuo and Kuo (1990).

Theorem 12.2.1 *Given the reliability of n components, the necessary conditions for the optimal (worst) arrangement for a Lin/Con/k/n:F or G system are to arrange*

  (i) *components from positions* 1 *to* $\min\{k, n-k+1\}$ *in nondecreasing (nonincreasing) order of component reliability;*
 (ii) *components from positions n to* $\max\{k, n-k+1\}$ *in nondecreasing (nonincreasing) order of component reliability; and*
(iii) *if* $n < 2k$, *the* $2k-n$ *most (least) reliable components from positions* $(n-k+1)$ *to k in any order.*

For the *Lin/Con/k/n*:F and G systems with $n > 2k$, the necessary conditions in Theorem 12.2.1 only specify the order of component reliability from positions 1 to k and positions $(n-k+1)$ to n. For the *Lin/Con/k/n*:G systems with $n \leq 2k$, the invariant optimal arrangement presented in Table 12.1 satisfies these necessary conditions. For the *Lin/Con/k/n*:F systems with $n < 2k$, Theorem 12.2.1 shows that the optimal arrangement should always put the $2k-n$ most reliable components in the middle (in any order) and then arrange the remaining components into the *Lin/Con/(n − k)/2(n − k)*:F system in an optimal way. However, the optimal arrangement for the *Lin/Con/(n − k)/2(n − k)*:F system is not invariant and depends on the reliability of those remaining components. For the *Lin/Con/k/2k*:F system, the optimal arrangement must satisfy the monotonicity of component reliability from positions 1 to k and positions $k + 1$ to n. Analogous results and illustrations apply to the case of minimizing systems reliability.

With the necessary conditions in Theorem 12.2.1, the enumeration method is only needed to evaluate the permutations satisfying these necessary conditions. The randomization method can also incorporate these necessary conditions. For an example of a *Lin/Con/k/n*:F system with $n > 2k$, after a permutation is randomly generated, rearrange the most left k numbers in ascending order and the most right k numbers in descending order to make the permutation satisfy the necessary conditions for the optimal arrangement, assuming that $\tilde{p}_1 \leq \tilde{p}_2 \leq \ldots \leq \tilde{p}_n$.

Essentially, the necessary conditions can drastically reduce the solution space for the CAP in *Lin/Con/k/n* systems. For example, to find the optimal arrangement that maximizes the reliability of a *Lin/Con/k/n*:F system, there are a total of $n!/2$ arrangements to compare in consideration of structural symmetry. When $n > 2k$, using the necessary conditions in Theorem 12.2.1, the number of arrangements to compare is reduced to $2^{-1}n!/(k!)^2$. When $n \leq 2k$, to construct the optimal arrangement, Theorem 12.2.1 indicates putting the $2k-n$ most reliable components in the middle (in any order), then partitioning the remaining $2n-2k$ components into two equal sets, and arranging the components in each set in the orderings of their reliability as described in Theorem 12.2.1. Thus, the problem is limited to finding the best partition, and this reduces to the amount of $\binom{2n-2k}{n-k}/2 = \binom{2n-2k-1}{n-k-1}$ arrangements to be checked for optimality. For an example in *Lin/Con/8/11*:F system, in order to find an optimal arrangement among the $11!/2 = 19,958,400$ possibilities, it suffices to check $\binom{6}{3}/2 = 10$ possibilities for *Lin/Con/3/6*:F system.

The following necessary condition can further reduce the solution space for the CAP in the $Lin/Con/k/2k$:F system (Malon 1985). We provide a proof as follows.

Theorem 12.2.2 *Considering a $Lin/Con/k/2k$:F system and assuming that component reliability satisfies $\tilde{p}_1 \leq \tilde{p}_2 \leq \ldots \leq \tilde{p}_{2k}$, under permutations in which $\pi_i < \pi_{2k+1-i}$ (or $\pi_i > \pi_{2k+1-i}$ by symmetry) for all $1 \leq i \leq k$, the systems reliability can be increased by exchanging the components in positions k and $k+1$.*

Proof. Let $\bar{R}(\mathbf{p})$ denote the unreliability of the $Lin/Con/k/2k$:F system. Then,

$$\bar{R}(\mathbf{p}) = q_k q_{k+1} \bar{R}(0_k, 0_{k+1}, \mathbf{p}^{(k,k+1)}) + p_k q_{k+1} \prod_{i=k+2}^{2k} q_i + q_k p_{k+1} \prod_{i=1}^{k-1} q_i.$$

Thus,

$$\bar{R}(\mathbf{p}_{\boldsymbol{\pi}}) - \bar{R}(\mathbf{p}_{\boldsymbol{\pi}(k,k+1)}) = \tilde{p}_{\pi_k} \tilde{q}_{\pi_{k+1}} \prod_{i=k+2}^{2k} \tilde{q}_{\pi_i} + \tilde{q}_{\pi_k} \tilde{p}_{\pi_{k+1}} \prod_{i=1}^{k-1} \tilde{q}_{\pi_i} - \tilde{p}_{\pi_{k+1}} \tilde{q}_{\pi_k} \prod_{i=k+2}^{2k} \tilde{q}_{\pi_i} - \tilde{q}_{\pi_{k+1}} \tilde{p}_{\pi_k} \prod_{i=1}^{k-1} \tilde{q}_{\pi_i}$$

$$= (\tilde{p}_{\pi_k} - \tilde{p}_{\pi_{k+1}})(\prod_{i=k+2}^{2k} \tilde{q}_{\pi_i} - \prod_{i=1}^{k-1} \tilde{q}_{\pi_i}),$$

which is greater than 0 because $\pi_i < \pi_{2k+1-i}$ implies $\tilde{p}_{\pi_i} < \tilde{p}_{\pi_{2k+1-i}}$ for $1 \leq i \leq k$. The proof is complete.

According to Theorem 12.2.2, the permutations in which $\pi_i < \pi_{2k+1-i}$ (or $\pi_i > \pi_{2k+1-i}$ by symmetry) for all $1 \leq i \leq k$ cannot be the optimal solution. There are $\binom{2k}{k}/(k+1)$ such permutations so that, for example, the optimal permutation for $Lin/Con/8/11$:F system, or for any $Lin/Con/k/k+3$:F system with $k \geq 3$, can be found after checking only $\binom{6}{3}/2 - \binom{6}{3}/4 = 5$ permutations for a related $Lin/Con/3/6$:F system.

For $Lin/Con/k/n$:F systems, Table 12.4 compares some uniform B-i.i.d. importance patterns in Theorem 8.2.3 and the necessary conditions for the optimal permutations of the CAP in Theorem 12.2.1, given that $\tilde{p}_1 \leq \tilde{p}_2 \leq \ldots \leq \tilde{p}_n$. In Table 12.4, notation "↗" ("↘") visually implies the increasing (decreasing) of the importance measures and reliability (equivalently, index of the assigned components) of components in the second and third rows, respectively. Note that these necessary conditions are derived from the permutation importance patterns, which, in turn, implies the corresponding uniform B-i.i.d. importance patterns. Thus, the necessary conditions for the optimal arrangements follow the uniform B-i.i.d. importance patterns. This observation encourages the conjecture that other B-i.i.d. importance patterns for $Lin/Con/k/n$ with $k < n/2$ in Theorems 8.2.3 and 8.2.4 may imply other necessary conditions.

Now, considering a $Cir/Con/k/n$ system, there exist $(n-1)!/2$ possible permutations to be optimal. For a $Cir/Con/k/n$:F system with $(n-1)/2 \leq k \leq n-2$, Theorem 12.2.3 (Tong 1986) provided a necessary condition for an optimal arrangement to maximize systems reliability. Without loss of generality, this result is presented in terms of the first two components for the sake of notation simplification (for a $Cir/Con/k/n$ system, components can be relabeled in a clockwise direction, so any component could be treated as component 1).

Table 12.4   The uniform B-i.i.d. importance and necessary conditions for the optimal arrangements for *Lin/Con/k/n*:F systems

$k \geq n/2$		$k < n/2$	
($I_i = I_{i+1}$)		(uncertain)	
$n-k+1$	k	k	$n-k+1$
$\nearrow I_i < I_{i+1}$ $I_i > I_{i+1} \searrow$		$\nearrow I_i < I_{i+1}$ $I_i > I_{i+1} \searrow$	
1	n	1	n
(the $2n-k$ most reliable components in any order)		(uncertain)	
$n-k+1$	k	k	$n-k+1$
$\nearrow \pi_i^* < \pi_{i+1}^*$ $\pi_i^* > \pi_{i+1}^* \searrow$		$\nearrow \pi_i^* < \pi_{i+1}^*$ $\pi_i^* > \pi_{i+1}^* \searrow$	
1	n	1	n

Notation I_i in the second row represents both the uniform B-i.i.d. importance and the permutation importance since they have the same orderings. The third row presents the necessary conditions.

Using Theorem 12.2.3, the reliability of a *Cir/Con/k/n*:F system may be increased by exchanging two consecutive components.

Theorem 12.2.3   *In a Cir/Con/k/n:F system with* $(n-1)/2 \leq k \leq n-2$, $R(\mathbf{p}_{\boldsymbol{\pi}(1,2)}) \geq R(\mathbf{p}_{\boldsymbol{\pi}})$ *if and only if under permutation* $\boldsymbol{\pi}$ *(then* $p_i = \tilde{p}_{\pi_i}$*),*

$$(p_1 - p_2)\left[p_{k+2}\left(\prod_{i=3}^{k+1} q_i\right) - p_{n-k+1}\left(\prod_{i=n-k+2}^{n} q_i\right)\right] \leq 0.$$

Furthermore, the inequality is strict if the reliability values of any two components are distinct.

12.3   Sequential Component Assignment Problems in *Con/2/n*:F Systems

The CAP described in Section 11.1 is to construct a system nonsequentially by specifying the complete assignment of components together. Differently, the sequential CAP studied in this section assumes that the state of a component becomes known once it is added to the system. The sequential CAP is to construct the system sequentially, adding components one by one and taking full advantage of the knowledge of the states of the components already added. In this case, a dynamic policy is a rule or strategy for deciding in what order to use the components. In this section, assume that only the ordering of reliability of *n* components is known and satisfies $\tilde{p}_1 \leq \tilde{p}_2 \leq \ldots \leq \tilde{p}_n$.

Derman et al. (1982) proposed a greedy algorithm in Table 12.5 for the sequential CAP in *Lin/Con/2/n*:F systems and proved that it generates invariant optimal assignments, as shown in Theorem 12.3.1.

Theorem 12.3.1   *For the sequential CAP in a Lin/Con/2/n:F system, the greedy algorithm in Table 12.5 generates the invariant optimal assignment.*

Table 12.5   The greedy algorithm for $Lin/Con/2/n$:F systems

1. Assign component 1 as the initial component to position 1.
2. Suppose component j is the last component assigned so far. If component j is functioning, assign the least reliable component remaining to the available neighbor of component j next. If component j fails, assign the most reliable component remaining to the available neighbor of component j next.

Note that the optimality of step 2 is independent of step 1. Namely, if component i is assigned first, the subsequent optimal strategy depends only on the state of that component, not on its index i. Let $\mathfrak{G}(n; i)$ be an extension of the greedy algorithm in Table 12.5 to the case that component i is assigned first. Then $\mathfrak{G}(n; i)$ is invariant among those algorithms where component i is assigned first. Note that the optimal assignment generated by the algorithm is invariant because the algorithm uses only the ordering of component reliability, not their actual values.

For $Cir/Con/2/n$:F systems, after the initial component is assigned to the system, both of its neighbors can receive the next assignment, and this duality of receivership exists until the assignment of the last component. Hwang and Pai (2000) modified $\mathfrak{G}(n; i)$ for the sequential CAP in $Cir/Con/2/n$:F systems as in Table 12.6 and proved Theorems 12.3.2 and 12.3.3.

Note that whenever two consecutive components both fail, the system fails no matter how the remaining components are assigned. In particular, if either component x_i or y_i fails in step 3 in Table 12.6, then the system fails and subsequent assignments are inconsequential.

Theorem 12.3.2   *Among the class of algorithms for the sequential CAP in $Cir/Con/2/n$:F systems that start with component i, $\mathfrak{G}(n; i)$ is invariant.*

Theorem 12.3.3   *For $n \geq 6$, an optimal assignment for the sequential CAP in $Cir/Con/2/n$:F systems must be among $\mathfrak{G}(n; i)$ for $i \in \{\lceil n/2 \rceil + 1, \lceil n/2 \rceil + 2, \ldots, n - 2\}$.*

For the sequential CAP in $Cir/Con/2/n$:F systems, Theorem 12.3.3 reduces the number of candidates for the optimal assignment from $n!$ to $\lfloor n/2 \rfloor - 2$. It is easy to verify that for $n = 1, 2$, and 3, all assignments are equivalent, and for $n = 4$ and 5, both $\mathfrak{G}(n; n - 1)$ and $\mathfrak{G}(n; n - 2)$ are invariant. By Theorem 12.3.3, for $n = 6$ and 7, $\mathfrak{G}(n; n - 2)$ is invariant. For $n \geq 8$, Hwang and Pai (2000) concluded that no invariant assignment exists in general.

Table 12.6   The greedy algorithm for $Cir/Con/2/n$:F systems

1. Assign component i as the initial component.
2. If component i is functioning, use $\mathfrak{G}(n - 1; 1)$ on the set of components excluding component i.
3. If component i fails, assign the most reliable component x_i from the remaining components to either neighbor of component i next. If component x_i is functioning, use $\mathfrak{G}(n - 2; y_i)$ on the remaining $n-2$ components (starting by assigning the currently most reliable component y_i to the other neighbor of component i).

12.4 Consecutive-2 Failure Systems on Graphs

A consecutive-2 failure system is a graph, in which vertices are subject to failure and an edge fails if both of its adjacent vertices fail. The system fails if and only if there exists a failing edge. Assume that the failures of the vertices are stochastically independent of each other. Then the CAP is to find the optimal arrangements of a given set of components of distinct reliability to the vertices in the system with the objective of maximizing systems reliability.

For a consecutive-2 failure system on a general graph, there is no guarantee of the existence of an invariant optimal arrangement. Du and Hwang (1987) proved a sufficient condition for the existence of an invariant optimal arrangement in terms of graph theory, conjecturing that this condition is also a necessary condition. However, on some special graphs, the consecutive-2 failure system admits the invariant optimal arrangement (Du and Hwang 1987). For example, $Lin/Con/2/n$:F and $Cir/Con/2/n$:F systems are two special cases in which the underlying graphs are a line and a circle, respectively. By Theorems 12.1.1 and 12.1.4, $Lin/Con/2/n$:F and $Cir/Con/2/n$:F systems admit the invariant optimal arrangements. A (3, 4)-caterpillar or a cube also admits the invariant optimal arrangement. A (k, m)-caterpillar is a graph containing a path of m vertices of degree k ($k \geq 2$) and having all other vertices of degree one. The next subsection focuses on another special case where the underlying graph is a tree.

12.4.1 Consecutive-2 Failure Systems on Trees

This subsection gives the definitions of various types of trees in graph theory and then presents the results on invariant optimal arrangements of the CAP for consecutive-2 failure systems on trees. The proofs are omitted and interested readers are referred to Santha and Zhang (1987).

Definition 12.4.1 *A tree is a graph in which any two vertices are connected by exactly one path. Alternatively, a tree is a connected graph with no cycles. A tree is called a rooted tree if one vertex has been designated the root, in which case the edges have a natural orientation toward or away from the root. A leaf is a vertex having no successors. The level of the root is 0, and the height of a tree is the maximum level of its leaves.*

Definition 12.4.2 *A complete binary tree is an ordered rooted tree in which every vertex except the leaves has exactly two successors and all leaves are at the same distance from the root.*

Theorem 12.4.3 *Complete binary trees admit invariant optimal arrangements if and only if they have one or two levels.*

Definition 12.4.4 *A complete k-regular tree is a rooted tree in which every nonleaf vertex has degree k and all leaves are at the same distance from the root.*

Theorem 12.4.5 *Complete k-regular trees with $k \geq 3$ admit invariant optimal arrangements.*

Theorem 12.4.6 *Arrangement $\mathfrak{A}$, defined as follows, is the invariant optimal arrangement for every complete 3-regular tree. $\mathfrak{A}$ assigns the set of the least reliable components on the leaves (i.e., vertices of level t) growing from left to right, then jumps two levels, on the vertices of level $t - 2$, assigning the next set of the least reliable components growing again from left to right, and keeps doing so on every second level until reaching the root or the level just below the root, depending on the parity of the height. $\mathfrak{A}$ then assigns components to those unassigned levels, starting from the level just below the root or from the root and working downward. At each level, $\mathfrak{A}$ assigns the next set of the least reliable components growing, this time, from right to left. In particular, the vertices adjacent to the leaves are assigned the set of the most reliable components.*

12.5   Series $Con/k/n$ Systems

A series $Con/k/n$ system is a system with a few embedded $Con/k_i/n_i$ systems functionally connected in series (these embedded systems may have different values of k_i and n_i). Examples 12.5.1 and 12.5.2 are two applications of series $Con/k/n$:F and G systems, provided by Shen and Zuo (1994).

Example 12.5.1   Consider the problem of monitoring the nuclear reactions in a reactor. To obtain three-dimensional snapshots of the reactions in a certain area, say A, in the reactor, four high-speed cameras (labeled 1, 2, 3, and 4) are mounted at slightly different angles, focusing on area A. At least two adjacent cameras have to function properly to obtain the required data or image regarding the reactions in area A. Then, the four cameras form $Lin/Con/2/4$:G system. For another area of interest, say B, in the reactor, three cameras (labeled 5, 6, and 7) are mounted for monitoring, and at least two adjacent cameras have to function properly to obtain the image in area B. Then, the three cameras form $Lin/Con/2/3$:G system. In order to analyze the reactions in the reactor, high-quality snapshots of both areas are needed. Thus, the whole monitoring system is a series $Lin/Con/k/n$:G system that has an embedded $Lin/Con/2/4$:G system of cameras 1, 2, 3, and 4 and an embedded $Lin/Con/2/3$:G system of cameras 5, 6, and 7 functionally connected in series.

Example 12.5.2   Consider a telecommunication system of 20 relay stations, where stations 9 and 10 are far apart from each other such that the only signal that can reach station 10 has to come from station 9. However, the distances between other neighboring stations are not too far from each other such that each station can transmit signal to the subsequent two stations. Then, the whole system has an embedded $Lin/Con/2/8$:F system with stations 1 through 8, a single station 9, and an embedded $Lin/Con/2/11$:F system with stations 10 through 20 functionally connected in series. Whenever at least two consecutive stations fail in an embedded system or station 9 fails, the whole system fails.

For a series $Con/k/n$ system to admit an invariant optimal arrangement, it is obvious that all of its embedded $Con/k_i/n_i$ systems must admit invariant optimal arrangements. Then, under this condition, the optimal design of a series $Con/k/n$ system is obtained once the available components are optimally partitioned among the embedded systems. Therefore, the invariant optimal arrangement for a series $Con/k/n$ system is the invariant optimal partition of components among the embedded $Con/k_i/n_i$ systems, given that these embedded systems themselves admit the invariant optimal arrangements. The next two subsections study the invariant optimal arrangements of the CAP in two special series $Con/k/n$ systems.

12.5.1 Series Con/2/n:F Systems

A series $Con/2/n$:F system consists of m embedded $Con/2/n_i$:F systems functionally connected in series. By the analysis in Section 12.1, the invariant optimal arrangements of the CAP in the $Lin/Con/2/n$ and $Cir/Con/2/n$ systems exist. Thus, the optimal solution of the CAP in a series $Con/2/n$:F system depends only on the partition of components into lines or circles; however, the invariant optimal design generally does not exist. Du and Hwang (1985) obtained the following two theorems about the invariant optimal arrangements (i.e., invariant optimal partitions among lines or circles) for some special cases.

Theorem 12.5.3 *For a series $Lin/Con/2/n$:F system with embedded $Lin/Con/2/n_i$:F systems of four values of n_i (i.e., four lengths of lines): $1, 3, r - 2$, and r, where r is even, the invariant optimal arrangement of the CAP exists. Let $\ell_1, \ell_3, \ell_{r-2}$, and ℓ_r denote the multiplicity of lines of length $1, 3, r - 2$, and r, respectively. Then, the optimal partition of components into lines is the following. The ℓ_1 least reliable components go to the ℓ_1 lines of length 1. Take the next two least reliable components and the most reliable components to form a line of length 3 and repeat this ℓ_3 times to form the ℓ_3 lines of length 3. Add ℓ_{r-2} perfect components and ℓ_{r-2} dummy components to the remaining components. Order the $r(\ell_{r-2} + \ell_r)$ components by treating the perfect components as the most reliable and the dummy components as the least reliable. Assign the jth least reliable component to set h if $j = h$ or $j = \ell_{r-2} + \ell_r + 1 - h$ (mod $\ell_{r-2} + \ell_r$) for $h = 1, 2, \ldots, \ell_{r-2} + \ell_r$. Delete the added perfect and dummy components from the sets after assignment to obtain ℓ_{r-2} lines of length $(r - 2)$ and ℓ_r lines of length r.*

Theorem 12.5.4 *Consider a series $Cir/Con/2/n$:F system with m embedded $Cir/Con/2/r$:F systems of the same length r, where r is even. The invariant optimal partition is to form a circle of length r by the $r/2$ least reliable components remaining and the $r/2$ most reliable components remaining, continuing until all circles are formed.*

12.5.2 Series Lin/Con/k/n:G Systems

Because a $Lin/Con/k/n$:G system with $n > 2k$ does not admit an invariant optimal arrangement, only the series $Lin/Con/k/n$:G system in which all embedded $Lin/Con/k_i/n_i$:G systems satisfy $k_i \le n_i \le 2k_i$ for $i = 1, 2, \ldots, m$ may admit an invariant optimal arrangement. When $k_i \le n_i \le 2k_i$, the $Lin/Con/k_i/n_i$:G system functions if and only if all components in the middle $2k_i - n_i$ positions function and the remaining $2(n_i - k_i)$ components constitute a functioning $Lin/Con/(n_i - k_i)/2(n_i - k_i)$:G system. Therefore, the $\sum_{i=1}^{m}(2k_i - n_i)$ most reliable components should be assigned to the middle positions in any order ($2k_i - n_i$ of them to line i), since the whole system fails if any one of them fails. After deleting these middle positions, the original system is reduced to a series system consisting of embedded $Lin/Con/(n_i - k_i)/2(n_i - k_i)$:G systems for $i = 1, 2, \ldots, m$. On the basis of this analysis, Chang and Hwang (1999) obtained the following theorem.

Theorem 12.5.5 *A series system consisting of embedded $Lin/Con/k_i/n_i$:G systems for $k_i \leq n_i \leq 2k_i$ and $i = 1, 2, \ldots, m$ admits an invariant optimal arrangement if and only if the series system consisting of embedded $Lin/Con/(n_i - k_i)/2(n_i - k_i)$:G systems for $i = 1, 2, \ldots, m$ admits an invariant optimal arrangement.*

Theorems 12.5.6 (Shen and Zuo 1994) and 12.5.7 (Chang and Hwang 1999) provide additional results on the invariant optimal arrangements for series $Lin/Con/k/n$:G systems consisting of embedded $Lin/Con/k_i/2k_i$:G systems for $i = 1, 2, \ldots, m$.

Theorem 12.5.6

(i) *For a series $Lin/Con/k/n$:G system consisting of two embedded $Lin/Con/k/2k$:G systems, S_1 and S_2, the invariant optimal partition*
   (a) *assigns the most reliable component remaining to S_1 and the next most reliable component to S_2;*
   (b) *assigns the most reliable component remaining to S_2 and the next most reliable component to S_1; and*
   (c) *repeats steps a and b until all components are assigned.*

(ii) *For a series $Lin/Con/k/n$:G system consisting of an embedded $Lin/Con/k/2k$:G system, S_1, and an embedded $Lin/Con/(k-1)/2(k-1)$:G system, S_2, the invariant optimal partition*
   (a) *assigns the most and the least reliable components to S_1;*
   (b) *assigns the most reliable component remaining to S_1 and the next most reliable component to S_2;*
   (c) *assigns the most reliable component remaining to S_2 and the next most reliable component to S_1; and*
   (d) *repeats steps b and c until all components are assigned.*

(iii) *For a series $Lin/Con/k/n$:G system consisting of m embedded $Lin/Con/k/2k$:G systems, denoted by $S_1, S_2, \ldots, S_m$, given that component reliability $\tilde{p}_1 < \tilde{p}_2 < \ldots < \tilde{p}_{2km}$, the invariant optimal partition*
   (a) *sets $i = 1$;*
   (b) *assigns $\tilde{p}_{2(i-1)m+1}, \tilde{p}_{2(i-1)m+2}, \ldots, \tilde{p}_{2(i-1)m+m}$ to $S_1, S_2, \ldots, S_m$, respectively;*
   (c) *assigns $\tilde{p}_{(2i-1)m+1}, \tilde{p}_{(2i-1)m+2}, \ldots, \tilde{p}_{2im}$ to $S_m, S_{m-1}, \ldots, S_1$, respectively; and*
   (d) *increases i by one and repeats steps b and c until all components are assigned.*

(iv) *For a series $Lin/Con/k/n$:G system consisting of ℓ embedded $Lin/Con/k/2k$:G systems and m embedded $Lin/Con/(k-1)/2(k-1)$:G systems, the invariant optimal partition*
   (a) *adds m perfect components as the most reliable components and m dummy components as the least reliable components to the set of available components;*
   (b) *does optimal design for the series $Lin/Con/k/n$:G system with $(m + \ell)$ embedded $Lin/Con/k/2k$:G systems following part (iii); and*
   (c) *deletes the added perfect and dummy components after assignment.*

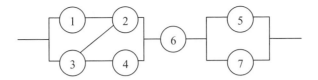

Figure 12.1   A diagram representing the camera system

Theorem 12.5.7   *For a series $Lin/Con/k/n$:G system consisting of embedded $Lin/Con/$ $k_i/2k_i$:G systems for $i = 1, 2, \ldots, m$ and $k_1 \leq k_2 \leq \ldots \leq k_m$, the system does not admit an invariant optimal arrangement if $k_m \geq k_1 + 2$.*

Example 12.5.8 (Example 12.5.1 continued)   Consider the camera system in Example 12.5.1. The embedded $Lin/Con/2/3$:G system of positions 5, 6, and 7 can be decomposed to a system where positions 5 and 7 are connected in parallel and position 6 is connected with them in series. Then the whole system can be represented as in Figure 12.1.

Suppose that seven cameras are to be assigned to seven positions in the system, and that the reliability of the seven cameras satisfies $\tilde{p}_1 < \tilde{p}_2 < \ldots < \tilde{p}_7$. Position 6 is more permutation important than any other position; thus, by Theorem 6.5.13, camera 7 is assigned to position 6. According to Theorem 12.5.6(ii), of the six remaining cameras, cameras 6, 1, 5, and 2 should be assigned to the embedded $Lin/Con/2/4$:G system (of positions 1, 2, 3, and 4) and cameras 3 and 4 should be assigned to the embedded $Lin/Con/1/2$:G system (of positions 5 and 7). According to Theorem 12.1.2, the optimal arrangement for the embedded $Lin/Con/2/4$:G system assigns cameras 1, 5, 6, and 2 to positions 1, 2, 3, and 4, respectively. Cameras 3 and 4 can be arbitrarily assigned to positions 5 and 7. Hence, the invariant optimal arrangement for the camera system is $(1, 5, 6, 2, 4, 7, 3)$.

12.6   Consecutive-*k*-out-of-*r*-from-*n* Systems

A consecutive-*k*-out-of-*r*-from-*n* failure system fails if and only if within any *r* consecutive components there are at least *k* failures. It was first introduced by Griffith (1986) but was originally named as a *k*-within-*r*-out-of-*n* failure system. Such a system is also referred to as a *k*-out-of-*r*-out-of-*n* (Chang et al. 1999), or *k*-within-consecutive-*r*-out-of-*n* (Koutras et al. 1994) system. Commonly, though, as used in this book, it is named as the consecutive-*k*-out-of-*r*-from-*n*:F system (Papastavridis and Sfakianakis 1991) and defined as follows.

Definition 12.6.1   *A consecutive-k-out-of-r-from-n:F (G) system with $k \leq r \leq n$, denoted by $Con/k/r/n$:F (G), consists of n ordered components such that the system fails (functions) if and only if there exists a set of r consecutive components that contains k or more failed (functioning) components.*

If the ordered components are arranged linearly (circularly), it is a linear (circular) consecutive-*k*-out-of-*r*-from-*n* system, denoted by $Lin/Con/k/r/n$ ($Cir/Con/k/r/n$). When $r = n$, the $Con/k/r/n$:G system becomes a *k*-out-of-*n* system, and the $Con/k/r/n$:F system becomes a $(n - k + 1)$-out-of-n system. When $k = r$, the $Con/k/r/n$ system becomes a $Con/k/n$ system. The $Con/k/r/n$:G and $Con/(r - k + 1)/r/n$:F systems are equivalent. The $Con/k/r/n$ systems can model quality-control problems and inspection procedures (Separstein 1973, 1975), radar detection problems (Nelson 1978), and so on. An example from Sfakianakis et al. (1992) is provided as follows.

Example 12.6.2   A telecommunication system uses n-byte messages. The last bit of every byte is a parity bit (1 when the parity of the byte is correct). An error detector indicates an error when it finds two or more errors in a "window" of width 4 in the parity bit sequence. This is a $Lin/Con/2/4/n$:F system.

Much research has focused on reliability evaluation of $Con/k/r/n$ systems (e.g., Cai 1994; Griffith 1986; Habib and Szántai 1997; Higashiyama et al. 1995, 1996, 1999; Separstein 1973, 1975); other research proposed some lower and upper bounds on the reliability of $Con/k/r/n$ systems (e.g., Cai 1994; Habib and Szántai 2000; Papastavridis and Koutras 1993; Sfakianakis et al. 1992). A few studies are related to importance measures and the CAP in $Con/k/r/n$ systems.

Theorem 12.6.3 presents the permutation importance patterns for a $Lin/Con/k/r/n$ system. According to Figure 6.3, the permutation importance implies the uniform B-i.i.d. importance; thus, following Theorem 12.6.3, Corollary 12.6.4 gives the uniform B-i.i.d. importance patterns.

Theorem 12.6.3   *In a $Lin/Con/k/r/n$:F or G system, $(i + 1) >_{\text{pe}} i$ for $1 \leq i \leq \min\{r - 1, n - r\}$, and $i >_{\text{pe}} (i + 1)$ for $\max\{r, n - r + 1\} \leq i \leq n - 1$. Moreover, when $n < 2r$, $i =_{\text{pe}} (i + 1)$ for $n - r + 1 \leq i \leq r - 1$; when $n > 2r$, the intermediate components $r, r + 1, \ldots, n - r + 1$ are not comparable according to the permutation importance.*

Proof. The result can be easily established through Theorem 6.5.7.

Corollary 12.6.4   *In a $Lin/Con/k/r/n$:F or G system, for all $0 < p < 1$, $I_{\text{B}}(i; p) \leq I_{\text{B}}(i + 1; p)$ for $1 \leq i \leq \min\{r - 1, n - r\}$, and $I_{\text{B}}(i; p) \geq I_{\text{B}}(i + 1; p)$ for $\max\{r, n - r + 1\} \leq i \leq n - 1$. Moreover, when $n < 2r$, $I_{\text{B}}(i; p) = I_{\text{B}}(i + 1; p)$ for $n - r + 1 \leq i \leq r - 1$.*

According to Theorem 6.5.13, Theorem 12.6.3 is equivalent to the statement that $R(\mathbf{p}_{\pi}) \geq R(\mathbf{p}_{\pi(i,j)})$ whenever (i) for $1 \leq i \leq \min\{r - 1, n - r\}$: $\tilde{p}_{\pi_{i+1}} \geq \tilde{p}_{\pi_i}$ or (ii) for $\max\{r, n - r + 1\} \leq i \leq n - 1$: $\tilde{p}_{\pi_{i+1}} \leq \tilde{p}_{\pi_i}$. Papastavridis and Sfakianakis (1991) directly proved this statement. Consequently, the necessary conditions for the optimal arrangement for a $Lin/Con/k/r/n$:F system are implied as in Corollary 12.6.5, which is similar to Theorem 12.2.1 for a $Lin/Con/k/n$ system. Sfakianakis (1993) gave an equivalent optimal arrangement policy for the $Lin/Con/k/r/n$:F system with $n < 2r$. When $n \geq 2r$, the situation is complicated for $i > r$ and $i < n - r + 1$ even in the simple case of $r = k$ (i.e., a $Lin/Con/k/n$:F system).

Corollary 12.6.5   *Given the reliability of n components, the necessary conditions for the optimal arrangement for a $Lin/Con/k/r/n$:F or G system are to arrange*

     (i) components from positions 1 to $\min\{r, n - r + 1\}$ in nondecreasing order of component reliability;

(ii) components from positions n to $\max\{r, n - r + 1\}$ in nondecreasing order of component reliability; and

(iii) if $n < 2r$, the $2r - n$ most reliable components from positions $(n-r+1)$ to r in any order.

The following corollary follows from Corollary 12.6.5.

Corollary 12.6.6 *For a $Lin/Con/k/r/n$:F or G system with $r = n - 1$, the invariant optimal arrangement is to assign the two least reliable components to positions 1 and n and the other components to positions 2 to r in any order.*

Sfakianakis (1993) stated that a $Lin/Con/k/r/n$:F system with $n \geq 2r$ admits an invariant optimal arrangement if and only if $k = r = 2$. However, his proof is incomplete. On the basis of the result in Sfakianakis (1993), we conjecture (characterize) all values of k, r, and n for which an invariant optimal arrangement exists for a $Lin/Con/k/r/n$:F system, as in Conjecture 12.6.7. The hard part is to show that the $Lin/Con/k/r/n$:F system with $1 < k < r < n$ does not admit an invariant optimal arrangement.

Conjecture 12.6.7 *The $Lin/Con/k/r/n$:F system with $1 < k < r < n$ does not admit an invariant optimal arrangement. When $k = 1$, any arrangement is optimal since a $Lin/Con/1/r/n$:F system is a series system. When $k = r$, an invariant optimal arrangement exists if and only if $k \in \{1, 2, n - 2, n - 1, n\}$ (Theorem 12.1.1). When $r = n$, any arrangement is optimal since a $Lin/Con/k/n/n$:F system is an $(n - k + 1)$-out-of-n system.*

For a $Lin/Con/2/r/n$:F system ($k = 2$), Chang et al. (1999) presented a closed form for the B-i.i.d. importance of components in terms of the reliability of the whole system and its subsystems; B-i.i.d. importance patterns; and a cut-path importance pattern, as in the following theorem. Let $R(n)$ denote the reliability of a $Lin/Con/2/r/n$:F system with i.i.d. components of a same reliability p.

Theorem 12.6.8 *In a $Lin/Con/2/r/n$:F system with i.i.d. components of a same reliability p,*

(i) $I_B(i; p) = [R(n) - p^{x_i}R(i - r)R(n - i - r + 1)]/p$, where $x_i = \min\{r - 1, i - 1\} + \min\{r - 1, n - i\} \geq r - 1$;

(ii) $R(x)p^{r-1}R(y) \leq R(x + r - 1 + y)$ for integers $x, y \geq 1$;

(iii) $I_B(1; p) \leq I_B(i; p)$ for $i = 1, 2, \ldots, n$;

(iv) $I_B(r; p) \geq I_B(i; p)$ for $i = 1, 2, \ldots, n$ when $n \geq 2r$; and

(v) $r \geq_{cp} i$ (cut-path importance) for $i = 1, 2, \ldots, n$ when $n \geq 2r$.

12.7   Two-dimensional and Redundant $Con/k/n$ Systems

Salvia and Lasher (1990) first introduced a two-dimensional $Con/k/n$:F system; Zuo (1993) then generalized it.

Definition 12.7.1 *A two-dimensional consecutive-$k_1 k_2$-out-of-mn:F (G) system with $2 \leq k_1 \leq m$ and $2 \leq k_2 \leq n$, denoted by $Con/(k_1, k_2)/(m, n)$:F (G), consists of mn ordered* *components such that the system fails (functions) if and only if there is at least one grid of dimension $k_1 \times k_2$ that contains all failed (functioning) components.*

Corresponding to (one-dimensional) linear and circular $Con/k/n$ systems, two-dimensional $Con/k/n$ systems can be defined and classified into rectangular (*Rec*) and cylindrical (*Cyl*) $Con/(k_1, k_2)/(m, n)$ systems. In a $Rec/Con/(k_1, k_2)/(m, n)$ system, mn components are arranged in a rectangular grid of dimension $m \times n$, that is, there are m rows of components, and each row has n components. In a $Cyl/Con/(k_1, k_2)/(m, n)$ system, mn components are arranged on the surface of a cylinder. There are m rows of components arranged along the axial direction of the cylinder and n columns of components along the circumferential direction of the cylinder. Columns n and 1 are neighbors. Salvia and Lasher (1990) provided two applications of two-dimensional $Con/k/n$ systems as follows.

Example 12.7.2 (a) (**Electronic device**) A group of connector pins for an electronic device includes some redundancy in its design such that the connection is good unless a square of side 2 (4 pins) is defective.

(b) (**Pattern detection**) The presence of a disease is diagnosed by reading an X-ray. An individual cell (or other small portion of the X-ray) is healthy with a certain probability. Unless diseased cells are aggregated into a sufficiently large pattern (say a k^2 square), the radiologist might not detect their presence.

As for the relationship between $Con/k/n$:F and G systems (Theorem 2.13.2 and Corollary 2.13.3), $Con/(k_1, k_2)/(m, n)$:F and G systems are dual and mirror images of each other (Zuo 1993). For a $Con/(k_1, k_2)/(m, n)$:F (G) system, any grid of size $k_1 \times k_2$ is a minimal cut (path). The component in row i and column j is labeled as (i, j), and its reliability is $p_{i,j}$.

Applying Theorem 6.5.7, Koutras et al. (1994) identified the permutation importance patterns in Theorem 12.7.3 for a $Rec/Con/(k, k)/(n, n)$:F or G system. These patterns can be used in the elimination procedure in Table 11.1 to find the optimal arrangements of the CAP.

Theorem 12.7.3 *In a $Rec/Con/(k, k)/(n, n)$:F or G system,*

(i) *if $1 \leq j < \min\{k, n - k + 1\}$, then for $i = 1, 2, \ldots, n$, $(i, j) <_{\mathrm{pe}} (i, j + 1)$, and $(j, i) <_{\mathrm{pe}} (j + 1, i)$;*

(ii) *if $\max\{k, n - k + 1\} \leq j < n$, then for $i = 1, 2, \ldots, n$, $(i, j + 1) <_{\mathrm{pe}} (i, j)$, and $(j + 1, i) <_{\mathrm{pe}} (j, i)$; and*

(iii) *all permutation patterns are the preceding ones and those deducible from them by the transitivity property of the permutation importance (Theorem 6.5.3).*

The concept of a two-dimensional $Con/k/n$ system can be extended to the multidimensional $Con/k/n$ system. For example, a three-dimensional $Con/k/n$:F system can be defined on a cubic of size $m \times n \times h$ (containing mnh components), and the system fails if and only if there exists at least a small cubic of size $k_1 \times k_2 \times k_3$ ($k_1 \leq m, k_2 \leq n, k_3 \leq h$) containing all failed components.

Redundancy has been introduced to $Con/k/n$ systems to enhance systems reliability. A redundant $Con/k/n$ system can be a redundant linear or circular $Con/k/n$:F or G system, depending on the type of the original (nonredundant) $Con/k/n$ system.

Definition 12.7.4 *A redundant $Con/k/n$ system is a $Con/k/n$ system where each component i, $i = 1, 2, \ldots, n$, is actually an embedded parallel system of r_i components. If $r_1 = r_2 = \ldots = r_n = r$, the system is known as a $Con/k/n$ system with redundancy r.*

Note that a $Con/k/n$ system with redundancy r is exactly a two-dimensional $Con/(r, k)/(r, n)$ system, which consists of nr components arranged either in a grid or on the surface of a cylinder of r rows and n columns. When $r = 1$, this system becomes a regular $Con/k/n$ system. The next subsection investigates the CAP in the $Con/(r, k)/(r, n)$ system for $r \geq 2$.

12.7.1 $Con/(r, k)/(r, n)$ Systems

For the CAP, suppose that nr components are available to be assigned to a $Con/(r, k)/(r, n)$ system, each one to a unique position. Since each embedded system has r components connected in parallel, the arrangement of components within an embedded system does not affect the systems reliability once these components are assigned to the embedded system.

Theorem 12.7.5 *In a $Rec/Con/(r, k)/(r, n)$:F or G system, the following permutation patterns exist.*

(i) $(i_1, j) =_{\mathrm{pe}} (i_2, j)$ *for* $i_1, i_2 \in \{1, 2, \ldots, r\}$ *and* $1 \leq j \leq n$.
(ii) $(i_1, j_1) <_{\mathrm{pe}} (i_2, j_2)$ *for* $i_1, i_2 \in \{1, 2, \ldots, r\}$ *and* $1 \leq j_1 < j_2 \leq \min\{k, n - k + 1\}$.
(iii) $(i_1, j_2) <_{\mathrm{pe}} (i_2, j_1)$ *for* $i_1, i_2 \in \{1, 2, \ldots, r\}$ *and* $\max\{k, n - k + 1\} \leq j_1 < j_2 \leq n$.
(iv) *If* $k < n < 2k$, $(i_1, j_1) =_{\mathrm{pe}} (i_2, j_2)$ *for* $i_1, i_2 \in \{1, 2, \ldots, r\}$ *and* $n - k + 1 \leq j_1, j_2 \leq k$.
(v) *All permutation patterns are the preceding ones and those deducible from them by the transitivity property of the permutation importance (Theorem 6.5.3).*

Assuming that the reliability values of all nr components are distinct and $\tilde{p}_1 < \tilde{p}_2 < \ldots < \tilde{p}_{nr}$, according to Theorems 6.5.13 and 12.7.5, the necessary conditions for the optimal arrangement of the CAP in a $Rec/Con/(r, k)/(r, n)$:F or G system are derived as in Corollary 12.7.6 (Zuo 1993). Corollary 12.7.6 can be illustrated similarly to Theorem 12.2.1, which pertains to a regular $Lin/Con/k/n$ system.

Corollary 12.7.6 *For a $Rec/Con/(r, k)/(r, n)$:F or G system, the necessary conditions for the optimal arrangement of the CAP are as follows:*

(i) $p_{i_1, j_1} < p_{i_2, j_2}$ *for* $i_1, i_2 \in \{1, 2, \ldots, r\}$ *and* $1 \leq j_1 < j_2 \leq \min\{k, n - k + 1\}$;
(ii) $p_{i_1, j_2} < p_{i_2, j_1}$ *for* $i_1, i_2 \in \{1, 2, \ldots, r\}$ *and* $\max\{k, n - k + 1\} \leq j_1 < j_2 \leq n$; *and*
(iii) *if* $k < n < 2k$, *the* $(2k - n)r$ *most reliable components should be arranged from columns* $(n - k + 1)$ *to k in any order.*

Table 12.7 Invariant optimal arrangements for $Con/(r, k)/(r, n)$:F systems

	$Rec/Con/(r, k)/(r, n)$:F	$Cyl/Con/(r, k)/(r, n)$:F
$r = 2$ and $k = n - 1$	1 (any arrangement) 2 4 (any arrangement) 3	Components 1 and 6, 2 and 5, and 3 and 4 are in the same column, respectively. These three columns are arranged next to each other in any order, and the other $2n - 6$ components are assigned in any order.
$r = 2$ and $k = 1$	Components s and $(2n - s + 1)$ are in the same column for $s = 1, 2, \ldots, n$, and these columns are arranged in any order.	
$k = n$	(any arrangement)	(any arrangement)

Hwang and Shi (1987) showed that a $Con/(r, k)/(r, n)$:F system with redundancy $r \geq 2$ and $2 \leq k \leq n - 1$ admits an invariant optimal arrangement if and only if $r = 2$ and $k = n - 1$. Additionally, when $k = n$, the system reduces to a parallel system; thus, any arrangement is optimal. When $k = 1$, the system becomes a series–parallel system, and by Theorem 11.5.3, the system admits the invariant optimal arrangement if $r = 2$. The following theorem includes these results.

Theorem 12.7.7 *For $r \geq 2$, both $Rec/Con/(r, k)/(r, n)$:F and $Cyl/Con/(r, k)/(r, n)$:F systems admit invariant optimal arrangements if and only if (i) $r = 2$ and $k = n - 1$; (ii) $r = 2$ and $k = 1$; or (iii) $k = n$. These invariant optimal arrangements are listed in Table 12.7, given that component reliability satisfies $\tilde{p}_1 < \tilde{p}_2 < \ldots < \tilde{p}_{rn}$.*

Zuo (1993) identified invariant optimal arrangements for $Rec/Con/(2, k)/(2, n)$:G systems (i.e., $r = 2$) with $k < n \leq 2k$ and $Cyl/Con/(2, k)/(2, n)$:G systems with $k < n \leq 2k + 1$ and proved that neither the $Rec/Con/(r, k)/(r, n)$:G system with $2 \leq k < n/2$ and $r \geq 2$ nor the $Cyl/Con/(r, k)/(r, n)$:G system with $2 \leq k < (n - 1)/2$ and $r \geq 2$ admits an invariant optimal arrangement, as presented in Theorem 12.7.8.

Theorem 12.7.8 *For a $Rec/Con/(2, k)/(2, n)$:G system with $k < n \leq 2k$, an invariant optimal arrangement exists, and, given that $\tilde{p}_1 < \tilde{p}_2 < \ldots < \tilde{p}_{2n}$, it is*

$$
\begin{array}{lllll}
1, 5, 9, & \ldots & 4(n - k) - 3, & (\text{any arrangement}), & 4(n - k) - 1, \quad \ldots, \quad 11, 7, 3 \\
2, 6, 10, & \ldots & 4(n - k) - 2, & (\text{any arrangement}), & 4(n - k), \qquad \ldots, \quad 12, 8, 4.
\end{array}
$$

A $Rec/Con/(r, k)/(r, n)$:G system with $2 \leq k < n/2$ and $r \geq 2$ does not admit an invariant optimal arrangement. For a $Cyl/Con/(2, k)/(2, n)$:G system with $k < n \leq 2k + 1$, an invariant optimal arrangement exists, and, given that $\tilde{p}_1 < \tilde{p}_2 < \ldots < \tilde{p}_{2n}$, it is

$$
\begin{array}{lllll}
1, 3, 7, 11, 15, & \ldots & 2n - 1, & \ldots, & 17, 13, 9, 5, 1 \\
2, 4, 8, 12, 16, & \ldots & 2n, & \ldots, & 18, 14, 10, 6, 2.
\end{array}
$$

A $Cyl/Con/(r, k)/(r, n)$:G system with $2 \leq k < (n - 1)/2$ and $r \geq 2$ does not admit an invariant optimal arrangement.

In Theorems 12.7.7 and 12.7.8, the arrangements obtainable by exchanging any two components in the same column or by flipping the left and right ends of the arrangement are not considered different because of symmetry. They result in the same systems reliability.

Additionally, Hwang and Shi (1987) investigated a redundancy design problem, in which n components are available to construct a relay system over a certain distance. A regular $Lin/Con/k/n$:F system could be built or, alternatively, $Rec/Con/(r, k/r)/(r, n/r)$:F systems for each r dividing both k and n could be built. The following theorem shows that the system with the largest redundancy always provides the maximum reliability.

Theorem 12.7.9 *For given $k_\ell r_\ell = k$ and $n_\ell r_\ell = n$, let $R_L^*(r_\ell, k_\ell, n_\ell)$ denote the reliability of the $Rec/Con/(r_\ell, k_\ell)/(r_\ell, n_\ell)$:F system under the optimal arrangement. Letting $\gcd(k, n) = g$, then $R_L^*(g, k/g, n/g)$ is maximum among $R_L^*(r_\ell, k_\ell, n_\ell)$ for all ℓ. The result also holds for the cylindrical case.*

12.8 Miscellaneous

Some other generalizations of $Con/k/n$ systems have been proposed. Lin et al. (2000) generalized the $Con/k/n$:G system to a so-called $Con/k^*/n$:G system. A $Con/k^*/n$:G system has n ordered components and functions if and only if k_i consecutive components that originate at component i function, where k_i is a function of i. They gave formulas to compute the reliability of linear and circular $Con/k^*/n$:G systems.

Griffith (1986) and Papastavridis (1990) studied reliability evaluation of a linear m-consecutive-k-out-of-n:F system. A linear m-consecutive-k-out-of-n:F system, which is an ordered linear sequence of n components, fails if and only if there are at least m nonoverlapping runs of k consecutive failures. A run of failures of length sk is treated as s runs of length k.

Makri and Psillakis (1996) studied a k-within two-dimensional consecutive-r-out-of-n failure system, which is a square grid of side n (containing n^2 components) and fails if and only if there is at least one square sub-grid of side r (containing r^2 components, $1 < r \leq n$) that has at least k ($1 \leq k \leq r^2$) failed components. For systems with i.i.d. components, a lower and an upper bound on systems reliability are derived using improved Bonferroni inequalities. For the system with $k = 2$ and i.i.d. components, Higashiyama (2004) and Higashiyama et al. (2006) gave algorithms for computing the systems reliability when the components are arranged in rectangular pattern and cylindrical pattern, respectively.

Some generalizations of $Lin/Con/k/r/n$ systems have also been proposed. Levitin (2003a) generalized a $Lin/Con/k/r/n$:F system to a system in which components have different reliability characteristics and performance rates. Each component can have two states: (i) complete failure with a performance rate of zero and (ii) perfect functioning with a nominal performance rate. The system fails if and only if the sum of the performance rates of any r consecutive components is lower than a specified level. Levitin (2003a) provided an algorithm to evaluate systems reliability that can also be used to evaluate the B-reliability importance of components according to Equation (4.11).

Levitin (2002, 2003b) further extended this system to the multistate case where the system is in one of two possible states (functioning or failed) but its components have more than two different states, from complete failure up to perfect functioning (see Sections 15.2–15.5). Levitin (2002) considered the CAP in this system, using a genetic algorithm as an optimization tool to find the approximate optimal arrangement and a universal generating function technique for systems reliability evaluation. Levitin (2003b) provided an algorithm for systems reliability evaluation using an extended universal moment-generating function.

Levitin (2004) generalized a $Lin/Con/k/r/n$:F system to a system with multiple failure criteria, and Levitin (2005) further generalized this multiple failure criteria system to a multistate case. The reliability of these $Lin/Con/k/r/n$:F related systems is evaluated using a universal generating function technique.

Habib et al. (2007) generalized a $Lin/Con/k/r/n$:G system to a multistate case and provided an algorithm for evaluating the reliability of a special multistate $Lin/Con/k/r/n$:G system. In this model, both the system and its components may be in one of $K+1$ possible states: $0, 1, \ldots$, and K. The system is in state μ or above if and only if at least k_μ components out of r consecutive components are in state μ or above.

Up to date, the research on these generalizations of $Con/k/n$ systems is limited and focuses on their reliability evaluation. For the CAP in these miscellaneous $Con/k/n$ related systems, the current research is almost blank and the future study could explore the special structures and properties of these systems for solving their CAP by means of importance measures.

References

Cai J. 1994. Reliability of a large consecutive-k-out-of-n:F system with unequal component-reliability. *IEEE Transactions on Reliability* **43**, 107–111.

Chang GJ, Cui L, and Hwang FK. 1999. New comparisons in Birnbaum importance for the consecutive-k-out-of-n system. *Probability in the Engineering and Informational Sciences* **13**, 187–192.

Chang HW and Hwang FK. 1999. Existence of invariant series consecutive-k-out-of-n:G systems. *IEEE Transactions on Reliability* **R-48**, 306–308.

Cui LR and Hawkes AG. 2008. A note on the proof for the optimal consecutive-k-out-of-n:G line for $n \le 2k$. *Journal of Statistical Planning and Inference* **138**, 1516–1520.

Derman C, Lieberman GJ, and Ross SM. 1982. On the consecutive-k-of-n:F system. *IEEE Transactions on Reliability* **R-31**, 57–63.

Du DZ and Hwang FK. 1985. Optimal consecutive-2 systems of lines and cycles. *Networks* **15**, 439–447.

Du DZ and Hwang FK. 1986. Optimal consecutive 2-out-of-n systems. *Mathematics of Operations Research* **11**, 187–191.

Du DZ and Hwang FK. 1987. Optimal assignments for consecutive-2 graphs. *SIAM Journal on Algebraic and Discrete Methods* **8**, 510–518.

Du D, Hwang FK, Jung Y, and Ngo HQ. 2001. Optimal consecutive-k-out-of-$(2k + 1)$:G cycle. *Journal of Global Optimization* **19**, 51–60.

Du D, Hwang FK, Jia X, and Ngo HQ. 2002. Optimal consecutive-k-out-of-n:G cycle for $n \le 2k + 1$. *SIAM Journal on Discrete Mathematics* **15**, 305–316.

Griffith W. 1986. On consecutive k-out-of-n:F failure system and their generations. In *Reliability and Quality Control* (ed. Basu AP). Elsevier (North-Holland), New York, pp. 157–165.

Habib A, Al-Seedy R, and Radwan T. 2007. Reliability evaluation of multi-state consecutive k-out-of-r-from-n:G system. *Applied Mathematical Modelling* **31**, 2412–2423.

Habib A and Szántai T. 1997. An algorithm evaluating the exact reliability of a consecutive k-out-of-r-from-n:F system. *Proceedings of 1997 IMACS Conference*, pp. 421–425.

Habib A and Szántai T. 2000. New bounds on the reliability of the consecutive k-out-of-r-from-n:F system. *Reliability Engineering and System Safety* **68**, 97–104.

Higashiyama Y. 2004. A method for exact reliability of consecutive 2-out-of-(r, r)-from-(n, n):F systems. *International Transactions in Operational Research* **11**, 217–224.

Higashiyama Y, Ariyoshi H, and Kraetzl M. 1995. Fast solutions for consecutive 2-out-of-r-from-n:F system. *IEICE Transactions on Fundamentals* **E-87A**, 680–684.

Higashiyama Y, Kraetzl M, and Caccetta L. 1996. Efficient algorithms for a consecutive 2-out-of-r-from-n:F system. *Australasian Journal of Combinatorics* **14**, 31–36.

Higashiyama Y, Kraetzl M, and Caccetta L. 1999. Formulas for the reliability of a consecutive k-out-of-r-from-n:F system. *Proceedings of SCI' 99*, pp. 131–134.

Higashiyama Y, Ohkura T, and Rumchev VG. 2006. Recursive method for reliability evaluation of circular consecutive 2-out-of-(r, r)-from-(n, n):F systems. *International Journal of Reliability, Quality and Safety Engineering* **13**, 355–363.

Hwang FK. 1982. Fast solutions for consecutive k-out-of-n:F systems. *IEEE Transactions on Reliability* **R-31**, 447–448.

Hwang FK. 1989. Invariant permutations for consecutive-k-out-of-n cycles. *IEEE Transactions on Reliability* **R-38**, 65–67.

Hwang FK and Pai CK. 2000. Sequential construction of a circular consecutive-2 system. *Information Processing Letters* **75**, 231–235.

Hwang FK and Shi D. 1987. Redundant consecutive-k-out-of-n:F systems. *Operations Research Letters* **6**, 293–296.

Jalali A, Hawkes AG, Cui LR, and Hwang FK. 2005. The optimal consecutive-k-out-of-n:G line for $n \leq 2k$. *Journal of Statistical Planning and Inference* **128**, 281–287.

Koutras MV, Papadopoylos G, and Papastavridis SG. 1994. Note: Pairwise rearrangements in reliability structures. *Naval Research Logistics* **41**, 683–687.

Kuo W, Zhang W, and Zuo MJ. 1990. A consecutive k-out-of-n:G: The mirror image of a consecutive k-out-of-n:F system. *IEEE Transactions on Reliability* **R-39**, 244–253.

Kuo W and Zuo MJ. 2003. *Optimal Reliability Modeling: Principles and Applications*. John Wiley & Sons, New York.

Levitin G. 2002. Optimal allocation of elements in a linear multi-state sliding window system. *Reliability Engineering and System Safety* **76**, 245–254.

Levitin G. 2003a. Element availability importance in generalized k-out-of-r-from-n systems. *IIE Transactions* **35**, 1125–1131.

Levitin G. 2003b. Linear multi-state sliding window systems. *IEEE Transactions on Reliability* **52**, 263–269.

Levitin G. 2004. Consecutive k-out-of-r-from-n system with multiple failure criteria. *IEEE Transactions on Reliability* **53**, 394–400.

Levitin G. 2005. Reliability of linear multistate multiple sliding window systems. *Naval Research Logistics* **52**, 212–223.

Lin MS, Chang MS, and Chen DJ. 2000. A generalization of consecutive k-out-of-n:G systems. *IEICE Transactions on Information and Systems* **E83-D**, 1309–1313.

Makri FS and Psillakis ZM. 1996. Bounds for reliability of k-within two-dimensional consecutive-r-out-of-n failure systems. *Microelectronics Reliability* **36**, 341–345.

Malon DM. 1984. Optimal consecutive 2-out-of-n:F component sequencing. *IEEE Transactions on Reliability* **R-33**, 414–418.

Malon DM. 1985. Optimal consecutive k-out-of-n:F component sequencing. *IEEE Transactions on Reliability* **R-34**, 46–49.

Nelson JB. 1978. Minimal-order models for false-alarm calculations on sliding windows. *IEEE Transactions on Aerospace Electron System* **ASE-14**, 351–363.

Papastavridis S. 1990. m-consecutive-k-out-of-n:F systems. *IEEE Transactions on Reliability* **39**, 386–388.

Papastavridis S and Hadzichristos I. 1988. Formulas for the reliability of a consecutive-k-out-of-n:F system. *Journal of Applied Probability* **26**, 772–779.

Papastavridis S and Koutras MV. 1993. Bounds for reliability of consecutive-k within-m-out-of-n systems. *IEEE Transactions on Reliability* **R-42**, 156–160.

Papastavridis SG and Sfakianakis M. 1991. Optimal-arrangement and importance of the components in a consecutive-k-out-of-r-from-n:F system. *IEEE Transactions on Reliability* **R-40**, 277–279.

Salvia AA and Lasher WC. 1990. 2-dimensional consecutive-k-out-of-n:F models. *IEEE Transactions on Reliability* **R-39**, 382–385.

Santha M and Zhang Y. 1987. Consecutive-2 systems on trees. *Probability in the Engineering and Informational Sciences* **1**, 441–456.

Separstein B. 1973. On the occurrence of n successes within n Bernouli trials. *Technometrics* **15**, 809–818.

Separstein B. 1975. Note on a clustering problem. *Journal of Applied Probability* **12**, 629–632.

Sfakianakis M. 1993. Optimal arrangement of components in a consecutive k-out-of-r-from-n:F system. *Microelectronics and Reliability* **33**, 1573–1578.

Sfakianakis M, Kounias S, and Hillaris A. 1992. Reliability of a consecutive-k-out-of-r-from-n:F system. *IEEE Transactions on Reliability* **R-41**, 442–447.

Shen J and Zuo MJ. 1994. Optimal design of series consecutive-k-out-of-n:G system with age-dependent minimal repair. *Reliability Engineering and System Safety* **45**, 277–283.

Tong YL. 1985. A rearrangement inequality for the longest run, with an application to network reliability. *Journal of Applied Probability* **22**, 386–393.

Tong YL. 1986. Some new results on the reliability of circular consecutive-k-out-of-n:F system. In *Reliability and Quality Control* (ed. Basu AP). Elsevier (North-Holland), New York, pp. 395–400.

Wei VK, Hwang FK, and Sös VT. 1983. Optimal sequencing of items in a consecutive-2-out-of-n system. *IEEE Transactions on Reliability* **R-32**, 30–33.

Zuo MJ. 1993. Reliability and design of 2-dimensional consecutive-k-out-of-n systems. *IEEE Transactions on Reliability* **R-42**, 488–490.

Zuo MJ and Kuo W. 1990. Design and performance analysis of consecutive k-out-of-n structure. *Naval Research Logistics* **37**, 203–230.

13

B-importance-based Heuristics for Component Assignment

As described in Section 11.1, CAP are combinatorial optimization problems and generally NP-hard. There is no study on exact optimal methods for the CAP except for the enumeration method as illustrated in Section 11.2. However, as n increases, the enumeration method is very time-consuming and may fail to find the solution (e.g., $n \geq 20$). Thus, fast and high-performance heuristics are in high demand.

Chapters 11 and 12 demonstrate the close relationship between the importance measures, especially the B-reliability importance, and the CAP in various systems. For some special systems (e.g., the systems admitting invariant optimal arrangements as in Theorem 11.6.3) or special structures of reliability of components (e.g., when the reliability values of available components are relatively close to each other as in Theorem 11.7.1), the B-i.i.d. importance can be used to achieve optimal arrangements of the CAP. However, for general coherent systems and general sets of component reliability values, the dependence of the optimal arrangements on the B-reliability importance is more intuitive than exact. The B-reliability importance is mostly used in devising heuristics for the CAP. A position with a larger B-reliability importance value indicates that it is more important than the other positions in the system and, thus, should get a more reliable component, at least on the first try.

This chapter presents the heuristics that use the B-reliability importance for solving the CAP. Both the theoretical analysis and computational evaluation of these heuristics are presented. Considering the fact that the component reliability values are estimated, these heuristics make more sense since they provide effective tools for generating good designs, given different estimates of component reliability. This chapter continues to use the notation from Chapter 11 and to assume that the reliability of available components satisfies

$$0 < \tilde{p}_1 \leq \tilde{p}_2 \leq \cdots \leq \tilde{p}_n < 1.$$

Sections 13.1–13.3 present three types of heuristics that, in different ways, try to match the arrangement of the CAP with the B-reliability importance ranking. Along with the

Importance Measures in Reliability, Risk, and Optimization: Principles and Applications, First Edition.
Way Kuo and Xiaoyan Zhu. © 2012 John Wiley & Sons, Ltd. Published 2012 by John Wiley & Sons, Ltd.

computational evaluation and results that benchmark the GAMS/CoinBonmin solver and randomization method, Section 13.4 presents the B-importance-based two-stage approach (BITS), with each stage using different B-reliability-importance-based heuristics. Section 13.5 presents the B-importance-based genetic local search approach (BIGLS), in which a local search based on the B-reliability importance is embedded into the genetic algorithm. Finally, Section 13.6 summarizes this chapter and initiates reflections on designing the algorithms for tackling problems in the associated areas.

13.1 The Kontoleon Heuristic

Kontoleon (1979) proposed the Kontoleon heuristic, in which the basic logic is the monotone property of the B-reliability importance as expressed in Equation (4.14). Maximum systems reliability is expected to be achieved by gradually improving the reliability of the position (i.e., by assigning a component of higher reliability) that has the largest B-reliability importance value.

Table 13.1 presents the Kontoleon heuristic. It starts with all positions having the same lowest reliability, $\tilde{p}_1$ (step 1). Then, the iterations proceed to assign available component of higher reliability to the system. At each iteration, the B-reliability importance values of the positions are calculated (step 3) and then two positions m and r are selected as candidates to receive the component of higher reliability (component j). The first candidate is position m in step 4, which has the largest B-reliability importance value and for which there exists a remaining component of higher reliability. The second candidate is position r in step 6, which has the highest reliability (assigned thus far) among the positions whose reliability is lower than component j. Component j in step 5 is the component to-be-assigned, which has the smallest reliability among the remaining components that have the reliability higher than the reliability of position m.

In step 7, if position r has the lowest reliability (i.e., $\tilde{p}_1$), then position m is chosen. Otherwise, the final decision about which position is chosen has to be made on the basis of

Table 13.1 The Kontoleon heuristic

---

1. Initially, set $p_i = \tilde{p}_1$ and $\pi_i = 1$ for $i = 1, 2, \ldots, n$.
2. Let $H = \{j : 1 < j \leq n, \tilde{p}_j \neq \tilde{p}_1\}$. ($H$ is the set of the indexes of components that have not yet been assigned.)
3. Calculate $I_B(i; \mathbf{p})$ for all positions i satisfying $p_i < \max_{\ell \in H} \tilde{p}_\ell$ according to Equation (4.11).
4. Obtain position m, which has the largest $I_B(\cdot; \mathbf{p})$ (break tie by choosing the first found with the lowest reliability).
5. Obtain component $j \in H$ such that $\tilde{p}_j$ is the smallest one satisfying $\tilde{p}_\ell > p_m$ for all $\ell \in H$.
6. Obtain position r such that p_r is the largest one satisfying $p_i < \tilde{p}_j$ for all $i = 1, 2, \ldots, n$ (break tie by choosing the first found).
7. If $p_r = \tilde{p}_1$, then set $\pi_m = j$, $H = H \setminus \{j\}$; otherwise
   (a) Set $I' = I_B(m; \mathbf{p})$, $z = \pi_m$, $p_m = p_r$, and recalculate $I_B(i; \mathbf{p})$ for $i = 1, 2, \ldots, n$.
   (b) If $I' \geq \max\{I_B(i; \mathbf{p}) : i = 1, 2, \ldots, n\}$, then set $\pi_m = j$; otherwise, set $\pi_m = \pi_r$, $\pi_r = j$.
   (c) If $\tilde{p}_z \neq \tilde{p}_1$, then set $H = H \cup \{z\} \setminus \{j\}$; otherwise, set $H = H \setminus \{j\}$.
8. STOP if $H = \emptyset$; otherwise, go to step 3.

---

further comparisons of the B-reliability importance. Setting the reliability of position r equal to that of position m and recalculating the B-reliability importance values of all positions, if any updated B-reliability importance value is larger than the original B-reliability importance value of position m, then position r is chosen; otherwise, position m is chosen. The iterations continue until all components are assigned to the system (step 8).

Example 13.1.1 (Example 6.14.5 continued)   Considering the CAP of assigning five components of reliability values $(\tilde{p}_1, \tilde{p}_2, \tilde{p}_3, \tilde{p}_4, \tilde{p}_5) = (0.1, 0.2, 0.3, 0.4, 0.5)$ to the system in Figure 6.4, Table 13.2 lists the procedure of the Kontoleon heuristic. The first column in Table 13.2 is the iteration number. The second column is current set H and the current reliability vector of positions, $\mathbf{p}$. Columns 3–7 are the values of the B-reliability importance, which are calculated using the current $\mathbf{p}$. Columns 8–10 give the values of positions m and r and component j, as illustrated in Table 13.1. The last column assigns component j of reliability $\tilde{p}_j$ to either position m or r.

For the set of component reliability values $(0.1, 0.2, 0.3, 0.4, 0.5)$, the Kontoleon heuristic generates the arrangement $(3, 1, 4, 5, 2)$ with systems reliability 0.3402, which is different from the optimal arrangement $(2, 1, 5, 4, 3)$ with systems reliability 0.3438. The relative error is 0.0036. For another set of component reliability values $(0.70, 0.73, 0.76, 0.79, 0.82)$, the Kontoleon heuristic finds the optimal arrangement $(2, 1, 3, 5, 4)$ with systems reliability 0.9282. The enumeration method is used to obtain the optimal arrangements.

The computational results in Yao et al. (2011) indicated that the Kontoleon heuristic never beats the LK-type heuristics, which are introduced in the next section.

13.2   The LK-Type Heuristics

13.2.1   The LKA Heuristic

Lin and Kuo (2002) proposed a greedy algorithm, namely, the LKA heuristic, with an analytical error term. Table 13.3 shows the LKA heuristic. The initial step has all positions assigned the component of the same lowest reliability. Then, at each iteration, step 3 calculates and orders the B-reliability importance values of positions in S, and allocates the most reliable component remaining (component k) to the position in S that has the largest B-reliability importance value (position m). The process is repeated until all components are assigned to the system.

Example 13.2.1 (Examples 6.14.5 and 13.1.1 continued)   For the system in Figure 6.4 and two sets of component reliability values $(0.1, 0.2, 0.3, 0.4, 0.5)$ and $(0.70, 0.73, 0.76, 0.79, 0.82)$, the LKA heuristic generates the same solutions as the Kontoleon heuristic in Example 13.1.1. Table 13.4 details the procedure of the LKA heuristic for the set of component reliability values $(0.1, 0.2, 0.3, 0.4, 0.5)$. The first column in Table 13.4 is the value of k in step 3 in Table 13.3. The second column is current reliability vector of positions, $\mathbf{p}$, and current set S. Columns 3–7 are the values of the B-reliability importance of each position in S which are calculated using current $\mathbf{p}$. Column 8 assigns component k of reliability $\tilde{p}_k$, that is, the most reliable component remaining, to the position that has the largest B-reliability importance value in this iteration.

Table 13.2 The Kontoleon heuristic procedure for the CAP in the system in Figure 6.4

Iter	$I_B(i;\mathbf{p})$					m	j	r	
	$i=1$	$i=2$	$i=3$	$i=4$	$i=5$				
1	$H = \{2,3,4,5\}$								
	0.089	0.089	0.170	0.179^a	0.179	4	2	1	$\pi_4 = 2, p_4 = 0.2$
2	$H = \{3,4,5\}$								
	$\mathbf{p} = (0.1,0.1,0.1,0.2,0.1)$ 0.178	0.088	0.250^a	0.179	0.168	3	3	4	$z = \pi_3 = 1, p_3 = p_4 = 0.2$
	$\mathbf{p} = (0.1,0.1,0.1,0.2,0.2,0.1)$ 0.158	0.078	0.250	0.259^a	0.238				$\pi_3 = \pi_4 = 2, p_3 = 0.2, \pi_4 = 3, p_4 = 0.3$
3	$H = \{4,5\}$								
	$\mathbf{p} = (0.1,0.1,0.2,0.3,0.1)$ 0.237	0.077	0.330^a	0.259	0.217	3	4	4	$z = \pi_3 = 2, p_3 = p_4 = 0.3$
	$\mathbf{p} = (0.1,0.1,0.3,0.3,0.1)$ 0.207	0.067	0.330	0.339^a	0.277				$\pi_3 = \pi_4 = 3, p_3 = 0.3, \pi_4 = 4, p_4 = 0.4$
4	$H = \{2,5\}$								
	$\mathbf{p} = (0.1,0.1,0.3,0.4,0.1)$ 0.277	0.067	0.410^a	0.339	0.247	3	5	4	$z = \pi_3 = 3, p_3 = p_4 = 0.4$
	$\mathbf{p} = (0.1,0.1,0.4,0.4,0.1)$ 0.237	0.057	0.410	0.419^a	0.297				$\pi_3 = \pi_4 = 4, p_3 = 0.4, \pi_4 = 5, p_4 = 0.5$
5	$H = \{2,3\}$								
	$\mathbf{p} = (0.1,0.1,0.4,0.5,0.1)$ 0.297^a	0.057			0.257	1	2	1	$\pi_1 = 2, p_1 = 0.2$
6	$H = \{3\}$								
	$\mathbf{p} = (0.2,0.1,0.4,0.5,0.1)$ 0.297^a	0.054	0.441	0.478^a	0.254	1	3	1	$z = \pi_1 = 2, p_1 = p_1 = 0.2$
	$\mathbf{p} = (0.2,0.1,0.4,0.5,0.1)$ 0.297	0.054			0.254				$\pi_1 = 3, p_1 = 0.3$
7	$H = \{2\}$								
	$\mathbf{p} = (0.3,0.1,0.4,0.5,0.1)$	0.051			0.251^a	5	2	1	$\pi_5 = 2, p_5 = 0.2$

^{a}The largest B-reliability importance value in current iteration

Table 13.3   The LKA heuristic

1. Initially, set $p_i = \tilde{p}_1$ and $\pi_i = 1$ for $i = 1, 2, \ldots, n$.
2. Let $S = \{1, 2, \ldots, n\}$. (S is the set of available positions that could receive other components.)
3. For $k = n$ to 2, do loop
   (a) calculate $I_B(i; \mathbf{p})$ for all $i \in S$ according to Equation (4.11);
   (b) find position $m \in S$ such that $I_B(m; \mathbf{p}) = \max_{i \in S} I_B(i; \mathbf{p})$ (break tie arbitrarily);
   (c) let $S = S \setminus \{m\}$; and
   (d) assign component k to position m, that is, set $\pi_m = k$.
4. STOP.

13.2.2   Another Three LK-type Heuristics

Continuing the schema of the LKA heuristic, Yao et al. (2011) derived the LKB, LKC, and LKD heuristics that implement the same rationale by using different initializations in step 1 and/or different assignment rules in step 3. The rationale of the LK-type heuristics (i.e., the LKA, LKB, LKC, and LKD heuristics) is to assign high reliable component to the position with a large B-reliability importance value. As shown in Table 13.5, each LK-type heuristic is delineated by replacing the partial contents in steps 1, 3, 3b, and 3d of the LKA heuristic in Table 13.3 with the corresponding contents in Table 13.5.

The LKB heuristic initializes all positions using the most reliable component (i.e., component n) and then constructs the assignment from the least reliable component, that is, assigning the least reliable component remaining to position m, which has the smallest B-reliability importance value among the positions currently assigned component n. The LKC heuristic starts from the same initialization as the LKA heuristic but uses a different assignment rule, which allocates the least reliable component remaining to all positions in S except position m, which has the smallest B-reliability importance value among positions in S. The LKD heuristic uses the same assignment rule as the LKC heuristic, but it iterates differently from the most reliable component to the least reliable component.

Table 13.4   The LKA heuristic procedure for the CAP in the system in Figure 6.4

k		$i = 1$	$i = 2$	$i = 3$	$i = 4$	$i = 5$	
		\multicolumn{5}{c}{$I_B(i; \mathbf{p})$}	
5	$\mathbf{p} = (0.1, 0.1, 0.1, 0.1, 0.1)$						
	$S = \{1, 2, 3, 4, 5\}$	0.089	0.089	0.170	0.179^a	0.179	$p_4 = 0.5, \pi_4 = 5$
4	$\mathbf{p} = (0.1, 0.1, 0.1, 0.5, 0.1)$						
	$S = \{1, 2, 3, 5\}$	0.445	0.085	0.490^a		0.135	$p_3 = 0.4, \pi_3 = 4$
3	$\mathbf{p} = (0.1, 0.1, 0.4, 0.5, 0.1)$						
	$S = \{1, 2, 5\}$	0.297^a	0.057			0.257	$p_1 = 0.3, \pi_1 = 3$
2	$\mathbf{p} = (0.3, 0.1, 0.4, 0.5, 0.1)$						
	$S = \{2, 5\}$		0.051			0.251^a	$p_5 = 0.2, \pi_5 = 2$
	$\mathbf{p} = (0.3, 0.1, 0.4, 0.5, 0.2)$						
	$S = \{2\}$						$p_2 = 0.1, \pi_2 = 1$

^{a}The largest B-reliability importance value in current iteration

Table 13.5 The LK-type heuristics

Heu.	step 1	step 3	step 3b	step 3d
LKA	$p_i = \tilde{p}_1$, that is, $\pi_i = 1$	n to 2	$I_B(m; \mathbf{p}) = \max\limits_{i \in S} I_B(i; \mathbf{p})$	component k to position m, that is, $\pi_m = k$
LKB	$p_i = \tilde{p}_n$, that is, $\pi_i = n$	1 to $n-1$	$I_B(m; \mathbf{p}) = \min\limits_{i \in S} I_B(i; \mathbf{p})$	component k to position m, that is, $\pi_m = k$ [a]
LKC	$p_i = \tilde{p}_1$, that is, $\pi_i = 1$ [a]	2 to n	$I_B(m; \mathbf{p}) = \min\limits_{i \in S} I_B(i; \mathbf{p})$	component k to positions in S, that is, $\pi_i = k$ for $i \in S$
LKD	$p_i = \tilde{p}_n$, that is, $\pi_i = n$	$n-1$ to 1	$I_B(m; \mathbf{p}) = \max\limits_{i \in S} I_B(i; \mathbf{p})$ [a]	component k to positions in S, that is, $\pi_i = k$ for $i \in S$

[a] No change is made in this step compared to the LKA heuristic

13.2.3 Relation to Invariant Optimal Arrangements

The existence of an invariant optimal arrangement for a given system enables the LK-type heuristics to generate the (invariant) optimal solution, as shown in Theorem 13.2.2. Furthermore, if a system admits an invariant optimal arrangement, the arrangement generated in each iteration of the LK-type heuristic is the optimal arrangement, assuming that the components have the reliability specified by current $\mathbf{p}$ and that no tie is present in step 3b in all proceeding iterations.

Theorem 13.2.2 *If a system admits an invariant optimal arrangement, the solution generated by each of the LKA, LKB, LKC, and LKD heuristics must be the invariant optimal arrangement, given that no tie is present in its step 3b in all iterations of the heuristic.*

The proof for Theorem 13.2.2 can be found in Lin and Kuo (2002) and Yao et al. (2011); however, the statement corresponding to the LKA heuristic and its proof in Lin and Kuo (2002) neglected the condition that no tie is present in step 3b in all iterations. Under the situation of a tie in step 3b in some iteration (i.e., more than one position achieves the same largest (or least) B-reliability importance value), the LK-type heuristics may arbitrarily choose a position that is not consistent with the invariant optimal arrangement and, consequently, the generated arrangement may be not the invariant optimal or even the optimal arrangement, as shown in Example 13.2.3.

Example 13.2.3 Consider the CAP instance of assigning seven components of reliability values (0.806, 0.809, 0.818, 0.833, 0.853, 0.925, 0.934) to the *Lin/Con/2/7*:F system that admits an invariant optimal arrangement $\boldsymbol{\pi}^* = (1, 7, 3, 5, 4, 6, 2)$ (treating the inverse (2, 6, 4, 5, 3, 7, 1) the same as the invariant optimal arrangement due to the symmetry of the *Lin/Con/2/7*:F system). Under the invariant optimal arrangement, the maximum systems reliability is 0.910892. To solve this instance using the LKA heuristic, any tie in step 3b is broken by choosing the position with the lowest index among tied positions. In step 3 of the LKA heuristic, at the first iteration ($k=7$), positions 2 and 6 have the same largest B-reliability importance value (i.e., a tie), and the rule of breaking ties chooses position 2 to receive the candidate component (i.e., component $k=7$). At the subsequent four iterations ($k = 6, 5, 4, 3$), there is no tie and the intermediate assignment after these five iterations is $(1, 7, 3, 5, 4, 6, 1)$, which is so far consistent with the invariant optimal arrangement π^*. At the last iteration ($k = 2$), a tie is present that positions 1 and 7 have the same B-reliability importance value, and the rule of breaking ties chooses position 1 to receive component $k=2$. Thus, the final generated arrangement is $(2, 7, 3, 5, 4, 6, 1)$ with systems reliability 0.910868, which is not even an optimal arrangement.

In fact, it is equivalent for the LK-type heuristics to use, instead of the B-reliability importance, some importance measures that are proportional to the B-reliability importance as later summarized in Table 14.1. For example, the criticality importance for system functioning of position i equals $p_i I_B (i; \mathbf{p})/R(\mathbf{p})$ as in Equation (4.20), and in step 3a, all positions $i \in S$ temporarily have the same reliability; therefore, the importance ranking of positions in S is the same according to the B-reliability importance or the criticality importance, and the heuristics choose the same position m in step 3b. In turn, even if the LK-type heuristics use the criticality importance, Theorem 13.2.2 still holds true. In solving the CAP, the B-reliability importance is better than those importance measures that are proportional to it because of its simplicity.

According to Theorems 13.2.2 and 11.6.3, if the LK-type heuristic, under the condition in Theorem 13.2.2, generates a solution that is not optimal or is different from the B-i.i.d. importance ordering, then the system does not admit an invariant optimal arrangement. Thus, the LK-type heuristics can be used to examine whether a system admits an invariant optimal arrangement, especially when the system has consistent B-i.i.d. importance ordering.

13.2.4 Numerical Comparisons of the LK-type Heuristics

For solving the CAP in the redundant $Lin/Con/k/n$:F with redundancy r (i.e., $Rec/Con/(r, k)/(r, n)$:F) systems, Zuo and Shen (1992) first proposed the LKA heuristic and conducted the computational tests on such the systems with redundancy $r = 2$, including the small instances ($k = 2, 4 \leq n \leq 7$; $k = 3, 4 \leq n \leq 6$; $k = 4, n = 6$) with sets of component reliability values randomly generated from uniform $[0, 1.0]$ and $[0.8, 1.0]$ distributions and the medium to large instances ($k = 2, 10 \leq n \leq 120$) with sets of component reliability values randomly generated from uniform $[0.8, 1.0]$ distribution. Their computational results show that the LKA heuristic generates solutions that are within 1.5% of the optimal solutions for the small instances and that it outperforms the randomization method with $10, 000$ random arrangements (see Section 11.2) for the medium to large instances in terms of solution quality.

More comprehensive computational tests were done by Yao et al. (2011) to compare the performance of four LK-type heuristics, which use 30 sets of the CAP instances as described in Remark 13.2.4. It is found that the best performance is always achieved by either the LKA or LKB heuristic. Although the LKC and LKD heuristics can outperform either of the LKA and LKB heuristics in some instances, they never outperform both. Moreover, the LKD heuristic is more likely to generate extremely bad solutions as compared to the others. These observations imply that in step 3d, updating the reliability of only one position (as in the LKA and LKB heuristics) is more effective than updating the reliability of multiple positions (as in the LKC and LKD heuristics).

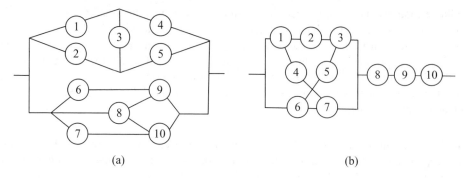

Figure 13.1   Two systems of ten components

Remark 13.2.4 (The CAP instances)   A CAP instance is defined in a given system and in-volves a set of available components that are of specific reliability values to be assigned to the system. For the computational tests, Yao et al. (2011) used 30 sets of the CAP instances in ten systems with three types of components. Ten systems include four general coherent systems in Figures 6.1, 6.4, and 13.1 (the system in Figure 13.1(b) originally from Prasad and Kuo (2000)) ranging from five to ten components; two $Lin/Con/3/n$:F systems with $n = 7, 8$; and four $Lin/Con/k/n$:G systems with $k = 2, 3$ and $n = 7, 8$. Note that the $Lin/Con/2/n$:F systems are not included because they have invariant optimal assignments as shown in Theorem 12.1.1. The three types of components are high-, low-, and arbitrary reliable components, whose reliability values are distributed from uniform $[0.8, 0.99]$, $[0.01, 0.2]$, and $[0.01, 0.99]$, respectively. To avoid biased results, for each system and each type of components, a set of 100 instances is generated, each instance corresponding to a different set of component reliability values. There-fore, there are 30 sets of the instances in which the number of components is not greater than ten.

In summary, regardless of component types and system structures, the LKA and LKB heuristics perform better than the LKC and LKD heuristics. On average, the LKA heuristic is recommended for systems with components of high reliability (≥ 0.8), and the LKB heuristic is recommended for systems with components of arbitrary or low reliability (≤ 0.2).

13.3   The ZK-Type Heuristics

13.3.1   Four ZK-type Heuristics

According to Theorems 8.2.1 and 12.2.1, the necessary conditions for an optimal arrangement of the CAP follow the B-reliability importance patterns. Motivated by this, Zuo and Kuo (1990) developed two ZK-type heuristics for $Lin/Con/k/n$ systems, the ZKA and ZKB heuristics, which can be extended to solve the CAP in any coherent system. On the basis of that, Yao et al. (2011) further proposed two derivative methods, the ZKC and ZKD heuristics. Table 13.6 details the ZKA heuristic. The ZKB, ZKC, and ZKD heuristics are delineated by replacing the partial contents in steps 3, 3a, and 3b of the ZKA heuristic with the corresponding contents in Table 13.7.

The ZK-type heuristics start with a feasible initial arrangement (i.e., each position is initially assigned a different component) (step 1) and try to improve the arrangement by pairwise

Table 13.6 The ZKA heuristic

1. Generate an initial arrangement, $\pi = (\pi_1, \pi_2, \ldots, \pi_n)$.
2. Calculate $I_B(i; \mathbf{p})$ for all positions $i = 1, 2, \ldots, n$ according to Equation (4.11).
3. For $k = 1$ to $n - 1$, do loop
   (a) Find positions m and r such that $\pi_m = k$ and $\pi_r = k + 1$.
   (b) If $I_B(m; \mathbf{p}) > I_B(r; \mathbf{p})$ and $R(\mathbf{p}\pi_{(m,r)}) > R(\mathbf{p}\pi)$, then exchange the assignments of components π_m and π_r, that is, set $\pi = \pi(m, r)$.
4. STOP if there is no exchange in step 3; otherwise, go to step 2.

exchanging the assignments of components to match the B-reliability importance ranking. Subsection 13.3.3 specializes the initial arrangements in the ZK-type heuristics.

Step 3 of the ZK-type heuristics conducts pairwise exchange iteratively until no more exchange can be made (step 4). In step 3, the ZKA and ZKB heuristics start pairwise exchange from the least reliable component, while the ZKC and ZKD heuristics start from the most reliable component. The four ZK-type heuristics use different methods to choose positions m and r for exchange. In loop k of step 3a, the ZKA heuristic compares component k in position m with the next more reliable component $k+1$ in position r; the ZKB heuristic compares component k in position m with the component π_r in position r that has the smallest B-reliability importance value among the positions whose reliability is higher than that of component k (i.e., p_m) (breaking the tie arbitrarily); the ZKC heuristic compares component $\pi_m = k$ with the next less reliable component $\pi_r = k - 1$; and the ZKD heuristic compares component $\pi_m = k$ with component π_r in position r that has the largest importance value among the positions whose reliability is lower than p_m (breaking the tie arbitrarily).

In step 3b, if the B-reliability importance value of the less reliable component is larger than that of the more reliable component and the exchange of these two components can improve the systems reliability, then the exchange is made. Note that, as shown in Example 11.7.3, the optimal arrangement does not always match the B-reliability importance ranking. Therefore, it is necessary that the ZK-type heuristics include the condition $R(\mathbf{p}\pi_{(m,r)}) > R(\mathbf{p}\pi)$ in step 3b for pairwise exchange to guarantee the improvement of the solution. Furthermore, in the case that the optimal arrangement matches the B-reliability importance ranking, the ZK-type heuristics are expected to obtain a good even optimal solution. Note that if an initial arrangement already satisfies that $I_B(m; \mathbf{p}) \leq I_B(r; \mathbf{p})$ whenever $p_m \leq p_r$ for any m and r, then the ZK-type heuristics are unable to improve the initial arrangement.

Example 13.3.1 (Examples 6.14.5 and 13.1.1 continued)   For the system in Figure 6.4 and five available components of reliability values $(0.1, 0.2, 0.3, 0.4, 0.5)$, the ZKA heuristic using

Table 13.7 The ZK-type heuristics

Heu.	step 3	step 3a	step 3b
ZKA	1 to $n - 1$	$\pi_r = k + 1$	$I_B(m; \mathbf{p}) > I_B(r; \mathbf{p})$
ZKB	1 to $n - 1$	$\pi_r : I_B(r; \mathbf{p}) = \min_{i:p_i > p_m} I_B(i; \mathbf{p})$	$I_B(m; \mathbf{p}) > I_B(r; \mathbf{p})$
ZKC	n to 2	$\pi_r = k - 1$	$I_B(m; \mathbf{p}) < I_B(r; \mathbf{p})$
ZKD	n to 2	$\pi_r : I_B(r; \mathbf{p}) = \max_{i:p_i < p_m} I_B(i; \mathbf{p})$	$I_B(m; \mathbf{p}) < I_B(r; \mathbf{p})$

Table 13.8 The ZKA heuristic procedure for the CAP in the system in Figure 6.4

Iter:	k	π	$R(\mathbf{p}_\pi)$	$i=1$	$i=2$	$i=3$	$i=4$	$i=5$	m	r	$\pi(m,r)$	$R(\mathbf{p}_{\pi(m,r)})$
				\multicolumn{5}{c}{$I_B(i;\mathbf{p})$}				
1:	1	(1,2,3,4,5)	0.3052	0.252	0.336	0.564	0.213	0.314	1	2	—	—
	2	(1,2,3,4,5)	0.3052	0.252	0.336	0.564	0.213	0.314	2	3	—	—
	3	(1,2,3,4,5)	0.3052	0.252	0.336	0.564	0.213	0.314	3	4	(1,2,4,3,5)	0.3362
	4	(1,2,4,3,5)	0.3362	0.162	0.291	0.523	0.254	0.396	3	5	(1,2,5,3,4)	0.3438
2:	1	(1,2,5,3,4)	0.3438	0.138	0.194	0.472	0.346	0.447	1	2	—	—
	2	(1,2,5,3,4)	0.3438	0.138	0.194	0.472	0.346	0.447	2	4	—	—
	3	(1,2,5,3,4)	0.3438	0.138	0.194	0.472	0.346	0.447	4	5	—	—
	4	(1,2,5,3,4)	0.3438	0.138	0.194	0.472	0.346	0.447	5	3	—	—

initial arrangement $(1, 2, 3, 4, 5)$ finds the optimal arrangement $(1, 2, 5, 3, 4)$ with systems reliability 0.3438. Table 13.8 details the procedure. In Table 13.8, the first two columns are the iteration number and k value in step 3 in Table 13.6, respectively; columns m and r are the index of positions such that $\pi_m = k$ and $\pi_r = k + 1$; and the other columns are self-illustrated.

As shown in Table 13.8, starting from initial arrangement $(1, 2, 3, 4, 5)$, the ZKA heuristic exchanges the assignments of positions 3 and 4 first and then positions 3 and 5 in the first iteration according to the criteria in step 3b in Table 13.6. After these exchanges, the ZKA heuristic ends up with arrangement $(1, 2, 5, 3, 4)$. In the second iteration, there is no exchange; thus, the ZKA heuristic stops. For another set of component reliability values $(0.70, 0.73, 0.76, 0.79, 0.82)$, the ZKA heuristic using initial arrangement $(1, 2, 3, 4, 5)$ generates the same optimal arrangement as in Example 13.1.1.

13.3.2   Relation to Invariant Optimal Arrangements

Different from the LK-type heuristics (as in Theorem 13.2.2), when a system admits an invariant optimal arrangement, the ZK-type heuristics may not generate the optimal solution, as shown in the following example from Yao et al. (2011).

Example 13.3.2   Considering $Lin/Con/2/7$:F system and seven available components of reliability values $(0.1, 0.2, 0.3, 0.4, 0.5, 0.6, 0.7)$, the invariant optimal arrangement exists, which is $\pi^* = (1, 7, 3, 5, 4, 6, 2)$ with systems reliability 0.2538. Using initial arrangement $(1, 3, 5, 7, 6, 4, 2)$ in step 1, both the ZKA and ZKB heuristics generate the optimal arrangement π^*. However, using initial arrangement $(1, 2, 3, 4, 5, 6, 7)$, the ZKA heuristic generates the arrangement $\pi^1 = (1, 7, 2, 6, 4, 5, 3)$ with systems reliability 0.2524, and the ZKB heuristic generates the arrangement $\pi^2 = (3, 4, 6, 1, 7, 2, 5)$ with systems reliability 0.1559. Both π^1 and π^2 differ from π^*.

13.3.3 Comparisons of Initial Arrangements

It is critical to choose appropriate initial arrangements in the ZK-type heuristics. Zuo and Kuo (1990) proposed four initial arrangements for $Lin/Con/k/n$ systems listed in the following text as initial arrangements 1–4. For tackling the CAP in any coherent system, Yao et al. (2011) proposed four instance-specified initial arrangements, listed in the following text as initial arrangements 5–8.

Initial arrangement 1: $(1, n-1, 3, n-3, \ldots, n-2, 4, n, 2)$ for the $Lin/Con/2/n$ systems only;

Initial arrangement 2: $(1, n, 3, n-2, \ldots, n-3, 4, n-1, 2)$, the invariant optimal arrangement for the $Lin/Con/2/n$:F systems only;

Initial arrangement 3: $(1, 3, 5, \ldots, n, \ldots, 6, 4, 2)$, a pattern found to be optimal in many cases;

Initial arrangement 4: $(1, 2, 3, 4, \ldots, n)$, a naturally ordered pattern;

Initial arrangements 5–8: arrangements generated by the LKA, LKB, LKC, and LKD heuristics, respectively.

Since initial arrangements 1 and 2 are designed for the $Lin/Con/2/n$ systems, only initial arrangements 3–8 are considered for general systems. As initial arrangements 5–8 are the approximate solutions generated by the LK-type heuristics, using them ensures that the worst performance of the ZK-type heuristics is not of bad quality. Thus, initial arrangements 5–8 would have stable performance in various instances. Although it takes time to generate initial arrangements 5–8 by the LK-type heuristics, their high quality may accelerate the termination of the ZK-type heuristics. Thus, compared to using fixed initial arrangements 3 and 4, using initial arrangements 5–8 does not increase the computation time of the ZK-type heuristics significantly.

Note that the ZK-type heuristics using an initial arrangement with high systems reliability may produce a worse final arrangement than using an initial arrangement with low systems reliability, as shown in Example 13.3.3.

Example 13.3.3 Considering the CAP in $Lin/Con/3/8$:F system and eight available components having reliability $\tilde{p}_i = 0.61 + 0.03 \times (i-1)$, $i = 1, 2, \ldots, 8$, Table 13.9 shows two sets of initial and final arrangements and the systems reliability values corresponding to these arrangements. In Table 13.9, the systems reliability associated with initial arrangement 4, $(1, 2, 3, 4, 5, 6, 7, 8)$, is lower than that associated with initial arrangement 5,

Table 13.9 Comparisons of the initial and final arrangements associated with the ZKA heuristic

Initial arrangement π	$R(\mathbf{p}_\pi)$	Final arrangement π'	$R(\mathbf{p}_{\pi'})$
4: (1, 2, 3, 4, 5, 6, 7, 8)	0.8865	(1, 5, 7, 4, 6, 8, 3, 2)	0.9207
5: (2, 5, 8, 3, 6, 7, 4, 1)	0.9200	(2, 3, 8, 4, 6, 7, 5, 1)	0.9205

$(2, 5, 8, 3, 6, 7, 4, 1)$. However, the systems reliability associated with the final arrangement generated by the ZKA heuristic using initial arrangement 4 is higher than that generated using initial arrangement 5.

Therefore, it is not sufficient to judge the performance of initial arrangements by simply comparing the systems reliability values associated with them. Yao et al. (2011) compared initial arrangements 3–8 by using all six initial arrangements in each of four ZK-type heuristics and concluded that on average, for any ZK-type heuristics, initial arrangement 6 is the first choice and initial arrangements 5 and 7 are the second and third choices, respectively.

A strategy to improve the performance of the ZK-type heuristics is to use multiple initial arrangements rather than a single one and then choose the best arrangement generated as the final solution. On the basis of the computational results in Yao et al. (2011), using initial arrangements 6 and 5 can improve the performance of all ZK-type heuristics significantly compared to using initial arrangement 6 alone; however, further using initial arrangement 7 cannot make any more improvement. Therefore, initial arrangements 6 and 5 are recommended for the ZK-type heuristics.

13.3.4   Numerical Comparisons of the ZK-type Heuristics

Computational tests have been conducted by Yao et al. (2011), Zuo and Kuo (1990), and Zuo and Shen (1992). The most comprehensive one is in Yao et al. (2011), which used the test instances in Remark 13.2.4. As discussed and suggested in Subsection 13.3.3, both initial arrangements 5 and 6 are used in each ZK-type heuristic, and the better arrangement generated is selected as the final solution from that heuristic. On this basis, the ZKB heuristic on average performs best for the CAP instances involving components of low reliability (≤ 0.2), and the ZKD heuristic performs best for the ones with high reliable (≥ 0.8) or arbitrary components. In addition, the ZKB and ZKD heuristics, which use the B-reliability importance to effectively find the pairs of components for exchange, are superior to the ZKA and ZKC heuristics.

13.4   The B-importance-based Two-stage Approach

As stated in Sections 13.2 and 13.3, the solutions from the LKA and LKB heuristics can be used as the initial arrangements in the ZK-type heuristics, and the ZKB heuristic should be used for solving the CAP with low reliable components and the ZKD heuristic for the CAP with other types of components. Integrating these results, Yao et al. (2011) proposed the BITS for solving the CAP as in Table 13.10.

Stage 2 can be regarded as an improvement stage that further refines the arrangements generated in stage 1. If computation time is restricted, use only stage 1 (i.e., the LKA and LKB heuristics) and then choose the better arrangement as the final solution or even use one single LK-type heuristic, according to the discussion in Subsection 13.2.4.

Example 13.4.1 (Examples 6.14.5 and 13.1.1 continued)   Consider the system in Figure 6.4 and five available components of reliability values $(0.1, 0.2, 0.3, 0.4, 0.5)$. In stage 1 of

Table 13.10   The BITS

Stage 1: Use both the LKA and LKB heuristics to generate two initial arrangements.

Stage 2: If the CAP instance involves only components of low reliability (≤ 0.2), choose the ZKB heuristic; otherwise, choose the ZKD heuristic. Solve the instance using the chosen ZK-type heuristic with two initial arrangements separately. STOP with the arrangement generated with higher systems reliability as the final solution.

---

the BITS, the LKA heuristic generates a nonoptimal arrangement $(3, 1, 4, 5, 2)$ with systems reliability 0.3402 as in Example 13.1.1, and the LKB heuristic generates another nonoptimal arrangement $(1, 2, 5, 3, 4)$ with systems reliability 0.3438. Under arrangements $(3, 1, 4, 5, 2)$ and $(1, 2, 5, 3, 4)$, the five positions have the B-reliability importance of values 0.294, 0.102, 0.433, 0.496, and 0.251 and values 0.138, 0.194, 0.472, 0.346, and 0.447, respectively, whose orderings match arrangements $(3, 1, 4, 5, 2)$ and $(1, 2, 5, 3, 4)$, respectively. Therefore, in stage 2 of the BITS, the ZKD heuristic does not make any pairwise exchange for any of these two arrangements, so return the better one among them as the final solution from the BITS. That is arrangement $(1, 2, 5, 3, 4)$ with systems reliability 0.3438.

Now assume that the reliability values of five available components are $(0.2, 0.3, 0.4, 0.5, 0.6)$. In stage 1 of the BITS, the LKA heuristic generates arrangement $\pi = (2, 1, 4, 5, 3)$ with systems reliability 0.5028, and the LKB heuristic generates arrangement $(1, 2, 5, 3, 4)$ with systems reliability 0.5072. Under arrangement π, the five positions have the B-reliability importance of values 0.276, 0.164, 0.514, 0.438, and 0.282, whose ordering does not match the arrangement. Therefore, in stage 2 of the BITS, the ZKD heuristic exchanges the assignments in positions 3 ($\pi_3 = 4$, $p_3 = 0.5$) and 4 ($\pi_4 = 5$, $p_4 = 0.6$) because $I_B(3; \mathbf{p}) = 0.514 > I_B(4; \mathbf{p}) = 0.438$ and $R(\mathbf{p}_{\pi(3,4)}) = 0.5072 > R(\mathbf{p}_\pi) = 0.5028$. After this exchange, the ZKD heuristic stops with arrangement $(2, 1, 5, 4, 3)$ with systems reliability 0.5072. In addition, under arrangement $(1, 2, 5, 3, 4)$, the B-reliability importance ranking of the five positions matches the arrangement, so no improvement is further made in stage 2 of the BITS. Thus, the final solution is arrangements $(1, 2, 5, 3, 4)$ and $(2, 1, 5, 4, 3)$ with systems reliability 0.5072. It can be verified by the enumeration method that both $(1, 2, 5, 3, 4)$ and $(2, 1, 5, 4, 3)$ are the optimal arrangements for this case.

Yao et al. (2011) conducted numerical testing on the BITS on a server with two Intel Dual-Core Xeon 5160 3.0 GHz processors and 8GB RAM, as shown in the following two subsections.

13.4.1   Numerical Comparisons with the GAMS/CoinBomin Solver and Enumeration Method

Note that a CAP can be formulated as an MINLP as in Section 11.1 and solved by the GAMS/CoinBonmin solver, which basically uses a branch-and-bound (B&B) method. The GAMS/CoinBonmin solver, however, does not guarantee the generation of optimal solutions for nonconvex problems like the CAP. Thus, the enumeration method is also used for finding optimal solutions.

The performance of the BITS was compared with the GAMS/CoinBonmin solver and enumeration method on solving 30 sets of the CAP instances, as described in Remark 13.2.4. The BITS achieves better solutions than the GAMS/CoinBonmin in 29 sets of instances and is also much faster than the GAMS/CoinBonmin by around 54–1268 times. It is concluded that the BITS has a clear advantage over the GAMS/CoinBonmin solver in terms of both solution quality and computation time, especially when the system size (i.e., the number of components) gets larger.

Furthermore, the BITS takes much less computation time than the enumeration method, especially for those systems with more than seven components. Although the enumeration method can find optimal solutions for small instances, it takes an extremely long time and may result in out-of-memory for medium to large instances, impeding its practical use. For example, to solve an instance in $Lin/Con/3/20$:F system, the enumeration method cannot achieve the optimal solution even after several days of computation, and the GAMS/CoinBonmin solver takes 49.09 seconds with a solution of systems reliability 0.99360; on the other hand, the BITS takes only 0.02 seconds with a solution of systems reliability 0.99366.

Therefore, for solving the large instances (with 20 or more components) in the next subsection, only the BITS deserves further study.

13.4.2 Numerical Comparisons with the Randomization Method

Benchmarking on the randomization method with 10,000 random arrangements, the performance of the BITS was evaluated on solving 84 sets of the CAP instances that are in 28 large $Lin/Con/k/n$ systems with three types of components. The 28 large systems include 12 $Lin/Con/k/n$:F systems with $k = 3, 4, 5$ and $n = 20, 30, 50, 100$ and 16 $Lin/Con/k/n$:G systems with $k = 2, 3, 4, 5$ and $n = 20, 30, 50, 100$. As in Remark 13.2.4, the three types of components are of high-, low-, and arbitrary reliability, with reliability values distributed from uniform $[0.8, 0.99]$, $[0.01, 0.2]$, and $[0.01, 0.99]$, respectively. Also, as in Remark 13.2.4, for each system and each type of components, a set of 100 instances is generated, each instance corresponding to a different set of component reliability values. Therefore, there are 84 sets of the large instances in which the number of components is not less than 20.

For the instances associated with high reliable components, the BITS generates better arrangements on average than the randomization method in 24 out of total 28 sets of instances with four exceptions in $Lin/Con/2/50$:G, $Lin/Con/2/100$:G, $Lin/Con/3/100$:G, and $Lin/Con/4/100$:G systems. For the instances associated with low reliable or arbitrary components, the BITS always generates better arrangements than the randomization method.

Furthermore, the BITS takes short computation time to solve the large CAP instances. For example, the longest average computation time is about 3.8 seconds for solving an instance in $Lin/Con/5/100$:F system. On average, the BITS takes only 1/6 of the time that the randomization method takes for the same instance. Solving even larger intances, for example, a set of 100 instances in $Lin/Con/3/300$:F system with high reliable components, the BITS takes 14 seconds per instance, and the randomization method takes 25 seconds per instance. That is, the BITS uses only about 3/5 the time of and obtains far better assignments than the randomization method.

Therefore, the BITS outperforms the randomization method and is capable of solving large instances with high accuracy in short computation time. The BITS is practical, friendly, and implementable.

13.5   The B-importance-based Genetic Local Search

Metaheuristics (e.g., genetic algorithm, simulated annealing, tabu search, and so on) have shown great potential to break local optimum compared to other heuristics and have been successfully applied to solve some difficult combinatorial optimization problems. A genetic algorithm and simulated annealing have been proposed by Shingyoch et al. (2009, 2010) for solving the CAP in $Cir/Con/k/n$:F systems. Note that for a $Cir/Con/k/n$ system, two assignments are equivalent if one can be obtained by rotating or reversing another because the components are arranged in a circle. Shingyoch et al. (2009) developed a genetic algorithm that modifies Grefenstette's ordinal representation schema (Grefenstette et al. 1985) by keeping only one of the equivalent assignments and eliminating the others. Their genetic algorithm always assigns the most reliable components to every kth position because the $Cir/Con/k/n$:F system functions if the components at every kth position function.

In addition to this genetic algorithm, Shingyoch et al. (2010) proposed two simulated annealing algorithms—a standard one and an improved one that further eliminates equivalent assignments as the genetic algorithm does. The numerical experiments showed that these two simulated annealing algorithms usually generate better solutions than the genetic algorithm in Shingyoch et al. (2009) but take longer computation time; the improved simulated annealing can generate similar quality solutions as the standard one but take only half of the computation time due to its reduced search space. However, all of these methods are designed specially for $Cir/Con/k/n$:F systems and do not utilize the importance measures, which can be used to improve the performance of heuristics.

Yao et al. (2012) proposed a genetic local search approach for solving the CAP in any coherent system. The genetic local search is a hybrid genetic algorithm that takes advantage of both population-based search and local optimization. The genetic operators explore the large search space to identify promising subregions; meanwhile, the local search exploits in-depth these small localized regions by iteratively moving from one solution to a better one in its neighborhood until a local optimum is reached. The genetic local search is a promising method and has been successfully applied to tackle traveling salesman problems (Freisleben and Merz 1996; Ulder et al. 1991), quadratic assignment problems (Lim et al. 2000, 2002; Merz and Freisleben 1997), CAP (Yao et al. 2012), and so on. The rest of this section describes the BIGLS, which is based on and similar to the genetic local search in Yao et al. (2012) but uses a different B-importance-based local search.

13.5.1   The Description of Algorithm

The BIGLS for solving the CAP uses the B-reliability importance in generating the initial population and directing local improvement. It applies the evolutionary mechanism of the genetic algorithm to suppress the premature phenomena that the local search might give rise to. The B-reliability importance is used in a pairwise-exchange local search that

Table 13.11 The BIGLS framework

---

1. Generate an initial population with s chromosomes. Perform the B-importance-based local search on each initial chromosome.
2. If the current population satisfies at least one termination condition, perform the B-importance-based local search on the best chromosome and STOP.
3. Evaluate the scaled fitness of each chromosome in the current population.
4. Perform elitism strategy, that is, the $s\beta$ best chromosomes in the current population are directly reproduced into the next population.
5. Perform crossover on the current population to generate $s(1 - \beta)$ offspring chromosomes into the next population.
6. Perform mutation on the selected offspring chromosomes and perform the B-importance-based local search on each mutated offspring chromosome and on the unmutated offspring chromosomes that are better than the best one in the current population.
7. Replace the current population with the next population and go to step 2.

---

greatly enhances the effectiveness of the local search, as demonstrated in Subsection 13.5.2. Table 13.11 presents the BIGLS framework.

As shown in Table 13.11, starting with the initial population of size s in step 1, an iterative procedure from steps 2 to 7 is performed on each generation of population until one of the termination conditions in step 2 is satisfied. After step 6, a new population of size s is generated, and step 7 replaces the current population with it. The BIGLS is detailed by specifying the operators in each step, as described in the rest of this subsection.

Gene, chromosome, and population

In the BIGLS, a chromosome is coded as an arrangement π, and each integer π_i in the chromosome is a gene, $i = 1, 2, \ldots, n$. Each population has s chromosomes. The initial population includes $s-1$ randomly generated chromosomes that maintain the diversity of the population and a special chromosome generated by the BITS that improves the quality of the initial population. According to Section 13.4, the BITS can quickly generate high-quality arrangements.

B-importance-based local search

A B-importance-based local search is used in steps 1, 2, and 6 of the BIGLS to boost a given arrangement π in its neighborhood. It is a pairwise-exchange method based on the B-reliability importance, as detailed in Table 13.12. Essentially, it is the inner loop of the ZKD heuristic (see Subsection 13.3.1). Note that the ZKD heuristic outperforms the other ZK-type heuristics in most cases, as discussed in Subsection 13.3.4. Taking only the inner loop of the ZKD heuristic as the local search in the BIGLS avoids the long computation time possibly incurred by the ZKD heuristic due to its multiple executions of this inner loop. This local search can easily start with a given arrangement as the initial arrangement.

Table 13.12 The B-importance-based local search

Given an arrangement π, for $k = n$ to 2, do loop
1. Calculate $I_B(i; \mathbf{p})$ for $i = 1, 2, \ldots, n$ according to Equation (4.11).
2. Find positions m and r such that $\pi_m = k$ and $I_B(r; \mathbf{p}) = \max_{i:p_i < p_m} I_B(i; \mathbf{p})$ (break tie arbitrarily).
3. If $I_B(m; \mathbf{p}) < I_B(r; \mathbf{p})$ and $R(\mathbf{p}\pi_{(m,r)}) > R(\mathbf{p}\pi)$, then exchange the assignments of components π_m and π_r, that is, set $\pi = \pi(m, r)$.

Fitness scaling

In the CAP, the objective is to maximize the systems reliability, which is bounded by zero and one. It is natural to use the systems reliability $R(\mathbf{p}\pi_t)$ as the fitness of chromosome π_t, denoted by $f_t = R(\mathbf{p}\pi_t)$, for $t = 1, 2, \ldots, s$. Meanwhile, step 3 calculates a scaled fitness of chromosome π_t, denoted by F_t, which is defined using a linear scaling function with coefficients α_1 and α_2 as

$$F_t = \alpha_1 f_t + \alpha_2. \tag{13.1}$$

Let f_{max}, f_{min}, and f_{avg} denote the maximum, minimum, and average fitness values of the s chromosomes in a population before the scaling, respectively, and F_{max}, F_{min}, and F_{avg} denote the maximum, minimum, and average fitness values after the scaling, respectively. The coefficients α_1 and α_2 are chosen such that $F_{avg} = f_{avg}$ and $F_{max} = S_f F_{min}$, which yield

$$\alpha_1 = \frac{(S_f - 1) f_{avg}}{S_f (f_{avg} - f_{min}) + (f_{max} - f_{avg})}, \tag{13.2}$$

$$\alpha_2 = (1 - \alpha_1) f_{avg},$$

where $S_f > 1$ is the scaling factor. Note that, because the fitness values are positive, α_1 is always positive but α_2 may not be.

The scaled fitness F_t in Equation (13.1) is used with the aim of balancing the premature and the divergence of the BIGLS. On one hand, with too much scaling ($S_f \gg 1$), a super-good chromosome can quickly dominate the population and result in a premature and suboptimal solution. On the other hand, with little scaling ($S_f \simeq 1$), the population may always keep large dispersion, converge slowly, and may even diverge. A proper scaling factor S_f can prevent a premature solution in the early stage and divergence in the late stage.

Termination conditions

The denominator in Equation (13.2) reflects the extent of convergence of the population, and it is thus used as one termination condition in step 2 of the BIGLS:

$$S_f (f_{avg} - f_{min}) + (f_{max} - f_{avg}) < \varepsilon, \tag{13.3}$$

where ε is a small positive tolerance. Another termination condition used in step 2 is that the number of generations cannot exceed a prespecified threshold (e.g., 200 generations). The BIGLS stops whenever one of the two termination conditions is met.

Elitism

An elitism strategy is used in step 4 of the BIGLS. A proportion (say, β) of the best chromosomes (i.e., those with high fitness values) in the current population are directly reproduced to the next population, where β is the elitist rate, $0 < \beta < 1$. These chromosomes are called elitists and carry the best genes from the current population to the next. This strategy can normally improve the performance of a genetic local search (Davis 1991). Recalling that the initial population includes the arrangement generated by the BITS, the solution from the BIGLS is never worse than that from the BITS because of the elitism strategy.

Crossover

Step 5 of the BIGLS executes the crossover in order to produce diverse offspring chromosomes for the next population. The biased roulette wheel criterion is used to select a pair of parent chromosomes from the current population; each chromosome π_t, $t = 1, 2, \ldots, s$, is selected as a parent with the probability $F_t / \sum_{j=1}^{s} F_j$, proportional to its scaled fitness value. Note that $F_{max} = S_f F_{min}$, which means that the best chromosome has S_f times probability to be selected as the worst chromosome. A chromosome can be selected more than one time and is replaced back into the current population each time for the next selection. Totally, $s(1 - \beta)$ pairs of parent chromosomes are selected for reproducing $s(1 - \beta)$ offsprings into the next population. With a probability P_{cro}, a crossover is executed on a pair of parent chromosomes to generate an offspring chromosome; with the probability $1 - P_{cro}$, no crossover is taken, and the first selected parent is reproduced to the next population.

The partial matched crossover (PMX) operator is used, under which two crossing points are uniformly picked at random. These two crossing points define a matching section between the two parent chromosomes, according to which a crossing of position-to-position exchanges is implemented in the first parent chromosome to obtain the offspring chromosome. As an example, consider two parent chromosomes π^a and π^b:

$$\pi^a = (8\ 7\ |\ 6\ 4\ 2\ |\ 1\ 5\ 3)$$
$$\pi^b = (2\ 5\ |\ 1\ 7\ 3\ |\ 8\ 4\ 6),$$

and two random crossing points 3 and 6 as indicated. Then the generated offspring chromosome is

$$\pi' = (8\ 4\ |\ 1\ 7\ 3\ |\ 6\ 5\ 2),$$

which is formed from chromosome π^a by exchanging positions of genes 6 and 1, 4 and 7, and 2 and 3.

Mutation

Following step 5 of crossover, step 6 of the BIGLS mutates each offspring chromosome with a probability P_{mut} to further diversify the population and expand the search region. The mutation operator randomly picks two positions and swaps the genes in these two positions. For example, if the aforementioned offspring chromosome π' is chosen for mutation and the second and fourth positions are randomly picked, then genes 4 and 7 in these two positions are swapped, producing a mutated chromosome

$$\pi'' = (8\ 7\ 1\ 4\ 3\ 6\ 5\ 2).$$

In step 6, the mutated offspring chromosomes and the promising unmutated ones that have fitness values higher than the best chromosome in the current population go through the B-importance-based local search. Then, the improved chromosomes replace the original ones to enter the next population. Note that the local search is performed only on the promising new chromosomes rather than all new ones because it is shown to be more computationally efficient in coping with quadratic assignment problems (Lim et al. 2000) and CAP (Yao et al. 2012).

13.5.2   Numerical Comparisons with the B-importance-based Two-stage Approach and a Genetic Algorithm

To demonstrate the effectiveness of the B-importance-based local search (Table 13.12) in the BIGLS, Yao et al. (2012) proposed a general genetic algorithm (GGA) that uses the same framework and parameter setting as the BIGLS but does not use local search at all. All of the BIGLS and GGA runs are performed with the population size $s = 50$, elitist rate $\beta = 0.04$, fitness scaling factor $S_f = 3$, crossover probability $P_{cro} = 0.8$, and mutation probability $P_{mut} = 0.05$. The BIGLS and GGA are terminated after 200 generations or when the convergence criterion in Equation (13.3) is satisfied with $\varepsilon = 0.0001$.

The performance of the BIGLS, GGA, and BITS was compared on both 30 sets of the small instances described in Remark 13.2.4 and 84 sets of the large systems described in Subsection 13.4.2. Note that the same computer resource as described at the end of Subsection 13.4.1 was used for these computational tests. Yao et al. (2012) presented the similar computational tests.

Computational tests on robustness

Because the execution of the BIGLS depends on the realization of its random operators, the robustness of the BIGLS to its random setting was first examined by executing ten random BIGLS runs on each of 84 selected instances. These 84 instances are from 84 distinct sets of the large instances, each from one set. Then, for each instance, the coefficient of variation (i.e., σ/μ) over the ten systems reliability values that correspond to the solutions of the ten runs is recorded, where σ is the standard deviation and μ is the mean. It is demonstrated that all of the coefficients of variation are extremely small with the largest one at 0.0026 for the instance in $Lin/Con/4/100:F$ system with arbitrary components. Therefore, the BIGLS is robust to its random setting, and there is no need to execute multiple BIGLS runs for solving a single instance.

Computational tests on small instances

For the small instances, the enumeration method can find the optimal solutions. On average over each of 30 sets of the small instances, the BIGLS always generates the solutions better than the

GGA and finds optimal solutions for more instances. Note that the initial populations of both the BIGLS and GGA include the arrangement generated by the BITS and the elitism strategy keeps the best chromosome; thus, the BIGLS and GGA always generate better solutions than the BITS. Over 30 sets of the small instances, the BITS finds the optimal arrangements for all of the 100 instances in 10 sets and 31–99 (over 100) instances in the other 20 sets. For most of the instances that the BITS fails to find the optimal solutions, both the BIGLS and GGA can improve the BITS solutions. Moreover, the BIGLS is more capable of improving the BITS solutions than the GGA. For example, in the set of 100 instances in $Lin/Con/2/7$:G system with low reliable components, the BITS does not find the optimal solutions for 48 instances; the GGA improves the BITS solutions for 38 instances among them and achieves the optimal solutions for 35 instances; the BIGLS improves the BITS solutions for 47 instances and achieves the optimal solutions for all of the 47 instances. Overall, the GGA finds the optimal arrangements for all of the 100 instances in 16 sets and 72–99 instances in other 14 sets, and the BIGLS finds the optimal arrangements for all of the 100 instances in 25 sets, 96–99 instances in four sets, and 81 instances in one set.

Over 30 sets of the small instances, the computation time of the BITS for a single instance is less than 0.019 seconds; that of the BIGLS ranges from 0.19 to 3.63 seconds; and that of the GGA ranges from 0.20 to 2.90 seconds. The BITS is always much faster than the BIGLS and GGA. Thus, including the BITS in constructing the initial population does not significantly increase the computational burden of the BIGLS and GGA. In addition, the BIGLS is faster than the GGA in solving 20 sets of instances, although the local search in the BIGLS consumes additional time in each generation. For the small instances, because the solution space containing $n!$ arrangements is relatively small, the BIGLS can find high-quality chromosomes more easily than the GGA with the aid of the B-importance-based local search. Therefore, the BIGLS may take fewer generations to converge and terminate; consequently, the overall time required for the BIGLS may be shorter than that for the GGA.

Computational tests on large instances

For the large instances for which the enumeration method cannot find the optimal solutions in a reasonable time, the BIGLS and GGA were evaluated on the capability of improving the solutions generated by the BITS. Over 84 sets of the large instances, the BIGLS improves the BITS solutions for more instances than the GGA except the set of instances in $Lin/Con/5/20$:F system with high reliable components, where the BIGLS improves the BITS solutions for 30 instances, while the GGA improves for 44 instances over the total 100 instances in the set. However, in most other sets, the BIGLS significantly outperforms the GGA. For example, in the set of instances in $Lin/Con/3/50$:F system with arbitrary components, the BIGLS improves the BITS solutions for 43 instances, while the GGA fails to improve any of them; in the set of instances in $Lin/Con/5/30$:G system with high reliable components, the BIGLS improves the BITS solutions for 79 instances, while the GGA improves only 19 instances. Note that, for the instances in the $Lin/Con/k/n$:G systems with arbitrary components, although the BIGLS works better than the GGA, neither of them can significantly improve the BITS solutions. A possible reason is that the BITS works very well on these instances and achieves optimal or near-optimal arrangements (see Subsection 13.4.2), leaving very limited room for further improvement to the BIGLS and GGA.

The computation time for a single instance over 84 sets of the large instances ranges from 0.04 to 3.95 seconds for the BITS, from 0.37 to 123.90 seconds for the BIGLS, and from 0.03 to 13.87 seconds for the GGA. Similar to the case of the small instances, the BITS is much faster than the BIGLS and GGA, again validating the reasonability of including the arrangement generated by the BITS in the initial populations of the BIGLS and GGA. However, different from the case of the small instances, the BIGLS is usually slower than the GGA with only a few exceptions on the instances of size $n = 20$. For the large instances, because the solution space of $n!$ arrangements is large, it is not easy to find high-quality arrangements by the B-importance-based local search over a local subregion. Consequently, the local search does not help a lot with convergence and the BIGLS takes as many generations as the GGA before termination. On the other hand, the local search consumes additional computation time in each generation of the BIGLS to improve the solution quality; thus, the overall time required for the BIGLS is longer than that for the GGA. Compared to the BITS and GGA, the longer (but still reasonable) computation time of the BIGLS is justified by its higher accuracy.

Summary

In summary, the B-importance-based local search, which is embedded in the BIGLS, increases the accuracy of the solutions and accelerates the convergence of the heuristic for the small instances but adds an additional computational burden for the large instances. The BIGLS can improve the solutions generated by the BITS, which, however, can solve the instances quickly.

13.6   Summary and Discussion

The use of B-importance

The heuristics for the CAP in this chapter are all based on the B-reliability importance in directing the assignment of components. The common intuitive rationale is that a position with a large B-reliability importance value should be assigned a component of high reliability. However, the way that each heuristic implements this rationale is different.

For the Kontoleon and LK-type heuristics, the algorithms initially assign all positions the available component of the same highest or lowest reliability. It is an infeasible solution because it is not an arrangement of all available components. Then, at each iteration, the B-reliability importance is calculated for certain positions, and the remaining components are allocated to the system gradually based on the ordering of the B-reliability importance of these positions. A feasible arrangement is obtained at the end of the procedure when all components are assigned to the system.

Unlike the Kontoleon and LK-type heuristics, the ZK-type heuristics start with a feasible initial arrangement and maintain the feasibility throughout the procedure. The solution is improved by pairwise exchange based on the B-reliability importance. The ZK-type heuristics are unified methods that can easily incorporate different initial arrangements and different strategies for pairwise exchange of components.

The BITS is built on the LK-type and ZK-type heuristics. Through numerical experiments involving both small and large instances, the BITS is shown to be efficient and capable of generating high-quality solutions. The BITS absolutely outperforms the GAMS/CoinBonmin solver, enumeration method, and randomization method, especially for large instances.

By using the B-reliability importance in the genetic algorithm, the BIGLS can generate solutions of higher quality than the GGA and BITS, yet consuming more time within a reasonable range, especially for solving large instances. The B-reliability importance is used in the local search of the BIGLS and in generating its initial population. For solving large instances, the computation time associated with the BIGLS is reasonable and relative small, taking into account that the enumeration method and the GAMS/CoinBonmin solver cannot generally solve large instances in a very long time.

Extension to $\mathbb{K}$-terminal network

The CAP described in Section 11.1 is defined on a reliability system in which the systems reliability is consistently defined and used as the objective function. As an extension, the CAP could be defined on a $\mathbb{K}$-terminal network (see Section 16.2) and have the objective function as the probability that all target vertices in some specified set $\mathbb{K}$ are joined by paths of functioning edges. Interestingly, there is another way to define the CAP in a $\mathbb{K}$-terminal network with the multiple objectives, each as the reliability of a pair of target vertices in $\mathbb{K}$. As shown in Section 13.5, the genetic algorithm in conjunction with importance measures successfully solves the CAP. It is expected that a genetic algorithm could address the CAP in a $\mathbb{K}$-terminal network with a single objective, and that a multiobjective genetic algorithm (e.g., NSGA and NSGA II) could address the multiobjective CAP in a $\mathbb{K}$-terminal network. Such an extension of CAP finds applications in airline network and communication network design.

Extension to continuous deterioration case

Consider the continuous degradation or deterioration of components that is further described in Section 15.6 for continuum model. Suppose that the functionally interchangeable components are allocated in the system according to their initial conditions. For example, they may simply have been arbitrarily assigned when new, each having the same reliability. Due to their different positions and uses in the system, the wear-out extents of these components are different in most cases. Thus, as time goes on, the components degrade differently, and their conditions and reliability become different. Rearranging them has the potential to improve the systems reliability, for example, exchanging the front and rear tires of a car. The corresponding CAP can provide a new arrangement of these components based on their current extents of degradation so that the systems reliability is improved to a possible maximum level.

However, rearrangement is costly and thus cannot be executed frequently. A maintenance policy for when rearrangement should be made is necessary. Simply, rearrangement could be made after a fixed time period or a certain provided service. For example, rotation of the front and rear tires of a car is recommended after running a car for a certain number of miles. Alternatively, one may decide to make a rearrangement when it can improve the systems reliability significantly. To implement this policy, it is necessary to monitor and acquire the reliability values of the current system and its components and then use efficient heuristics (e.g., the BIGLS) to approximate the systems reliability after near-optimal rearrangement that is determined on the basis of the current component reliability values. One other consideration is to replace some components whose reliability has fallen below some threshold value with new ones. Then, the CAP is to redesign the system subject to this updated set of components.

Extension to multiple types of components

One limitation of the CAP in Section 11.1 is that only one type of functionally interchangeable components is allowed. Although it is usual for a system to need multiple components (units) that perform the same function, many systems consist of more than one type of functionally interchangeable components, similar as in Section 12.8. One way to deal with multiple types of components is to assign components of the same type, type-wise one-by-one in the order of their importance to the system until all types of components are assigned. For each type of components, the CAP is to determine the allocation of these functionally interchangeable components to the system so as to maximize the systems reliability; the determination is based on the assignments of the allocated types of components and the assumption that all types of components unassigned are perfect, with a reliability value of one.

Another method is to develop an importance-measure-based genetic algorithm to simultaneously make an arrangement of all types of components. Through appropriate coding, the genetic algorithm can easily restrict the components to their allowed positions (which are determined by their types). Meanwhile the importance measures can be applied to all types of components since the importance measures are determined by the positions and reliability values of the components irrespective of their types.

Inspiration for designing solution methods

The outstanding performance of B-importance-based heuristics on solving the CAP inspires research in integrating importance measures into the design of various heuristics. The findings and tests in this chapter will benefit any more complicated importance-measure-based heuristics for the CAP. The importance-measure-based metaheuristics (e.g., genetic algorithm and simulated annealing) have the potential to break local optimal arrangements and thus are promising.

In broader research, importance-measure-based methods for solving various hard problems in the fields of reliability, risk, and mathematical programming deserve investigation since importance-measure-based methods such as those described in this chapter and those presented in Part Five are probably the most practical decision-making tools.

References

Davis L. 1991. *Handbook of Genetic Algorithms*. Van Nostrand Reinhold, New York.

Freisleben B and Merz P. 1996. A genetic local search algorithm for solving symmetric and asymmetric traveling salesman problems. In *Proceedings of the 1996 IEEE International Conference on Evolutionary Computation* (ed. Grefenstette JJ). IEEE Press, New York, pp. 616–621.

Grefenstette JJ, Gopal R, Rosmaita B, and Cucht DV. 1985. Genetic algorithms for the traveling salesman problem. *Proceedings of the 1st International Conference on Genetic Algorithms and Their Applications*. Erlbaum, Hillsdale, NJ, pp. 160–168.

Kontoleon JM. 1979. Optimal link allocation of fixed topology networks. *IEEE Transactions on Reliability* **28**, 145–147.

Lim MH, Yuan Y, and Omatu S. 2000. Efficient genetic algorithms using simple genes exchange local search policy for the quadratic assignment problem. *Computational Optimization and Applications* **15**, 249–268.

Lim MH, Yuan Y, and Omatu S. 2002. Extensive testing of a hybrid genetic algorithm for solving quadratic assignment prolems. *Computational Optimization and Applications* **23**, 47–64.

Lin FH and Kuo W. 2002. Reliability importance and invariant optimal allocation. *Journal of Heuristics* **8**, 155–171.

Merz P and Freisleben B. 1997. A genetic local search approach to the quadratic assignment problem. In *Proceedings of the 7th International Conference on Genetic Algorithms* (ed. Grefenstette JJ). Morgan Kaufmann, San Francisco, CA, pp. 465–472.

Prasad VR and Kuo W. 2000. Reliability optimization of coherent systems. *IEEE Transactions on Reliability* **49**, 323–330.

Shingyoch K, Yamamoto H, Tsujimura Y, and Akiba T. 2010. Proposal of simulated annealing algorithms for optimal arrangement in a circular consecutive-k-out-of-n:F system. *Quality Technology and Quantitative Management* **7**, 395–405.

Shingyoch K, Yamamoto H, Tsujimura Y, and Kambayashi Y. 2009. Improvement of ordinal representation scheme for solving optimal component arrangement problem of circular consecutive-k-out-of-n:F system. *Quality Technology and Quantitative Management* **6**, 11–22.

Ulder NLJ, Aarts EHL, Bandelt HJ, van Laarhoven PJM, and Pesch E. 1991. Genetic local search algorithms for the traveling salesman problem. In *Parallel Problem Solving from Nature I* (ed. Schwefel H and Manner R). Springer-Verlag, Berlin, pp. 109–116.

Yao Q, Zhu X, and Kuo W. 2011. Heuristics for component assignment problems based on the Birnbaum importance. *IIE Transactions* **43**, 1–14.

Yao Q, Zhu X, and Kuo W. 2012. Importance-measure based genetic local search for component assignment problems. *Annals of Operations Research*, In review.

Zuo MJ and Kuo W. 1990. Design and performance analysis of consecutive k-out-of-n structure. *Naval Research Logistics* **37**, 203–230.

Zuo MJ and Shen J. 1992. System reliability enhancement through heuristic design. *1992 OMAE Volume II, Safety and Reliability, ASME*, pp. 301–304.

Part Four

Relations and Generalizations

Introduction

Parts Two and Three present many types of importance measures, which are actually related to each other in some respects. Part Four summarizes, compares, and generalizes these importance measures in different aspects, establishing the relationships among them and deepening the understanding of them. In particular, this part focuses on the importance measures of individual components in reliability and covers the importance measures of pairs and groups of components when appropriate. For the details relating to importance measures of pairs and groups of components, readers should refer to Chapter 7.

Part Four consists of two chapters. Chapter 14 presents the relationships of all of the importance measures to the B-importance, demonstrating the central role of the B-importance. It also summarizes the possible judgment of the reliability importance measures of components by means of the structure importance measures, compares the importance measures in some typical structure systems, and discusses the relevant computation methods. Chapter 15 generalizes the importance measures for some uncommon but nonetheless important reliability systems. These two chapters extend some of the highlights provided in Kuo and Zhu (2012) on the relations and generalizations of the importance measures.

Reference

Kuo W and Zhu X. 2012. Relations and generalizations of importance measures in reliability. *IEEE Transactions on Reliability* **61**.

14

Comparisons of Importance Measures

The first observation is that excluding the structure importance measures that are defined by means of comparisons, all other types of importance measures of individual components are nonnegative, and almost all of them are less than or equal to one, except for the RAW, RRW, first-term importance, and rare-event importance (Kuo and Zhu 2012). The RAW and RRW are always greater than or equal to one according to their definitions in Equations (7.15) and (7.16).

A second observation is that all types of importance measures hold the transitivity property, as stated in the following theorem.

Theorem 14.0.1 *The transitivity property holds for every importance measure. That is, according to a given importance measure, if component i is more important than component j, which is more important than component k, then component i is more important than component k.*

Proof. Of course, the transitivity property holds for importance measures that are defined as numerical values. Among importance measures that are defined by means of comparisons, the transitivity property can be easily verified for the absoluteness importance, cut-path importance, min-cut importance, and min-path importance. In addition, the transitivity properties of the cut importance and path importance follow from the transitivity of lexicographic order. Finally, the transitivity properties of the permutation importance and domination importance have been proved in Theorems 6.5.3 and 6.6.5, respectively.

In this chapter, Section 14.1 presents the relations of all of the importance measures to the B-importance, which can be viewed as a general summary of various importance measures. Section 14.2 explains how to select the potentially most B-important components using some structure importance measures and based on the orderings or ranges of component reliability. Section 14.3 demonstrates the performance of importance measures for some systems

Importance Measures in Reliability, Risk, and Optimization: Principles and Applications, First Edition.
Way Kuo and Xiaoyan Zhu. © 2012 John Wiley & Sons, Ltd. Published 2012 by John Wiley & Sons, Ltd.

with special structures. Finally, Section 14.4 discusses issues relating to the computation of importance measures.

14.1   Relations to the B-importance

The B-importance plays an extremely important role as many importance measures are defined by or have relationships to it. Table 14.1 summarizes the relations of all other importance measures to the B-importance for binary coherent systems (Kuo and Zhu 2012). The first column in Table 14.1 gives the types and notation of importance measures, and the second column indicates the corresponding relationships.

From Table 14.1, it is easy to see the essential role of the B-importance. Some importance measures, including the B-importance for system functioning and one for system failure, criticality importance, DIM^I, and redundancy reliability importance, are proportional to the B-reliability or B-TDL importance. The Bayesian importance, RAW, and RRW are functions of and increasing with the B-reliability importance. Others, including the BP importance, redundancy TIL importance, L_1–L_6 importance measures, and yield importance, are weighted averages of the B-i.i.d. or B-TDL importance with different weight functions. The TOI^k is an integrated importance measure of the JRI of all orders from one up to k. The first-order JRI is indeed the B-reliability importance. Note that the FV importance does not have much of a relation to the B-importance because the former takes into account the contribution of a component to the system without being critical, while the latter considers the case of being critical (see Section 5.6).

Among structure importance measures, the cut importance and path importance have the closest relationship to the B-i.i.d. importance; they are equivalent to $I_B(\cdot; p)$ as p approaches 1 and 0, respectively (Theorems 6.7.4 and 6.7.6). The first-term importance and rare-event importance are the counterparts of each other; they are relevant as p approaches 1 and 0, respectively (see Section 6.10). The absoluteness importance, permutation importance, domination importance, and cut-path importance are all stronger than the uniform B-i.i.d. importance. That is, for example, if component i is absolutely more important than component j, then component i is more uniformly B-i.i.d. important than component j (see Section 6.13). However, the min-cut importance and min-path importance do not have a clear relationship to the B-importance. Apart from the importance measures listed in Table 14.1, a series of structure importance measures can be defined by integrating the B-i.i.d. importance over p (see Section 6.4), in which a prior distribution of component reliability p needs to be prescribed. Example 6.14.4 demonstrates the effect of the ranges of component reliability on the B-i.i.d. importance and the differences among the B-structure importance, BP structure importance, cut importance, and path importance.

As can be seen from the preceding chapters, various importance measures have been developed for dealing with different problems and situations in which the B-importance may not be suitable for direct use. However, almost all of these importance measures are in one way or another related to the B-importance, which is essentially the sensitivity analysis.

Table 14.1 not only presents the relations of various importance measures to the B-importance but also demonstrates the overall relations and differences among the importance measures. For example, it is easy to see that the BP importance and L_6 importance are special cases of the yield importance with yield function $Y(t) = t$ and $Y(t) = t^2$, respectively (see the

Table 14.1 Relations to the B-importance

Importance measure	Relation to $I_B(\cdot;\cdot)$	
B-reliability for system functioning (improvement potential) and B-reliability for system failure		
$\quad I_{Bs}(i;\mathbf{p})$	$q_i I_B(i;\mathbf{p})$	$\propto I_B(i;\mathbf{p})$
$\quad I_{Bf}(i;\mathbf{p})$	$p_i I_B(i;\mathbf{p})$	$\propto I_B(i;\mathbf{p})$
Bayesian $I_{Bay}(i;\mathbf{p})$	$q_i\left(1+p_i I_B(i;\mathbf{p})/(1-R(\mathbf{p}))\right)$	$\propto I_B(i;\mathbf{p})$
Criticality for system functioning and criticality for system failure		
$\quad I_{Cs}(i;\mathbf{p})$	$p_i I_B(i;\mathbf{p})/R(\mathbf{p})$	$\propto I_B(i;\mathbf{p})$
$\quad I_{Cf}(i;\mathbf{p})$	$q_i I_B(i;\mathbf{p})/(1-R(\mathbf{p}))$	$\propto I_B(i;\mathbf{p})$
$\quad I_{Cf}(i;\overline{\mathbf{F}}(t))$	$F_i(t)I_B(i;\overline{\mathbf{F}}(t))/(1-R(\overline{\mathbf{F}}(t)))$	$\propto I_B(i;\overline{\mathbf{F}}(t))$
FV		
$\quad I_{FV^c}(i;\overline{\mathbf{F}}(t)), I_{FV^c}(i;\mathbf{p}), I_{FV^c}(i;\phi), I_{FV^p}(i;\overline{\mathbf{F}}(t)), I_{FV^p}(i;\mathbf{p}), I_{FV^p}(i;\phi)$	—	
BP		
$\quad I_{BP}(i;\overline{\mathbf{F}}(t))$	$\int_0^t I_B(i;\overline{\mathbf{F}}(u))\mathrm{d}F_i(u)$	Integral of $I_B(i;\overline{\mathbf{F}}(t))$
$\quad I_{BP}(i;\overline{\mathbf{F}})$	$\int_0^\infty I_B(i;\overline{\mathbf{F}}(t))\mathrm{d}F_i(t)$	Integral of $I_B(i;\overline{\mathbf{F}}(t))$
$\quad I_{BP}(i;\phi)$	$\int_0^1 I_B(i;p)\mathrm{d}p$	Integral of $I_B(i;p)$
Permutation $>_{pe}$	Stronger than $I_B(i;p)$ for all $0<p<1$	
Internal domination $>_{in}$	Stronger than $I_B(i;p)$ for all $0<p<1$	
External domination $>_{ex}$	Stronger than $I_B(i;p)$ for all $0<p<1$	
Cut $>_c$	Consistent with $I_B(i;p)$ as p approaches 1	
Path $>_p$	Consistent with $I_B(i;p)$ as p approaches 0	
Absoluteness $>_{a^c}, >_{a^p}$	Both c-type and p-type stronger than $I_B(i;p)$ for all $0<p<1$	
Cut-path $>_{cp}$	Stronger than $I_B(i;p)$ for all $0<p<1$	
Min-cut $>_{mc}$ and min-path $>_{mp}$	—	
First-term $I_{FT}(i;\phi)$	Relevant as p approaches 1	
Rare-event $I_{RE}(i;\phi)$	Relevant as p approaches 0	
DIM^I $I_{DIM^I}(i;\mathbf{p})$	$I_B(i;\mathbf{p})\mathrm{d}p_i/\sum_{j=1}^n I_B(j;\mathbf{p})\mathrm{d}p_j$	$\propto I_B(i;\mathbf{p})$
TOI $I_{TOI^k}(i;\mathbf{p})$	Equation (7.14) integrating $I_B(i;\mathbf{p})$ and JRI of orders from two to k	
RAW $I_{RAW}(i;\mathbf{p})$	$1+q_i I_B(i;\mathbf{p})/R(\mathbf{p})$	$\propto I_B(i;\mathbf{p})$
RRW $I_{RRW}(i;\mathbf{p})$	$1/\left(1-p_i I_B(i;\mathbf{p})/R(\mathbf{p})\right)$	$\propto I_B(i;\mathbf{p})$
Redundancy (parallel and series)		
$\quad I_{PR}(i;\mathbf{p})$	$p_i^* q_i I_B(i;\mathbf{p})$	$\propto I_B(i;\mathbf{p})$
$\quad I_{SR}(i;\mathbf{p})$	$q_i^* p_i I_B(i;\mathbf{p})$	$\propto I_B(i;\mathbf{p})$
$\quad I_{PR}(i;\overline{\mathbf{F}})$	$\int_0^\infty \overline{F}_i^*(t)F_i(t)I_B(i;\overline{\mathbf{F}}(t))\mathrm{d}t$	Integral of $I_B(i;\overline{\mathbf{F}}(t))$
L_1-L_6 importance measures		
$\quad I_{L_1}(i;\overline{\mathbf{F}})$	$\propto \int_0^\infty \overline{F}_i(t)\left[-\ln\overline{F}_i(t)\right]I_B(i;\overline{\mathbf{F}}(t))\mathrm{d}t$	Integral of $I_B(i;\overline{\mathbf{F}}(t))$
$\quad I_{L_2}(i;\overline{\mathbf{F}})$	$\propto \int_0^\infty \lambda_i^2 h(t)\exp(-\lambda_i h(t))I_B(i;\overline{\mathbf{F}}(t))\mathrm{d}t$	Integral of $I_B(i;\overline{\mathbf{F}}(t))$
$\quad I_{L_3}(i;\overline{\mathbf{F}})$	$\propto \int_0^\infty \int_0^t f_i(t-u)\overline{F}_i(u)\mathrm{d}u\, I_B(i;\overline{\mathbf{F}}(t))\mathrm{d}t$	Integral of $I_B(i;\overline{\mathbf{F}}(t))$
$\quad I_{L_4}(i;\overline{\mathbf{F}})$	$\propto \int_0^\infty F_i(t)I_B(i;\overline{\mathbf{F}}(t))\mathrm{d}t$	Integral of $I_B(i;\overline{\mathbf{F}}(t))$
$\quad I_{L_5}(i;\overline{\mathbf{F}})$	$\propto \int_0^\infty \left(\overline{F}_i(t)^\alpha - \overline{F}_i(t)\right)I_B(i;\overline{\mathbf{F}}(t))\mathrm{d}t$	Integral of $I_B(i;\overline{\mathbf{F}}(t))$
$\quad I_{L_6}(i;\overline{\mathbf{F}})$	$\propto \int_0^\infty t I_B(i;\overline{\mathbf{F}}(t))\mathrm{d}F_i(t)$	Integral of $I_B(i;\overline{\mathbf{F}}(t))$
yield $I_Y(i;\overline{\mathbf{F}})$	$\propto \int_0^\infty Y'(t)I_B(i;\overline{\mathbf{F}}(t))\mathrm{d}F_i(t)$	Integral of $I_B(i;\overline{\mathbf{F}}(t))$

Notation $\propto$ denotes proportional to or increasing with.

discussion following Corollary 10.3.4). Hence, Table 14.1 can be used as a general reference guideline for importance measures.

14.2 Rankings of Reliability Importance Measures

The reliability importance measures are believed to give generally superior results to the structure importance measures. However, when probabilistic information is not available or the computation involved is prohibitively extensive, structure importance measures must be employed.

Using the structure importance measures alone or in conjunction with the known orderings or ranges of component reliability can help determine the ranking of components relative to the reliability importance measures (e.g., the B-reliability importance) without computing their exact values, and thus can help find the components that have the greatest effect on system performance. Subsection 14.2.1 gives the relationships of the permutation importance to the B-reliability importance, improvement potential importance, RAW, and RRW. Subsection 14.2.2 presents the relationships between the permutation importance, B-reliability importance, and JRI, which is an extension of the B-reliability importance for pairs of components. Subsection 14.2.3 shows that the internal and external domination importance measures can also be used to judge the B-reliability importance ranking of two components. Finally, Subsection 14.2.4 summarizes all of these relationships.

14.2.1 Using the Permutation Importance

Theorem 14.2.1 (Meng 1994) shows that the permutation importance implies the ranking induced by the B-reliability importance when the reliability values of all components are equal or *nearly* equal. Similarly but differently, Theorem 14.2.2 (Meng 1995) shows that the permutation importance ranking of two components implies their B-reliability importance ranking when these two components have the same reliability. Note that Theorems 14.2.1 and 14.2.2 provide stronger results than that $i >_{\text{pe}} j$ implies $I_B(i; p) > I_B(j; p)$ for all p, which can be induced from Figure 6.3.

Theorem 14.2.1 *Assume that* $\mathbf{p} : \mathbb{R} \mapsto \mathbb{R}^n$ *satisfies* $0 < \mathbf{p}(\varepsilon) < 1$ *for all* $\varepsilon \in \mathbb{R}$ *and that* $\lim_{\varepsilon \to 0} p_i(\varepsilon) = \lim_{\varepsilon \to 0} p_j(\varepsilon) = p$ *for some* $0 < p < 1, i \neq j$. *Then there exists a* ε_0 *such that for all* $\varepsilon < \varepsilon_0, i >_{\text{pe}} j$ *implies that* $I_B(i; \mathbf{p}(\varepsilon)) > I_B(j; \mathbf{p}(\varepsilon))$.

Proof. From Equation (4.8) that $I_B(i; \mathbf{p}(\varepsilon)) = \mathbb{E}(\phi(1_i, \mathbf{X})) - \mathbb{E}(\phi(0_i, \mathbf{X}))$, it is easy to see that

$$I_B(i; \mathbf{p}(\varepsilon)) = p_j(\varepsilon)\mathbb{E}(\phi(1_i, 1_j, \mathbf{X}^{(ij)})) + (1 - p_j(\varepsilon))\mathbb{E}(\phi(1_i, 0_j, \mathbf{X}^{(ij)}))$$
$$- p_j(\varepsilon)\mathbb{E}(\phi(0_i, 1_j, \mathbf{X}^{(ij)})) - (1 - p_j(\varepsilon))\mathbb{E}(\phi(0_i, 0_j, \mathbf{X}^{(ij)})).$$

Hence,

$$I_B(i; \mathbf{p}(\varepsilon)) - I_B(j; \mathbf{p}(\varepsilon))$$
$$= \mathbb{E}(\phi(1_i, 0_j, \mathbf{X}^{(ij)}) - \phi(0_i, 1_j, \mathbf{X}^{(ij)})) + (p_j(\varepsilon) - p_i(\varepsilon))$$
$$\times \mathbb{E}(\phi(1_i, 1_j, \mathbf{X}^{(ij)}) - \phi(1_i, 0_j, \mathbf{X}^{(ij)}) - \phi(0_i, 1_j, \mathbf{X}^{(ij)}) + \phi(0_i, 0_j, \mathbf{X}^{(ij)})),$$

and

$$\lim_{\varepsilon \to 0}[I_{\mathrm{B}}(i; \mathbf{p}(\varepsilon)) - I_{\mathrm{B}}(j; \mathbf{p}(\varepsilon))] = \mathbb{E}(\phi(1_i, 0_j, \mathbf{X}^{(ij)}) - \phi(0_i, 1_j, \mathbf{X}^{(ij)})).$$

Because $i >_{\mathrm{pe}} j$, $\mathbb{E}(\phi(1_i, 0_j, \mathbf{X}^{(ij)})) - \mathbb{E}(\phi(0_i, 1_j, \mathbf{X}^{(ij)})) > 0$. Then, the result under the assumptions of $\mathbf{p}(\varepsilon)$ in the statement follows from the continuity of $I_{\mathrm{B}}(\cdot; \mathbf{p})$.

Theorem 14.2.2 $i >_{\mathrm{pe}} j$ *if and only if* $I_{\mathrm{B}}(i; \mathbf{p}) > I_{\mathrm{B}}(j; \mathbf{p})$ *for all* $\mathbf{0} < \mathbf{p} < \mathbf{1}$ *satisfying* $p_i = p_j$.

Proof. From Equation (4.11) that $I_{\mathrm{B}}(i; \mathbf{p}) = R(1_i, \mathbf{p}) - R(0_i, \mathbf{p})$ and using pivotal decomposition (2.10),

$$I_{\mathrm{B}}(i; \mathbf{p}) = p_j R(1_i, 1_j, \mathbf{p}^{(ij)}) + (1 - p_j)R(1_i, 0_j, \mathbf{p}^{(ij)})$$
$$- p_j R(0_i, 1_j, \mathbf{p}^{(ij)}) - (1 - p_j)R(0_i, 0_j, \mathbf{p}^{(ij)}). \tag{14.1}$$

Suppose that α is the reliability of both components i and j. Then,

$$I_{\mathrm{B}}(i; \alpha_j, \mathbf{p}^{(ij)}) - I_{\mathrm{B}}(j; \alpha_i, \mathbf{p}^{(ij)}) = R(1_i, 0_j, \mathbf{p}^{(ij)}) - R(0_i, 1_j, \mathbf{p}^{(ij)})$$
$$= \mathbb{E}(\phi(1_i, 0_j, \mathbf{X}^{(ij)}) - \phi(0_i, 1_j, \mathbf{X}^{(ij)})).$$

Furthermore, following Definition 6.5.1 of the permutation importance, $i >_{\mathrm{pe}} j$ if and only if $\mathbb{E}(\phi(1_i, 0_j, \mathbf{X}^{(ij)}) - \phi(0_i, 1_j, \mathbf{X}^{(ij)})) > 0$ for all $\mathbf{0} < \mathbf{p}^{(ij)} < \mathbf{1}$. This completes the proof.

Theorems 14.2.3 and 14.2.4 specify four structural conditions under which comparisons of the B-reliability importance can be made based on the permutation importance. The proofs of these two theorems are here omitted but can be found in Meng (2004). Under the similar structural conditions, the permutation importance of two components can be used to determine the relative c-FV importance ranking of the two components as shown in Theorems 6.13.5 and 6.13.7 and the relative p-FV importance ranking as shown in Theorems 6.13.10 and 6.13.11.

Theorem 14.2.3 *Supposing that $i =_{\mathrm{pe}} j$, then the following hold:*

(i) $\overline{\mathscr{C}}_i = \overline{\mathscr{C}}_j$ *if and only if* $I_{\mathrm{B}}(i; \mathbf{p}) \geq I_{\mathrm{B}}(j; \mathbf{p})$ *for all* $\mathbf{0} < \mathbf{p} < \mathbf{1}$ *satisfying* $p_i \geq p_j$;
(ii) $\overline{\mathscr{P}}_i = \overline{\mathscr{P}}_j$ *if and only if* $I_{\mathrm{B}}(i; \mathbf{p}) \geq I_{\mathrm{B}}(j; \mathbf{p})$ *for all* $\mathbf{0} < \mathbf{p} < \mathbf{1}$ *satisfying* $p_i \leq p_j$.

Theorem 14.2.4 *Supposing that $i >_{\mathrm{pe}} j$, then the following hold:*

(i) $\overline{\mathscr{C}}_i \subset \overline{\mathscr{C}}_j$ *if and only if* $I_{\mathrm{B}}(i; \mathbf{p}) > I_{\mathrm{B}}(j; \mathbf{p})$ *for all* $\mathbf{0} < \mathbf{p} < \mathbf{1}$ *satisfying* $p_i \geq p_j$;
(ii) $\overline{\mathscr{P}}_i \subset \overline{\mathscr{P}}_j$ *if and only if* $I_{\mathrm{B}}(i; \mathbf{p}) > I_{\mathrm{B}}(j; \mathbf{p})$ *for all* $\mathbf{0} < \mathbf{p} < \mathbf{1}$ *satisfying* $p_i \leq p_j$.

In addition, Freixas and Pons (2008b) showed the relationships of the permutation importance to the B-reliability importance for system functioning (improvement potential importance), RAW, and RRW, as follows.

Theorem 14.2.5 *Suppose that $i \geq_{\text{pe}} j$. If $p_i \leq p_j$, then $I_{\text{Bs}}(i; \mathbf{p}) \geq I_{\text{Bs}}(j; \mathbf{p})$, and $I_{\text{RRW}}(i; \mathbf{p}) \geq I_{\text{RRW}}(j; \mathbf{p})$. If $p_i \geq p_j$, then $I_{\text{RAW}}(i; \mathbf{p}) \geq I_{\text{RAW}}(j; \mathbf{p})$. Note that the strictly greater relation holds for the above importance measures of components i and j when $i >_{\text{pe}} j$.*

14.2.2   Using the Permutation Importance and Joint Reliability Importance

Using the JRI in addition to the permutation importance, Meng (2000, 2004) found necessary and sufficient conditions, as in Theorems 14.2.6 and 14.2.7, under which the B-reliability importance of two components may be ranked given certain orderings of the reliability of the two components.

Theorem 14.2.6 *Suppose that $i =_{\text{pe}} j$. Then $I_{\text{JR}^{\prime\prime}}(i, j; \mathbf{p}) \geq (\leq) 0$ for all $0 < \mathbf{p} < 1$ if and only if $I_{\text{B}}(i; \mathbf{p}) \geq I_{\text{B}}(j; \mathbf{p})$ for all $0 < \mathbf{p} < 1$ satisfying $p_i \leq (\geq) p_j$.*

Proof. By Equations (7.3) and (14.1), $I_{\text{B}}(i; \mathbf{p})$ can be expressed using $I_{\text{JR}^{\prime\prime}}(i, j; \mathbf{p})$ as

$$I_{\text{B}}(i; \mathbf{p}) = p_j I_{\text{JR}^{\prime\prime}}(i, j; \mathbf{p}) + R(1_i, 0_j, \mathbf{p}^{(ij)}) - R(0_i, 0_j, \mathbf{p}^{(ij)}).$$

It is then obtained that

$$I_{\text{B}}(i; \mathbf{p}) - I_{\text{B}}(j; \mathbf{p}) = (p_j - p_i) I_{\text{JR}^{\prime\prime}}(i, j; \mathbf{p}) + R(1_i, 0_j, \mathbf{p}^{(ij)}) - R(0_i, 1_j, \mathbf{p}^{(ij)})$$

$$= (p_j - p_i) I_{\text{JR}^{\prime\prime}}(i, j; \mathbf{p}) + \mathbb{E}(\phi(1_i, 0_j, \mathbf{X}^{(ij)}) - \phi(0_i, 1_j, \mathbf{X}^{(ij)})).$$

$$(14.2)$$

When $i =_{\text{pe}} j$, $\mathbb{E}(\phi(1_i, 0_j, \mathbf{X}^{(ij)}) - \phi(0_i, 1_j, \mathbf{X}^{(ij)})) = 0$, and thus,

$$I_{\text{B}}(i; \mathbf{p}) - I_{\text{B}}(j; \mathbf{p}) = (p_j - p_i) I_{\text{JR}^{\prime\prime}}(i, j; \mathbf{p}).$$

Hence, $I_{\text{JR}^{\prime\prime}}(i, j; \mathbf{p}) \geq (\leq) 0$ for all $0 < \mathbf{p} < 1$ if and only if $I_{\text{B}}(i; \mathbf{p}) \geq I_{\text{B}}(j; \mathbf{p})$ for all $0 < \mathbf{p} < 1$ satisfying $p_i \leq (\geq) p_j$.

Theorem 14.2.7 *Suppose that $i >_{\text{pe}} j$. Then $I_{\text{JR}^{\prime\prime}}(i, j; \mathbf{p}) \geq (\leq) 0$ for all $0 < \mathbf{p} < 1$ if and only if $I_{\text{B}}(i; \mathbf{p}) > I_{\text{B}}(j; \mathbf{p})$ for all $0 < \mathbf{p} < 1$ satisfying $p_i \leq (\geq) p_j$.*

Proof. ("Only if") The result follows immediately from Equation (14.2).

("If") To show that $I_{\text{JR}^{\prime\prime}}(i, j; \mathbf{p}) \geq 0$ for all $0 < \mathbf{p} < 1$, it is equivalent to prove that $\phi(1_i, 1_j, \mathbf{x}^{(ij)}) - \phi(1_i, 0_j, \mathbf{x}^{(ij)}) - \phi(0_i, 1_j, \mathbf{x}^{(ij)}) + \phi(0_i, 0_j, \mathbf{x}^{(ij)}) \geq 0$ holds for all $\mathbf{x}^{(ij)}$. Supposing that the claim is not true, then there exists an $\tilde{\mathbf{x}}$ such that $\phi(1_i, 0_j, \tilde{\mathbf{x}}^{(ij)}) = \phi(0_i, 1_j, \tilde{\mathbf{x}}^{(ij)}) = 1$ and $\phi(0_i, 0_j, \tilde{\mathbf{x}}^{(ij)}) = 0$. Choosing a probability vector $\tilde{\mathbf{p}}$ such that $\Pr\{(\cdot_i, \cdot_j, \mathbf{X}) = (\cdot_i, \cdot_j, \tilde{\mathbf{x}})\}$ approaches 1, then, when $\mathbf{p} = \tilde{\mathbf{p}}$ in Equation (14.2), $R(1_i, 0_j, \mathbf{p}^{(ij)}) - R(0_i, 1_j, \mathbf{p}^{(ij)})$ approaches 0 and $I_{\text{JR}^{\prime\prime}}(i, j; \mathbf{p})$ approaches -1. Hence, $I_{\text{B}}(i; \mathbf{p}) < I_{\text{B}}(j; \mathbf{p})$ holds for some $p_i < p_j$, which contradicts the assumption.

A similar argument can be applied to show the case $I_{\text{JR}^{\prime\prime}}(i, j; \mathbf{p}) \leq 0$.

The following two corollaries follow from Theorems 14.2.3, 14.2.4, 14.2.6, and 14.2.7 and indicate the relationships between the permutation importance and JRI.

Corollary 14.2.8 *Supposing that $i =_{\text{pe}} j$, then the following hold:*

(i) $\overline{\mathscr{C}}_i = \overline{\mathscr{C}}_j$ *if and only if* $I_{\text{JR}^{II}}(i, j; \mathbf{p}) \leq 0$ *for all* $0 < \mathbf{p} < \mathbf{1}$;
(ii) $\overline{\mathscr{P}}_i = \overline{\mathscr{P}}_j$ *if and only if* $I_{\text{JR}^{II}}(i, j; \mathbf{p}) \geq 0$ *for all* $0 < \mathbf{p} < \mathbf{1}$.

Corollary 14.2.9 *Supposing that $i >_{\text{pe}} j$, then the following hold:*

(i) $\overline{\mathscr{C}}_i \subset \overline{\mathscr{C}}_j$ *if and only if* $I_{\text{JR}^{II}}(i, j; \mathbf{p}) \leq 0$ *for all* $0 < \mathbf{p} < \mathbf{1}$;
(ii) $\overline{\mathscr{P}}_i \subset \overline{\mathscr{P}}_j$ *if and only if* $I_{\text{JR}^{II}}(i, j; \mathbf{p}) \geq 0$ *for all* $0 < \mathbf{p} < \mathbf{1}$.

14.2.3   Using the Domination Importance

Freixas and Pons (2008a) proved the relationship between the domination importance and B-reliability importance as follows.

Theorem 14.2.10

(i) If $i \geq_{\text{in}} j$, then $I_{\text{B}}(i; \mathbf{p}) < I_{\text{B}}(j; \mathbf{p})$ for all $0 < \mathbf{p} < \mathbf{1}$ satisfying $p_i < p_j$.
(ii) If $i \geq_{\text{ex}} j$, then $I_{\text{B}}(i; \mathbf{p}) > I_{\text{B}}(j; \mathbf{p})$ for all $0 < \mathbf{p} < \mathbf{1}$ satisfying $p_i > p_j$.

Proof. To prove part (i), considering Theorem 6.6.6, $R(1_i, 0_j, \mathbf{p}^{(ij)}) = R(1_i, 1_j, \mathbf{p}^{(ij)})$. Thus, by Equation (14.1),

$$I_{\text{B}}(i; \mathbf{p}) - I_{\text{B}}(j; \mathbf{p}) = (p_j - p_i)(R(0_i, 0_j, \mathbf{p}^{(ij)}) - R(0_i, 1_j, \mathbf{p}^{(ij)}))$$
$$+(R(1_i, 1_j, \mathbf{p}^{(ij)}) - R(0_i, 1_j, \mathbf{p}^{(ij)})).$$

By Theorem 6.6.4, if $i \geq_{\text{in}} j$, then $i \not\geq_{\text{ex}} j$. Thus, according to Theorem 6.6.6, $R(0_i, 0_j, \mathbf{p}^{(ij)}) \neq R(0_i, 1_j, \mathbf{p}^{(ij)})$; consequently, $R(0_i, 0_j, \mathbf{p}^{(ij)}) < R(0_i, 1_j, \mathbf{p}^{(ij)})$. Finally, noting that $p_i < p_j$ and $R(1_i, 1_j, \mathbf{p}^{(ij)}) - R(0_i, 1_j, \mathbf{p}^{(ij)}) \geq 0$, thus $I_{\text{B}}(i; \mathbf{p}) - I_{\text{B}}(j; \mathbf{p}) < 0$.
   Part (ii) can be proved similarly.

14.2.4   Summary

The theorems in this section explain all possible ways to determine which one of two components has the larger B-reliability importance value by means of either the permutation importance or the domination importance in conjunction with reliability orderings between the two components. These types of judgement are attractive because using them does not require exact component reliability values and avoids the possibly extensive computation of the reliability importance measures. The component having the smaller B-reliability importance can be removed from the set of potentially important components. Given the set of minimal paths and the particular ordering of reliability of components with tie allowed, Freixas and Pons (2008a) constructed an algorithm that uses these theorems to select a list of potentially important components.

14.3   Importance Measures for Some Special Systems

Summarizing the related analysis in the preceding chapters, this section presents the importance measures for several systems with special structures, including systems in which a component is in series or parallel with the rest of the system, k-out-of-n systems, series systems, and parallel systems.

Suppose that component i is in series with the rest of a system and that $j \neq i$. Considering the structure importance measures, $i \geq_{\text{pe}} j$ (Theorem 6.5.5), $i \geq_{\text{c}} j$ (Corollary 6.7.10), $i \geq_{\text{a}^p} j$ (Theorem 6.8.4), $i \geq_{\text{ex}} j$ (Theorem 6.13.1), and $i \geq_{\text{cp}} j$ (Theorem 6.9.5). Furthermore, letting $F_i(t) \geq F_j(t)$ for all $t \geq 0$, then $I_{\text{B}}(i; \overline{\mathbf{F}}(t)) \geq I_{\text{B}}(j; \overline{\mathbf{F}}(t))$ (Theorem 5.1.4), $I_{\text{BP}}(i; \overline{\mathbf{F}}) \geq I_{\text{BP}}(j; \overline{\mathbf{F}})$ (Theorem 5.3.9), $I_{\text{FV}^c}(i; \overline{\mathbf{F}}(t)) \geq I_{\text{FV}^c}(j; \overline{\mathbf{F}}(t))$ (Theorem 4.2.6), and $I_{\text{L}_1}(i; \overline{\mathbf{F}}) \geq I_{\text{L}_1}(j; \overline{\mathbf{F}})$ (Theorem 10.2.7). In addition, $I_{\text{Y}}(i; \overline{\mathbf{F}}) \geq I_{\text{Y}}(j; \overline{\mathbf{F}})$ under the condition $r_i(t) \geq r_j(t)$ for all $t \geq 0$ (Theorem 10.3.11(iv)). Note that when component i is in series with a module containing component j, these results still hold for the B-importance, BP importance, L_1 importance, and yield importance.

Suppose that component i is in parallel with the rest of a system and that $j \neq i$. Considering the structure importance measures, $i \geq_{\text{pe}} j$ (Theorem 6.5.5), $i \geq_{\text{p}} j$ (Corollary 6.7.13), $i \geq_{\text{a}^c} j$ (Theorem 6.8.4), $i \geq_{\text{in}} j$ (Theorem 6.13.1), and $i \geq_{\text{cp}} j$ (Theorem 6.9.5). Furthermore, letting $\overline{F}_i(t) \geq \overline{F}_j(t)$ for all $t \geq 0$, then $I_{\text{B}}(i; \overline{\mathbf{F}}(t)) \geq I_{\text{B}}(j; \overline{\mathbf{F}}(t))$ (Theorem 5.1.4), $I_{\text{BP}}(i; \overline{\mathbf{F}}) \geq I_{\text{BP}}(j; \overline{\mathbf{F}})$ (Theorem 5.3.9), $I_{\text{FV}^p}(i; \overline{\mathbf{F}}(t)) \geq I_{\text{FV}^p}(j; \overline{\mathbf{F}}(t))$ (Theorem 4.2.6), and $I_{\text{L}_1}(i; \overline{\mathbf{F}}) \geq I_{\text{L}_1}(j; \overline{\mathbf{F}})$ (Theorem 10.2.7). Note that when component i is in parallel with a module containing component j, these results still hold for the B-importance, BP importance, and L_1 importance.

For a k-out-of-n system, all components are structurally symmetric (see Example 2.2.9), and thus, equivalently important according to any type of structure importance measure except for the absoluteness and the domination importance measures (according to Corollary 6.14.2). For a series system (n-out-of-n system), $i =_{\text{a}^c} j$, and $i =_{\text{in}} j$; for a parallel system (1-out-of-n system), $i =_{\text{a}^p} j$, and $i =_{\text{ex}} j$; and for a general k-out-of-n system ($1 < k < n$), the components cannot be compared according to the c-absoluteness, p-absoluteness, internal domination, or external domination importance measure.

For a general k-out-of-n system ($1 < k < n$), the c-FV importance always assigns the largest value to the component of the lowest reliability, while the p-FV importance always assigns the largest value to the component of the highest reliability (see Example 6.13.9). For the other nonstructure importance measures, there are no such simple rules to follow, and more information is needed to determine the ranking of the importance of components in a general k-out-of-n system.

Table 14.2 summarizes the performance of reliability and lifetime importance measures in parallel and series systems. For parallel systems, "HR" represents the component of the highest reliability being the most important since the system is at least as strong as its most reliable component; "E" represents all components being equally important (to the value in parenthesis), since the system functions if any component functions. For series systems, "LR" represents the component of the lowest reliability being the most important, reflecting the principle that a chain is only as strong as its weakest link; "E" represents all components being equally important (to the value in parenthesis), since the failure of any component results in the failure of the system.

The term "N/A" in Table 14.2 means that conclusions cannot be drawn based solely on the knowledge of the relative ordering of component reliability. To make comparisons of

Table 14.2 Importance measures for parallel and series systems

Importance measure	Parallel system	Series system
B-importance, BP, L_1, L_5	HR	LR
B-importance for system functioning	E $(=1-R(\mathbf{p}))$	LR
Bayesian, criticality for system failure, c-FV	E$(=1)$	LR
RAW	E $(=1/R(\mathbf{p}))$	LR
L_4	E $(=1/n)$	LR
B-importance for system failure	HR	E $(=R(\mathbf{p}))$
criticality for system functioning, p-FV	HR	E $(=1)$
RRW	HR	E[a]
L_2	N/A	LR
L_3, L_6	N/A	N/A

[a]$I_{\text{RRW}}(i; \mathbf{p}) = R(\mathbf{p})/R(0_i, \mathbf{p})$ and denominator $R(0_i, \mathbf{p})$ is zero for series systems

component importance, other component reliability information is needed. Additionally, the DIM^I, redundancy importance, and yield importance are defined using other specified properties beyond the reliability of components, such as the changes of component reliability, the types of redundancy, and the yield function; thus, Table 14.2 does not include these importance measures.

14.4 Computation of Importance Measures

Importance measures can be calculated by means of analytical approaches and numerical methods, which are designed on the basis of the properties of particular importance measures as shown in the preceding as well as the following chapters. These methods might not be detailed in this book due to the limit of the scope of the book. Interested readers can find descriptions and examples in the original work referenced. In general, the calculations of reliability and lifetime importance measures for complex systems involve a large amount of computation and must sometimes rely on numerical approximations. For example, Aven (1986) established some bounds on the numerators of $I_{\text{BP}}(\cdot; \overline{\mathbf{F}}(t))$ in Equation (5.21) and of $I_{L_1}(\cdot; \overline{\mathbf{F}})$ in Equation (10.14) by restricting to a finite interval of time, although the original definition of $I_{L_1}(\cdot; \overline{\mathbf{F}})$ is for time from zero to infinity. For highly reliable systems, these bounds give good approximations when the lifetimes of components are assumed to be Weibull distributed as in Equation (2.17). Song and Der Kiureghian (2005) proposed a bounding method to approximate the B-importance, FV importance, RAW, and RRW by solving a linear programming that is modified or parameterized from the one used for estimating the systems unreliability.

Dutuit and Rauzy (2001) proposed an efficient methodology based on binary decision diagrams (BDD), and Chang et al. (2004) extended it to an algorithm based on ordered BDD to compute system failure frequency and the TDL importance measures, including the B-importance, criticality importance, RAW, and RRW for system risk evaluation. In addition, Chang et al. (2005) later dealt with the case of multistate systems (see Sections 15.2–15.5). Shrestha et al. (2010) presented an analytical method based on multistate multivalued decision diagrams for multistate component importance analysis, which is a natural extension of BDD for the multivalued case. The B-structure importance has been used in converting the fault tree structure into the BDD format by ranking the basic events of the tree (Bartlett and Andrews 2001). The ordering of the basic events is critical to the resulting size of the BDD, and ultimately affects the performance and benefits of this technique.

Another well known implementable method is simulation, especially when the form of systems reliability function does not exist. For example, Monte Carlo methods (Remark 5.3.5) have been used to calculate the BP lifetime importance for general systems with arbitrary lifetime distributions of components, especially for complex systems with a large number of components. Gertsbakh and Shpungin (2008) presented a Monte Carlo model based on systems reliability gradient vectors for evaluating the B-reliability importance of nonidentical components as well as a Monte Carlo model based on network combinatorial spectrum for the B-reliability importance of identical components. Marseguerra and Zio (2004) performed a reliability/availability analysis based on the DIM^I by means of the Monte Carlo simulation method. Zio et al. (2004) estimated the importance measures of multistate components by Monte Carlo simulation and tested the approach on a multistate transmission system of literature. Various Monte Carlo methods (e.g., Coyle et al. 2003; Zio et al. 2006, 2007) have been recently developed for computing various importance measures. Section 18.1 discusses the perturbation-analysis-based simulation for calculating the sensitivity measures and their related importance measures such as the DIM^I and TOI^O.

Most standard reliability and risk software tools, such as ReliaSoft (ReliaSoft 2003), SAPHIRE (Borgonovo and Smith 1999, 2000), and SIMLAB (Borgonovo 2007; SIMLAB 2007), produce importance measures (e.g., the B-importance, c-FV importance, RAW, and RRW) along with systems reliability.

References

Aven T. 1986. On the computation of certain measures of importance of system components. *Microelectronics and Reliability* **26**, 279–281.

Bartlett LM and Andrews JD. 2001. An ordering heuristic to develop the binary decision diagram based on structural importance. *Reliability Engineering and System Safety* **72**, 31–38.

Borgonovo E. 2007. A new uncertainty importance measure. *Reliability Engineering and System Safety* **92**, 771–784.

Borgonovo E and Smith C. 1999. A case study: Two components in parallel with epistemic uncertainty, Part I. *SAPHIRE Facets INEEL 1999*.

Borgonovo E and Smith C. 2000. A case study: Two components in parallel with epistemic uncertainty, Part II. *SAPHIRE Facets INEEL 2000*.

Chang YR, Amari SV, and Kuo SY. 2004. Computing system failure frequencies and reliability importance measures using OBDD. *IEEE Transactions on Computers* **53**, 54–68.

Chang YR, Amari SV, and Kuo SY. 2005. OBDD-based evaluation of reliability and importance measures for multistate systems subject to imperfect fault coverage. *IEEE Transactions on Dependable and Secure Computing* **2**, 336–347.

Coyle D, Buxton MJ, and O'Brien BJ. 2003. Measures of importance for economic analysis based on decision modeling. *Journal of Clinical Epidemiology* **56**, 989–997.

Dutuit Y and Rauzy A. 2001. Efficient algorithms to assess component and gate importance in fault tree analysis. *Reliability Engineering and System Safety* **72**, 213–222.

Freixas J and Pons M. 2008a. Identifying optimal components in a reliability system. *IEEE Transactions on Reliability* **57**, 163–170.

Freixas J and Pons M. 2008b. The influence of the node criticality relation on some measures of component importance. *Operations Research Letters* **36**, 557–560.

Gertsbakh I and Shpungin Y. 2008. Network reliability importance measures: Combinatorics and Monte Carlo based computations. *WSEAS Transactions on Computers* **7**, 216–227.

Kuo W and Zhu X. 2012. Relations and generalizations of importance measures in reliability. *IEEE Transactions on Reliability* **61**.

Marseguerra M and Zio E. 2004. Monte Carlo estimation of the differential importance measure: Application to the protection system of a nuclear reactor. *Reliability Engineering and System Safety* **86**, 11–24.

Meng FC. 1994. Comparing criticality of nodes via minimal cut (path) sets for coherent systems. *Probability in the Engineering and Informational Sciences* **8**, 79–87.

Meng FC. 1995. Some further results on ranking the importance of system components. *Reliability Engineering and System Safety* **47**, 97–101.

Meng FC. 2000. Relationships of Fussell-Vesely and Birnbaum importance to structural importance in coherent systems. *Reliability Engineering and System Safety* **67**, 55–60.

Meng FC. 2004. Comparing Birnbaum importance measure of system components. *Probability in the Engineering and Informational Sciences* **18**, 237–245.

ReliaSoft. 2003. Using reliability importance measures to guide component improvement efforts. Available at http://www.maintenanceworld.com/Articles/reliasoft/usingreliability.html.

Shrestha A, Xing L, and Coit DW. 2010. An efficient multistate multivalued decision diagram-based approach for multistate system sensitivity analysis. *IEEE Transactions on Reliability* **59**, 581–592.

SIMLAB. 2007. SIMLAB *Reference Manual* POLIS-JRC ISIS. Joint Research Center of the European Community.

Song J and Der Kiureghian A. 2005. Component importance measures by linear programming bounds on system reliability. *Proceedings of the 9th International Conference on Structural Safety and Reliability*, pp. 19–23.

Zio E, Marella M, and Podofillini L. 2007. Importance measures-based prioritization for improving the performance of multi-state systems: Application to the railway industry. *Reliability Engineering and System Safety* **92**, 1303–1314.

Zio E, Podofillini L, and Levitin G. 2004. Estimation of the importance measures of multi-state elements by Monte Carlo simulation. *Reliability Engineering and System Safety* **86**, 191–204.

Zio E, Podofillini L, and Zille V. 2006. A combination of Monte Carlo simulation and cellular automata for computing the availability of complex network systems. *Reliability Engineering and System Safety* **91**, 181–190.

15

Generalizations of Importance Measures

Chapters 2–14 are based on general assumptions about a system and its components, as presented in Section 2.14. These assumptions pertain to the coherence of the system, binary states of components and the system, the nonrepairable components and system, and so on. They hold true unless stated otherwise.

This chapter investigates importance measures when one or more of these assumptions in Section 2.14 are released. Importance measures are either generalizations of the ones introduced for binary coherent systems or new ones proposed for special components and systems. In particular, Section 15.1 discusses the importance measures for binary noncoherent systems, Sections 15.2 those for multistate coherent systems (MCS), Section 15.3 those for multistate monotone systems (MMS), Section 15.4 those for binary-type multistate monotone system (BTMMS), Section 15.6 those for continuum systems, and Section 15.7 those for repairable systems. Section 15.8 gives one application in the power industry, in which the system is repairable multistate, and system performance is measured as outage rate.

Section 15.5 summarizes and gives a comprehensive discussion of importance measures for multistate systems based on Sections 15.2–15.4. Kuo and Zuo (2003, Chapter 12) have already presented theory and reliability evaluation for multistate systems, especially for $Con/k/n$ systems. This chapter does not purport to study the theory of various extended systems and their reliability evaluation in depth; instead, this chapter focuses on importance measures for these systems.

15.1 Noncoherent Systems

A system is noncoherent because (i) its structure function is decreasing in terms of component states, and/or (ii) some components are irrelevant. Cause (i) implies that improvement of components may degrade the system performance. A typical noncoherent system is a k-to-ℓ-out-of-n:G system, which functions if and only if no fewer than k and no more than ℓ out of n components function ($1 \leq k \leq \ell \leq n$). Such a system is noncoherent because the systems reliability can decrease when the number of good components increases. Jain and

Importance Measures in Reliability, Risk, and Optimization: Principles and Applications, First Edition.
Way Kuo and Xiaoyan Zhu. © 2012 John Wiley & Sons, Ltd. Published 2012 by John Wiley & Sons, Ltd.

Gopal (1985), Pham (1991), and Upadhyaya and Pham (1993) gave some examples of the k-to-ℓ-out-of-n:G systems in communication, multiprocessor, and transportation systems. One example from Pham (1991) is presented as follows.

Example 15.1.1 A multiprocessor system shares resources, such as memory, I/O units, and buses, among n processors. If fewer than k processors are being used, the system does not function to its maximum capacity, and system efficiency is poor. On the other hand, if more than ℓ processors are being used, then efficiency is still poor due to the traffic congestion caused by having too many processors in use. Both of these cases need to be avoided; this can be accomplished by modeling the system as failed for these cases. This multiprocessor system is thus the k-to-ℓ-out-of-n:G system. Any system with limited resources, for example, a computer network or transportation system, can be modeled in a similar manner.

If a system satisfies the monotonicity requirement on structure function (i.e., nondecreasing) but contains some irrelevant component(s), then the system is specifically identified as a (binary) monotone system. Subsection 15.1.1 focuses on the importance measures for binary monotone systems.

On the importance measures for general noncoherent systems, the following gives a brief review rather than a comprehensive investigation, since most studies have only preliminary results. Early related research includes Andrews and Beeson (1983), Bossche (1987), Gandini (1990), and Zhang and Mei (1985). By utilizing the calculation procedure of Inagaki and Henley (1980), Andrews and Beeson (2003) recently extended the B-TDL importance to a noncoherent system by separately considering the contribution of component failure to system failure and the contribution of component repair to system failure. In their work, the B-importance in a noncoherent system is obtained by separate differentiations of the systems reliability function with respect to component reliability and unreliability, since the reliability of a noncoherent system is a function of both component reliability and unreliability. Along with this idea, Beeson and Andrews (2003) provided the extensions of the criticality importance, BP importance, FV importance, and an importance measure proposed by Lambert (1975a, 1975b) to a noncoherent system.

Lu and Jiang (2007) investigated the JRI and JFI in a noncoherent fault tree, which includes at least one NOT operation. They showed that Theorems 7.1.3, 7.1.4 (using the noncoherent B-reliability importance in Equations (7.5)–(7.8)) and 7.1.7 pertain to noncoherent systems and demonstrated the results in some real systems. Borgonovo (2010) defined the TOI in a unified way for both coherent and noncoherent systems. For a noncoherent system, the changes of both component reliability Δp_i and unreliability Δq_i are included in Equation (7.14) in defining the TOI. Theorem 7.3.3 for the TOI also pertains to noncoherent systems since it is induced from Theorem 2.10.3, which shows that the reliability functions of both coherent and noncoherent systems are multilinear that is the basic assumption of defining the TOI.

15.1.1 Binary Monotone Systems

In a coherent system, all components are relevant to the operation of the system. However, in a practical system, irrelevant components may exist and be important even though they never directly cause failure of the system. As an example in Natvig (1979), consider a condenser being in parallel with an electrical device in a large engine. The task of the condenser is to prevent high voltage from destroying the electrical device. Hence, although being irrelevant, the condenser can be very important in increasing the lifetime of the device, and consequently, the lifetime of the entire engine.

Definition 15.1.2 *A system* (N, ϕ) *is a monotone (semicoherent) system if (i)* $\phi(\mathbf{0}) = 0$ *and* $\phi(\mathbf{1}) = 1$, *and (ii)* $\phi(\mathbf{x}) \leq \phi(\mathbf{y})$ *for all* $\mathbf{x} \leq \mathbf{y}$.

A monotone system (also known as a semicoherent system) satisfies the monotonicity requirement for the structure function. The monotone system in Definition 15.1.2 was initially called a "coherent system" in Birnbaum (1969), but it is different with the coherent system in Definition 2.3.2, which is the one that has been commonly recognized. Note that a coherent system in Definition 2.3.2 is a special monotone system; however, a monotone system may not be coherent because the monotone system may contain irrelevant components. For an example of a two-component system where $\phi(1, 1) = 1$, $\phi(0, 1) = 1$, $\phi(1, 0) = 0$, and $\phi(0, 0) = 0$, the system is monotone but not coherent since component 1 is irrelevant. The monotone system is coherent if it does not have any irrelevant components.

In a monotone system, the B-importance can be defined exactly the same as in a coherent system. Then, the following theorem holds for irrelevant component i because $\phi(1_i, \mathbf{x}) = \phi(0_i, \mathbf{x})$ for all vectors $(\cdot_i, \mathbf{x})$.

Theorem 15.1.3 *In a monotone system, if component i is irrelevant,* $I_{\mathrm{B}}(i, \overline{\mathbf{F}}(t)) = 0$ *for* $t \geq 0$.

Due to Theorem 15.1.3, any importance measure that is either proportional to or a weighted average of the B-importance is zero for the irrelevant components; such importance measures are listed in Table 14.1. Apart from those, the FV importance of irrelevant component is also zero, but the Bayesian importance of irrelevant component i is q_i and the RAW and RRW of irrelevant components are one, achieving their smallest values. That is, the irrelevant components are the least important.

Considering structure importance measures, for example, the cut-path importance, by Definition 6.9.1, if component i is irrelevant, then for every cut $\{i\} \cup S$ where $j \notin S$, $\{j\} \cup S$ is also a cut since S is. Therefore, $|\mathscr{C}_i(d)| \leq |\mathscr{C}_j(d)|$ for all d, and component i is the least important. Similarly, it can be verified that the irrelevant component is the least important in a binary monotone system according to the min-cut importance, min-path importance, permutation importance, cut importance, and path importance. But the c-absoluteness and p-absoluteness importance measures cannot evaluate irrelevant components. For the internal and external domination importance measures, Freixas and Pons (2008) showed a characteristic of irrelevant components as follows. Note that by Theorem 6.6.4, for two distinct relevant components i and j, $i \geq_{\mathrm{in}} j$ and $i \geq_{\mathrm{ex}} j$ cannot coexist.

Theorem 15.1.4 *Component i is irrelevant if and only if $j \geq_{\mathrm{in}} i$ and $j \geq_{\mathrm{ex}} i$ for any component $j \neq i$.*

Proof. Using Definition 6.6.1, the relations $j \geq_{\mathrm{in}} i$ and $j \geq_{\mathrm{ex}} i$ hold for all $j \neq i$ if and only if $i \in S \in \mathscr{P}$ implies $S \setminus \{i\} \in \mathscr{P}$.

15.2 Multistate Coherent Systems

The importance measures introduced in the preceding chapters are for binary systems in which the system and its components are either functioning or failed. In most cases, binary models are adequate. However, they may not be sufficient to represent some situations such as manufacturing production, power generation, and transportation systems. In these situations, some components may experience several *degraded* states before they completely fail, and consequently the system may function at less than full capacity. Thus, multistate models, in which both the system and its components may be in any of *a finite number of states*, become necessary.

Let $S = \{0, 1, \ldots, K\}$ and $S^n = \{\mathbf{x} \in \mathbb{R}^n : x_i \in S\}$ where $K \geq 2$. Then the state, ϕ, of the general $(K + 1)$-state system is given by the structure function $\phi = \phi(\mathbf{x})$, $\mathbf{x} \in S^n$ and could be in any state of $\{0, 1, \ldots, K\}$, ranging from the complete failure state 0 up to perfect functioning state K. Butler (1979) defined a relevant component and an MCS as follows.

Definition 15.2.1 *Component i is relevant if $\phi(K_i, \mathbf{x}) \neq \phi(0_i, \mathbf{x})$ for some $\mathbf{x} \in S^n$. Otherwise, component i is irrelevant. Component i is fully relevant if for each $\mu = 1, 2, \ldots, K, \phi(\mu_i, \mathbf{x}) \neq \phi((\mu - 1)_i, \mathbf{x})$ for some $\mathbf{x} \in S^n$.*

Definition 15.2.2 *A system (N, ϕ) is a multistate coherent system (MCS) if (i) $\phi(\mathbf{0}) = 0$ and $\phi(\mathbf{K}) = K$ ($\mathbf{K} = (K, K, \ldots, K)$), (ii) $\phi(\mathbf{x})$ is nondecreasing in $\mathbf{x}$, and (iii) each component is relevant.*

If a component is not fully relevant, then only less than $K + 1$ states are required to describe its status. Such components are permissible in an MCS to allow for a mixture of components with states of $K + 1$ or less.

Define the matrix $\mathbf{P} = [p_{i\mu}]$ by

$$p_{i\mu} = \Pr\{\text{component } i \text{ is in state } \mu\}, \quad i = 1, 2, \ldots, n, \quad \mu = 0, 1, \ldots, K,$$

and $\sum_{\mu=0}^{K} p_{i\mu} = 1$. Let $(\kappa_j, \mathbf{P})$ denote the matrix whose (i, μ)th entry is given by

$$(\kappa_j, \mathbf{P})_{i\mu} = \begin{cases} p_{i\mu} & i \neq j \\ 1 & i = j, \mu = \kappa \\ 0 & i = j, \mu \neq \kappa. \end{cases}$$

In the following, whenever a vector $\mathbf{p} \in \mathbb{R}^{(K+1)}$ appears in an expression normally involving the matrix $\mathbf{P}$, $\mathbf{P}$ should be understood to be the matrix of all whose rows are equal to $\mathbf{p}$.

The reliability function of the system with respect to system state η, $R(\mathbf{P}; \eta)$, is defined as

$$R(\mathbf{P}; \eta) = \Pr\{\phi(\mathbf{X}) \geq \eta\}, \quad \eta = 1, 2, \ldots, K.$$

This generalized reliability function is to consider the system whose components have several states but whose structure function has two effective states ($\geq \eta$ or $< \eta$). *All subsequent definitions and results in this section are for a fixed value of η. For simplification, the dependence of $R(\cdot; \eta)$ upon η is suppressed in the notation.*

Given an MCS (N, ϕ) and a partition $\mathcal{C} = (N_0, N_1, \ldots, N_K)$ of N into $K + 1$ sets, define vector $\mathbf{x}(\mathcal{C}) \in S^n$ as

$$(\mathbf{x}(\mathcal{C}))_i = \begin{cases} 0 & i \in N_0 \\ 1 & i \in N_1 \\ \vdots & \vdots \\ K & i \in N_K. \end{cases}$$

The function $\mathbf{x}(\mathcal{C})$ shows how a partition $\mathcal{C}$ determines the states of all components. On the basis of that, Butler (1979) defined a cut and a minimal cut in an MCS as follows.

Definition 15.2.3 *A partition $\mathcal{C} = (N_0, N_1, \ldots, N_K)$ of N is a cut in an MCS (N, ϕ) if $\phi(\mathbf{x}(\mathcal{C})) < \eta$. A cut $\mathcal{C}$ is minimal if $\phi(\mathbf{y}) \geq \eta$ for all $\mathbf{y} \in S^n$ such that $\mathbf{y} \geq \mathbf{x}(\mathcal{C}), \mathbf{y} \neq \mathbf{x}(\mathcal{C})$.*

15.2.1 The μ, ν B-importance

For the MCS, Butler (1979) studied the B-importance with respect to any two distinct component states, as presented in this subsection.

Definition 15.2.4 *A vector $(\cdot_i, \mathbf{x}) \in S^n$ is μ, ν critical for component i in an MCS (N, ϕ) if $\phi(\mu_i, \mathbf{x}) \geq \eta$ and $\phi(\nu_i, \mathbf{x}) < \eta$, $0 \leq \nu < \mu \leq K$, given that the system function has two effective states: $\geq \eta$ or $< \eta$. That is, under $(\cdot_i, \mathbf{x})$, the system is in state η or better, given that component i is in state μ, and the system is in state worse than η, given that component i is in state ν.*

Definition 15.2.5 *The μ, ν B-reliability importance of component i in an MCS, denoted by $I_B^{\mu,\nu}(i; \mathbf{P})$, is defined as the probability that the system is in state η or better, given that component i is in state μ minus the probability that the system is in state η or better, given that component i is in state ν. Mathematically,*

$$I_B^{\mu,\nu}(i; \mathbf{P}) = R(\mu_i, \mathbf{P}) - R(\nu_i, \mathbf{P}),$$

where $0 \leq \nu < \mu \leq K$. The $K, 0$ B-reliability importance is sometimes simply called the B-reliability importance and is denoted by $I_B(i; \mathbf{P})$.

Definition 15.2.6 *Let $m_\eta^{\mu,\nu}(i) = |\{(\cdot_i, \mathbf{x}) \in S^n : \mathbf{x} \text{ is } \mu, \nu \text{ critical for component } i\}|$, given that the system has two effective states: $\geq \eta$ or $< \eta$, $0 \leq \nu < \mu \leq K$. The μ, ν B-structure importance of component i in an MCS, denoted by $I_B^{\mu,\nu}(i; \phi)$, is defined as*

$$I_B^{\mu,\nu}(i; \phi) = (K + 1)^{-(n-1)} m_\eta^{\mu,\nu}(i).$$

Theorem 15.2.7

(i) $I_B^{\mu,\nu}(i; \mathbf{P}) = I_B^{\mu,\kappa}(i; \mathbf{P}) + I_B^{\kappa,\nu}(i; \mathbf{P})$ for $0 \le \nu < \kappa < \mu \le K$.

(ii) $I_B^{\mu,\nu}(i; \phi) = I_B^{\mu,\kappa}(i; \phi) + I_B^{\kappa,\nu}(i; \phi)$ for $0 \le \nu < \kappa < \mu \le K$.

(iii) $I_B^{\mu,\nu}(i; \phi) = I_B^{\mu,\nu}\left(i; \left(\frac{1}{K+1}, \frac{1}{K+1}, \dots, \frac{1}{K+1}\right)\right)$ for $0 \le \nu < \mu \le K$.

Proof. The proofs of parts (i) and (ii) are trivial. To prove part (iii), note that by summing over the $(K+1)^{n-1}$ possible values for $(\cdot_i, \mathbf{x})$,

$$R\left(\kappa_i, \left(\tfrac{1}{K+1}, \tfrac{1}{K+1}, \dots, \tfrac{1}{K+1}\right)\right) = (K+1)^{-(n-1)} |\{(\kappa_i, \mathbf{x}) \in S^n : \phi(\kappa_i, \mathbf{x}) \ge \eta\}|$$

Thus,

$$I_B^{\mu,\nu}\left(i; \left(\tfrac{1}{K+1}, \tfrac{1}{K+1}, \dots, \tfrac{1}{K+1}\right)\right)$$
$$= (K+1)^{-(n-1)} \left[|\{(\mu_i, \mathbf{x}) \in S^n : \phi(\mu_i, \mathbf{x}) \ge \eta\}| - |\{(\nu_i, \mathbf{x}) \in S^n : \phi(\nu_i, \mathbf{x}) \ge \eta\}| \right]$$
$$= (K+1)^{-(n-1)} m_\eta^{\mu,\nu}(i) = I_B^{\mu,\nu}(i; \phi).$$

As shown in Theorem 15.2.7(i) and (ii), the μ, ν B-importance can decompose into the sum of the μ, κ, and κ, ν B-importance measures for $\mu > \kappa > \nu$. The generalized cut importance, to be defined in the next subsection, has a similar property. In practice, it is likely that the K, 0 measure would be the most commonly used, while the measures at other states can be useful in providing more detailed information about which states are the most relevant in determining a given component's ranking (see Example 15.2.11).

15.2.2   The μ, ν Cut Importance

While in principle it is possible to develop a complete cut importance for an MCS, in practice the calculation of the entire generalized $\mathbf{b}_i$ vector for each component i is too complex to be feasible (see Definition 6.7.1 for vector $\mathbf{b}_i$ in a binary coherent system). However, a partial ranking of the components involving few calculations can be developed by generalizing Theorem 6.7.9 appropriately. To present this generalization (Butler 1979), first define the size of a partition.

Definition 15.2.8   *The size of a partition* $C = (N_0, N_1, \dots, N_K)$, *denoted by* $z(C)$, *is* $\sum_{\mu=0}^{K-1} w_\mu |N_\mu|$, *where* w_μ, $\mu = 0, 1, \dots, K-1$, *are arbitrary constants satisfying* $w_0 > w_1 > \dots > w_{K-1} > 0$.

Definition 15.2.9   *Consider an MCS with minimal cuts* $C^k = \left(N_0^k, N_1^k, \dots, N_K^k\right)$, $k = 1, 2, \dots, u$, *where* u *is the number of minimal cuts (with respect to a fixed value of* η). *For each component* i, *let*

$$e_i^{\mu+1,\mu} = \min_{1 \le k \le u} \{z(C^k) : i \in N_\mu^k\} - w_\mu$$

Table 15.1 States of the system

$\phi(0,0) = 0$	$\phi(1,0) = 1$	$\phi(2,0) = 2$
$\phi(0,1) = 0$	$\phi(1,1) = 1$	$\phi(2,1) = 2$
$\phi(0,2) = 1$	$\phi(1,2) = 2$	$\phi(2,2) = 2$

and

$$\ell_i^{\mu+1,\mu} = \left| \{ C^k : i \in N_\mu^k, z(C^k) - w_\mu = e_i^{\mu+1,\mu}, 1 \le k \le u \} \right|.$$

(By convention $e_i^{\mu+1,\mu} = +\infty$ if $i \notin N_\mu^k$ for all $1 \le k \le u$; $e_i^{\mu+1,\mu}$ is just the size of the "smallest" minimal cut which contains component i in the μth set of the partition less w_μ, and $\ell_i^{\mu+1,\mu}$ is the number of such minimal cuts.) For $0 \le v < \mu \le K$, define

$$e_i^{\mu,v} = \min_{v \le \kappa < \mu} \{ e_i^{\kappa+1,\kappa} \}$$

and

$$\ell_i^{\mu,v} = \min_{\substack{v \le \kappa < \mu \\ e_i^{\kappa+1,\kappa} = e_i^{\mu,v}}} \{ \ell_i^{\kappa+1,\kappa} \}.$$

Component i is more μ, v cut important than component j, denoted by $i >_c^{\mu,v} j$, if either (i) $e_i^{\mu,v} < e_j^{\mu,v}$ or (ii) $e_i^{\mu,v} = e_j^{\mu,v}$ and $\ell_i^{\mu,v} > \ell_j^{\mu,v}$. The $K, 0$ cut importance is sometimes simply called the cut importance and denoted by $>_c$.

As in the binary case (Theorem 6.7.4), the μ, v cut importance is consistent with the ranking induced by the corresponding μ, v B-importance as the reliability of components in state K approaches one. The next theorem shows a special case in this regard.

Theorem 15.2.10 *Let $K = 2$ and $\mathbf{p}(\varepsilon) = (\varepsilon^{w_0}, \varepsilon^{w_1}, 1 - \varepsilon^{w_0} - \varepsilon^{w_1})$. For ε sufficiently close to zero, the component ranking induced by $I_B^{\mu,v}(\cdot; \mathbf{p}(\varepsilon))$ is consistent with the ranking of the μ, v cut importance, $0 \le v < \mu \le K$.*

The proof of Theorem 15.2.10 is lengthy and complicated; it is, thus, omitted to conserve space. Interested readers can find the proof in Butler (1979). For the special case in Theorem 15.2.10, as ε approaches 0, $\mathbf{p}(\varepsilon)$ puts almost all of its mass on the best state, K. Of the mass left over, the ratio of the mass put on state 0 to that put on state 1 approaches zero. Thus, the parameters w_0 and w_1 give the relative weights put on components in state 0 versus components in state 1 in the cut importance ranking, and they also determine the relative likelihood of a component fully failing (state 0) or partially failing (state 1). The following example (Butler 1979) calculates the μ, v B-importance and μ, v cut importance for an MCS.

Example 15.2.11 Table 15.1 shows the states of a ternary system ($K = 2$) consisting of two components and with $\eta = 2$, $w_0 = 2$, and $w_1 = 1$. Furthermore, $\mathbf{p}(\varepsilon) = (\varepsilon^2, \varepsilon, 1 - \varepsilon^2 - \varepsilon)$.
   First, to calculate the μ, v B-importance,

$$R(P) = 1 - p_{10}p_{20} - p_{10}p_{21} - p_{11}p_{20} - p_{11}p_{21} - p_{10}p_{22} = 1 - p_{10} - p_{11}(p_{20} + p_{21}).$$

Then,

$$I_B^{2,1}(1; \mathbf{p}(\varepsilon)) = \varepsilon^2 + \varepsilon, \quad I_B^{1,0}(1; \mathbf{p}(\varepsilon)) = 1 - \varepsilon - \varepsilon^2, \quad I_B^{2,0}(1; \mathbf{p}(\varepsilon)) = 1;$$

$$I_B^{2,1}(2; \mathbf{p}(\varepsilon)) = \varepsilon, \quad\quad I_B^{1,0}(2; \mathbf{p}(\varepsilon)) = 0, \quad\quad\quad I_B^{2,0}(2; \mathbf{p}(\varepsilon)) = \varepsilon.$$

Because the minimal cuts are $C^1 = (\emptyset, \{1, 2\}, \emptyset)$ and $C^2 = (\{1\}, \emptyset, \{2\})$, the cut importance can be calculated as follows:

$$e_1^{2,1} = 1, \quad \ell_1^{2,1} = 1, \quad e_1^{1,0} = 0, \quad\quad \ell_1^{1,0} = 1, \quad e_1^{2,0} = 0, \quad \ell_1^{2,0} = 1;$$

$$e_2^{2,1} = 1, \quad \ell_2^{2,1} = 1, \quad e_2^{1,0} = +\infty, \quad \ell_2^{1,0} = 0, \quad e_2^{2,0} = 1, \quad \ell_2^{2,0} = 1.$$

Thus, $1 >_c^{1,0} 2$, and $1 >_c^{2,0} 2$. However, components 1 and 2 are not comparable under the 2, 1 cut importance. Hence, component 1 is more important than component 2 in an overall sense (i.e., according to the 2, 0 cut importance). Moreover, the 2, 1 and 1, 0 cut importance measures show that it is the state 1 to state 0 transaction of the components that determines the 2, 0 cut importance.

As is the case for binary coherent systems, analogous results based on minimal paths can be developed for an MCS composed of very unreliable components.

15.3   Multistate Monotone Systems

An MMS is a generalization of a binary monotone system. Continue to use the notation from Section 15.2. Let $S = \{0, 1, ..., K\}$ represent the state space of a multistate system and its components and $P = (\mathbf{p}_1, \mathbf{p}_2, \ldots, \mathbf{p}_n)$, where $\mathbf{p}_i = (p_{i0}, p_{i1}, \ldots, p_{iK})$ is the reliability vector of component i, $i = 1, 2, \ldots, n$. According to Natvig (1985), an MMS can be defined as follows.

Definition 15.3.1   *A system (N, ϕ) is a multistate monotone system (MMS) if (i) $\phi(\mathbf{0}) = 0$ and $\phi(\mathbf{K}) = K$, and (ii) $\phi(\mathbf{x})$ is nondecreasing in $\mathbf{x}$.*

The MCS is a special MMS, but the MMS may not be coherent since it may contain some irrelevant components. The relationship between the MMS and MCS is the same as that between the binary monotone system and binary coherent system.

As discussed following Definition 15.2.2, the set of states of components may be different from each other and different from the state space of the system. For the sake of notation simplification, this section assumes that the state spaces for all components and the system are the same as $S = \{0, 1, \ldots, K\}$ unless stated otherwise. Although this assumption is made, most of the definitions and results in this section allow differences in the state spaces for the system and its components.

For a multistate system, Griffith (1980) proposed the following performance utility function:

$$U = \sum_{\kappa=1}^{K} a_\kappa \Pr\{\phi(\mathbf{X}) = \kappa\}, \tag{15.1}$$

where a_κ is the utility level of the system when it is at state κ, and $a_K \geq a_{K-1} \geq \cdots \geq a_1 \geq a_0 = 0$. Note that when $a_\kappa = \kappa$ for $\kappa = 0, 1, \ldots, K$, the system performance utility is just the expected system state, that is, $U = \mathbb{E}(\phi(\mathbf{X}))$. Such defined system performance utility can represent satisfied demand, expected maintenance cost but possibly with the reversed order of a_κ, and so on.

Subsection 15.3.1 extends the permutation importance to evaluate the importance of positions in an MMS. Then, Subsections 15.3.2, 15.3.3, and 15.3.4 present the importance measures with respect to the performance utility function in Equation (15.1).

15.3.1   The Permutation Importance

Definition 15.3.2   *Component i is more permutation important than component j in an MMS if $\phi(\mu_i, \nu_j, \mathbf{x}^{(ij)}) \geq \phi(\nu_i, \mu_j, \mathbf{x}^{(ij)})$ for all $0 \leq \nu < \mu \leq K$ and all $\mathbf{x}^{(ij)}$, and strict inequality holds for some $\nu < \mu$ and $\mathbf{x}^{(ij)}$.*

For the permutation importance in an MMS, Meng (1996) proved Theorem 15.3.3 that provides a principle for pairwise exchange of components to improve systems reliability. In Theorem 15.3.3, let X and Y be two discrete random variables taking values on S with probability mass functions $\mathbf{f}^X$ and $\mathbf{f}^Y$, respectively. Notation $(\mathbf{f}^X{}_i, \mathbf{f}^Y{}_j, \mathbf{P}^{(ij)})$ denotes the reliability matrix with the reliability vector $\mathbf{f}^X$ in row i, the reliability vector $\mathbf{f}^Y$ in row j, and the same reliability vectors as $\mathbf{P}$ in the other $(n-2)$ rows; $(\mathbf{f}^Y{}_i, \mathbf{f}^X{}_j, \mathbf{P}^{(ij)})$ is defined similarly. Theorem 15.3.3 corresponds to Theorem 6.5.13, which pertains to the permutation importance in a binary system.

Theorem 15.3.3   *In an MMS, component i is more permutation important than component j if and only if $R(\mathbf{f}^X{}_i, \mathbf{f}^Y{}_j, \mathbf{P}^{(ij)}; \eta) \geq R(\mathbf{f}^Y{}_i, \mathbf{f}^X{}_j, \mathbf{P}^{(ij)}; \eta)$ for all $\mathbf{f}^X \geq^{lr} \mathbf{f}^Y$ (see Section 2.8), all states $\eta \geq 1$, and all $\mathbf{P}^{(ij)}$, with strict inequality for some $\mathbf{f}^X \geq^{lr} \mathbf{f}^Y$, some state $\eta \geq 1$, and some $\mathbf{P}^{(ij)}$.*

The following example from Meng (1996) illustrates the permutation importance in an MMS.

Example 15.3.4   Consider an MMS composed of n positions where each position is occupied by a module. Furthermore, suppose that each one of the n modules is composed of K i.i.d. binary components and that the state of a module is represented by the number of functioning components within it. The structure function of the system, denoted by ϕ, is a mapping from $\{0, 1, \ldots, K\}^n$ to $\{0, 1, \ldots, \overline{K}\}$, where the perfect system state $\overline{K}$ can be any nonzero positive integer in general. Let g_i be the reliability of the components within the ith module, $i = 1, 2, \ldots, n$. Then the reliability vector of the ith module $\mathbf{p}_i = (p_{i0}, p_{i1}, \ldots, p_{iK})$ follows the binomial density, that is, $p_{i\mu} = \binom{K}{\mu} g_i^\mu (1 - g_i)^{K-\mu}$, $\mu = 0, 1, \ldots, K$, $i = 1, 2, \ldots, n$. It is easily verified that $\mathbf{p}_i \geq^{lr} \mathbf{p}_j$ holds if and only if $g_i \geq g_j$. Suppose that the goal is to maximize the probability (at a given time) that the system performs at state η or better ($1 \leq \eta \leq \overline{K}$), that is, to maximize $\Pr\{\phi(\mathbf{X}) \geq \eta\}$. It then follows from Theorem 15.3.3 that if $g_i > g_j$ and position i is more permutation important than position j, then the ith module should be allocated to position i rather than position j.

15.3.2 The Utility-decomposition Reliability Importance

Wu and Chan (2003) proposed a utility-decomposition importance as follows, which evaluates the contribution of the particular states of a component to the system performance utility defined in Equation (15.1).

Definition 15.3.5 *The utility-decomposition reliability importance of state μ of component i with respect to the performance utility function (15.1) in an MMS, denoted by $I_{U\text{-}D}^{\mu}(i; \boldsymbol{P})$, for $\mu = 0, 1, \ldots, K$, is defined as*

$$I_{U\text{-}D}^{\mu}(i; \boldsymbol{P}) = \sum_{\kappa=0}^{K} a_{\kappa} \Pr\{\phi(\mathbf{X}) = \kappa, X_i = \mu\}$$

$$= \sum_{\kappa=0}^{K} a_{\kappa} \Pr\{\phi(\mathbf{X}) = \kappa | X_i = \mu\} \Pr\{\phi(X_i) = \mu\}$$

$$= p_{i\mu} \sum_{\kappa=0}^{K} a_{\kappa} \Pr\{\phi(\mu_i, \mathbf{X}) = \kappa\}.$$

The significance of $I_{U\text{-}D}^{\mu}(i; \boldsymbol{P})$ is that for each component i, the system performance utility function U can be expressed in terms of $I_{U\text{-}D}^{\mu}(i; \boldsymbol{P})$ for $\mu = 0, 1, \ldots, K$:

$$U = \sum_{\kappa=0}^{K} a_{\kappa} \Pr\{\phi(\mathbf{X}) = \kappa\}$$

$$= \sum_{\kappa=0}^{K} a_{\kappa} \sum_{\mu=0}^{K} \Pr\{\phi(\mathbf{X}) = \kappa | X_i = \mu\} \Pr\{\phi(X_i) = \mu\}$$

$$= \sum_{\mu=0}^{K} \sum_{\kappa=0}^{K} a_{\kappa} \Pr\{\phi(\mu_i, \mathbf{X}) = \kappa\} p_{i\mu}$$

$$= \sum_{\mu=0}^{K} I_{U\text{-}D}^{\mu}(i; \boldsymbol{P}).$$

This shows that a state μ of a component i with larger $I_{U\text{-}D}^{\mu}(i; \boldsymbol{P})$ contributes more to the system performance utility.

The following theorem (Wu and Chan 2003) gives expressions of $I_{U\text{-}D}^{\mu}(i; \boldsymbol{P})$ for a series and a parallel system.

Theorem 15.3.6 *In a series system whose structure function $\phi(\mathbf{x}) = \min\{x_1, x_2, \ldots, x_n\}$,*

$$I_{\text{U-D}}^{\mu}(i; \mathbf{P}) = \begin{cases} p_{i\mu}\left[\sum_{\kappa=0}^{\mu-1} a_\kappa \left(\prod_{\substack{j=1 \\ j\neq i}}^{n} \Pr\{X_j \geq \kappa\} - \prod_{\substack{j=1 \\ j\neq i}}^{n} \Pr\{X_j \geq \kappa+1\}\right)\right. \\ \left. + a_\mu \prod_{\substack{j=1 \\ j\neq i}}^{n} \Pr\{X_j \geq \mu\}\right] & \mu = 1, 2, \ldots, K \\ a_0\, p_{i0} & \mu = 0. \end{cases}$$

In a parallel system whose structure function $\phi(\mathbf{x}) = \max\{x_1, x_2, \ldots, x_n\}$,

$$I_{\text{U-D}}^{\mu}(i; \mathbf{P}) = \begin{cases} p_{i\mu}\left[\sum_{\kappa=\mu+1}^{K} a_\kappa \left(\prod_{\substack{j=1 \\ j\neq i}}^{n} \Pr\{X_j < \kappa+1\} - \prod_{\substack{j=1 \\ j\neq i}}^{n} \Pr\{X_j < \kappa\}\right)\right. \\ \left. + a_\mu \prod_{\substack{j=1 \\ j\neq i}}^{n} \Pr\{X_j < \mu+1\}\right] & \mu = 0, 1, \ldots, K-1 \\ a_K\, p_{iK} & \mu = K. \end{cases}$$

15.3.3   The Utility B-reliability Importance

With respect to the performance utility function (15.1), Griffith (1980) generalized the B-reliability importance as follows, which is referred to as the utility B-reliability importance and can be interpreted as the change of the system performance utility caused by the deterioration of a particular component from state μ to state $\mu - 1$.

Definition 15.3.7 *The utility B-reliability importance of state μ of component i with respect to the performance utility function (15.1) in an MMS, denoted by $I_{\text{U-B}}^{\mu}(i; \mathbf{P})$, for $\mu = 1, 2, \ldots, K$, is defined as*

$$I_{\text{U-B}}^{\mu}(i; \mathbf{P}) = \sum_{\kappa=1}^{K} (a_\kappa - a_{\kappa-1})\left(\Pr\{\phi(\mu_i, \mathbf{X}) \geq \kappa\} - \Pr\{\phi((\mu-1)_i, \mathbf{X}) \geq \kappa\}\right). \tag{15.2}$$

When $a_\kappa = \kappa$ for $\kappa = 0, 1, \ldots, K$, that is, $U = \mathbb{E}(\phi(\mathbf{X}))$, the utility B-reliability importance of state μ of component i is

$$I_{\text{U-B}}^{\mu}(i; \mathbf{P}) = \mathbb{E}\left(\phi(\mu_i, \mathbf{X}) - \phi((\mu-1)_i, \mathbf{X})\right).$$

It preserves some properties of the B-reliability importance. For example, the vector of importance $\left(I_{\text{U-B}}^{1}(i; \mathbf{P}), I_{\text{U-B}}^{2}(i; \mathbf{P}), \ldots, I_{\text{U-B}}^{K}(i; \mathbf{P})\right)$ is the gradient of the systems reliability $(\mathbb{E}(\phi(\mathbf{X})))$ with respect to $\mathbf{p}_i$, when $\mathbf{p}_j$ are fixed for all $j \neq i$, thus preserving the chain rule property. Theorem 4.1.13 states the chain rule of the B-importance in a binary case. Bueno (1989) presented a similar chain rule of the utility B-reliability importance in a multistate case, as shown in Theorem 15.3.6.

Theorem 15.3.8 *Let an MMS (N, ϕ) have a module (M, χ). Assume that $a_\kappa = \kappa$ for $\kappa = 0, 1, \ldots, K$ in the performance utility function (15.1). For $i \in M$,*

$$I_{\text{U-B}}^{\mu}(i; \mathbf{P}, \phi) = \sum_{\kappa=1}^{K} I_{\text{U-B}}^{M,\kappa}(M; \mathbf{P}^{M^c}, \psi)\left[\Pr\{\chi(\mu_i, \mathbf{X}^M) \geq \kappa\} - \Pr\{\chi((\mu-1)_i, \mathbf{X}^M) \geq \kappa\}\right],$$

where ψ is the organizing structure, $\boldsymbol{P}^{M^c}$ is the vector with rows $j \in M^c$ $(M^c = N \setminus M)$, $I_{\text{U-B}}^{M,\kappa}(M; \boldsymbol{P}^{M^c}, \psi)$ denotes the utility B-reliability importance of state κ of modular set M for the system ψ, and $\mathbf{X}^M$ is the vector with rows $j \in M$.

Note that the relationship between $I_{\text{U-B}}^{\mu}(i; \boldsymbol{P})$ in Equation (15.2) and $I_{\text{U-D}}^{\mu}(i; \boldsymbol{P})$ in Definition 15.3.5 is specified as:

$$I_{\text{U-B}}^{\mu}(i; \boldsymbol{P}) = \frac{\partial I_{\text{U-D}}^{\mu}(i; \boldsymbol{P})}{\partial p_{i\mu}} - \frac{\partial I_{\text{U-D}}^{\mu-1}(i; \boldsymbol{P})}{\partial p_{i,\mu-1}}, \quad \mu = 1, 2, \ldots, K.$$

15.3.4 The Utility B-structure Importance, Joint Structure Importance, and Joint Reliability Importance

Consider an MMS where a component can only degrade one state each time (i.e., from states μ to $\mu - 1$), whereas the system may degrade more than one state at a time. For this MMS, Wu (2005) defined a critical path vector and, based on that, proposed a generalized B-structure importance, joint structure importance (JSI), and JRI with respect to the system performance utility function (15.1), as presented in the following text.

Definition 15.3.9 *A vector $(\mu_i, \mathbf{x}) \in S^n$ is a critical path vector for state μ of component i in an MMS (N, ϕ) given that the system is at state η, if $\phi(\mu_i, \mathbf{x}) = \eta$ and $\phi((\mu - 1)_i, \mathbf{x}) \leq \eta - 1$.*

Definition 15.3.10 *The utility B-structure importance of component i with respect to the performance utility function (15.1) in an MMS, denoted by $I_{\text{U-B}}(i; \phi)$, is defined as*

$$I_{\text{U-B}}(i; \phi) = \frac{\sum_{\mu=1}^{K} \sum_{\eta=1}^{K} \sum_{\kappa=1}^{\eta} (a_\eta - a_{\eta-\kappa}) |\{(\cdot_i, \mathbf{x}) : \phi(\mu_i, \mathbf{x}) = \eta \text{ and } \phi((\mu - 1)_i, \mathbf{x}) = \eta - \kappa\}|}{(K + 1)^{n-1}}.$$

Definition 15.3.11 *The JSI of components i and j, $i \neq j$ with respect to the performance utility function (15.1) in an MMS, denoted by $I_{\text{U-JS}}(i, j; \phi)$, is defined as*

$$I_{\text{U-JS}}(i, j; \phi) = \sum_{\mu=1}^{K} \sum_{\nu=1}^{K} \left(I_{\text{U-JS}}^{\mu,\nu}(i, j; \phi) - I_{\text{U-JS}}^{\mu,\nu-1}(i, j; \phi) \right)$$

$$= \sum_{\mu=1}^{K} \left(I_{\text{U-JS}}^{\mu,K}(i, j; \phi) - I_{\text{U-JS}}^{\mu,0}(i, j; \phi) \right),$$

where $I_{\text{U-JS}}^{\mu,\nu}(i, j; \phi) =$

$$\frac{\sum_{\eta=1}^{K} \sum_{\kappa=1}^{\eta} (a_\eta - a_{\eta-\kappa}) \left|\{\mathbf{x}^{(ij)} : \phi(\mu_i, \nu_j, \mathbf{x}^{(ij)}) = \eta \text{ and } \phi((\mu - 1)_i, \nu_j, \mathbf{x}^{(ij)}) = \eta - \kappa\}\right|}{(K + 1)^{n-2}}.$$

In Definitions 15.3.10 and 15.3.11, if the system steps from state η down to state $\eta - \kappa$, then the reduction of the system performance level is $(a_\eta - a_{\eta-\kappa})$. $I_{\text{U-JS}}(i, j; \phi)$ is the sum of the change of the system performance utility caused by component j stepping from states K to 0 while component i changes from states 1 to K. $I_{\text{U-JS}}(i, j; \phi)$ indicates the change of importance of component i with respect to the system performance utility when the state of component j changes from K to 0.

Definition 15.3.12 *The JRI of state μ of component i and state v of component j, $i \neq j$, with respect to the performance utility function* (15.1) *in an MMS, denoted by $I_{U\text{-}JR}^{\mu,v}(i, j; P)$, for $i, j \in \{1, 2, \ldots, n\}$ and $\mu, v \in \{1, 2, \ldots, K\}$, is defined as*

$$I_{U\text{-}JR}^{\mu,v}(i, j; P) = \frac{\partial^2 U}{\partial R_i^\mu \partial R_j^v},$$

where $R_i^\mu = \Pr\{X_i \geq \mu\}$. The JRI of components i and j is then defined as

$$I_{U\text{-}JR}(i, j; P) = \sum_{\mu=1}^{K} \sum_{v=1}^{K} I_{U\text{-}JR}^{\mu,v}(i, j; P).$$

Similar to $I_{U\text{-}B}^{\mu}(i; P)$ in Equation (15.2), which can be interpreted as the change of the system performance utility caused by component i deteriorating from states μ to $\mu - 1$, $I_{U\text{-}JR}^{\mu,v}(i, j; P)$ represents the change of the importance of state μ of component i caused by component j deteriorating from states v to $v - 1$. $I_{U\text{-}JR}(i, j; P)$ indicates how the system performance utility changes with the change of states of one component when the other functions.

Lemma 15.3.13 gives an expression of $I_{U\text{-}JR}(i, j; P)$. Theorem 15.3.14 gives expressions of $I_{U\text{-}JR}^{\mu,v}(i, j; P)$ for a series and a parallel system. Chacko and Manoharan (2011) generalized the JRI similarly and used the universal generating function technique for numerical calculation of the JRI.

Lemma 15.3.13

$$I_{U\text{-}JR}(i, j; P) = \sum_{\kappa=1}^{K} (a_\kappa - a_{\kappa-1}) \left(\Pr\{\phi(K_i, K_j, \mathbf{X}^{(ij)}) \geq \kappa\} - \Pr\{\phi(0_i, K_j, \mathbf{X}^{(ij)}) \geq \kappa\} \right.$$

$$\left. - \Pr\{\phi(K_i, 0_j, \mathbf{X}^{(ij)}) \geq \kappa\} + \Pr\{\phi(0_i, 0_j, \mathbf{X}^{(ij)}) \geq \kappa\} \right).$$

Theorem 15.3.14 *In a series system whose structure function $\phi(\mathbf{x}) = \min\{x_1, x_2, \ldots, x_n\}$,*

$$I_{U\text{-}JR}^{\mu,v}(i, j; P) = \begin{cases} 0 & v \neq \mu \\ (a_\mu - a_{\mu-1}) \prod_{s=1, s \neq i, j}^{n} \Pr\{X_s > \mu\} & v = \mu. \end{cases}$$

In a parallel system whose structure function $\phi(\mathbf{x}) = \max\{x_1, x_2, \ldots, x_n\}$,

$$I_{U\text{-}JR}^{\mu,v}(i, j; P) = \begin{cases} 0 & v \neq \mu \\ -(a_\mu - a_{\mu-1}) \prod_{s=1, s \neq i, j}^{n} \Pr\{X_s \leq \mu\} & v = \mu. \end{cases}$$

Lemma 15.3.15 indicates the relationship between the JRI of two states and the utility B-reliability importance of a state.

Lemma 15.3.15

$$I_{\text{U-JR}}^{\mu,v}(i, j; \boldsymbol{P}) = I_{\text{U-B}}^{\mu}(i; \boldsymbol{P}|v_j) - I_{\text{U-B}}^{\mu}(i; \boldsymbol{P}|(v-1)_j),$$

where $I_{\text{U-B}}^{\mu}(i; \boldsymbol{P}|v_j) = \sum_{\kappa=1}^{K}(a_\kappa - a_{\kappa-1})(\Pr\{\phi(\mu_i, v_j, \mathbf{X}^{(ij)}) \geq \kappa\} - \Pr\{\phi((\mu-1)_i, v_j, \mathbf{X}^{(ij)}) \geq \kappa\}).$

The following theorem shows the relationship between the JSI and JRI. Because of the effect of the system performance utility, under the equal probability condition, the JRI is not necessarily equal to the JSI.

Theorem 15.3.16   *If* $\Pr\{X_i = \mu\} = 1/(K+1)$ *for* $i = 1, 2, \ldots, n$ *and* $\mu = 0, 1, \ldots, K$, *then for* $i, j \in \{1, 2, \ldots, n\}$, $I_{\text{U-JR}}(i, j; (1/(K+1), 1/(K+1), \ldots, 1/(K+1))) \leq I_{\text{U-JS}}(i, j; \phi)$.

Note that if $K = 1$ (binary), $a_1 = 1$ and $a_0 = 0$, then the utility B-structure importance in Definition 15.3.10 is just the B-structure importance in Definition 6.1.5, and the JRI of components i and j in Definition 15.3.12 is just the JRI in Definition 7.1.1.

In this subsection, the system and all of its components are assumed to have the same number of states. If this assumption is relaxed, the JSI and JRI can be extended by replacing the fixed number of states $K + 1$ with different ones. Another assumption in this subsection is that only one-state degradation for components is allowed. If a component can degrade more than one state, the aforementioned definitions become very complicated but can still be defined similarly (Wu 2005).

15.4   Binary-Type Multistate Monotone Systems

Natvig (1985) studied a BTMMS and generalized the B-TDL importance, BP TIL importance, and L_1 TIL importance. This section first introduces the BTMMS and then presents the extensions of the importance measures for the BTMMS in Subsection 15.4.1.

Let $S = \{0, 1, \ldots, K\}$ be the set of states of the system and S_i be the set of states of component i. Assume that $\{0, K\} \subseteq S_i \subseteq S$. Let indicator vector $\boldsymbol{\mathcal{I}}_\mu(\mathbf{x}) = (\mathcal{I}(x_1 \geq \mu), \mathcal{I}(x_2 \geq \mu), \ldots, \mathcal{I}(x_n \geq \mu))$, $\mu = 1, 2, \ldots, K$, where the indicator function $\mathcal{I}(x_i \geq \mu) = 1$ if $x_i \geq \mu$ and 0 otherwise. If the states $\{0, 1, \ldots, \mu - 1\}$ correspond to the failure state when a binary approach is applied, then $\boldsymbol{\mathcal{I}}_\mu(\mathbf{x})$ is the corresponding vector of binary component states.

Definition 15.4.1   *A system* (N, ϕ) *is a binary-type multistate monotone system (BTMMS) if there exist binary structures* φ_μ, $\mu = 1, 2, \ldots, K$, *which are nondecreasing with* $\varphi_\mu(\mathbf{0}) = 0$ *and* $\varphi_\mu(\mathbf{1}) = 1$ *such that the multistate structure function* ϕ *satisfies* $\phi(\mathbf{x}) \geq \mu$ *if and only if* $\varphi_\mu(\boldsymbol{\mathcal{I}}_\mu(\mathbf{x})) = 1$ *for all* $\mu = 1, 2, \ldots, K$ *and all* $\mathbf{x}$.

A BTMMS is always an MMS because (i) $\phi(\mathbf{0}) < 1$ (i.e., $\phi(\mathbf{0}) = 0$) induced by $\varphi_1(\boldsymbol{\mathcal{I}}_1(\mathbf{0})) = \varphi_1(\mathbf{0}) = 0$, (ii) $\phi(\mathbf{K}) = K$ induced by $\varphi_K(\boldsymbol{\mathcal{I}}_K(\mathbf{K})) = \varphi_K(\mathbf{1}) = 1$, and (iii) ϕ is a nondecreasing function induced by the facts that for any two component state vectors, $\mathbf{x} \geq \mathbf{y}$ implies $\boldsymbol{\mathcal{I}}_\mu(\mathbf{x}) \geq \boldsymbol{\mathcal{I}}_\mu(\mathbf{y})$ and that all φ_μ are nondecreasing functions, $\mu = 1, 2, \ldots, K$.

With the binary property of the BTMMS, the binary structures φ_μ uniquely determine the system's binary state from the components' binary states. It also follows that $\min_{1 \leq i \leq n} x_i \leq \phi(\mathbf{x}) \leq \max_{1 \leq i \leq n} x_i$ and that the binary structures φ_μ must satisfy $\varphi_\mu(\mathbf{z}) \geq \varphi_{\mu+1}(\mathbf{z})$ for all

$\mu = 1, 2, \ldots, K - 1$ and all binary vector $\mathbf{z}$. These results make it possible to analyze the system using the binary approach.

Example 15.4.2 As an example of the BTMMS, consider the system depicted in Figure 6.2 as a two-terminal network, for instance an oil pipeline network. The oil flows in from the left, goes through the system, and flows out from the right.

Assume for simplicity that the flow through component (edge) i is either integer-valued y_i or 0, $i = 1, 2, \ldots, n$. From the max-flow min-cut theorem of graph theory (Jensen and Barnes 1980), the maximum flow capacity of the system is

$$M = M(\mathbf{y}) = \min_{1 \le k \le u} \sum_{1 \in C_k} y_i,$$

where $C_1, C_2, \ldots, C_u$ are the minimal cuts of the graph. Now let $x_i = M(0)$, $i = 1, 2, \ldots, n$, if the flow through component i is $y_i(0)$. Then the max flow through the system, as a function of $\mathbf{x}$, is given by the structure function

$$\phi(\mathbf{x}) = \min_{1 \le k \le u} M^{-1} \sum_{1 \in C_k} y_i x_i.$$

It is easily seen that the system corresponding to ϕ is a BTMMS. Finally, for each edge i, assume that there exists vector $(\cdot_i, \mathbf{x})$ such that $\phi(M_i, \mathbf{x}) > \phi(0_i, \mathbf{x})$ (otherwise remove the edge from the graph). Hence, all components are relevant to at least one state of the system.

Assume that the marginal performance processes of the components are mutually independent in $[0, \infty)$. The performance process of component i, $i = 1, 2, \ldots, n$, is a stochastic process $\{X_i(t), t \ge 0\}$, where for each fixed $t \ge 0$, $X_i(t)$ is a random variable that takes values in S_i. The marginal performance processes $\{X_i(t), t \ge 0\}$, $i = 1, 2, \ldots, n$, are independent in $[0, \infty)$ if and only if for any integer m and $\{t_1, t_2, \ldots, t_m\} \subset [0, \infty)$ the random vectors $(X_1(t_1), X_1(t_2), \ldots, X_1(t_m))$, $(X_2(t_1), X_2(t_2), \ldots, X_2(t_m))$, $\ldots$, $(X_n(t_1), X_n(t_2), \ldots, X_n(t_m))$ are independent.

Consider a case where the components cannot be repaired. Introducing $F_i^{\mu}(t)$ as the lifetime distribution in the states $\{\mu, \mu + 1, \ldots, K\}$ of component i, mathematically,

$$\Pr\{X_i(t) \ge \mu\} = 1 - F_i^{\mu}(t) = \overline{F}_i^{\mu}(t), \quad i = 1, 2, \ldots, n, \ \mu = 1, 2, \ldots, K.$$

Furthermore, assume that $\mathbf{X}(0) = \mathbf{K}$ and that $F_i^{\mu}(t)$ has a density $f_i^{\mu}(t)$. Then, for a BTMMS,

$$\Pr\{\phi(\mathbf{X}(t)) \ge \mu\} = \Pr\{\varphi_{\mu}(\mathcal{I}_{\mu}(\mathbf{X}(t))) = 1\} = R_{\varphi_{\mu}}(\overline{\mathbf{F}}^{\mu}(t)),$$

where $R_{\varphi_{\mu}}$ is the reliability function of the binary structure φ_{μ} and $\overline{\mathbf{F}}^{\mu}(t) = (\overline{F}_1^{\mu}(t), \overline{F}_2^{\mu}(t), \ldots, \overline{F}_n^{\mu}(t))$.

15.4.1 The B-TDL Importance, BP TIL Importance, and L_1 TIL Importance

Definition 15.4.3 *Component i is critical at time t for the system being in states $\{\mu, \mu + 1, \ldots, K\}$ if putting component i in states $\{0, 1, \ldots, \mu - 1\}$ causes the system to be in states $\{0, 1, \ldots, \mu - 1\}$, while putting component i in states $\{\mu, \mu + 1, \ldots, K\}$ causes the system to be in states $\{\mu, \mu + 1, \ldots, K\}$.*

Definition 15.4.4 *A generalization of the B-TDL importance of state μ of component i in a BTMMS, denoted by $I_B^\mu(i; \overline{\mathbf{F}}^\mu(t))$, for $\mu = 1, 2, \ldots, K$, is defined as the probability that component i is critical at time t for the system being in states $\{\mu, \mu + 1, \ldots, K\}$. Mathematically,*

$$I_B^\mu(i; \overline{\mathbf{F}}^\mu(t)) = R_{\varphi_\mu}(1_i, \overline{\mathbf{F}}^\mu(t)) - R_{\varphi_\mu}(0_i, \overline{\mathbf{F}}^\mu(t)).$$

Furthermore, let c_μ be the average loss in utility when the system leaves the states $\{\mu, \mu + 1, \ldots, K\}$, and assume $\sum_{\mu=1}^K c_\mu = 1$. Then, the generalized B-TDL importance of component i in a BTMMS is defined as

$$I_B(i; t) = \sum_{\mu=1}^K \frac{c_\mu I_B^\mu(i; \overline{\mathbf{F}}^\mu(t))}{\sum_{j=1}^n I_B^\mu(j; \overline{\mathbf{F}}^\mu(t))}.$$

Note that $\sum_{i=1}^n I_B(i; t) = 1$, which is different from the B-TDL importance in a binary coherent system.

Definition 15.4.5 *A generalization of the BP TIL importance of state μ of component i in a BTMMS, denoted by $I_{BP}^\mu(i; \overline{\mathbf{F}}^\mu)$, for $\mu = 1, 2, \ldots, K$, is defined as the probability that component i is critical for the system being in states $\{\mu, \mu + 1, \ldots, K\}$. Mathematically,*

$$I_{BP}^\mu(i; \overline{\mathbf{F}}^\mu) = \int_0^\infty I_B^\mu(i; \overline{\mathbf{F}}^\mu(t)) f_i^\mu(t) dt. \tag{15.3}$$

Then, the generalized BP TIL importance of component i in a BTMMS is defined as

$$I_{BP}(i) = \sum_{\mu=1}^K c_\mu I_{BP}^\mu(i; \overline{\mathbf{F}}^\mu).$$

Definition 15.4.6 *Let Z_i^μ be the reduction of remaining system time in states $\{\mu, \mu + 1, \ldots, K\}$ due to the departure of the same states of component i. Then, the generalized L_1 TIL importance of component i in a BTMMS is defined as*

$$I_{L_1}(i) = \sum_{\mu=1}^K \frac{c_\mu \mathbb{E}(Z_i^\mu)}{\sum_{j=1}^n \mathbb{E}(Z_j^\mu)}.$$

The L_1 importance and random variable Z_i for a binary coherent system are defined in Subsection 10.2.2. Different interpretations of Z_i^μ can be given, corresponding to the ones in Subsection 10.2.2 for a binary case. This leads to

$$\mathbb{E}(Z_i^\mu) = \int_0^\infty \overline{F}_i^\mu(t) \left[-\ln \overline{F}_i^\mu(t) \right] I_B^\mu(i; \overline{\mathbf{F}}^\mu(t)) dt.$$

If component i is not relevant to state μ, that is, there is no binary vector $(\cdot_i, \mathbf{z})$ such that $\varphi_\mu(1_i, \mathbf{z}) - \varphi_\mu(0_i, \mathbf{z}) = 1$, then immediately $I_B^\mu(i; \overline{\mathbf{F}}^\mu(t)) = I_{BP}^\mu(i; \overline{\mathbf{F}}^\mu) = \mathbb{E}(Z_i^\mu) = 0$ because of the independence of the marginal performance processes of the components, which is exactly right.

Remark 15.4.7   Note that $I_{BP}(i)$ and $I_{L_1}(i)$ are based firmly on $I_B^\mu(i; \overline{\mathbf{F}}^\mu(t))$. For example, using different generalizations of the B-TDL importance in Equation (15.3) results in different generalizations of the BP TIL importance. With respect to the performance utility function (15.1) in an MMS, Bueno (1989) defined an extension of the BP TIL importance of state μ of component i as

$$I_{U\text{-}BP}^\mu(i; \overline{\mathbf{F}}^\mu) = \int_0^\infty I_{U\text{-}B}^\mu(i; t)\, f_i^\mu(t)\mathrm{d}t,$$

where the utility B-TDL importance $I_{U\text{-}B}^\mu(i; t)$ is defined in Equation (15.2) by using $\overline{\mathbf{F}}^\mu(t)$, $\mu = 1, 2, \ldots, K$, instead of $\mathbf{P}$.

It is not obvious how to extend these ideas from the BTMMS to the MMS, which is a generalized BTMMS. However, Natvig (1985) gave some preliminary suggestions.

15.5   Summary of Importance Measures for Multistate Systems

As demonstrated, various importance measures and their extensions have been proposed for different multistate systems. Note that an importance measure can be generalized in multiple ways even for one multistate system. According to the discussion in Sections 15.2–15.4, importance measures for multistate systems can be classified into two types (Ramirez-Marquez and Coit 2005):

Type 1 (composite importance measures): Measures how a specific *component* affects system performance.

Type 2: Measures how a particular *state* or set of *states* of a specific component affects system performance.

Since the importance measure of a component, inclusive all of its states, can be understood as a composite of the importance of all states of the component, type 1 importance measures are also referred to as composite importance measures. Some generalizations of importance measures introduced in Subsections 15.3.1, 15.3.4, and 15.4.1 are type 1 importance measures. The importance measures that are later discussed in Section 16.1 are also type 1 importance measures for the multistate systems in which the multistates represent the different levels of network capacity. The other importance measures in Sections 15.2–15.4 are type 2 importance measures.

Type 1 importance measures evaluate the overall contribution of a component to system performance. Type 2 importance measures characterize, for a given component, the most important component state with regard to its impact on system performance. The most important component state may not necessarily correspond to the most important component. For some system design and reliability problems, it is necessary to identify the most important component as all, inclusive all of its states. For example, it may be better to use type 1 importance measures in determining which component should be replaced with a new one. In other applications, determination of the most important component state is the primary concern, and type 2 importance measures should be applied. For example, it is of interest to know how an increase or a decrease in the state of a component affects system performance.

In addition to the work introduced in Sections 15.2–15.4, there has been other related work on generalizing and evaluating reliability and importance measures in multistate systems, for example, Abouammoh and Al-Kadi (1991), Barlow and Wu (1978), Block and Savits (1982), Finkelstein (1994), Levitin (2004), Levitin and Lisnianski (1999), Levitin et al. (2003), Meng (1993), Natvig (1982, 2011), and Ramirez-Marquez and Coit (2005). Levitin et al. (2003) discussed the B-reliability importance, FV reliability importance, RAW, and RRW in an MMS, focusing on an approach based on the universal generating function technique for the evaluation of these importance measures. Levitin (2004) discussed a protection survivability importance, which is an extension of the B-reliability importance, and used the universal generating function technique to evaluate it. Levitin and Lisnianski (1999) used the universal generating function technique for their sensitivity analysis of system output performance (see also Subsection 18.1.1).

Zio and Podofillini (2003b) proposed similar importance measures to those presented by Levitin et al. (2003). Zio and Podofillini (2003b) and Zio et al. (2004) performed Monte Carlo simulation methods to evaluate these importance measures. Ramirez-Marquez and Coit (2005) used Monte Carlo simulation to compute their proposed composite importance measures.

Zio et al. (2007) gave an application of a multistate RAW in the railway industry where each rail section is treated as a component. For prioritizing the rail sections to most effectively improve the performance of the rail network, the RAW is defined in terms of a decrease in the overall delay of trains and is calculated using the Monte Carlo simulation. The method in this application may be used to address the problem in Example 1.0.3.

Ramirez-Marquez et al. (2006) introduced some type 1 and type 2 importance measures for an MMS, including unsatisfied demand index providing insight regarding a component or component state contribution to unsatisfied demand, multistate failure frequency index quantifying the contribution of a particular component or component state to system failure, and multistate redundancy importance identifying where to allocate component redundancy as to improve systems reliability. All of these importance measures are defined on the basis of systems reliability, and conditional systems reliability conditioned on each component being in each state.

Some importance measures for multistate systems have been well defined for binary coherent systems in the preceding chapters while others have been developed specifically for multistate systems, for example, the performance utility importance, which deals with the system performance utility. The increase in the states of components or the system adds to the sophistication of the importance measures. In a practical model, the number of states of components and the system must be kept small to make the model manageable.

15.6   Continuum Systems

Baxter (1984) and Baxter (1986) first introduced the standard continuum structure model, which allows a system or component's state at time t to assume any value in a segment of the real number line: $\phi(t) \in [0, 1]$ and $X_i(t) \in [0, 1]$. After that, a few supplements (e.g., Baxter and Kim, 1986, 1987; Kim and Baxter, 1987a; Montero et al., 1990) have complemented the continuum structure model. The review paper of Brunelle and Kapur (1999) clarified the difference between binary, multistate, continuum, and mixed (continuous and discrete) systems and proposed a classification schema for reliability measures that can be defined equally for these systems. The continuum structure enables explicitly modeling of the continuous performance degradation (Yang and Xue 1996) and deterioration of components and systems. An example of a continuum system is an automobile tire, the performance of which degrades continuously as the tread wears (Brunelle and Kapur 1999). Another example is a passive system whose utilization is growlingly advocated in the new generation of nuclear power plants and whose physical behavior dictates the system performance in a continuous fashion (Zio and Podofillini 2003a). Other examples include a coal-fired power station, fueling a gas turbine, and a manufacturing process (Baxter and Kim 1986, 1987).

Example 15.6.1 (A manufacturing process) Suppose that a certain manufactured product is assembled from k distinct parts, and that each part is the end result of a different sequence of manufacturing stages. At each stage in each sequence, a certain proportion of the input is damaged by defects in the machinery; let x_{ij} denote the proportion of its input successfully processed by the jth machine in the ith sequence ($j = 1, 2, \ldots, n_i$; $i = 1, 2, \ldots, k$). Clearly, x_{ij} takes values in the unit interval. Then the overall operational level of the factory is given by the structure function $\phi : [0, 1]^{n_1 + n_2 + \cdots + n_k} \mapsto [0, 1]$ defined by $\phi(\mathbf{x}) = \min_{1 \leq i \leq k} \prod_{j=1}^{n_i} x_{ij}$.

A continuum structure function is a continuous function of the states of its n components. One criticism about continuum system models is that, though well understood for binary and multistate systems, the method of constructing a continuum structure function through a continuous analog to minimal cuts and minimal paths (or lower and upper boundary points) appears to be generally impractical. Baxter and Kim (1986, 1987) determined bounds on the distribution of the continuum system structure for some special cases, and Block and Savits (1984) and Montero et al. (1990) studied bounds on the reliability of continuum systems. There are several simplifying techniques for continuum modeling including subsystem identification (Brunelle and Kapur 1999), scattered data interpolation (Brunelle and Kapur 1998), partial-information bounds (Brunelle and Kapur 1999), and estimation of boundary points based on the universal generating function technique (Lisnianski 2001). Recently, Gámiz and Martínez Miranda (2010) proposed a regression method to build a structure function for a continuum system by using multivariate nonparametric regression techniques, which can then be used to estimate the systems reliability.

For the rest of this section, assume that the continuum structure function is a mapping from the unit hypercube to the unit interval, that is, $\phi : [0, 1]^n \mapsto [0, 1]$, which is nondecreasing in each argument and satisfies $\phi(\mathbf{0}) = 0$ and $\phi(\mathbf{1}) = 1$. For any component i and any subset $A \subset [0, 1]^n$, define $A^i = \{\mathbf{x} \in [0, 1]^n : (\cdot_i, \mathbf{x}) = (\cdot_i, \mathbf{z})$ for some $\mathbf{z} \in A\}$; thus, $A \subset A^i$. Let λ denote Lebesgue measure on n-dimensional real number space $\mathbb{R}^n$, and $U_\alpha = \{\mathbf{x} \in [0, 1]^n : \phi(\mathbf{x}) \geq \alpha\}$ for $\alpha \in [0, 1]$. Baxter and Lee (1989) defined an almost irrelevant component with respect to the continuum structure function ϕ.

Definition 15.6.2 *Let ϕ be a continuum structure function and let $\alpha \in [0, 1]$. Component i is said to be almost irrelevant to ϕ at level α if there exists a subset $V_\alpha \subset [0, 1]^n$ such that $\lambda(V_\alpha^c) = 0$ ($V_\alpha^c = [0, 1]^n \setminus V_\alpha$) and $U_\alpha \cap V_\alpha = (U_\alpha \cap V_\alpha)^i \cap V_\alpha$. Furthermore, if component i is almost irrelevant to ϕ at level α for all $\alpha \in [0, 1]$, component i is said to be almost irrelevant to ϕ.*

Work on importance measures for continuum systems is very limited. Finkelstein (1994) suggested extending the importance measures in the binary and multistate models by performing a limit transition (where the number of states of a system and components is unlimited within the given state intervals) but had no conclusive results. In particular, Kim and Baxter (1987b) extended the B-reliability importance as in Definition 15.6.3, which is based on vector δ_α that denotes the intersection of the boundary of U_α in $[0, 1]^n$ and the diagonal of $[0, 1]^n$ for $\alpha \in [0, 1]$. δ_α is known as the key vector of U_α, and δ_α denotes the corresponding key element. For any continuum system structure ϕ, the key vector always exists and is unique, and, if ϕ is continuous, $\phi(\delta_\alpha) = \alpha$ for all $\alpha \in (0, 1]$. Baxter and Lee (1989) showed that if ϕ is continuous at $\mathbf{0}$ and $\mathbf{1}$, then $0 < \delta_\alpha < 1$ for all $\alpha \in (0, 1]$.

Definition 15.6.3 *For a continuum structure function ϕ, the B-reliability importance of component i at level $\alpha \in (0, 1]$, denoted by $I_B(i; \alpha)$, is defined as*

$$I_B(i; \alpha) = \Pr\{\phi(\mathbf{X}) \geq \alpha | X_i \geq \delta_\alpha\} - \Pr\{\phi(\mathbf{X}) \geq \alpha | X_i < \delta_\alpha\}, \tag{15.4}$$

where δ_α is the key element of U_α.

This extension of the B-reliability importance to continuum systems is motivated by the expression of the B-reliability importance in Equation (4.4) for binary systems. It regards part of the unit interval, say $[0, \alpha)$, as corresponding to the failure state of the system and $[\alpha, 1]$ as the functioning state. $I_B(i; \alpha)$ can be interpreted as the probability that $\phi(\mathbf{X}) \geq \alpha$ if and only if $X_i \geq \delta_\alpha$.

In Theorem 15.6.4, notation $H_\alpha = \{\mathbf{x} \in [0, 1]^n : \phi(\mathbf{x}) \geq \alpha \text{ whereas } \phi(\mathbf{y}) < \alpha \text{ for all } \mathbf{y} < \mathbf{x}\}$ for $\alpha \in (0, 1]$, and $K_\alpha = \{\mathbf{x} \in [0, 1]^n : \phi(\mathbf{x}) \leq \alpha \text{ whereas } \phi(\mathbf{y}) > \alpha \text{ for all } \mathbf{y} > \mathbf{x}\}$ for $\alpha \in [0, 1)$. Kim and Baxter (1987b) derived conditions (i)–(iii) under which $\lim_{\alpha \to 1} I_B(i; \alpha) = 0$, $\lim_{\alpha \to 0} I_B(i; \alpha) = 0$, and $I_B(i; \alpha) > 0$, respectively. Baxter and Lee (1989) derived conditions (iv) and (v) under which $I_B(i; \alpha) > 0$ and $I_B(i; \alpha)$ is continuous, respectively. The proofs are lengthy and therefore omitted.

Theorem 15.6.4 *Suppose that ϕ is a continuum structure function and that $X_1, X_2, \ldots, X_n$ are independent, absolutely continuous random variables. Let $i = 1, 2, \ldots, n$.*

(i) *If ϕ is continuous and for all $\mathbf{x} \in H_1$, $x_j = 1$ for some $j \neq i$, then $\lim_{\alpha \to 1} I_B(i; \alpha) = 0$.*

(ii) *If ϕ is continuous and for all $\mathbf{x} \in K_0$, $x_j = 0$ for some $j \neq i$, then $\lim_{\alpha \to 0} I_B(i; \alpha) = 0$.*

(iii) *Suppose that ϕ is continuous such that $\lambda(U_\alpha) > 0$ for all $\alpha \in (0, 1)$ and the support of each of $X_1, X_2, \ldots, X_n$ is the unit interval. Then $I_B(i; \alpha) > 0$ for $\alpha \in (0, 1)$ if and only if $x_i \neq 0$ for some $\mathbf{x} \in H_\alpha$ for which $\lambda(\{\mathbf{y} \in [0, 1]^n : \mathbf{y} \geq \mathbf{x}\}) > 0$.*

(iv) *Suppose that ϕ is continuous at $\mathbf{0}$ and $\mathbf{1}$ and the support of each of $X_1, X_2, \ldots, X_n$ is the unit interval. Then $I_B(i; \alpha) > 0$ for $\alpha \in (0, 1)$ if and only if component i is almost irrelevant to ϕ at level α.*

(v) *Suppose that ϕ is continuous at $\mathbf{0}$ and $\mathbf{1}$ and strongly increasing (i.e., $\phi(\mathbf{x}) > \phi(\mathbf{y})$ whenever $\mathbf{x} \geq \mathbf{y}$ and $\mathbf{x} \neq \mathbf{y}$) and the support of each of $X_1, X_2, \ldots, X_n$ is the unit interval. Then $I_B(i; \alpha)$ is continuous on $(0, 1)$.*

Baxter and Lee (1989) compared the B-reliability importance of two components for all α for the special cases in which a component is in series or parallel with the rest of a system as shown in Theorem 15.6.5.

Theorem 15.6.5 *Suppose that a continuum system structure ϕ is continuous at $\mathbf{0}$ and $\mathbf{1}$ and that $X_1, X_2, \ldots, X_n$ are random variables, the support of each of which is the unit interval. If $U_\alpha \subset (\supset) \{\mathbf{x} \in [0, 1]^n : x_i \geq \delta_\alpha\}$ (i.e., component i is in series (parallel) with the rest of the system) for all $\alpha \in [0, 1]$ and if $X_i \leq^{st} (\geq^{st}) X_j$ (see Definition 2.8.1), then $I_B(i; \alpha) \geq I_B(j; \alpha)$ for all $\alpha \in (0, 1)$, $j \neq i$.*

Baxter and Lee (1989) and Kim and Baxter (1997) proposed a numerical approximation method for evaluating the B-reliability importance in Equation (15.4). They showed that a sequence $\{I_B^{(m)}(i; \alpha), m = 1, 2, \ldots\}$ converges to $I_B(i; \alpha)$ uniformly as m approaches infinity, where each $I_B^{(m)}(i; \alpha)$ is readily calculated. However, this approximation procedure is only practicable for the system of small or moderate number of components, since the computational complexity of the calculation grows rapidly with the number of components.

The continuum model successfully describes a nuclear waste repository (Eisenberg and Sagar 2000) in which the performance of the system is represented by a scalar variable, such as the individual dose or cumulative release of radionuclides, and the components of the repository system also perform on a continuous scale. For example, a waste package is intended to entirely contain the waste for a long period of time, but will eventually degrade and allow the release of radionuclides to the geosphere. Such release begins some time after water is first able to penetrate into the waste package and is expected initially to increase in time. The repository performance depends, in a complex fashion, not only on the time at which the waste package integrity is initially lost but also on the rate at which water is able to contact the waste and move radionuclides from the degraded waste package into the geosphere. To evaluate the contributions of a repository component to system risk, the RRW is used by means of determining how the repository system functions without the component (i.e., neutralizing the functioning of the component from the system) and comparing the performance of the systems before and after the change. The generalization of RRW into the continuum model is straightforward since it depends on the evaluation of system performance.

15.7 Repairable Systems

For a repairable system in which one or more components can be repaired after failure, the system may experience several failure-and-functioning cycles before it finally breaks down. Thus, for a repairable system, system performance is often measured by availability function, which gives more information than the reliability function. Consequently, the importance measures, especially the classes of reliability and lifetime importance measures, should be defined using availability. Note that for nonrepairable systems, reliability is identical in value to availability. In the following, Subsections 15.7.1, 15.7.2, 15.7.3, and 15.7.4 illustrate the B-importance, c-FV importance, BP importance, and L$_1$ importance of repairable components in terms of availability, respectively.

For complex repairable systems, the systems reliability and availability depend not only on the reliability/availability of components but also on other characteristics of the components and the system, such as component time-to-failure distributions, component time-to-restore distributions, maintenance practices, spare availability, and so on. Thus, exact analytical solutions to systems reliability and availability become intractable. Moreover, analytical approaches for computing importance measures may be intractable. To resolve this intractability, Wang et al. (2004) turned to simulation to define and compute some importance measures. Subsection 15.7.5 presents these importance measures.

Although in most cases the availability function is hard to obtain, the steady-state availability could be known if it exists. For the steady state of a repairable system, the B-TDL importance, criticality TDL importance, FV TDL importance, RAW, and RRW can be assessed simply by substituting the steady-state unavailability $\bar{A}_i(t)$ for $F_i(t)$ or q_i without change in probabilistic interpretations. Section 15.8 discusses the applications of these importance measures in the outage rate analysis of the power plant. The outage rate is a measure of steady-state unavailability in a repairable power plant.

Markov models are frequently used in reliability analysis to assess different measures of interest, for example, systems reliability, availability, and maintainability. Within the Markov modeling framework, the repair rates of components and the intercomponent and functional dependencies can be easily included. In this context, Subsection 18.1.2 introduces an importance measure that offers a promising tool for sensitivity analysis of Markov processes in reliability studies.

15.7.1   The B-availability Importance

Extending the B-reliability importance for a nonrepairable system, Barabady and Kumar (2007) proposed the B-availability importance for a repairable system under the perfect repair and the steady state of the system. Similar to Equation (4.10), the B-availability importance of component i is defined as

$$\frac{\partial A_\phi}{\partial A_i},$$

recalling that A_ϕ is the system availability and A_i is the availability of component i. The analytical results can only be obtained for the simple systems such as parallel, series, series–parallel, and parallel-series systems.

15.7.2   The c-FV Unavailability Importance

Vesely (1970) and Fussell (1975) introduced the c-FV unavailability importance for a repairable system, which is defined as the probability (a function of time) that failure of component i contributes to system failure given that the system fails. The c-FV unavailability counts the contribution of a component to system failure without requiring it to be critical for the system. Let $\bar{A}_{C_k}$ be the unavailability of minimal cut C_k and $\bar{A}_\phi$ the unavailability of the system. Then, the c-FV unavailability importance of component i is upper bounded by $\sum_{C_k \in \mathscr{C}_i} \bar{A}_{C_k} / \bar{A}_\phi$, where the sum is taken over all minimal cuts containing component i.

15.7.3   The BP Availability Importance

This subsection assumes that whenever component i fails, a (perfect) repair is conducted immediately. Assume that repair times of components follow continuous distributions and let

$M_i(t)$ denote the expected number of failures of component i in $[0, t]$. Given system failure at time t, the probability that component i causes system failure is

$$\frac{[R(1_i, \mathbf{A}(t)) - R(0_i, \mathbf{A}(t))]\, dM_i(t)}{\sum_{j=1}^{n} [R(1_j, \mathbf{A}(t)) - R(0_j, \mathbf{A}(t))]\, dM_j(t)},$$

where $dM_i(t)$ is called the renewal density, that is, the probability that a renewal occurs in some differential time interval. $dM_i(t)$ is analogous to $dF_i(t)$ in the nonrepairable case. Note that the numerator represents the probability that component i is critical for the system at time t and that component i then fails at time t (i.e., in the small element of time following the instant t), and the denominator represents the probability that some component is critical for the system at time t and that it fails at time t (i.e., the system fails at time t).

The expected number of times during $[0, t]$ that component i causes system failure, denoted by $\widetilde{M}_i(t)$, is then

$$\widetilde{M}_i(t) = \int_0^t [R(1_i, \mathbf{A}(u)) - R(0_i, \mathbf{A}(u))]\, dM_i(u). \tag{15.5}$$

Formally, Barlow and Proschan (1975) proved Equation (15.5) by forming a partition of $[0, t]$: $0 = t_0 < t_1 < \ldots < t_s = t$. Observe that

$$\big[R(1_i, \mathbf{A}(\xi_j)) - R(0_i, \mathbf{A}(\xi_j))\big] \big[M_i(t_{j+1}) - M_i(t_j)\big]$$

is approximately the probability that component i causes system failure in (t_j, t_{j+1}) and $\xi_j \in (t_j, t_{j+1})$. Summing these probabilities and then letting s approach infinity yield Equation (15.5).

Then, the BP time-dependent availability importance is defined as

$$I_{\mathrm{BP}}(i; \mathbf{A}(t)) = \frac{\widetilde{M}_i(t)}{\sum_{j=1}^{n} \widetilde{M}_j(t)}.$$

Letting time t approach infinity, the BP time-independent availability importance is defined as

$$I_{\mathrm{BP}}(i; \mathbf{A}) = \lim_{t \to \infty} I_{\mathrm{BP}}(i; \mathbf{A}(t)).$$

As in the case of no repair,

$$\sum_{i=1}^{n} I_{\mathrm{BP}}(i; \mathbf{A}(t)) = \sum_{i=1}^{n} I_{\mathrm{BP}}(i; \mathbf{A}) = 1.$$

The process generated by component failures and repairs forms an alternating renewal process. Considering component i, $i = 1, 2, \ldots, n$, let α_i be the MTBF representing the average continuous functioning duration, β_i the MTTR representing the average amount of time needed to repair a failure, and A_i the stationary availability. Applying the elementary renewal theorem (Barlow and Proschan 1965, Chapter 3), that is,

$$\lim_{t \to \infty} \frac{M_i(t)}{t} = \frac{1}{\alpha_i + \beta_i} \quad \text{and} \quad A_i = \lim_{t \to \infty} A_i(t) = \frac{\alpha_i}{\alpha_i + \beta_i},$$

it is shown that

$$I_{\text{BP}}(i; \mathbf{A}) = \frac{(\alpha_i + \beta_i)^{-1}(R(1_i, \mathbf{A}) - R(0_i, \mathbf{A}))}{\sum_{j=1}^{n}(\alpha_j + \beta_j)^{-1}(R(1_j, \mathbf{A}) - R(0_j, \mathbf{A}))}.$$

$\alpha_i + \beta_i$ is the average amount of time between failures for component i, that is, the average length of time for a renewal cycle. $1/(\alpha_i + \beta_i)$ is the average rate at which the renewal process takes place for component i. Asymptotically, the system failure probability is time-invariant. Thus, $I_{\text{BP}}(i; \mathbf{A})$ represents the stationary probability that the failure of component i causes system failure, given that the system eventually fails.

Example 15.7.1 (Example 2.2.1 continued)   For a series system with perfect repair of components,

$$I_{\text{BP}}(i; \mathbf{A}) = \frac{\prod_{j \neq i} \alpha_j}{\sum_{k=1}^{n} \prod_{j \neq k} \alpha_j}.$$

Note that the BP availability importance does not depend on the MTTR of components.

Example 15.7.2 (Example 2.2.2 continued)   For a parallel system with perfect repair of components,

$$I_{\text{BP}}(i; \mathbf{A}) = \frac{\prod_{j \neq i} \beta_j}{\sum_{k=1}^{n} \prod_{j \neq k} \beta_j}.$$

Thus, the BP availability importance does not depend on the MTBF of components.

15.7.4   The L_1 TIL Importance

Assume that, while repair of one component is occurring, the remaining components continue to function. For $i = 1, 2, \ldots, n$ and $j = 1, 2, \ldots$, let random variables

$$T_{ij} = \text{length of the } j\text{th functioning period for component } i,$$

$$D_{ij} = \text{length of the } j\text{th repair period for component } i,$$

and assume the T_{ij} to be independent with the lifetime distribution function $F_i(t)$ and the D_{ij} to be independent with the repair time distribution function $G_i(t)$.

To apply Definition 10.2.3 of the L_1 TIL importance for a repairable component, Z_i needs to be specified under the case of repair. Letting random variables

$$Z_i^1 = \text{reduction in time until a functioning system fails due to the failure of component } i,$$

$$Z_i^2 = \text{increase in time until a failed system functions due to the failure of component } i,$$

then $Z_i = Z_i^1 + Z_i^2$. Natvig and Gåsemyr (2009) derived the expression of the stationary L_1 TIL importance of repairable component i as

$$I_{L_1}(i; \mathbf{A}) = \frac{\mathbb{E}(Z_i)}{\sum_{j=1}^{n} \mathbb{E}(Z_j)} = \frac{(\alpha_i + \beta_i)^{-1}\left(R(1_i, \mathbf{A}) - R(0_i, \mathbf{A})\right) \int_0^\infty \overline{F}_i(t)(-\ln \overline{F}_i(t)) \mathrm{d}t}{\sum_{j=1}^{n}(\alpha_j + \beta_j)^{-1}\left(R(1_j, \mathbf{A}) - R(0_j, \mathbf{A})\right) \int_0^\infty \overline{F}_j(t)(-\ln \overline{F}_j(t)) \mathrm{d}t}.$$

Natvig and Gåsemyr (2009) further extended the L_1 TIL importance to take into account the probability that the repair of component i is the cause of system repair given that system repair has occurred. Letting $\overline{G}_i(t) = 1 - G_i(t)$, the extended version is

$$\frac{(\alpha_i + \beta_i)^{-1} \left(R(1_i, \mathbf{A}) - R(0_i, \mathbf{A})\right) \int_0^\infty [\overline{F}_i(t)(-\ln \overline{F}_i(t)) + \overline{G}_i(t)(-\ln \overline{G}_i(t))] dt}{\sum_{j=1}^n (\alpha_j + \beta_j)^{-1} \left(R(1_j, \mathbf{A}) - R(0_j, \mathbf{A})\right) \int_0^\infty [\overline{F}_j(t)(-\ln \overline{F}_j(t)) + \overline{G}_j(t)(-\ln \overline{G}_j(t))] dt}.$$

Natvig et al. (2009) applied these measures to an offshore oil and gas production system. According to the extended version of the L_1 importance, component is important if both by failing it strongly reduces the expected system uptime and by being repaired it strongly reduces the expected system downtime. The results show that different distributions affect the ranking of the components. All numerical results were computed using discrete event simulation (Huseby and Natvig 2010).

Natvig et al. (2011) extended the work in Natvig et al. (2009) to multistate systems. In the multistate case, the L_1 TIL importance of a component is calculated separately for each component state, and the importance of a component is evaluated by adding up the importance measures for each individual state. The simulations of this case study show that the results based on the original L_1 importance and its extended version are almost identical because the fictive prolonged repair times are much shorter than the fictive prolonged times spent in each of the noncomplete failure states.

15.7.5   Simulation-based Importance Measures

Unlike the importance measures of repairable components in Subsections 15.7.2–15.7.4, importance measures in this subsection are defined and directly calculated through simulation outcomes over the simulation time period $[0, t]$. The simulation is set up by various component and system characteristics such as system structure, component lifetime distributions, types of repair and maintenance, and repair time distributions of components.

Definition 15.7.3   *A failure criticality importance of component i is defined as the percentage of times that a system failure event is caused by a failure of this component in the simulation time $[0, t]$, that is,*

$$I_{\mathrm{FCI}}(i; t) = \frac{\textit{number of system failures caused by component } i \textit{ in } [0, t]}{\textit{number of system failures in } [0, t]}. \tag{15.6}$$

Alternatively, it can be defined as the percentage of times that a failure of component i causes a system failure in $[0, t]$, that is,

$$I_{\mathrm{FCI}}(i; t) = \frac{\textit{number of system failures caused by component } i \textit{ in } [0, t]}{\textit{number of component } i \textit{ failures in } [0, t]}.$$

Intuitively, the importance measure (15.6) would have the same meaning, and therefore, the same use as the BP availability importance in Subsection 15.7.3. As in Remark 5.3.5, an efficient method for computing the BP importance is Monte Carlo simulation.

Definition 15.7.4   *A restore criticality importance of component i is defined as the percentage of times that system restoration results from the restoration of component i in* $[0, t]$, *that is,*

$$I_{\text{RCI}}(i; t) = \frac{number\ of\ actions\ on\ component\ i\ that\ restore\ system\ in\ [0, t]}{number\ of\ times\ the\ system\ is\ restored\ in\ [0, t]}.$$

Alternatively, it can be defined as the percentage of times that a restoration of component i results in system restoration from a failure state in $[0, t]$, *that is,*

$$I_{\text{RCI}}(i; t) = \frac{number\ of\ actions\ on\ component\ i\ that\ restore\ system\ in\ [0, t]}{number\ of\ times\ component\ i\ is\ restored\ in\ [0, t]}.$$

Definition 15.7.5   *An operational criticality importance of component i is defined as the percentage of downtime of component i over the system downtime in* $[0, t]$, *that is,*

$$I_{\text{OCI}}(i; t) = \frac{total\ downtime\ of\ component\ i\ when\ system\ down\ in\ [0, t]}{total\ system\ downtime\ in\ [0, t]}.$$

Alternatively, it can be defined as the percentage of uptime of component i over the system uptime in $[0, t]$, *that is,*

$$I_{\text{OCI}}(i; t) = \frac{total\ uptime\ of\ component\ i\ when\ system\ up\ in\ [0, t]}{total\ system\ uptime\ in\ [0, t]}.$$

These simulation methods for defining importance measures are straightforward to implement once the system simulation model is constructed.

15.8   Applications in the Power Industry

In the power industry, electricity systems require high reliability and low outage. Supposing that limited resources are available to maintain and improve the performance of an electricity system, managers and administrators want to invest resources to upgrade old and aging components and to guarantee the maximum increase of system performance. Importance measures of components can be used for this purpose to identify the weakest areas of the system.

Before designing the importance measures of components for an electricity system, it is necessary first to analyze the system's mechanisms. Take the example of an electricity transmission system (ETS) in the power industry, which is composed mainly of components such as lines, transformers, breakers, and buses. All of these components are interconnected with the aim of transporting electrical energy from the bulk transmission system to various load points. An ETS has no definite lifetime period and is expected to function endlessly without any service interruptions. In addition, it is a repairable system, and the components are characterized by outage rates and repair rates instead of probabilities of failure. An ETS exhibits failures as a stochastic process in terms of outage rates. Thus, to be used in an ETS, reliability importance measures, which assume a fixed mission time, are not appropriate; importance measures need to be revised in terms of outage.

Espiritu et al. (2007) revised the B-importance, criticality importance, FV importance, RAW, and RRW for use in an ETS. Assume that (i) component failures are statistically independent and failure at one component does not impact the outage rate of the other components; (ii) time between component outages and repair durations are distributed in accordance with known exponential distributions; and (iii) the system is in steady state. Thus, let α_i denote the sustained outage rate of component i and let β_i denote the average repair time of component i. Let $\bar{A}$ denote the system steady-state unavailability; $\bar{A}$ is a function of $\boldsymbol{\alpha} = (\alpha_1, \alpha_2, \ldots, \alpha_n)$ and $\boldsymbol{\beta} = (\beta_1, \beta_2, \ldots, \beta_n)$. In order to adopt the original ideas of the relevant importance measures, a lower limit l_i and an upper limit u_i of outage rate specification are selected for each individual component in the system. The selection of these values depends on each individual component, and past data or experience may be useful. Typically, u_i could be a very large value, and l_i could be zero.

Using this notation, the transformed B-importance according to Equation (4.11) is

$$I_{B^T}(i) = \bar{A}(\boldsymbol{\alpha}, \boldsymbol{\beta}|\alpha_i = u_i) - \bar{A}(\boldsymbol{\alpha}, \boldsymbol{\beta}|\alpha_i = l_i),$$

where $\bar{A}(\boldsymbol{\alpha}, \boldsymbol{\beta}|\alpha_i = u_i)$ is the system unavailability when the sustained outage rate of component i is u_i, and $\bar{A}(\boldsymbol{\alpha}, \boldsymbol{\beta}|\alpha_i = l_i)$ is the system unavailability when the sustained outage rate of component i is l_i. $I_{B^T}(i)$ represents the maximum change in system unavailability when component i switches from the condition of highest possible availability to the condition of lowest possible availability.

Similarly, the transformed criticality importance for system failure is

$$I_{Cf^T}(i) = \left(\bar{A}(\boldsymbol{\alpha}, \boldsymbol{\beta}|\alpha_i = u_i) - \bar{A}(\boldsymbol{\alpha}, \boldsymbol{\beta}|\alpha_i = l_i)\right) \frac{\alpha_i}{\bar{A}(\boldsymbol{\alpha}, \boldsymbol{\beta})}.$$

The transformed FV importance (approximately) is

$$I_{FV^T}(i) = \frac{\bar{A}(\boldsymbol{\alpha}, \boldsymbol{\beta}|\alpha_i = u_i) - \bar{A}(\boldsymbol{\alpha}, \boldsymbol{\beta})}{\bar{A}(\boldsymbol{\alpha}, \boldsymbol{\beta})}.$$

The transformed RAW is

$$I_{RAW^T}(i) = \frac{\bar{A}(\boldsymbol{\alpha}, \boldsymbol{\beta})}{\bar{A}(\boldsymbol{\alpha}, \boldsymbol{\beta}|\alpha_i = l_i)}.$$

Finally, the transformed RRW is

$$I_{RRW^T}(i) = \frac{\bar{A}(\boldsymbol{\alpha}, \boldsymbol{\beta}|\alpha_i = u_i)}{\bar{A}(\boldsymbol{\alpha}, \boldsymbol{\beta})}.$$

Espiritu et al. (2007) demonstrated the applications of these importance measures on some commonly used electrical configurations, such as breaker-and-a-half, breaker-and-a-third, and dual-element-spot-network for an ETS. Assessment of reliability and importance measures has a variety of applications in the power system planning, operation, and maintenance (Akhavein and Fotuhi-Firuzabad 2011; Hong and Lee 2009).

References

Abouammoh AM and Al-Kadi MA. 1991. On measures of importance for components in multistate coherent systems. *Microelectronics and Reliability* **31**, 109–122.

Akhavein A and Fotuhi-Firuzabad M. 2011. A heuristic-based approach for reliability importance assessment of energy producers. *Energy Policy* **39**, 1562–1568.

Andrews JD and Beeson S. 1983. On the s-importance of elements and prime implicants of noncoherent systems. *IEEE Transactions on Reliability* **R-32**, 21–25.

Andrews JD and Beeson S. 2003. Birnbaum's measure of component importance for noncoherent systems. *IEEE Transactions on Reliability* **52**, 213–219.

Barabady J and Kumar U. 2007. Availability allocation through importance measures. *International Journal of Quality and Reliability Management* **24**, 643–657.

Barlow R and Wu A. 1978. Coherent systems with multistate components. *Mathematics of Operations Research* **3**, 275–281.

Barlow RE and Proschan F. 1965. *Mathematical Theory of Reliability*. John Wiley & Sons, New York.

Barlow RE and Proschan F. 1975. Importance of system components and fault tree events. *Statistic Processes and Their Applications* **3**, 153–172.

Baxter LA. 1984. Continuum structures I. *Journal of Applied Probability* **21**, 802–815.

Baxter LA. 1986. Continuum structures II. *Mathematical Proceedings of the Cambridge Philosophical Society* **99**, 331–338.

Baxter LA and Kim C. 1986. Bounding the stochastic performance of continuum structure functions I. *Journal of Applied Probability* **23**, 660–669.

Baxter LA and Kim C. 1987. Bounding the stochastic performance of continuum structure functions II. *Journal of Applied Probability* **24**, 609–618.

Baxter LA and Lee SM. 1989. Further properties of reliability importance for continuum structure functions. *Probability in the Engineering and Informational Sciences* **3**, 237–246.

Beeson S and Andrews JD. 2003. Importance measures for non-coherent system analysis. *IEEE Transactions on Reliability* **52**, 301–310.

Birnbaum ZW. 1969. On the importance of different components in a multicomponent system. In *Multivariate Analysis, Vol. 2* (ed. Krishnaiah PR). Academic Press, New York, pp. 581–592.

Block, HW and Savits TH. 1982. A decomposition for multistate monotone system. *Journal of Applied Probability* **19**, 391–402.

Block HW and Savits TH. 1984. Continuous multistate structure functions. *Operations Research* **32**, 703–714.

Borgonovo E. 2010. The reliability importance of components and prime implicants in coherent and non-coherent systems including total-order interactions. *European Journal of Operational Research* **204**, 485–495.

Bossche A. 1987. Calculation of critical importance for multi-state components. *IEEE Transactions on Reliability* **R-36**, 247–249.

Brunelle RD and Kapur KC. 1998. Continuous-state system reliability: An interpolation approach. *IEEE Transactions on Reliability* **47**, 181–187.

Brunelle RD and Kapur KC. 1999. Review and classification of reliability measures for multistate and continuum models. *IIE Transactions* **31**, 1171–1180.

Bueno VC. 1989. On the importance of components for multistate monotone systems. *Statistics and Probability Letters* **7**, 51–59.

Butler DA. 1979. A complete importance ranking for components of binary coherent systems with extensions to multi-state systems. *Naval Research Logistics* **4**, 565–578.

Chacko VM and Manoharan M. 2011. Joint importance measures in network system. *Reliability: Theory & Applications #04 (23)* **2**, 129–139.

Eisenberg NA and Sagar B. 2000. Importance measures for nuclear waste repositories. *Reliability Engineering and System Safety* **70**, 217–239.

Espiritu JF, Coit DW, and Prakash U. 2007. Component criticality importance measures for the power industry. *Electric Power Systems Research* **77**, 407–420.

Finkelstein MS. 1994. Once more on measures of importance of system components. *Microelectronics and Reliability* **34**, 1431–1439.

Freixas J and Pons M. 2008. Identifying optimal components in a reliability system. *IEEE Transactions on Reliability* **57**, 163–170.

Fussell JB. 1975. How to hand-calculate system reliability and safety characteristics. *IEEE Transactions on Reliability* **R-24**, 169–174.

Gámiz ML and Martínez Miranda MD. 2010. Regression analysis of the structure function for reliability evaluation of continuous-state system. *Reliability Engineering and System Safety* **95**, 134–142.

Gandini A. 1990. Importance & sensitivity analysis in assessing system reliability. *IEEE Transactions on Reliability* **39**, 61–70.

Griffith WS. 1980. Multistate reliability models. *Journal of Applied Probability* **17**, 735–744.

Hong Y and Lee L. 2009. Reliability assessment of generation and transmission systems using fault-tree analysis. *Energy Conversion and Management* **50**, 2810–2817.

Huseby AB and Natvig B. 2010. Advanced discrete simulation methods applied to repairable multistate systems. In *Reliability, Risk and Safety: Theory and Applications, Vol. 1* (eds. Bris R, Guedes Soares C, and Martorell S). CRC Press, London, pp. 659–666.

Inagaki T and Henley EJ. 1980. Probabilistic evaluation of prime implicants and top-events for non-coherent systems. *IEEE Transactions on Reliability* **R-29**, 361–367.

Jain SP and Gopal K. 1985. Reliability of k-to-ℓ-out-of-n systems. *Reliability Engineering* **12**, 175–179.

Jensen PA and Barnes JW. 1980. *Network Flow Programming*. John Wiley & Sons, New York.

Kim C and Baxter LA. 1987a. Axiomatic characterizations of continuum structure functions. *Operations Research Letters* **6**, 297–300.

Kim C and Baxter LA. 1987b. Reliability importance for continuum structure functions. *Journal of Applied Probability* **24**, 779–785.

Kim C and Baxter LA. 1997. Approximation of reliability importance for continuum structure functions. *Kangweon-Kyungki Mathematical Journal* **5**, 55–60.

Kuo W and Zuo MJ. 2003. *Optimal Reliability Modeling: Principles and Applications*. John Wiley & Sons, New York.

Lambert HE. 1975a. *Fault Trees for Decision Making in System Safety and Availability*. PhD Thesis, University of California, Berkeley.

Lambert HE. 1975b. Measure of importance of events and cut sets in fault trees. In *Reliability and Fault Tree Analysis* (eds. Barlow RE, Fussell JB, and Singpurwalla ND). Society for Industrial and Applied Mathematics, Philadelphia, pp. 77–100.

Levitin G. 2004. Protection survivability importance in systems with multilevel protection. *Quality and Reliability Engineering International* **20**, 727–738.

Levitin G and Lisnianski A. 1999. Importance and sensitivity analysis of multi-state systems using the universal generating function method. *Reliability Engineering and System Safety* **65**, 271–282.

Levitin G, Podofillini L, and Zio E. 2003. Generalised importance measures for multi-state elements based on performance level restrictions. *Reliability Engineering and System Safety* **82**, 287–298.

Lisnianski A. 2001. Estimation of boundary points for continuum-state system reliability measures. *Reliability Engineeging and System Safety* **74**, 81–88.

Lu L and Jiang J. 2007. Joint failure importance for noncoherent fault trees. *IEEE Transactions on Reliability* **56**, 435–443.

Meng FC. 1993. Component-relevancy and characterization results in multistate systems. *IEEE Transactions on Reliability* **42**, 478–483.

Meng FC. 1996. More on optimal allocation of componenets in coherent systems. *Journal of Applied Probability* **33**, 548–556.

Montero J, Tejada J, and Yáñez J. 1990. Structural properties of continuum systems. *European Journal of Operational Research* **45**, 231–240.

Natvig B. 1979. A suggestion of a new measure of importance of system component. *Stochastic Processes and Their Applications* **9**, 319–330.

Natvig B. 1982. Two suggestions of how to define a multistate coherent system. *Advances in Applied Probability* **14**, 434–455.

Natvig B. 1985. Recent developments in multistate reliability theory. In *Probabilistic Methods in the Mechanics of Solids and Structures* (eds. Eggwertz S and Lind NC). Springer Verlag, Berlin, pp. 385–393.

Natvig B. 2011. *Multistate Systems Reliability Theory with Applications*. Wiley Series in Probability and Statistics. John Wiley & Sons, West Sussex, UK.

Natvig B, Eide KA, Gåsemyr J, Huseby AB, and Isaksen SL. 2009. Simulation based analysis and an application to an offshore oil and gas production system of the Natvig measures of component importance in repairable systems. *Reliability Engineering and System Safety* **94**, 1629–1638.

Natvig B and Gåsemyr J. 2009. New results on the Barlow-Proschan and Natvig measures of component importance in nonrepairable and repairable systems. *Methodology and Computing in Applied Probability* **11**, 603–620.

Natvig B, Huseby AB, and Reistadbakk MO. 2011. Measures of component importance in repairable multistate systems—a numerical study. *Reliability Engineering and System Safety* **96**, 1680–1690.

Pham H. 1991. Optimal design for a class of noncoherent systems. *IEEE Transactions on Reliability* **40**, 361–363.

Ramirez-Marquez JE and Coit DW. 2005. Composite importance measures for multi-state systems with multi-state components. *IEEE Transactions on Reliability* **54**, 517–529.

Ramirez-Marquez JE, Rocco CM, Gebre BA, Coit DW, and Tortorella M. 2006. New insights on multi-state component criticality and importance. *Reliability Engineering and System Safety* **91**, 894–904.

Upadhyaya SJ and Pham H. 1993. Analysis of noncoherent systems and an architecture for the computation of the system reliability. *IEEE Transactions on Computers* **42**, 484–493.

Vesely WE. 1970. A time dependent methodology for fault tree evaluation. *Nuclear Engineering and Design* **13**, 337–360.

Wang W, Loman J, and Vassiliou P. 2004. Reliability importance of componetns in complex system. *Proceedings of Annual Reliability and Maintainability Symposium*, pp. 6–8.

Wu S. 2005. Joint importance of multistate systems. *Computers and Industrial Engineering* **49**, 63–75.

Wu S and Chan LY. 2003. Performance utility-analysis of multi-state systems. *IEEE Transactions on Reliability* **52**, 14–21.

Yang K and Xue J. 1996. Continuous state reliability analysis. *Processings of Annual Reliability and Maintainability Symposium*, pp. 251–257.

Zhang Q and Mei Q. 1985. Element importance and system failure frequency of a 2-state system. *IEEE Transactions on Reliability* **R-34**, 308–313.

Zio E, Marella M, and Podofillini L. 2007. Importance measures-based prioritization for improving the performance of multi-state systems: Application to the railway industry. *Reliability Engineering and System Safety* **92**, 1303–1314.

Zio E and Podofillini L. 2003a. Importance measures of multi-state components in multi-state systems. *International Journal of Reliability Quality and Safety Engineering* **10**, 289–310.

Zio E and Podofillini L. 2003b. Monte-Carlo simulation analysis of the effects on different system performance levels on the importance on multi-state components. *Reliability Engineering and System Safety* **82**, 63–73.

Zio E, Podofillini L, and Levitin G. 2004. Estimation of the importance measures of multi-state elements by Monte Carlo simulation. *Reliability Engineering and System Safety* **86**, 191–204.

Part Five

Broad Implications to Risk and Mathematical Programming

Introduction

As partially demonstrated in Chapter 1, the concept of importance measures has entered various disciplines. This part investigates the broad implications of importance measures to mathematical programming and risk, for example, (i) network flow, (ii) $\mathbb{K}$-terminal network, (iii) mathematical programming, (iv) sensitivity analysis including software reliability, and (v) PRA and PSA as well as fault diagnosis and maintenance. Chapters 16–18 investigate the applications (i)–(iv) in mathematical programming. Chapter 19 addresses the importance measures in application (v) of PRA and PSA and fault diagnosis.

It is interesting to see the connections of importance measures from reliability to other disciplines as pointed out in Zhu and Kuo (2012) which reviewed the implications of importance measures in broader areas. Part Five studies and extends these implications. For example, Freixas and Puente (2002) listed the basic analogies between reliability and game theory (e.g., component to player; functioning to winning coalition; failure to losing coalition; system structure function to simple game; monotone system to monotonic simple game; reliability function to multilinear extension). The B-importance, BP importance, and permutation importance in reliability, respectively, coincide with the well known Banzhaf index (Banzhaf 1965), Shapley-Shubik index (Shapley and Shubik 1954), and desirability relation (Isbell 1956; Taylor and Zwicker 1999) in game theory. The connections between different disciplines promote the transdisciplinary applications of various importance measures and the methodologies.

References

Banzhaf JF. 1965. Weighted voting doesn't work: A mathematical analysis. *Rutges Law Review* **19**, 317–343.

Freixas J and Puente MA. 2002. Reliability importance measures of the components in a system based on semivalues and probabilistic values. *Annals of Operations Research* **109**, 331–342.

Isbell J. 1956. A class of majority games. *Quarterly Journal of Mathematics. Oxford Series* **7**, 183–187.

Shapley LS and Shubik M. 1954. A method for evaluating the distribution of power in a committee system. *American Political Science Review* **48**, 787–792.

Taylor A and Zwicker W. 1999. *Simple Games: Desirability Relations, Trading, and Pseudoweightings*. Princeton Universtiy Press, Princeton, NJ.

Zhu X and Kuo W. 2012. Importance measures in reliability and mathematical programming. *Annals of Operations Research*, Accepted.

16

Networks

This chapter has two sections, discussing importance measures in network flow systems and $\mathbb{K}$-terminal networks, respectively. These importance measures evaluate the relative importance of edges with respect to the performance of various networks and could be used to address transportation problems such as in Examples 1.0.3 and 1.0.7.

16.1 Network Flow Systems

Consider a single-commodity network flow system on a directed graph with a single input and a single output vertex. The network has n edges, numbering from 1 to n, and *edge i represents component i*. For edge (component) i, $S_i = \{s_{i0}, s_{i1}, \ldots, s_{iK_i}\}$ is a set of finite capacity states, where $K_i + 1$ is the number of capacity states of edge i, $1 \leq K_i < \infty$, and $s_{i\mu}$ is the μth capacity state of edge i, $0 \leq \mu \leq K_i$. Assume that $0 = s_{i0} < s_{i1} < \ldots < s_{iK_i}$, $i = 1, 2, \ldots, n$, where $s_{i0} = 0$ is the state of complete failure, and let $S^n = \{\mathbf{y} = (y_1, y_2, \ldots, y_n) : y_i \in S_i, i = 1, 2, \ldots, n\}$ be the component state space.

Let Y_i be a random variable indicating the capacity state of edge (component) i (at a given time point), $i = 1, 2, \ldots, n$. Y_i could be one of the values in S_i. Let $p_{i\mu} = \Pr\{Y_i = s_{i\mu}\}$ and $\mathbf{P} = [p_{i\mu}]$. The flow through each edge cannot exceed its capacity state Y_i, and flow is conserved at each vertex except the input and output vertices. Introduce random vector $\mathbf{Y} = (Y_1, Y_2, \ldots, Y_n)$.

The capacity state of the system, $\phi(\mathbf{y})$, is defined to be the maximum flow that can be transmitted from input vertex to output vertex when the capacity state vector of edges is specified by $\mathbf{y} \in S^n$. For a required capacity level of the system, c, the system functions if and only if $\phi(\mathbf{y}) \geq c$. The *system availability* is defined to be $\Pr\{\phi(\mathbf{Y}) \geq c\}$, the probability that at least c units of flow can be transmitted through the network at a given time point. *Assume that the required capacity c is constant and fixed.*

The system might, for example, be a gas/oil production system or a power transmission system (Aven and Ostebo 1986). Usually, system availability is calculated for various capacity levels (c). In some cases, this involves all possible capacity levels; in other cases, only the level or levels specified in, for example, the design or sales contract, are taken into account. For such

Importance Measures in Reliability, Risk, and Optimization: Principles and Applications, First Edition.
Way Kuo and Xiaoyan Zhu. © 2012 John Wiley & Sons, Ltd. Published 2012 by John Wiley & Sons, Ltd.

a network flow system, different importance measures are developed to evaluate the relative importance of edges with respect to different criteria. Subsection 16.1.1 focuses on a series of edge importance measures that were proposed by Aven and Ostebo (1986). Subsection 16.1.2 demonstrates the relations of these importance measures to those in a binary monotone reliability system.

Essentially, such a defined probabilistic network flow system is an MMS (see Section 15.3), with each edge of multiple capacity states representing a multistate component. Subsection 16.1.3 generalizes the B-importance, FV importance, RAW, and RRW for such an MMS in the context of network flow. Subsection 16.1.4 presents the importance measures of an edge that explicitly take into account the disruptions of other edges.

16.1.1   The Edge Importance Measures in a Network Flow System

Definition 16.1.1   *The availability improvement potential of edge i, denoted by $I_{E_1}(i; P)$, is defined as*

$$I_{E_1}(i; P) = \Pr\{\phi(\infty_i, \mathbf{Y}) \geq c\} - \Pr\{\phi(\mathbf{Y}) \geq c\}.$$

Using the expectation instead of probability generates the importance measure

$$I_{E_2}(i; P) = \mathbb{E}(\phi(\infty_i, \mathbf{Y})) - \mathbb{E}(\phi(\mathbf{Y})),$$

assuming $\mathbb{E}(\phi(\mathbf{Y})) < \infty$.

For a component in a binary system, the improvement potential importance in Subsection 4.1.1 and the L_4 TIL importance in Subsection 10.2.5 assume that the component is replaced with a perfect one of reliability one. Similar in spirit to them, $I_{E_1}(i; P)$ ($I_{E_2}(i; P)$) equals the improvement of system availability (expected system capacity) obtained by setting the capacity of edge i equal to infinity with probability one, that is, assuming that edge i does not restrict the flow in the system. Note that $I_{E_1}(i; P) = \Pr\{\phi(\infty_i, \mathbf{Y}) \geq c$ and $\phi(\mathbf{Y}) < c\}$; hence, $I_{E_1}(i; P)$ equals the probability that edge i acts as a bottleneck in the system. To calculate $I_{E_1}(i; P)$ and $I_{E_2}(i; P)$, it is necessary to calculate $\Pr\{\phi(\infty_i, \mathbf{Y}) \geq c\}$, $\Pr\{\phi(\mathbf{Y}) \geq c\}$, $\mathbb{E}(\phi(\infty_i, \mathbf{Y}))$, and $\mathbb{E}(\phi(\mathbf{Y}))$, which can be dealt with by an efficient algorithm in Doulliez and Jamoulle (1972) when Y_i are independent.

Recall Example 1.0.7 where each flight leg corresponds to an edge in an airline network. If a flight leg is down, then the capacity of the corresponding edge is zero. It might be desirable to know the effect of the failure of the flight leg on the system survivability, which could be addressed by the next importance measure.

Definition 16.1.2   *The survivability potential   of edge i, denoted by $I_{E_3}(i; P)$, is defined as*

$$I_{E_3}(i; P) = \Pr\{\phi(\mathbf{Y}) \geq c\} - \Pr\{\phi(0_i, \mathbf{Y}) \geq c\}.$$

Suppose that a flight leg i is down because the airplane for that leg suddenly experiences mechanical trouble. Instead of canceling flight leg i, another airplane, originally assigned to flight leg j, $j \neq i$, might be dispatched to flight leg i. The to-be-canceled flight leg j can be determined such that $I_{E_3}(j; P) < I_{E_3}(i; P)$, then the cancellation of flight leg j has less impact on overall system survivability than that of flight leg i.

A *cut* is a set of edges that failing together (i.e., whose capacity states equal zero) is sufficient to result in zero system capacity. A cut is *minimal* if it cannot be reduced and still be a cut. Let $\overline{\mathcal{C}}_i$ be the set of minimal cuts containing edge i. On the basis of the cuts, another edge importance measure is defined as follows.

Definition 16.1.3 *The expected unutilized capacity of edge i, denoted by $I_{E_4}(i; P)$, is defined as*

$$I_{E_4}(i; P) = \sum_{y \in S^n : \phi(y) \geq c} \min\{y_i, \phi_i(y) - \phi(y)\} \Pr\{Y = y\},$$

where $\phi_i(y) = \min_{C \in \overline{\mathcal{C}}_i} \sum_{j \in C} y_j$.

Note that by the max-flow min-cut theorem (Jensen and Barnes 1980), $\phi(y) = \min_{1 \leq k \leq u} \sum_{j \in C_k} y_j$, where u is the number of minimal cuts and C_k is the kth minimal cut for $k = 1, 2, \ldots, u$. It can be observed that $\min\{y_i, \phi_i(y) - \phi(y)\}$ equals the unutilized capacity of edge i, that is, the maximum amount of capacity state of edge i that can be reduced without changing the system capacity state. Thus, $I_{E_4}(i; P)$ is the expectation of the unutilized capacity of edge i that is taken over all $y \in S^n$ for which the system functions, that is, $\phi(y) \geq c$.

When a limited budget is available to improve system availability or expected system capacity by increasing the capacity of edges, $I_{E_1}(\cdot; P)$ or $I_{E_2}(\cdot; P)$ may be used as a good indicator to locate the budget to the most important edges; it is also a bottleneck identifier. In another situation, if some of the edge capacities must be reduced for some reason (e.g., budget issues), $I_{E_3}(\cdot; P)$ may be used to see how much system availability will be reduced if an edge totally fails, and $I_{E_4}(\cdot; P)$ may be used to indicate which edges and how much capacity can be cut off without changing the current system capacity.

The following numerical example illustrates the calculations of $I_{E_k}(\cdot; P)$, $k = 1, 2, 3, 4$.

Example 16.1.4 Considering the network shown in Figure 16.1, the system has $n = 5$ edges (components). The minimal cuts are $\{1, 2\}$, $\{1, 5\}$, $\{4, 5\}$, and $\{2, 3, 4\}$. Assume that $S_1 = S_2 = S_3 = S_4 = \{0, 10\}$ and $S_5 = \{0, 20\}$; Y_i are independent and $\Pr\{Y_i = 0\} = 0.1, i = 1, 2, \ldots, 5$; and $c = 20$. Then, $\Pr\{\phi(Y) \geq 20\} = 0.722$; $\mathbb{E}(\phi(Y)) = 16.929$; and

$I_{E_1}(1; P) = 0.153$	$I_{E_2}(1; P) = 8.271$	$I_{E_3}(1; P) = 0.722$	$I_{E_4}(1; P) = 0.00$
$I_{E_1}(2; P) = 0.178$	$I_{E_2}(2; P) = 9.171$	$I_{E_3}(2; P) = 0.722$	$I_{E_4}(2; P) = 0.00$
$I_{E_1}(3; P) = 0.007$	$I_{E_2}(3; P) = 0.081$	$I_{E_3}(3; P) = 0.066$	$I_{E_4}(3; P) = 5.91$
$I_{E_1}(4; P) = 0.007$	$I_{E_2}(4; P) = 0.171$	$I_{E_3}(4; P) = 0.066$	$I_{E_4}(4; P) = 5.91$
$I_{E_1}(5; P) = 0.080$	$I_{E_2}(5; P) = 0.981$	$I_{E_3}(5; P) = 0.722$	$I_{E_4}(5; P) = 6.56.$

Edges 1 and 2 are of the largest availability improvement potential, the largest survivability potential, and the smallest expected unutilized capacity (this fact does not come as any surprise). With a capacity of 20 for edge 1 (edge 2) (because $c = 20$), the system availability would increase with 0.153 (0.178) from 0.722 to 0.875 (0.900). Note that the rankings of edges induced by the availability improvement potential in terms of both probability ($I_{E_1}(\cdot; P)$) and expectation ($I_{E_2}(\cdot; P)$) are almost the same.

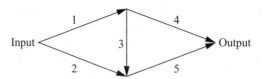

Figure 16.1   An example of a network flow system

Observing the survivability potential, the failure of edge 1, 2, or 5 causes the system to fail (i.e., $\Pr\{\phi(0_i, \mathbf{Y}) \geq 20\} = 0$, $i = 1, 2, 5$); however, the failure of edge 3 or 4 would not result in the failure of the system, although it would reduce the system availability. Thus, in the case of the airline system as in Example 1.0.7, if the airplane for flight leg 1 is down, then the airplane for flight leg 3 or 4 might be dispatched to flight leg 1, if possible, so that the system can still survive.

Since $\phi(\mathbf{Y}) \geq 20$ implies $y_1 = 10$ and $y_2 = 10$, it is clear that the expected unutilized capacity for each of edges 1 and 2 is zero. Edge 5 has the largest expected unutilized capacity. If edges 1, 2, 4, and 5 function, then the unutilized capacity of edge 5 is 10.

Additionally, Whitson and Ramirez-Marquez (2009) defined a static resiliency for a group of simultaneous failures of size α. The static resiliency allows recognizing the overall sensitivity of the network capacity to external failures and can be evaluated by Monte Carlo simulation.

Definition 16.1.5   *The expected resiliency of the network in the presence of α failures, denoted by $I_{\text{Resiliency}}(\alpha)$, is defined as*

$$I_{\text{Resiliency}}(\alpha) = \sum_{\Im} \Pr\{\phi(\mathbf{y}) \geq c | \alpha, \Im\} \, \Pr\{\Im\},$$

where $\Pr\{\phi(\mathbf{y}) \geq c | \alpha, \Im\}$ is the probability density function of network reliability when α external failures affect the network, and the index $\Im$ describes the specific α-failure case scenario.

16.1.2   The Edge Importance Measures for a Binary Monotone System

In this subsection, specifically, assume that each edge (component) in the network flow system has exactly two states: high s_{i1} and low s_{i0} with $s_{i1} > s_{i0} = 0$. Define random variable $X_i = 1$ if component i is in high state ($Y_i = s_{i1}$) and is 0 otherwise, and $p_i = \Pr\{X_i = 1\} = \Pr\{Y_i = s_{i1}\}$, $i = 1, 2, \ldots, n$. A component fails if its capacity is zero, and no flow is then allowed through it. The inspiration is to show the connection between the edge importance measures in a network flow system with the component importance measures in a reliability system.

Let structure function $\varphi(\mathbf{x}) = 1$ if and only if $\phi(\mathbf{y}) \geq c$ with respect to a given required system capacity c; thus, $R(\mathbf{p}) = \Pr\{\varphi(\mathbf{X}) = 1\} = \Pr\{\phi(\mathbf{Y}) \geq c\}$. Assume that $\phi(\mathbf{y}) \geq c$ when all components are in their high capacity states, that is, $\mathbf{y} = (s_{11}, s_{21}, \ldots, s_{n1})$. Of course, $\phi(\mathbf{y}) = 0$ when all components fail, that is, $\mathbf{y} = (s_{10}, s_{20}, \ldots, s_{n0}) = \mathbf{0}$. Then, the system of n components with structure function φ forms a binary monotone system.

Definition 16.1.6 *In a binary monotone (network flow) system, two importance measures of component i, denoted by $I_{E_5}(i; \mathbf{p})$ and $I_{E_6}(i; \mathbf{p})$, are defined as*

$$
\begin{aligned}
I_{E_5}(i; \mathbf{p}) &= \Pr\{\phi(s_{i1}, \mathbf{Y}) \geq c\} - \Pr\{\phi(\mathbf{Y}) \geq c\} \\
&= \Pr\{\varphi(1_i, \mathbf{X}) = 1\} - \Pr\{\varphi(\mathbf{X}) = 1\} \\
&= R(1_i, \mathbf{p}) - R(\mathbf{p})
\end{aligned}
$$

and

$$
I_{E_6}(i; \mathbf{p}) = \Pr\{\varphi(1_i, \mathbf{X}) = 1 \text{ and } \varphi(0_i, \mathbf{X}) = 1\} = \Pr\{\varphi(0_i, \mathbf{X}) = 1\} = R(0_i, \mathbf{p}).
$$

Similar to but different from $I_{E_1}(i; \mathbf{p})$, in which the capacity of component i is assumed to be increased to infinity, $I_{E_5}(i; \mathbf{p})$ indicates the improvement of system availability obtained by setting the capacity of component i equal to high level with probability one. Note that

$$
I_{E_5}(i; \mathbf{p}) = (1 - p_i)I_B(i; \mathbf{p}) = I_{Bs}(i; \mathbf{p}).
$$

Similarly, $I_{E_3}(i; \mathbf{p})$ for the binary monotone system can be derived as follows:

$$
\begin{aligned}
I_{E_3}(i; \mathbf{p}) &= \Pr\{\phi(\mathbf{Y}) \geq c\} - \Pr\{\phi(s_{i0}, \mathbf{Y}) \geq c\} \\
&= \Pr\{\varphi(\mathbf{X}) = 1\} - \Pr\{\varphi(0_i, \mathbf{X}) = 1\} \\
&= R(\mathbf{p}) - R(0_i, \mathbf{p}) \\
&= p_i I_B(i; \mathbf{p}) = I_{Bf}(i; \mathbf{p}).
\end{aligned}
$$

Thus, $I_{E_5}(i; \mathbf{p}) + I_{E_3}(i; \mathbf{p}) = I_B(i; \mathbf{p})$. $I_{Bs}(i; \mathbf{p})$ and $I_{Bf}(i; \mathbf{p})$ are the B-reliability importance measure for system functioning and the one for system failure, respectively, as introduced in Definition 4.1.1 and in Subsection 4.1.1.

Similar in spirit to $I_{E_4}(i; \mathbf{p})$, $I_{E_6}(i; \mathbf{p})$ evaluates the expected "excess capacity" of component i in the binary monotone system as the probability that when component i and the system are functioning, if component i were to fail, the system would continue functioning. If component i is in series with the rest of the system, then there is no unutilized capacity for component i, and $I_{E_6}(i; \mathbf{p}) = 0$. For a k-out-of-n system, $I_{E_6}(i; \mathbf{p})$ is equal to the reliability of the corresponding k-out-of-$(n-1)$ system excluding component i. Therefore, it is easy to see that in a k-out-of-n system, $p_i \leq p_j$ implies $I_{E_6}(i; \mathbf{p}) \leq I_{E_6}(j; \mathbf{p})$.

In addition, for the binary scenario in this subsection, Rocco et al. (2010) defined a flow importance measure of a group of edges as the reduction of maximum flow of the network when the set of edges fail simultaneously. A multiobjective evolutionary heuristic is used to identify the most vulnerable group of edges in a network in terms of the flow importance measure. Rocco et al. (2010) tested the flow importance measure and its calculation heuristic in an Italian high-voltage electrical transmission network.

16.1.3   The B-importance, FV Importance, Reliability Achievement Worth, and Reliability Reduction Worth

Representing each edge of multiple capacity states as a multistate component, Ramirez-Marquez and Coit (2005) explicitly treated the network flow system as an MMS (see Section 15.3) and generalized the B-reliability importance, FV reliability importance, RAW, and RRW for such an MMS. For each of these importance measures, there are two generalizations; one only considers the possible states and not the probability of a component being in a state, which, however, another one includes.

In a binary case where $K_i = 1$, the B-reliability importance can be written as

$$I_{\mathrm{B}}(i; \mathbf{p}) = \frac{1}{K_i} \sum_{\mu=0}^{K_i} |\Pr\{\phi(\mathbf{X}) = 1 | X_i = \mu\} - \Pr\{\phi(\mathbf{X}) = 1\}|.$$

Analogous to this expression, Definition 16.1.7 gives the generalizations of the B-reliability importance.

Definition 16.1.7 *In a network flow system, a generalization of the B-reliability importance of component i, denoted by $I_{\mathrm{NF\text{-}B}}(i; \boldsymbol{P})$, is defined as*

$$I_{\mathrm{NF\text{-}B}}(i; \boldsymbol{P}) = \frac{1}{K_i} \sum_{\mu=0}^{K_i} \left| \Pr\{\phi(\mathbf{Y}) \geq c | Y_i = s_{i\mu}\} - \Pr\{\phi(\mathbf{Y}) \geq c\} \right|.$$

An alternative generalization of the B-reliability importance is expressed as

$$I_{\mathrm{NFA\text{-}B}}(i; \boldsymbol{P}) = \mathbb{E}\left(|\Pr\{\phi(\mathbf{Y}) \geq c | Y_i\} - \Pr\{\phi(\mathbf{Y}) \geq c\}|\right)$$

$$= \sum_{\mu=0}^{K_i} p_{i\mu} |\Pr\{\phi(\mathbf{Y}) \geq c | Y_i = s_{i\mu}\} - \Pr\{\phi(\mathbf{Y}) \geq c\}|$$

They are also referred to as average of the sum of absolute deviations and mean absolute deviation, respectively (Shrestha et al. 2010).

To generalize the FV importance, RAW, and RRW, let

$$\beta_{i\mu} = \frac{\Pr\{\phi(\mathbf{Y}) \geq c | Y_i = s_{i\mu}\} - \Pr\{\phi(\mathbf{Y}) \geq c\}}{\Pr\{\phi(\mathbf{Y}) \geq c\}}.$$

Definition 16.1.8 *In a network flow system, a generalization of the FV reliability importance of component i, denoted by $I_{\mathrm{NF\text{-}FV}}(i; \boldsymbol{P})$, is defined as*

$$I_{\mathrm{NF\text{-}FV}}(i; \boldsymbol{P}) = \frac{1}{K_i} \sum_{\mu=0}^{K_i} \max\{0, -\beta_{i\mu}\}.$$

An alternative generalization of the FV importance is expressed as

$$I_{\mathrm{NFA\text{-}FV}}(i; \boldsymbol{P}) = \mathbb{E}(\max\{0, -\beta_{i\mu}\}) = \sum_{\mu=0}^{K_i} p_{i\mu} \max\{0, -\beta_{i\mu}\}.$$

In the binary case, the RAW of component i is

$$I_{\text{RAW}}(i; \boldsymbol{P}) = \frac{\Pr\{\phi(1_i, \boldsymbol{X}) = 1\}}{\Pr\{\phi(\boldsymbol{X}) = 1\}}$$

$$= 1 + \max\left\{0, \frac{\Pr\{\phi(1_i, \boldsymbol{X}) = 1\}}{\Pr\{\phi(\boldsymbol{X}) = 1\}} - 1\right\}$$

$$= 1 + \sum_{\mu=0}^{1} \max\left\{0, \frac{\Pr\{\phi(\boldsymbol{X}) = 1 | X_i = \mu\}}{\Pr\{\phi(\boldsymbol{X}) = 1\}} - 1\right\}.$$

Thus, the following generalizations result.

Definition 16.1.9 *In a network flow system, a generalization of the RAW of component i, denoted by $I_{\text{NF-RAW}}(i; \boldsymbol{P})$, is defined as*

$$I_{\text{NF-RAW}}(i; \boldsymbol{P}) = 1 + \frac{1}{K_i} \sum_{\mu=0}^{K_i} \max\{0, \beta_{i\mu}\}.$$

An alternative generalization of the RAW is expressed as

$$I_{\text{NFA-RAW}}(i; \boldsymbol{P}) = 1 + \mathbb{E}(\max\{0, \beta_{i\mu}\}) = 1 + \sum_{\mu=0}^{K_i} p_{i\mu} \max\{0, \beta_{i\mu}\}.$$

Definition 16.1.10 *In a network flow system, two generalizations of the RRW of component i, denoted by $I_{\text{NF-RRW}}(i; \boldsymbol{P})$ and $I_{\text{NFA-RRW}}(i; \boldsymbol{P})$, are defined as*

$$I_{\text{NF-RRW}}(i; \boldsymbol{P}) = \frac{1}{1 - I_{\text{NF-FV}}(i; \boldsymbol{P})} \quad and \quad I_{\text{NFA-RRW}}(i; \boldsymbol{P}) = \frac{1}{1 - I_{\text{NFA-FV}}(i; \boldsymbol{P})}.$$

Note that these generalizations of importance measures are for a component rather than a state of a component. Thus, according to Section 15.5, these generalized importance measures are composite importance measures (i.e., type 1 generalization). To complement the analysis, Ramirez-Marquez and Coit (2007) developed a greedy heuristic exploiting these composite importance measures for redundancy allocation subject to a budget constraint. This greedy method is similar to the one in Table 9.3 for the binary system but based on the ratio of the importance measure to cost and checking the budget constraint. An analytical method based on multistate multivalued decision diagrams can be used to calculate these importance measures (Shrestha et al. 2010).

Now, suppose that required capacity varies with the time. Let vectors $\boldsymbol{T} = (T_1, T_2, \ldots, T_\varpi)$ and $\boldsymbol{c} = (c_1, c_2, \ldots, c_\varpi)$ define the duration T_ω and required capacity level c_ω of interval $\omega = 1, 2, \ldots, \varpi$. Then the system availability can be understood as the probability that during the total time interval $\sum_{\omega=1}^{\varpi} T_\omega$, the capacity of the system can meet a required capacity of c_ω during the duration T_ω for $\omega = 1, 2, \ldots, \varpi$. Thus, the system availability is defined to be

$$\frac{\sum_{\omega=1}^{\varpi} \Pr\{\phi(\boldsymbol{Y}) \geq c_\omega\} T_\omega}{\sum_{\omega=1}^{\varpi} T_\omega}.$$

For this system, Ramirez-Marquez and Coit (2005) extended all the earlier importance measures in this subsection to include this additional information by taking the sum over the time intervals $\omega = 1, 2, \ldots, \varpi$.

Now, assume that the required minimum level of system performance is a function of time t, $c(t)$. Then, the system availability at time t is the probability that at that time the components in the system are in the states with performance $\phi(\mathbf{Y}) \geq c(t)$, that is, $\Pr\{\phi(\mathbf{Y}) \geq c(t)\}$. For a given required performance function, $c(t)$, $t \in [0, \tau]$, and τ being a fixed mission time, the mean multistate unavailability over τ is defined as

$$\bar{A} = \frac{1}{\tau} \int_0^\tau [1 - \Pr\{\phi(\mathbf{Y}) \geq c(t)\}] dt. \tag{16.1}$$

Furthermore, introduce

$\bar{A}_i^{\leq \mu}$ = system mean unavaliability when component i is always in state μ or worse in $[0, \tau]$,

$\bar{A}_i^{> \mu}$ = system mean unavaliability when component i is always in state above μ in $[0, \tau]$.

Using these quantities, Zio and Podofillini (2003) extended the B-importance, FV importance, RAW, and RRW to a multistate component i at state μ as $\bar{A}_i^{\leq \mu} - \bar{A}_i^{> \mu}$, $(\bar{A} - \bar{A}_i^{> \mu})/\bar{A}$, $\bar{A}_i^{\leq \mu}/\bar{A}$, and $\bar{A}/\bar{A}_i^{> \mu}$, respectively.

However, the mean unavailability in Equation (16.1) does not relate to the actual performance of the system over τ. In order to measure the importance of the components with respect to the overall system performance, introduce

$$\bar{W}_i^{\leq \mu} = \text{system mean performance over } \tau \text{ when component } i$$
$$\text{is always in state } \mu \text{ or worse in } [0, \tau],$$

$$\bar{W}_i^{> \mu} = \text{system mean performance over } \tau \text{ when component } i$$
$$\text{is always in state above } \mu \text{ in } [0, \tau].$$

Correspondingly, Zio and Podofillini (2003) extended the B-importance, FV importance, RAW, and RRW to a multistate component i at state μ as $\bar{W}_i^{> \mu} - \bar{W}_i^{\leq \mu}$, $(\bar{W} - \bar{W}_i^{\leq \mu})/\bar{W}$, $\bar{W}_i^{> \mu}/\bar{W}$, and $\bar{W}/\bar{W}_i^{\leq \mu}$, respectively.

16.1.4 The Flow-based Importance and Impact-based Importance

As in Examples 1.0.3 and 1.0.7, Jenelius (2010) considered the importance of road links (edges) as rerouting alternatives when other links in the network are disrupted (due to events such as snow, floods, landslides, and car accidents). He proposed two importance measures based on traffic flow and disruption impacts, respectively. The traffic-flow-based importance and the disruption-impact-based importance are different from the edge importance measures in Subsection 16.1.1 and 16.1.3 that capture a link's role for transport efficiency under normal conditions.

Let i index the edge for which the importance is calculated, and let ℓ index some other edge that is closed. Further, let f_i^0 denote the flow across edge i when the entire network is fully operational (normal conditions) and let f_i^ℓ denote the same quantity when edge ℓ is closed, all other edges remaining open. Both f_i^0 and f_i^ℓ are quantities that are easily obtained from model-based traffic assignments and simulations. The flow-based importance of edge i with respect to edge ℓ is defined as

$$(f_i^\ell - f_i^0)_+,$$

where $(x)_+$ means that $(x)_+ = x$ if $x > 0$ and 0 otherwise. The change in flow $f_i^\ell - f_i^0$ captures the net change in flow across edge i when edge ℓ is closed. The global flow-based importance of edge i is obtained by a weighted sum over closures on every other edge $\ell \neq i$,

$$I_{\text{Flow}}(i) = \sum_{\ell \neq i} w_\ell (f_i^\ell - f_i^0)_+.$$

The weights w_ℓ reflect the significance of being an alternative to edge i. For example, they may represent the relative probabilities that each edge fails.

Although the flow-based importance captures how many users take advantage of edge i as a rerouting alternative for the closed edges, it does not consider how much worse the impact would have been, both for the redirected travelers themselves and for other travelers affected by congestion, if edge i had not been available for rerouting. Let set H_i contain all origin-destination pairs for which at least one used route contains the considered edge i. Let ΔT_i^ℓ denote the total impact of a closure of edge ℓ assuming that only travelers between origin-destination pairs in H_i are allowed to use edge i, and ΔT^ℓ denote the total impact of a closure of ℓ when all other edges including edge i are open for any travelers. The quantities ΔT_i^ℓ should be possible to calculate with any traffic assignment model that allows multiclass assignments, since one can then prohibit the use of edge i for origin-destination pairs not in H_i. The impact-based importance of edge i with respect to edge ℓ is defined as

$$\Delta T_i^\ell - \Delta T^\ell.$$

Then, the global impact-based importance of edge i is as the weighted sum

$$I_{\text{Impact}}(i) = \sum_{\ell \neq i} w_\ell (\Delta T_i^\ell - \Delta T^\ell).$$

Jenelius (2010) and its references provided other importance measures used in network flow and transportation.

16.2 $\mathbb{K}$-terminal Networks

Reliability systems can be modeled as networks representing communication, computers, transportation, and so on. The mathematical model of a reliability network is a graph composed of vertices and edges. Assume that all vertices are failure-free and that all edges are subject to failure at a certain probability. The following are widely cited reliability networks:

- a two-terminal reliability network: the graph reliability is the probability that a signal can be transmitted between a given source and sink in the graph,
- an all-terminal reliability network: the graph reliability is the probability that all vertices are joined to all other vertices by paths of functioning edges, and
- a $\mathbb{K}$-terminal reliability network: the graph reliability is the probability that all vertices in some specified set $\mathbb{K}$ of target vertices are all joined by paths of functioning edges.

All of these three reliability networks are intractable (Ball 1980). Apparently, the $\mathbb{K}$-terminal reliability network is more general. Typical all-terminal reliability networks include the airline network, in which vertices represent locations and edges represent the flight legs, for example, in Example 1.0.7, and the transport network (MacGregor et al. 1993) that entails the set of real physical facilities that provide the bandwidth used by large numbers of data sessions or voice calls. In a transport network, an adapted reliability concept, *connectability*, which is the probability that a path exists between two vertices, is used to characterize the ability of the transport network to satisfy various requests. Examples of $\mathbb{K}$-terminal networks are the power grids (Kuo and Zuo 2003) and electrical networks (e.g., Hilber and Bertling 2007), which contain more than one input vertex (supply point) and/or more than one output vertex (load point).

It is worthwhile to evaluate an edge in terms of its contribution to the overall network performance, which can be the graph reliability as defined earlier, average reliability between every pair of target vertices in $\mathbb{K}$ (Zhang et al. 2011), and a weighted average over the performance of all pairs of target vertices as in Example 16.2.6. Subsection 16.2.1 introduces three importance measures of edges that are used for this purpose. Subsection 16.2.2 discusses the use of importance measures in a $\mathbb{K}$-terminal optimization problem and its potential extension to operations research.

Nomenclature

- Cutset: a set of edges whose removal disconnects the specified set $\mathbb{K}$ of target vertices
- Pathset: a set of consecutive edges connecting the specified set $\mathbb{K}$ of target vertices
- Minimum cutset: the cutset of the smallest size among all cutsets
- Minimum pathset: the pathset of the smallest size among all pathsets

Notation

- G: a graph representing a network
- $G - e$: graph with edge e deleted from graph G
- G^*e: graph obtained from graph G with edge e contracted
- $\mathbb{K}$: a specified set of target vertices in graph G
- $nc(G, \mathbb{K})$: the number of minimum cutsets of graph G with respect to $\mathbb{K}$
- $np(G, \mathbb{K})$: the number of minimum pathsets of graph G with respect to $\mathbb{K}$
- p: edge reliability when all edges have the same reliability p, $0 < p < 1$
- $\mathbf{p}$: vector of edge reliability
- $\mathscr{P}$: set of pathsets
- p_e: reliability of edge e
- $Rel(G, \mathbb{K}, \mathbf{p})$: reliability of graph G with respect to $\mathbb{K}$ and the reliability of edges specified by vector $\mathbf{p}$
- $Rel(G, \mathbb{K}, p)$: reliability of graph G with respect to $\mathbb{K}$, assuming that all edges have the same reliability p
- $sc(G, \mathbb{K})$: size of a minimum cutset of graph G with respect to $\mathbb{K}$
- $sp(G, \mathbb{K})$: size of a minimum pathset of graph G with respect to $\mathbb{K}$

16.2.1   Importance Measures of an Edge

Definition 16.2.1   *In a $\mathbb{K}$-terminal reliability network G, the B-reliability importance of edge e is adapted as*

$$I_\mathrm{B}(e; \mathbf{p}) = \frac{\partial Rel(G, \mathbb{K}, \mathbf{p})}{\partial p_e} = Rel(G^*e, \mathbb{K}, \mathbf{p}) - Rel(G - e, \mathbb{K}, \mathbf{p}). \quad (16.2)$$

Then the B-i.i.d. importance of edge e when all edges have the same reliability p is expressed as

$$I_\mathrm{B}(e; p) = Rel(G^*e, \mathbb{K}, p) - Rel(G - e, \mathbb{K}, p).$$

The second equality in expression (16.2) holds because by pivoting on edge e,

$$Rel(G, \mathbb{K}, \mathbf{p}) = p_e Rel(G^*e, \mathbb{K}, \mathbf{p}) + (1 - p_e)Rel(G - e, \mathbb{K}, \mathbf{p}),$$

which results in Equation (16.2) by taking a partial derivative of $Rel(G, \mathbb{K}, \mathbf{p})$ with respect to p_e.

Definition 16.2.2   *Let e_1 and e_2 be two distinct edges in a $\mathbb{K}$-terminal reliability network G. Then edge e_1 is more link important than edge e_2, denoted by $e_1 >_\mathrm{li} e_2$, if*

$$Rel(G - e_2, \mathbb{K}, p) \geq Rel(G - e_1, \mathbb{K}, p) \text{ for all } 0 < p < 1 \quad (16.3)$$

or, equivalently,

$$Rel(G^*e_2, \mathbb{K}, p) \leq Rel(G^*e_1, \mathbb{K}, p) \text{ for all } 0 < p < 1, \quad (16.4)$$

and the strict inequality holds for some $0 < p < 1$. If equality holds for all $0 < p < 1$, then $e_1 =_\mathrm{li} e_2$.

Intuitively, edge e_1 is more link important than edge e_2 if edge e_1 makes a greater difference to graph reliability both by its connecting and by its disconnecting. However, the link importance is a partial ranking in that some pairs of components cannot be compared, because conditions (16.3) and (16.4) may hold for certain values of p but not for all p values between 0 and 1.

The link importance was first proposed by Page and Perry (1994), requiring that *both* conditions (16.3) and (16.4) hold. However, these two conditions are indeed equivalent. To see that, by pivoting on edges e_1 and e_2, respectively, then

$$Rel(G, \mathbb{K}, p) = pRel(G^*e_1, \mathbb{K}, p) + (1 - p)Rel(G - e_1, \mathbb{K}, p) \quad (16.5)$$

and

$$Rel(G, \mathbb{K}, p) = pRel(G^*e_2, \mathbb{K}, p) + (1 - p)Rel(G - e_2, \mathbb{K}, p).$$

Thus, if $Rel(G - e_2, \mathbb{K}, p) \geq Rel(G - e_1, \mathbb{K}, p)$, then $Rel(G^*e_2, \mathbb{K}, p) \leq Rel(G^*e_1, \mathbb{K}, p)$, and vice versa. Therefore, this importance measure should be defined as in Definition 16.2.2.

Note that from Equation (16.5),

$$Rel(G, \mathbb{K}, p) = Rel(G^*e, \mathbb{K}, p) - (1 - p)I_\mathrm{B}(e; p)$$
$$= Rel(G - e, \mathbb{K}, p) + pI_\mathrm{B}(e; p);$$

thus, $Rel(G^*e_1, \mathbb{K}, p) \geq Rel(G^*e_2, \mathbb{K}, p)$ and $Rel(G - e_1, \mathbb{K}, p) \leq Rel(G - e_2, \mathbb{K}, p)$ if and only if $I_B(e_1; p) \geq I_B(e_2; p)$. In other words, the link importance is equivalent to the uniform B-i.i.d. importance as shown in Theorem 16.2.3. In this regard, Armstrong (1997) indicated that the holding of a uniform B-i.i.d. importance pattern is a necessary condition for that of the corresponding link importance pattern.

Theorem 16.2.3 $e_1 >_{li} e_2$ *if and only if* $I_B(e_1; p) \geq I_B(e_2; p)$ *for all* $0 < p < 1$, *and the strict inequality holds for some* $0 < p < 1$. $e_1 =_{li} e_2$ *if and only if* $I_B(e_1; p) = I_B(e_2; p)$ *for all* $0 < p < 1$.

For dependent components, the link importance can be defined in terms of expectations rather than reliability and then Theorem 16.2.3 is still valid for the relation between the link importance and the uniform B-i.i.d. importance.

To compute the B-reliability importance and link importance, it is necessary to evaluate graph reliability. One method is to enumerate all pathsets; then,

$$Rel(G, \mathbb{K}, \mathbf{p}) = \sum_{P \in \mathscr{P}} \left(\prod_{e \in P} p_e \prod_{e \notin P} (1 - p_e) \right).$$

For a reducible network that can be fully reduced to a two-terminal network using seven simple reduction axioms, Hsu and Yuang (1999) proposed a two-phase algorithm that has a complexity of $\mathcal{O}(m)$ for the computation of the B-reliability importance of all links, where m is the total number of links in a network. The first phase performs network reduction and the second phase backtracks the reduction steps and computes the B-reliability importance. Other computation algorithms exist for deriving graph reliability (e.g., Bailey and Kulkarni 1986). All of these methods make the approaches using the B-reliability importance and link importance implementable.

Page and Perry (1994) defined a contract-delete importance measure of edges as in Definition 16.2.4. Apparently, the contract-delete importance is also a partial ranking.

Definition 16.2.4   *Let e_1 and e_2 be two distinct edges in a $\mathbb{K}$-terminal reliability network G. Edge e_1 is more contract-delete important than edge e_2, denoted by $e_1 >_{cd} e_2$, if the following four conditions are true.*

(i) $sp(G^*e_1, \mathbb{K}) \leq sp(G^*e_2, \mathbb{K})$. If the equality holds, then $np(G^*e_1, \mathbb{K}) \geq np(G^*e_2, \mathbb{K})$.
(ii) $sc(G^*e_1, \mathbb{K}) \geq sc(G^*e_2, \mathbb{K})$. If the equality holds, then $nc(G^*e_1, \mathbb{K}) \leq nc(G^*e_2, \mathbb{K})$.
(iii) $sp(G - e_1, \mathbb{K}) \geq sp(G - e_2, \mathbb{K})$. If the equality holds, then $np(G - e_1, \mathbb{K}) \leq np(G - e_2, \mathbb{K})$.
(iv) $sc(G - e_1, \mathbb{K}) \leq sc(G - e_2, \mathbb{K})$. If the equality holds, then $nc(G - e_1, \mathbb{K}) \geq nc(G - e_2, \mathbb{K})$.

Furthermore, at least one strict inequality in these four conditions must hold. If all of these four conditions hold equality, then $e_1 =_{cd} e_2$.

The contract-delete importance of edges in a $\mathbb{K}$-terminal reliability network is similar in spirit to the cut importance and path importance of components in a reliability system.

As shown in Example 16.2.6, the link importance can provide more precise information than the contract-delete importance because the former is defined on edge reliability p. Theorem 16.2.5 indicates the relationship between these two importance measures.

Theorem 16.2.5   *If $e_1 \geq_{li} e_2$, then $e_1 \geq_{cd} e_2$.*

Example 16.2.6 (Example 1.0.7 continued)   This example presents an application in an airline network in which the vertices represent locations and the edges represent the flight legs. For the airline network, a weight, w_{ij}, is associated with each pair of target vertices i and j to assess their relative significance, for example, representing the average passenger flow volume from locations i to j. Letting $\rho_{ij}(\mathbf{p})$ denote the probability that vertices i and j are connected, the expected weight for the route from vertices i to j is $w_{ij}\rho_{ij}(\mathbf{p})$, and the expected total weight for the entire network is $\sum_{i \neq j, i, j \in \mathbb{K}} w_{ij}\rho_{ij}(\mathbf{p})$. Thus, the ratio

$$\eta(\mathbf{p}) = \frac{\sum_{i \neq j, i, j \in \mathbb{K}} w_{ij}\rho_{ij}(\mathbf{p})}{\sum_{i \neq j, i, j \in \mathbb{K}} w_{ij}}$$

represents the expected percentage of the weight that can be realized. To analyze the effect of an edge on $\eta(\mathbf{p})$, an importance measure of edge e is designed as

$$\frac{\partial \eta(\mathbf{p})}{\partial p_e}, \tag{16.6}$$

which is essentially the sensitivity analysis. For a two-terminal reliability network, it turns out to be the B-reliability importance.

In an airline network, passenger flow volumes vary on different routes. The importance of a flight leg (i.e., an edge) should consider this factor. The importance measure (16.6) evaluates the relative importance of a flight leg to the overall system performance. For instance, if the airline company has to temperately cancel a flight leg as illustrated in Example 1.0.7, considering the effect of the cancellation on the overall operations, the flight leg with the smallest value of importance measure (16.6) should be canceled. Other restrictions are easily included in the decision. For example, if the flight leg to be canceled or delayed must start from location i, then the candidate flight legs are restricted to the outgoing edges connected to vertex i, and the candidate flight leg with the smallest value of importance measure (16.6) has to be chosen for cancellation.

16.2.2   A $\mathbb{K}$-terminal Optimization Problem

Chiu et al. (2001) investigated a $\mathbb{K}$-terminal reliability network design problem, which is to choose a set, $\mathbb{K}$, of vertices from a given undirected graph to maximize the graph reliability subject to $|\mathbb{K}| = \kappa$, where κ denotes a required order, $2 \leq \kappa \leq n$, and n is the number of vertices. They proposed a simple greedy heuristic that uses the weights of vertices to guide the selection of a new vertex (with the largest weight) to $\mathbb{K}$ at each iteration until $|\mathbb{K}| = \kappa$. Serving as an importance measure, the weights of vertices are approximated on the basis of the reliability of edges and the degree of the vertices (i.e., topology of the graph) and are recalculated at each iteration based on the current selection of $\mathbb{K}$.

The concepts in this $\mathbb{K}$-terminal reliability network design can be extended to a distribution system. In a distribution system, the $\mathbb{K}$-terminal optimization problem can model the choice of a subset of warehouses among many potential warehouses with the restriction of a fixed number of warehouses, possibly due to the limited resources. The goal is to maximize the reliability of the resulting system so that the chosen warehouses are maximally connected and the transportation among them is maximally available. Rather than using the reliability of edges, using the cost of transporting through the edges can model the problem of minimizing the total system cost. Shen et al. (2003) studied the design of such a distribution system that is to locate the warehouses and allocate the retailers to them by means of integer programming. These network topologies can also be characterized by their message-delay in communication, network capacity in network flow, and so on. Thus, typical applications exist in operations research.

In addition, a multiobjective optimization problem could arise from a $\mathbb{K}$-terminal network in which each pair of target vertices in $\mathbb{K}$ could serve as an objective and is expected to perform well, for example, to be reliably connected.

References

Armstrong MJ. 1997. Reliability-importance and dual failure-mode components. *IEEE Transactions on Reliability* **46**, 212–221.

Aven T and Ostebo R. 1986. Two new component importance measures for a flow network system. *Reliability Engineering* **14**, 75–80.

Bailey MP and Kulkarni VG. 1986. A recursive algorithm for computing exact reliability measures. *IEEE Transactions on Reliability* **R-35**, 36–40.

Ball MO. 1980. Complexity of network reliability computations. *Networks* **10**, 153–165.

Chiu CC, Yeh YS, and Chou JS. 2001. An effective algorithm for optimal k-terminal reliability of distributed systems. *Malaysian Journal of Library & Information Science* **6**, 101–118.

Doulliez P and Jamoulle E. 1972. Transportation networks with random arc capacities. *RAIRO* **3**, 45–60.

Hilber P and Bertling L. 2007. Component reliability importance indices for electrical networks. *Proceedings of the 8th International Power Engineering Conference, IPEC, Singapore.*

Hsu SJ and Yuang MC. 1999. Efficient computation of marginal reliability importance for reducible networks in network management. *Proceedings of the 1999 IEEE International Conference on Communications*, pp. 1039–1045.

Jenelius E. 2010. Redundancy importance: Links as rerouting alternatives during road network disruptions. *Procedia Engineering* **3**, 129–137.

Jensen PA and Barnes JW. 1980. *Network Flow Programming.* John Wiley & Sons, New York.

Kuo W and Zuo MJ. 2003. *Optimal Reliability Modeling: Principles and Applications.* John Wiley & Sons, New York.

MacGregor M, Grover WD, and Maydell UM. 1993. Connectability: A performance metric for reconfigurability transport networks. *IEEE Journal on Selected Areas in Communications* **11**, 1461–1468.

Page LB and Perry JE. 1994. Reliability polynomials and link importance in networks. *IEEE Transactions on Reliability* **43**, 51–58.

Ramirez-Marquez JE and Coit DW. 2005. Composite importance measures for multi-state systems with multi-state components. *IEEE Transactions on Reliability* **54**, 517–529.

Ramirez-Marquez JE and Coit DW. 2007. Multi-state component criticality analysis for reliability improvement in multi-state systems. *Reliability Engineering and System Safety* **92**, 1608–1619.

Rocco C, Ramirez-Marquez JE, Salazar D, and Zio E. 2010. A flow importance measure with application to an Italian transmission power system. *International Journal of Performability Engineering* **6**, 53–61.

Shen ZJM, Coullard C, and Daskin MS. 2003. A joint location-inventory model. *Transportation Science* **37**, 40–55.

Shrestha A, Xing L, and Coit DW. 2010. An efficient multistate multivalued decision diagram-based approach for multistate system sensitivity analysis. *IEEE Transactions on Reliability* **59**, 581–592.

Whitson JC and Ramirez-Marquez JE. 2009. Resiliency as a component importance measure in network reliability. *Reliability Engineering and System Safety* **94**, 1685–1693.

Zhang C, Ramirez-Marquez JE, and Sanseverino CMR. 2011. A holistic method for reliability performance assessment and critical components detection in complex networks. *IIE Transactions* **43**, 661–675.

17

Mathematical Programming

In mathematics, the terms "mathematical programming" and "mathematical optimization" refer to the study of problems in which one seeks to minimize or maximize a real function by systematically prescribing the values of real or integer variables within an allowed set. This problem can be represented in the following way:

Given: a function $f : \mathbf{S} \mapsto \mathbb{R}$ from some set $\mathbf{S} \in \mathbb{R}^n$
Sought: an element $\mathbf{x}^*$ in $\mathbf{S}$ such that $f(\mathbf{x}^*) \leq f(\mathbf{x})$ for all $\mathbf{x}$ in $\mathbf{S}$ ("minimization") or such that $f(\mathbf{x}^*) \geq f(\mathbf{x})$ for all $\mathbf{x}$ in $\mathbf{S}$ ("maximization").

Such a formulation is called an optimization problem or a mathematical programming problem. Many real and theoretical problems may be modeled from this general framework. Depending on the types of objective functions, constraints, and decision variables, optimization problems have several major subfields including linear programming, integer programming, quadratic programming, nonlinear programming, convex programming, semidefinite programming, and so on.

Linear programming studies the case in which the objective function f is linear and the set $\mathbf{S}$ is specified using only linear equalities and inequalities. Integer programming refers to the case in which all or some of the decision variables are restricted to be integer. Sections 17.1 and 17.2 investigate linear programming and integer programming, respectively.

Note that this chapter is not intended to provide a deep study of linear programming and integer programming or broad mathematical programming. Instead, it aims to relate the basic concepts and general methods of linear programming and integer programming to importance measures and to illustrate applications of the importance measures in these contexts. Similar studies may be developed for other types of optimization problems by elaborating on the insights and properties of the problems. For example, one way to solve a problem approximately is to develop heuristics using importance measures rather than directly dealing with the NP-hard character of a complex mathematical programming problem.

Importance Measures in Reliability, Risk, and Optimization: Principles and Applications, First Edition.
Way Kuo and Xiaoyan Zhu. © 2012 John Wiley & Sons, Ltd. Published 2012 by John Wiley & Sons, Ltd.

17.1 Linear Programming

First, Subsection 17.1.1 gives the basic concepts and notation that are used in linear programming and defines two types of importance measures for decision variables. The simplex algorithm has been a viable and popular tool for solving linear programming problems since its first invention in the summer of 1947, although several variants of the simplex algorithm have evolved and other new computing algorithms have been proposed. Thus, Subsection 17.1.2 restates the simplex algorithm using the importance measures defined in Subsection 17.1.1. Finally, Subsection 17.1.3 conducts sensitivity analysis of linear programming using these importance measures.

17.1.1 Basic Concepts

Let $x_1, x_2, \ldots, x_n$ be n decision variables. Introduce an n-dimensional column vector $\mathbf{x} = (x_1, x_2, \ldots, x_n)^T$. (Notation $(\cdot)^T$ denotes the transpose of matrix $\cdot$.) Consider the system $A\mathbf{x} = \mathbf{b}$ and $\mathbf{x} \geq \mathbf{0}$, where A is an $m \times n$ matrix and $\mathbf{b}$ is an m-dimensional column vector. Letting vector $\mathbf{a}_j$ be the jth column of matrix A, then $A = [\mathbf{a}_1, \mathbf{a}_2, \ldots, \mathbf{a}_n]$. Suppose that $n \geq m$ and $rank(A, \mathbf{b}) = rank(A) = m$. After possibly rearranging the columns of A, let $A = [B, N]$, where B is an $m \times m$ invertible matrix and N is an $m \times (n - m)$ matrix. In addition, let $\mathbf{x}_B$ and $\mathbf{x}_N$ denote the set of variables corresponding to the given matrices B and N, respectively. The solution $\mathbf{x} = \binom{\mathbf{x}_B}{\mathbf{x}_N}$ to the equations $A\mathbf{x} = \mathbf{b}$, where

$$\mathbf{x}_B = B^{-1}\mathbf{b} \quad \text{and} \quad \mathbf{x}_N = \mathbf{0},$$

is called a *basic solution* of the system. (Notation $(\cdot)^{-1}$ denotes the inverse of matrix $\cdot$.) If $\mathbf{x}_B \geq \mathbf{0}$, then $\mathbf{x}$ is called a *basic feasible solution* of the system. Here, B is called the *basic matrix* (or simply the *basis*) and N is called the *nonbasic matrix*. The elements of $\mathbf{x}_B$ are called *basic variables* (or dependent variables), and the elements of $\mathbf{x}_N$ are called *nonbasic variables* (or independent variables). Let W be the current set of the indices of the nonbasic variables. If $\mathbf{x}_B > \mathbf{0}$, then $\mathbf{x}$ is called a *nondegenerate basic feasible solution*, and if at least one element of $\mathbf{x}_B$ is zero, then $\mathbf{x}$ is called a *degenerate basic feasible solution*.

Now, consider a linear programming problem

$$
\begin{aligned}
\min \quad & z = \mathbf{c}\mathbf{x} \\
\text{subject to} \quad & A\mathbf{x} = \mathbf{b} \\
& \mathbf{x} \geq \mathbf{0},
\end{aligned}
\tag{17.1}
$$

where z denotes the objective function value. Note that if the original problem has inequality constraints, they can be changed to equality constraints by adding a nonnegative slack (additional) variable to each inequality constraint.

Suppose that there exist a basis B and a corresponding basic feasible solution $\binom{B^{-1}\mathbf{b}}{0}$ whose objective value is given by $z_0 = (\mathbf{c}_B, \mathbf{c}_N)\binom{B^{-1}\mathbf{b}}{0} = \mathbf{c}_B B^{-1}\mathbf{b}$. The feasibility of problem (17.1) requires that $\mathbf{x}_B \geq \mathbf{0}$, $\mathbf{x}_N \geq \mathbf{0}$, and that $\mathbf{b} = A\mathbf{x} = B\mathbf{x}_B + N\mathbf{x}_N$. Multiplying the last equation by B^{-1} and rearranging the terms,

$$\mathbf{x}_B = B^{-1}\mathbf{b} - B^{-1}N\mathbf{x}_N = B^{-1}\mathbf{b} - \sum_{j \in W} B^{-1}\mathbf{a}_j x_j,$$

and

$$z = \mathbf{cx} = \mathbf{c}_B\mathbf{x}_B + \mathbf{c}_N\mathbf{x}_N$$

$$= \mathbf{c}_B\left(B^{-1}\mathbf{b} - \sum_{j \in W} B^{-1}\mathbf{a}_j x_j\right) + \sum_{j \in W} c_j x_j$$

$$= z_0 - \sum_{j \in W} (\mathbf{c}_B B^{-1}\mathbf{a}_j - c_j)x_j.$$

Using these transformations, the linear programming problem (17.1) can be written as

$$\min \quad z = z_0 - \sum_{j \in W}(\mathbf{c}_B B^{-1}\mathbf{a}_j - c_j)x_j$$

$$\text{subject to} \quad \sum_{j \in W} B^{-1}\mathbf{a}_j x_j + \mathbf{x}_B = B^{-1}\mathbf{b} \tag{17.2}$$

$$x_j \geq 0,\ j \in W \quad \text{and} \quad \mathbf{x}_B \geq \mathbf{0}.$$

Without loss of generality, assume that no row in Equation (17.2) has all zeros in the columns of the nonbasic variables x_j, $j \in W$. Otherwise, the basic variable in such a row is known in value, and this row can be deleted from the problem. Note that the linear programming problem can equivalently be written in the nonbasic variable space, that is, in terms of the nonbasic variables, as follows.

$$\min \quad z = z_0 - \sum_{j \in W}(\mathbf{c}_B B^{-1}\mathbf{a}_j - c_j)x_j$$

$$\text{subject to} \quad \sum_{j \in W} B^{-1}\mathbf{a}_j x_j \leq B^{-1}\mathbf{b} \tag{17.3}$$

$$x_j \geq 0,\ j \in W.$$

In expression (17.3), the objective function z and the basic variables $\mathbf{x}_B$ have been represented in terms of the nonbasic variables. The key result now simply follows:

If $\mathbf{c}_B B^{-1}\mathbf{a}_j - c_j \leq 0$ for all $j \in W$, then the current basic feasible solution is optimal.

This should be clear by noting that since $\mathbf{c}_B B^{-1}\mathbf{a}_j - c_j \leq 0$ for all $j \in W$, $z \geq z_0$ for any feasible solution, and $z = z_0$ for the current basic feasible solution because $x_j = 0$ for all $j \in W$.

Expression (17.3) expresses the linear programming problem (17.1) in some $(n-m)$-dimensional space because the number of nonbasic variables is $n - m$. This is to be expected since there are $n - m$ independent variables or $n - m$ degrees of freedom in the constraint system. The values $c_j - \mathbf{c}_B B^{-1}\mathbf{a}_j$ are sometimes referred to as *reduced cost coefficients*, since they are the coefficients of the (nonbasic) variables in this reduced space. The quantity $\mathbf{c}_B B^{-1}\mathbf{a}_j - c_j$ for $j \in W$ plays a crucial role; thus, we define an importance measure for nonbasic variables using it.

Definition 17.1.1 *Consider the linear programming problem* (17.1) *and a given basic* **B** *that corresponds to a basic feasible solution. Then, for each nonbasic variable* x_j, *define a nonbasic variable importance measure with respect to* **B** *as*

$$I_{\text{NBV}}(j; \boldsymbol{B}) = \mathbf{c}_B \boldsymbol{B}^{-1} \mathbf{a}_j - c_j.$$

The objective function z in expression (17.3) changes with nonbasic variable x_j at a rate of

$$\frac{\partial z}{\partial x_j} = c_j - \mathbf{c}_B \boldsymbol{B}^{-1} \mathbf{a}_j = -I_{\text{NBV}}(j; \boldsymbol{B}). \tag{17.4}$$

From this point of view, this nonbasic variable importance in the linear programming problem is similar in spirit to the B-reliability importance of components in the reliability system.

If $I_{\text{NBV}}(\cdot; \boldsymbol{B}) > 0$ for some nonbasic variable x_k, the objective function value can be improved (i.e., decreased) by increasing x_k. However, if the vector $\boldsymbol{B}^{-1} \mathbf{a}_k$ contains at least one positive element, the increase motion of x_k will be blocked by some nonnegative constraint on the basic variables so that the increase of x_k starting from the current basic feasible solution has to be stopped. A first basic variable x_r that drops to zero is called the blocking variable because it blocks further increase of x_k. The detection of which basic variable blocks the increase of x_k depends on the quantity of $(\boldsymbol{B}^{-1}\mathbf{b})_i/(\boldsymbol{B}^{-1}\mathbf{a}_k)_i$, given that $(\boldsymbol{B}^{-1}\mathbf{a}_k)_i > 0$ (see constraints in expression (17.3)), where $(\cdot)_i$ denotes the ith element in the vector $\cdot$ and $1 \leq i \leq m$. Thus, we define an importance measure for basic variables using this quantity.

Definition 17.1.2 *Considering the linear programming problem* (17.1), *suppose that a basis* **B**, *which corresponds to a basic feasible solution, and a nonbasic variable* x_k *are given. Then, for each basic variable* x_i, *define a basic variable importance measure with respect to* **B** *and* x_k *as*

$$I_{\text{BV}}(i; \boldsymbol{B}, k) = \begin{cases} \dfrac{(\boldsymbol{B}^{-1}\mathbf{b})_i}{(\boldsymbol{B}^{-1}\mathbf{a}_k)_i} & \text{if } (\boldsymbol{B}^{-1}\mathbf{a}_k)_i > 0, \\ +\infty & \text{otherwise.} \end{cases}$$

The minimum ratio test determines that the basic variable x_r blocks the increase of the nonbasic variable x_k if

$$\frac{(\boldsymbol{B}^{-1}\mathbf{b})_r}{(\boldsymbol{B}^{-1}\mathbf{a}_k)_r} = \min_{1 \leq i \leq m} \frac{(\boldsymbol{B}^{-1}\mathbf{b})_i}{(\boldsymbol{B}^{-1}\mathbf{a}_k)_i}, \quad \text{that is, } I_{\text{BV}}(r; \boldsymbol{B}, k) = \min_{1 \leq i \leq m} I_{\text{BV}}(i; \boldsymbol{B}, k).$$

Thus, x_k enters the basis, x_r leaves the basis, and the new basis is generated.

Note that the nonbasic variable importance is defined with respect to the basis **B**, while the basic variable importance is defined with respect to the basis **B** and the selecting nonbasic variable x_k.

17.1.2 The Simplex Algorithm

On the basis of Subsection 17.1.1, given a basic feasible solution and a corresponding basis, the simplex algorithm either improves the solution if $I_{\text{NBV}}(k; \boldsymbol{B}) > 0$ for some nonbasic variable x_k,

Table 17.1 The simplex algorithm (minimization problem)

Initialization step

Start with an initial basic feasible solution with basis B. (Here, the procedure for finding an initial basis is ignored; interested readers can find them in Bazaraa et al. (1990).)

Main steps

1. Calculate $I_{NBV}(\cdot; B)$ for all nonbasic variables. Choose the nonbasic variable x_k with the largest value of $I_{NBV}(\cdot; B)$ among current nonbasic variables.
2. If $I_{NBV}(k; B) \leq 0$, then stop with the current basic feasible solution $\binom{B^{-1}b}{0}$ as an optimal solution. If $B^{-1}a_k \leq 0$, then stop with the conclusion that the optimal solution is unbounded along the ray

$$\left\{ \binom{B^{-1}b}{0} + x_k \binom{-B^{-1}a_k}{e_k} : x_k \geq 0 \right\},$$

where e_k is an $(n-m)$-dimensional vector of zeros except for a one at the kth position.
3. Calculate $I_{BV}(\cdot; B, k)$ for all basic variables. Choose the basic variable x_r with the smallest value of $I_{BV}(\cdot; B, k)$ among current basic variables.
4. Update the basis B by letting x_k enter the basis and x_r leave the basis (i.e., replace a_k with a_r). Go to step 1.

or justifies the current solution as optimal and stops if $I_{NBV}(j; B) \leq 0$ for all nonbasic variables $j \in W$. If $I_{NBV}(k; B) > 0$ and the vector $B^{-1}a_k$ contains at least one positive element, then the increase of x_k will be blocked by one of the current basic variables that drops to zero and leaves the basis. On the other hand, if $I_{NBV}(k; B) > 0$ and $B^{-1}a_k \leq 0$, then x_k can be increased indefinitely, and the optimal solution value is unbounded $(-\infty)$. This discussion reflects the simplex algorithm exactly. Table 17.1 gives a summary of the simplex algorithm in terms of the nonbasic variable importance and basic variable importance for solving the linear programming problem (17.1).

Note that the criterion for a nonbasic variable x_k to enter the basis in step 1 is known as Dantzig's rule. In the absence of degeneracy (and assuming feasibility), the simplex algorithm stops in a finite number of iterations, either with an optimal basic feasible solution or with the conclusion that the optimal value is unbounded (Bazaraa et al. 1990, Chapter 3).

Example 17.1.3 Consider the following problem:

$$
\begin{array}{rrrrl}
\min & -2x_1 & +x_2 & -x_3 & \\
\text{subject to} & x_1 & +x_2 & +x_3 & \leq 6 \\
& -x_1 & +2x_2 & & \leq 4 \\
& x_1, x_2, x_3 \geq 0. &
\end{array}
$$

After introducing the nonnegative slack variables x_4 and x_5, the equivalent problem is obtained as

$$
\begin{array}{rrrrrrl}
\min & -2x_1 & + x_2 & -x_3 & & & \\
\text{subject to} & x_1 & + x_2 & +x_3 & +x_4 & & = 6 \\
& -x_1 & +2x_2 & & & +x_5 & = 4 \\
& x_1, x_2, x_3, x_4, x_5 \geq 0. &
\end{array}
$$

Table 17.2 The procedure of the simplex algorithm for the problem in Example 17.1.3

Iter	Basis B	B^{-1}	Nonbasic variables	Basic variables	
1	$\begin{bmatrix} 1 & 0 \\ 0 & 1 \end{bmatrix}$	$\begin{bmatrix} 1 & 0 \\ 0 & 1 \end{bmatrix}$	x_1, x_2, x_3 $I_{\text{NBV}}(1; B) = 2$ $I_{\text{NBV}}(2; B) = -1$ $I_{\text{NBV}}(3; B) = 1$ Thus, $x_k = x_1$	x_4, x_5 $I_{\text{BV}}(4; B, 1) = 6$ $I_{\text{BV}}(5; B, 1) = \infty$ Thus, $x_r = x_4$	x_1 enters the basis x_4 leaves the basis
2	$\begin{bmatrix} 1 & 0 \\ -1 & 1 \end{bmatrix}$	$\begin{bmatrix} 1 & 0 \\ 1 & 1 \end{bmatrix}$	x_4, x_2, x_3 $I_{\text{NBV}}(4; B) = -2$ $I_{\text{NBV}}(2; B) = -3$ $I_{\text{NBV}}(3; B) = -1$ Thus, $x_k = x_3$	x_1, x_5	$I_{\text{NBV}}(3; B) = -1 < 0$; thus stop with the optimal solution $= \left(\begin{matrix} B^{-1}b \\ 0 \end{matrix} \right)^T$ $(x_1, x_2, x_3, x_4, x_5)$ $= (6, 0, 0, 0, 10)$

Table 17.2 lists the procedure of the simplex algorithm, which initializes $B = [\mathbf{a}_4, \mathbf{a}_5] = \begin{bmatrix} 1 & 0 \\ 0 & 1 \end{bmatrix}$.

Example 17.1.4 Consider the problem

$$
\begin{aligned}
\min \quad & -x_1 \quad -3x_2 \\
\text{subject to} \quad & 2x_1 \quad +3x_2 \quad \leq 6 \\
& -x_1 \quad +x_2 \quad \leq 1 \\
& x_1, x_2 \geq 0.
\end{aligned}
$$

After introducing the nonnegative slack variables x_3 and x_4, the equivalent problem is obtained as

$$
\begin{aligned}
\min \quad & -x_1 \quad -3x_2 \\
\text{subject to} \quad & 2x_1 \quad +3x_2 \quad +x_3 \qquad = 6 \\
& -x_1 \quad +x_2 \qquad \quad +x_4 = 1 \\
& x_1, x_2, x_3, x_4 \geq 0.
\end{aligned}
$$

Table 17.3 lists the procedure of the simplex algorithm, which initializes $B = [\mathbf{a}_3, \mathbf{a}_4] = \begin{bmatrix} 1 & 0 \\ 0 & 1 \end{bmatrix}$.

17.1.3 Sensitivity Analysis

In most practical applications, some of the problem data are not known exactly and hence are estimated as well as possible. It is important to be able to find the new optimal solution as other estimates of some of the data become available without extensively resolving the problem from scratch. Also, at early stages of problem formulation, some factors may be overlooked. It is

Table 17.3 The procedure of the simplex algorithm for the problem in Example 17.1.4

Iter	Basis B	B^{-1}	Nonbasic variables	Basic variables	
1	$\begin{bmatrix} 1 & 0 \\ 0 & 1 \end{bmatrix}$	$\begin{bmatrix} 1 & 0 \\ 0 & 1 \end{bmatrix}$	x_1, x_2 $I_{\text{NBV}}(1; B) = 1$ $I_{\text{NBV}}(2; B) = 3$ Thus, $x_k = x_2$	x_3, x_4 $I_{\text{BV}}(3; B, 2) = 2$ $I_{\text{BV}}(4; B, 2) = 1$ Thus, $x_r = x_4$	x_2 enters the basis x_4 leaves the basis
2	$\begin{bmatrix} 1 & 3 \\ 0 & 1 \end{bmatrix}$	$\begin{bmatrix} 1 & -3 \\ 0 & 1 \end{bmatrix}$	x_1, x_4 $I_{\text{NBV}}(1; B) = 4$ $I_{\text{NBV}}(4; B) = -3$ Thus, $x_k = x_1$	x_3, x_2 $I_{\text{BV}}(3; B, 1) = \dfrac{3}{5}$ $I_{\text{BV}}(2; B, 1) = \infty$ Thus, $x_r = x_3$	x_1 enters the basis x_3 leaves the basis
3	$\begin{bmatrix} 2 & 3 \\ -1 & 1 \end{bmatrix}$	$\begin{bmatrix} \frac{1}{5} & \frac{3}{5} \\ \frac{1}{5} & \frac{2}{5} \end{bmatrix}$	x_3, x_4 $I_{\text{NBV}}(3; B) = -\dfrac{4}{5}$ $I_{\text{NBV}}(4; B) = -\dfrac{3}{5}$ Thus, $x_k = x_4$	x_1, x_2	$I_{\text{NBV}}(4; B) = -\dfrac{3}{5} < 0;$ thus stop with the optimal solution $(x_1, x_2, x_3, x_4) = \begin{pmatrix} B^{-1}\mathbf{b} \\ \mathbf{0} \end{pmatrix}^T$ $= \left(\dfrac{3}{5}, \dfrac{8}{5}, 0, 0 \right)$

important to update the current solution in a way that takes these factors into account. These and other related topics constitute sensitivity analysis (Bazaraa et al. 1990).

Consider the linear programming problem (17.1) and suppose that the simplex algorithm produces an optimal basis B. This subsection describes how to make use of the optimality conditions and importance measures to find the new optimal solution without resolving the problem from scratch if there is a change in some of the problem data. According to Equation (17.4), $I_{\text{NBV}}(j; B) = -\partial z/\partial x_j$; thus, the nonbasic variable importance is appropriate for the sensitivity analysis.

Adding a new activity

Suppose that a new activity x_{n+1} with unit cost c_{n+1} and consumption column $\mathbf{a}_{n+1}$ is added to the linear programming problem (17.1). Without resolving the problem, it can easily be determined whether adding x_{n+1} changes the optimal solution. First calculate $I_{\text{NBV}}(n+1; B)$. If $I_{\text{NBV}}(n+1; B) \leq 0$, then $x_{n+1} = 0$ (nonbasic variable), and the current solution is still optimal. On the other hand, if $I_{\text{NBV}}(n+1; B) > 0$, then x_{n+1} is introduced into the current basis, and the simplex algorithm continues to find the new optimal solution.

Example 17.1.5 (Example 17.1.3 continued)

Consider the linear programming problem in Example 17.1.3 and its optimal solution. To find the new optimal solution for the case that a new activity $x_6 \geq 0$ with $c_6 = 1$ and $\mathbf{a}_6 = \binom{-1}{2}$ is introduced, calculate $I_{\text{NBV}}(6; B) = c_B B^{-1}\mathbf{a}_6 - c_6 = (-2, 0)\binom{-1}{2} - 1 = 1$. Therefore, x_6 enters the basis, and the simplex algorithm continues until the new optimal solution is obtained.

Change in the objective function coefficient of a nonbasic variable

Suppose that c_k is changed to c'_k and that x_k is a nonbasic variable with respect to the current optimal basis $\boldsymbol{B}$. In this case, $\mathbf{c}_B$ is not affected, and $I_{\mathrm{NBV}}(k; \boldsymbol{B}) = \mathbf{c}_B \boldsymbol{B}^{-1} \mathbf{a}_k - c_k$ is replaced with

$$I'_{\mathrm{NBV}}(k; \boldsymbol{B}) = \mathbf{c}_B \boldsymbol{B}^{-1} \mathbf{a}_k - c'_k = I_{\mathrm{NBV}}(k; \boldsymbol{B}) + (c_k - c'_k).$$

Note that $I_{\mathrm{NBV}}(k; \boldsymbol{B}) \leq 0$ because the current solution is optimal for the original problem. If $I'_{\mathrm{NBV}}(k; \boldsymbol{B}) \leq 0$, then the current solution is still optimal for the new problem. Otherwise, x_k is introduced into the basis, and the simplex algorithm continues as usual.

Example 17.1.6 (Example 17.1.3 continued)   Consider the linear programming problem in Example 17.1.3 and its optimal solution. Suppose that $c_2 = 1$ is replaced with -3. Since x_2 is a nonbasic variable, then $I'_{\mathrm{NBV}}(2; \boldsymbol{B}) = I_{\mathrm{NBV}}(2; \boldsymbol{B}) + (c_2 - c'_2) = -3 + (1 - (-3)) = 1 > 0$, and all other $I_{\mathrm{NBV}}(j; \boldsymbol{B})$ are unaffected. Hence, x_2 enters the basis, and the simplex algorithm continues to find the new optimal solution.

Change in the objective function coefficient of a basic variable

Now suppose that c_r is changed to c'_r and that x_r is a basic variable with respect to the current optimal basis $\boldsymbol{B}$. Then $\mathbf{c}_B$ is changed to $\mathbf{c}'_B$. For each nonbasic variable x_j, recalculate

$$\begin{aligned}
I'_{\mathrm{NBV}}(j; \boldsymbol{B}) &= \mathbf{c}'_B \boldsymbol{B}^{-1} \mathbf{a}_j - c_j \\
&= \mathbf{c}_B \boldsymbol{B}^{-1} \mathbf{a}_j - c_j + (c'_r - c_r)(\boldsymbol{B}^{-1} \mathbf{a}_j)_r \\
&= I_{\mathrm{NBV}}(j; \boldsymbol{B}) + (c'_r - c_r)(\boldsymbol{B}^{-1} \mathbf{a}_j)_r.
\end{aligned}$$

If $I'_{\mathrm{NBV}}(j; \boldsymbol{B}) \leq 0$ for all nonbasic variable x_j, then the current solution is still optimal, and the optimal objective value is changed to be $\mathbf{c}'_B \boldsymbol{B}^{-1} \mathbf{b}$. Otherwise, choose the nonbasic variable x_k with the largest positive value of $I'_{\mathrm{NBV}}(\cdot; \boldsymbol{B})$ as the entering variable, and the simplex algorithm continues as usual.

Example 17.1.7 (Example 17.1.3 continued)   Consider the linear programming problem in Example 17.1.3 and its optimal solution. Suppose that $c_1 = -2$ is replaced with zero. Since x_1 is a basic variable, it needs to recalculate the nonbasic variable importance for all nonbasic variables using new $\mathbf{c}'_B = (0, 0)$. Then, $I'_{\mathrm{NBV}}(4; \boldsymbol{B}) = 0$, $I'_{\mathrm{NBV}}(2; \boldsymbol{B}) = -1$, and $I'_{\mathrm{NBV}}(3; \boldsymbol{B}) = 1$. Hence, x_3 enters the basis, and the simplex algorithm continues to find the new optimal solution.

17.2   Integer Programming

The general approach for solving integer programming is the enumeration-based B&B algorithm, which has been used in addressing redundancy allocation (Kuo et al. 2006). Subsection 17.2.1 gives the descriptions of general integer programming and the B&B algorithm and discusses applications of importance measures in the B&B algorithm. Subsection 17.2.2 studies importance measures in the B&B algorithm for solving pure integer linear programming.

Subsection 17.2.3 reviews the importance measures in MINLP for reliability-redundancy problems.

17.2.1 Basic Concepts and Branch-and-bound Algorithm

Consider the general mixed integer programming (MIP) problem (maximization case)

$$z_{\text{MIP}} = \max\{f(\mathbf{x}) : \mathbf{x} = (\mathbf{x}', \mathbf{x}'') \in \mathbf{S}\} \text{ with } \mathbf{S} = \{\mathbf{x}' \in \mathbb{Z}_+^n, \mathbf{x}'' \in \mathbb{R}_+^m : \mathbf{g}(\mathbf{x}) \le \mathbf{0}\}, \quad (17.5)$$

where $\mathbb{Z}_+^n$ represents the set of nonnegative integer n-dimensional vectors; $\mathbb{R}_+^m$ represents the set of nonnegative real m-dimensional vectors; $\mathbf{x}' = (x_1, x_2, \ldots, x_n)$ and $\mathbf{x}'' = (x_{n+1}, x_{n+2}, \ldots, x_{n+m})$ are the decision variable vectors; $f(\mathbf{x})$ is the objective function; and $\mathbf{g}(\mathbf{x})$ is the vector of constraint functions. When all decision variables are integers (i.e., $\mathbf{x} = \mathbf{x}''$), it is called pure integer programming.

A general and important type of algorithms for solving the MIP in (17.5) is based on an enumerative approach, which is, in turn, based on the division of solution set $\mathbf{S}$. Note that $\{\mathbf{S}^k\}_{k=1}^{\bar{k}}$ is a division of $\mathbf{S}$ if $\bigcup_{k=1}^{\bar{k}} \mathbf{S}^k = \mathbf{S}$, where $\bar{k}$ denotes the number of subsets in the division. A division is called a partition if $\mathbf{S}^k \cap \mathbf{S}^j = \emptyset$ for any $k, j \in \{1, 2, \ldots, \bar{k}\}, k \ne j$. Relative to the MIP in (17.5) and a division of $\mathbf{S}$, a set of subproblems MIP^k, $k = 1, 2, \ldots, \bar{k}$, can be defined as

$$z_{\text{MIP}}^k = \max\{f(\mathbf{x}) : \mathbf{x} \in \mathbf{S}^k\}. \quad (17.6)$$

Then, $z_{\text{MIP}} = \max_{k=1}^{\bar{k}} z_{\text{MIP}}^k$. That is, if the MIP in (17.5) is hard to optimize over $\mathbf{S}$, it can be solved by optimizing subproblem MIP^k in (17.6) over subset $\mathbf{S}^k$, $k = 1, 2, \ldots, \bar{k}$ and then selecting the maximum z_{MIP}^k as the solution for the original MIP in (17.5).

The division is usually done recursively. As an example, the tree in Figure 17.1 shows a simple recursive division in which $\mathbf{S} = \{0, 1\}^n$ (with $n = 2$) and $\mathbf{S}^{\delta^1 \delta^2 \ldots \delta^{i_0}} = \mathbf{S} \cap \{\mathbf{x} \in \{0, 1\}^n : x_i = \delta^i \in \{0, 1\}, i = 1, 2, \ldots, i_0\}$ for $i_0 \le n$. Note that $\{0, 1\}^n$ denotes the set of binary n-dimensional vectors. Then, the division $\{\mathbf{S}^{\delta^1 \delta^2 \ldots \delta^n} : (\delta^1, \delta^2, \ldots, \delta^n) \in \{0, 1\}^n\}$ is a partition of $\mathbf{S}$.

In an extreme case, the division can turn into total enumeration of the elements of $\mathbf{S}$. The corresponding total enumeration is not viable for solving problems with more than a very small number of variables. To avoid brute force enumerating many subsets of $\mathbf{S}$, the enumeration tree is pruned at the enumeration node corresponding to a subset of $\mathbf{S}$ if it is justified that the optimal solution does not lie in the division of this subset. To accomplish the pruning, the

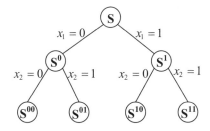

Figure 17.1   A simple division of $\mathbf{S} = \{0, 1\}^2$

Table 17.4   The general B&B algorithm

---

1. (Initialization) $\mathcal{L} = \{\text{MIP in (17.5)}\}$, $\mathbf{S}^0 = \mathbf{S}$, $\bar{z}^0 = \infty$, and $\underline{z}_{\text{MIP}} = -\infty$.
2. (Termination) If $\mathcal{L} = \emptyset$, then the solution $\mathbf{x}^*$ that yields $\underline{z}_{\text{MIP}} = f(\mathbf{x}^*)$ is optimal.
3. (Subproblem/node selection) Select and delete a subproblem MIP^k (i.e., a B&B node k) from $\mathcal{L}$.
4. (Relaxation problem) Design and solve the relaxation problem, RP^k.
5. (Pruning)
   (a) If RP^k is infeasible, go to step 2;
   (b) otherwise, if $z_{\text{RP}}^k \leq \underline{z}_{\text{MIP}}$, go to step 2;
   (c) otherwise, if $\mathbf{x}_{\text{RP}}^k \in \mathbf{S}^k$ and $z_{\text{RP}}^k > \underline{z}_{\text{MIP}}$, let $\underline{z}_{\text{MIP}} = z_{\text{RP}}^k$ and $\bar{z}^k = z_{\text{RP}}^k$. Delete from $\mathcal{L}$ all
       subproblems k' with $\bar{z}^{k'} \leq \underline{z}_{\text{MIP}}$ and go to step 2; and
   (d) otherwise, if $\mathbf{x}_{\text{RP}}^k \notin \mathbf{S}^k$, go to step 6.
6. (Division/branching) Select branching variable(s). On the basis of the branching variable(s), create a
   division of $\mathbf{S}^k$, denoted by $\{\mathbf{S}^{k_j}\}_{j=1}^{n_k}$ with n_k denoting the number of subsets in the division; add the
   corresponding subproblems $\{\text{MIP}^{k_j}\}_{j=1}^{n_k}$ to $\mathcal{L}$; and set $\bar{z}^{k_j} = z_{\text{RP}}^k$ for $j = 1, 2, \ldots, n_k$. Go to step 2.

---

upper bound is calculated at each node, for example, by solving a relaxation problem of MIP^k in (17.6), denoted by RP^k, as

$$z_{\text{RP}}^k = \max\{f(\mathbf{x}) : \mathbf{x} \in \mathbf{S}_{\text{RP}}^k \supseteq \mathbf{S}^k\}.$$

Set $\mathbf{S}_{\text{RP}}^k$ is a feasible region for the relaxation problem and contains $\mathbf{S}^k$; thus, $z_{\text{RP}}^k \geq z_{\text{MIP}}^k$. Let $\mathbf{x}_{\text{RP}}^k$ be an optimal solution of RP^k, if one exists, and let $\underline{z}_{\text{MIP}}$ be the lower bound for the MIP, normally, equal to the objective value of the best feasible solution found for the MIP. Then, the enumeration node corresponding to $\mathbf{S}^k$ can be pruned if one of the following three conditions holds:

(i) RP^k is infeasible, implying that MIP^k is infeasible;
(ii) $z_{\text{RP}}^k \leq \underline{z}_{\text{MIP}}$, implying that MIP^k cannot generate a better solution to the best one currently found;
(iii) $\mathbf{x}_{\text{RP}}^k \in \mathbf{S}^k$, implying $\mathbf{x}_{\text{RP}}^k$ is the optimal solution of MIP^k.

Such an enumerative relaxation approach is commonly referred to as B&B or implicit enumeration (Spielberg 1979). Integrating the aforementioned considerations, Table 17.4 presents a general B&B algorithm for solving the MIP in (17.5). In the description of the algorithm, $\mathcal{L}$ is a collection of subproblems MIP^k that have not been processed, and $\bar{z}^k$ is an upper bound for subproblem MIP^k in $\mathcal{L}$, that is, $\bar{z}^k \geq z_{\text{MIP}}^k$.

This general B&B algorithm has six steps. Step 1 creates the root node (node 0); step 2 tests the termination condition; step 3 selects a subproblem to process; step 4 designs and solves the relaxation problem and step 5 specifies the pruning conditions based on the result of the relaxation problem; and step 6 selects branching variable(s) and creates new subproblems accordingly. The B&B algorithm generates a B&B tree with each B&B node representing a unique subproblem, as later shown in Figure 17.2.

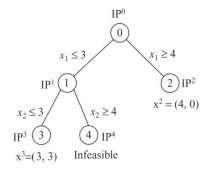

Figure 17.2 Procedure of the B&B algorithm for the problem in Example 17.2.3

Implementing importance measures in the B&B algorithm

In the B&B algorithm in Table 17.4, steps 1, 2, and 5 are standard, while steps 3, 4, and 6 leave flexibility for designing the B&B algorithm to solve a particular MIP problem based on its structure and properties. Designing and solving (exactly or heuristically) the relaxation problems in step 4 are much different for distinct MIP. Normally, the *linear* relaxation problems (i.e., $\mathbf{x} \in \mathbb{R}_+^{n+m}$) can be solved efficiently. The next subsection illustrates this typical relaxation in the context of pure integer linear programming. There is no unified way to incorporate the concept of importance measures to design and solve the relaxation problems, although for a particular problem, importance-measure-based methods could be developed to solve the relaxation problems. However, steps 3 and 6 can be well realized through the concepts of importance measures. Steps 3 and 6 directly affect the efficiency of pruning in step 5 and are vital for the success of the B&B algorithm.

For step 3, various subproblem (i.e., B&B node) selection rules such as depth-first, breadth-first, and best-first selections have been proposed (Beale 1979; Nemhauser and Wolsey 1988). According to a particular rule, at each iteration, the subproblems in $\mathcal{L}$ are ordered, and the one with the highest priority is processed by steps 4, 5, and 6 next. In this context, the importance measures can be designed to prioritize the subproblems in $\mathcal{L}$, and the subproblem with the highest importance is selected for processing. Such an importance measure of a subproblem could be defined on the basis of the estimated possibility of achieving an optimal solution for the original MIP from this subproblem, the expected computational effort to reach an integer solution from this subproblem, the improvement of upper and/or lower bounds, or other heuristics.

For step 6, various division rules, such as variable dichotomy (Dakin 1965) and generalized upper bound dichotomy (Tomlin 1970), have been proposed. The division usually involves the selection of integer variable(s) for branching and is then created by branching the region of the selected variable(s) into subregions, which depends on the solution of the relaxation problem at the B&B node. The most common way of choosing a branching variable is to specify priorities of the integer variables that are eligible for branching (e.g., the integer variables of fractional values in the solution of relaxation problems). Achterberg et al. (2005) summarized the rules for selecting branching variables (i.e., branching rules), including the most infeasible branching, strong branching, pseudo-cost branching, and reliability branching. Playing the same role as the division and branching rules, the importance measures of eligible integer variables or sets of integer variables can be defined. Then, the eligible integer variable(s) with the highest importance is (are) selected as the branching variable(s) for division.

The principle of designing the related importance measures is the same as the principle of designing the node selection rules and division/branching rules, that is, to facilitate the convergence of the B&B algorithm by quickly improving the lower ($\underline{z}_{\text{MIP}}$) and/or upper ($z_{\text{RP}}^k$) bounds and by efficient pruning. Techniques of tightening bounds for reliability optimization were studied by Kuo et al. (2006). Different importance measures can be defined according to various node selection rules and division/branching rules. During the B&B procedure, the choices of the importance measures can be changed in the distinct B&B nodes, equivalent to using the different rules. Subsection 17.2.2 presents several importance measures and their uses in this regard.

17.2.2   Branch-and-bound Using Linear Programming Relaxations

To detail a possible application of importance measures in the B&B algorithm in Table 17.4, this subsection uses pure integer linear programming as an example:

$$z_{\text{IP}} = \max\{\mathbf{c}\mathbf{x} : \mathbf{x} \in \mathbf{S}\} \text{ with } \mathbf{S} = \{\mathbf{x} \in \mathbb{Z}_+^n : A\mathbf{x} \leq \mathbf{b}\}, \tag{17.7}$$

where $\mathbf{c} \in \mathbb{R}^n$ is the objective coefficient vector. Essentially, however, all of the ideas carry over unchanged to the general MIP. To solve the integer programming in (17.7), a B&B algorithm with linear relaxation is widely used including in commercial software (Land and Powell 1979). That is, at B&B node k corresponding to an integer subproblem IP^k as

$$z_{\text{IP}}^k = \max\{\mathbf{c}\mathbf{x} : \mathbf{x} \in \mathbf{S}^k = \{\mathbf{x} \in \mathbb{Z}_+^n : A^k\mathbf{x} \leq \mathbf{b}^k\}\},$$

a linear relaxation problem LP^k is defined as

$$z_{\text{LP}}^k = \max\{\mathbf{c}\mathbf{x} : \mathbf{x} \in \mathbf{S}_{\text{LP}}^k = \{\mathbf{x} \in \mathbb{R}_+^n : A^k\mathbf{x} \leq \mathbf{b}^k\}\}$$

which is linear programming and can be solved efficiently, for example, by the simplex algorithm in Subsection 17.1.2. Here, A^k and $\mathbf{b}^k$ are extended matrix and vector from A and $\mathbf{b}$, respectively, which are unique for each subproblem and can be constructed as follows.

The B&B algorithm in Table 17.4 starts from the root node in step 1, in which $\mathbf{S}_{\text{LP}}^0 = \{\mathbf{x} \in \mathbb{R}_+^n : A\mathbf{x} \leq \mathbf{b}\}$. If the solution of LP^k, $\mathbf{x}_{\text{LP}}^k \notin \mathbf{S}^k$, a branching variable is selected in step 6 among the variables that have the fractional values in $\mathbf{x}_{\text{LP}}^k$. Suppose that x_j is selected and $x_j = \alpha$ in $\mathbf{x}_{\text{LP}}^k$ where α is a fractional value. Then, the division rule of variable dichotomy is usually used to generate two children (two branches) of B&B node k. The division is done by adding linear constraints $x_j \leq \lfloor \alpha \rfloor$ and $x_j \geq \lfloor \alpha \rfloor + 1$ to the constraint sets of the left and right children, respectively, so that $\mathbf{S}^k = \mathbf{S}^{k_1} \bigcup \mathbf{S}^{k_2}$ with $\mathbf{S}^{k_1} = \mathbf{S}^k \bigcap \{\mathbf{x} \in \mathbb{R}_+^n : x_j \leq \lfloor \alpha \rfloor\}$ and $\mathbf{S}^{k_2} = \mathbf{S}^k \bigcap \{\mathbf{x} \in \mathbb{R}_+^n : x_j \geq \lfloor \alpha \rfloor + 1\}$. Note that if x_j is binary, then it is equivalent to add constraints $x_j = 0$ and $x_j = 1$ to the left and right children, respectively. The rest of this subsection shows how appropriate importance measures can be defined for branching variable selection in step 6 and B&B node selection in step 3.

Branching variable selection

In the B&B algorithm in Table 17.4, at B&B node k, step 6 chooses an integer variable of the fractional value in the solution for LP^k (i.e., $\mathbf{x}_{\text{LP}}^k$) for branching and constructing the division. Denote $\mathbf{x}_{\text{LP}}^k = (\alpha_1^k, \alpha_2^k, \ldots, \alpha_n^k)$ and $\alpha_j^k = \lfloor \alpha_j^k \rfloor + f_j^k$ for $j = 1, 2, \ldots, n$. Let index set $N^k = \{j : f_j^k \neq 0\}$. One criterion of choosing a branching variable is based on the degradations of each variable in N^k. The degradation of a variable at B&B node k attempts to estimate the decrease in the upper bounds for the two children of node k that is caused by adding the new linear constraint on the variable. Supposing that the branching variable is x_j at B&B node k, estimate the decreases for the left and right children as $\rho_j^{-k} f_j^k$ and $\rho_j^{+k}(1 - f_j^k)$, respectively. The coefficients ρ_j^{-k} and ρ_j^{+k} are either known as a prior or are estimated using the in-process information (e.g., dual variable values at the node or information on previous branchings involving x_j) (Nemhauser and Wolsey 1988). On the basis of the degradations, we define two degradation importance measures of variables in N^k for directing the selection of branching variables in N^k.

Definition 17.2.1 *Given a B&B node k with $\mathbf{x}_{\text{LP}}^k = (\alpha_1^k, \alpha_2^k, \ldots, \alpha_n^k)$ and $N^k = \{j : f_j^k = \alpha_j^k - \lfloor \alpha_j^k \rfloor \neq 0, 1 \leq j \leq n\}$, the maximum and minimum degradation importance measures of integer variable x_j, $j \in N^k$, denoted by $I_{\text{MaxD}}(j; k)$ and $I_{\text{MinD}}(j; k)$, respectively, are defined as*

$$I_{\text{MaxD}}(j; k) = \max\{\rho_j^{-k} f_j^k, \rho_j^{+k}(1 - f_j^k)\} \ \text{ and } \ I_{\text{MinD}}(j; k) = \min\{\rho_j^{-k} f_j^k, \rho_j^{+k}(1 - f_j^k)\}.$$

Then, the branching variable should be the variable in N^k that has the largest value of the degradation importance measure. When $\rho_j^{+k} = \rho_j^{-k} = 1$, using the maximum degradation importance implies choosing the variable with the *minimum integer infeasibility* to the next branch, that is, the variable whose value is closest to an integer value but not equal to the integer value. The rationale of the maximum degradation importance is that the branch generated by branching the variable whose larger degradation is largest is most likely to be pruned. When $\rho_j^{+k} = \rho_j^{-k} = 1$, using the minimum degradation importance implies choosing the variable with the *maximum integer infeasibility* to the next branch, that is, the variable whose decimal part is closest to 0.5. The rationale of the minimum degradation importance is that the variable whose smaller degradation is largest is most likely to achieve integrality.

Node selection

In the B&B algorithm in Table 17.4, step 3 selects a subproblem IP^k (i.e., B&B node k) from list $\mathcal{L}$ for processing when $\mathcal{L}$ has more than one B&B node. In Definition 17.2.2, we define three node selection importance measures of B&B nodes in $\mathcal{L}$, which correspond to three different node selection rules. With respect to each type of importance measure, the B&B node with the largest value of importance measure should be selected.

Definition 17.2.2 *For a B&B algorithm with current collection of integer subproblems (B&B nodes), $\mathcal{L}$, three node selection importance measures of B&B node k in $\mathcal{L}$ are defined as*

$$I_{\text{NS}_1}(k) = z_{\text{LP}}^k \qquad \text{(best upper bound rule)}$$

$$I_{\text{NS}_2}(k) = \bar{z}^k \qquad \text{(best estimate rule)}$$

$$I_{\text{NS}_3}(k) = \frac{z_{\text{LP}}^k - z_{\text{IP}}}{z_{\text{LP}}^k - \bar{z}^k} \qquad \text{(quick improvement rule)},$$

where $\tilde{z}^k$ is an estimate of z_{IP}^k and can be computed (assuming that the degradations for each variable are independent) as

$$\tilde{z}^k = z_{LP}^k - \sum_{j \in N^k} \min\{\rho_j^{-k} f_j^k, \rho_j^{+k}(1 - f_j^k)\}.$$

Three node selection importance measures define different node selection rules. $I_{NS_1}(\cdot)$ corresponds to the *best upper bound rule* of selecting the node that has the largest objective value of the linear programming relaxation, z_{LP}^k; the rationale is that $z_{LP}^k \geq z_{IP}^k$ and the higher upper bound (i.e., z_{LP}^k) intuitively implies the higher z_{IP}^k. $I_{NS_2}(\cdot)$ corresponds to the *best estimate rule* of selecting the node that has the largest estimated objective value of z_{IP}^k and is thus believed to be most likely to contain an optimal or a good integer solution to obtain the largest possible value of $\underline{z}_{IP}$. $I_{NS_3}(\cdot)$ corresponds to the *quick improvement rule* of selecting the node that can most quickly yield a feasible integer solution and improve the lower bound $\underline{z}_{IP}$. $I_{NS_3}(\cdot)$ measures the potential improvement to the lower bound $\underline{z}_{IP}$ provided by each node and is used in some commercial software as the default option once a feasible solution is known.

Example 17.2.3   Consider a integer linear programming problem

$$
\begin{array}{rlrl}
z_{IP} = \min & 6x_1 & +2x_2 & \\
\text{subject to} & -x_1 & +2x_2 & \leq 4 \\
& 5x_1 & +x_2 & \leq 20 \\
& -2x_1 & -2x_2 & \leq -7 \\
& x_1, x_2 & \in \mathbb{Z}_+^1. &
\end{array}
\tag{17.8}
$$

At the root node (B&B node 0 and $\mathcal{L} = \{IP^0\}$), solve the linear relaxation of the integer problem (17.8) with $x_1, x_2 \geq 0$ instead of $x_1, x_2 \in \mathbb{Z}_+^1$, obtaining $z_{LP}^0 = 296/11$ and $\mathbf{x}_{LP}^0 = (36/11, 40/11) \notin \mathbb{Z}_+^2$ (step 5(d)). Then, step 6 needs to select either x_1 or x_2 as a branching variable. For that purpose, use the maximum degradation importance measure of variables with $\rho_j^{+k} = \rho_j^{-k} = 1$ and obtain $I_{MaxD}(1; k = 0) = \max\{3/11, 8/11\} = 8/11$ and $I_{MaxD}(2; k = 0) = \max\{7/11, 4/11\} = 7/11$. Hence, x_1 is chosen as the branching variable since $I_{MaxD}(1; k = 0) > I_{MaxD}(2; k = 0)$, and two new B&B nodes, indexed as nodes 1 and 2, are then created as shown in Figure 17.2. The subproblem at node 1 (node 2) has the added linear constraint $x_1 \leq 3$ ($x_1 \geq 4$) and the original constraints in the problem (17.8).

Now, step 3 uses $I_{NS_1}(\cdot)$ to select the next B&B node from $\mathcal{L} = \{IP^1, IP^2\}$ for processing. To calculate $I_{NS_1}(\cdot)$, solve the linear programming relaxation of both IP^1 and IP^2 and obtain $I_{NS_1}(1) = z_{LP}^1 = 25$ (at $\mathbf{x}_{LP}^1 = (3, 7/2)$) and $I_{NS_1}(2) = z_{LP}^2 = 24$ (at $\mathbf{x}_{LP}^2 = (4, 0)$). Thus, node 1 (i.e., IP^1) is selected, and the lower bound is updated to $\underline{z}_{IP} = z_{LP}^2 = 24$. In step 6, $x_2 = 7/2$ is the only integer variable of fractional value in $\mathbf{x}_{LP}^1$ and is thus the branching variable. Then, the two new nodes 3 and 4 are created by adding linear constraints $x_2 \leq 3$ and $x_2 \geq 4$ to IP^1, respectively, so $\mathcal{L} = \{IP^2, IP^3, IP^4\}$.

To proceed, select one node from $\mathcal{L} = \{IP^2, IP^3, IP^4\}$ based on $I_{NS_1}(\cdot)$. To calculate $I_{NS_1}(\cdot)$, solve the linear relaxation of both IP^3 and IP^4. Then, $I_{NS_1}(3) = z_{LP}^3 = 24$ (at $\mathbf{x}_{LP}^3 = (3, 3)$), and IP^4 is infeasible and thus pruned. Because $z_{LP}^3 = \underline{z}_{IP} = 24$, node 3 is pruned. Similarly, $z_{LP}^2 = \underline{z}_{IP} = 24$; thus, node 2 is pruned as well. The B&B algorithm is terminated because

$\mathcal{L} = \emptyset$. The optimal solutions for IP are $\mathbf{x}^* = (3, 3)$ and $(4, 0)$ with $z_{IP} = 24$. Figure 17.2 summarizes the B&B algorithm for this instance.

17.2.3   Mixed Integer Nonlinear Programming

In addition to the MINLP model for the CAP in Section 11.1, the mathematical formulation of an MINLP can be that the continuous variables represent component reliability and the integer variables represent the levels of redundancy. This is a typical reliability-redundancy allocation problem. Some methods, such as the B&B, generalized Benders decomposition, outer approximations, and so on, are useful in solving an MINLP. Tillman et al. (1977) were the first to report on the application of importance measures to solve a mixed integer reliability optimization problem.

In solving reliability-redundancy problems, different importance measures, including some heuristics-based approaches, have been adopted. A recent example of using an importance measure is in Ha and Kuo (2005) in solving such an MINLP problem, which is relatively unknown to the operations research community.

References

Achterberg T, Kocha T, and Martinb A. 2005. Branching rules revisited. *Operations Research Letters* **33**, 42–54.

Bazaraa MS, Jarvis JJ, and Sherali HD. 1990. *Linear Programming and Network Flows, 2nd edn.* John Wiley & Sons, Hoboken, New Jersey.

Beale EML. 1979. Branch and bound methods for mathematical programming systems. *Annals of Discrete Mathematics* **5**, 201–219.

Dakin RJ. 1965. A tree search algorithm for mixed integer programming problems. *Computer Journal* **8**, 250–255.

Ha C and Kuo W. 2005. Multi-path approach for reliability-redundancy allocation using a scaling method. *Journal of Heuristics* **11**, 201–217.

Kuo W, Prasad VR, Tillman FA, and Hwang CL. 2006. *Optimal Reliability Design: Fundamentals and Applications, 2nd edn.* Cambridge University Press, Cambridge, UK.

Land AH and Powell S. 1979. Computer codes for problems of integer programming. *Annals of Discrete Mathematics* **5**, 221–269.

Nemhauser GL and Wolsey LA. 1988. *Integer and Combinatorial Optimization.* Wiley-Interscience Series in Discrete Mathematics and Optimization. John Wiley & Sons, New York.

Spielberg K. 1979. Enumerative methods in integer programming. *Annals of Discrete Mathematics* **5**, 139–183.

Tillman FA, Hwang CL, and Kuo W. 1977. Determining component reliability and redundancy for optimum system reliability. *IEEE Transactions on Reliability* **R-26**, 162–165.

Tomlin JA. 1970. Branch and bound methods for integer and non-convex programming. In *Integer and Nonlinear Programming* (ed. Abadie J). Elsevier (North-Holland), New York, pp. 437–450.

18

Sensitivity Analysis

Sensitivity analysis is very useful in determining the impact of a particular input on an output under a set of assumptions, such as the effect that changes in interest rates have on a bond's price. Importance measures are close to sensitivity analysis, both of which provide a quantitative framework for investigating a system's mechanisms. They are the studies of how the variation (uncertainty) in the output of a mathematical model can be apportioned to different sources of variations in the inputs of the model. Sensitivity analysis has the complementary role of ordering by importance measures the strength and relevance of the inputs in determining the variation in the output. Sections 18.1 and 18.2 give an introductory and illuminative discussion on general optimization problems, focusing on importance measures in local sensitivity and perturbation analysis, and global sensitivity and uncertainty analysis, respectively. Both of them help in understanding the insights of a problem. Section 18.3 specially introduces the variance-based uncertainty importance measures that are used in the sensitivity analysis of uncertainty in systems reliability, including software reliability. Finally, Section 18.4 briefly describes the applications of sensitivity analysis, especially global sensitivity measures, in various areas.

18.1 Local Sensitivity and Perturbation Analysis

Assume that a mathematical model under investigation is described by a function $f(\mathbf{x})$, where the input $\mathbf{x} = (x_1, x_2, \ldots, x_n)$ is a point inside a certain region and the output $f(\mathbf{x})$ is a real number. For example, in reliability, the input region could be an n-dimensional unit hypercube of component reliability vectors, and $f(\mathbf{x})$ could be the systems reliability. Let $f(\mathbf{x}^*)$ be the required solution at the *nominal point* $\mathbf{x}^*$. In most papers, the sensitivity of the solution $f(\mathbf{x}^*)$ with respect to x_i is considered by estimating the partial derivative $(\partial f / \partial x_i)_{\mathbf{x}=\mathbf{x}^*}$. This approach is called local sensitivity. Local methods are commonly used on steady-state models or on addressing the stability of a nominal point.

18.1.1 The B-reliability Importance

According to Equation (4.10), the B-reliability importance evaluates the rate at which systems reliability improves with the reliability of a component. In this sense, the B-reliability

Importance Measures in Reliability, Risk, and Optimization: Principles and Applications, First Edition.
Way Kuo and Xiaoyan Zhu. © 2012 John Wiley & Sons, Ltd. Published 2012 by John Wiley & Sons, Ltd.

importance is a typical local sensitivity measure where $f(\mathbf{x})$ is the systems reliability and $\mathbf{x}$ is the component reliability vector.

Levitin and Lisnianski (1999) extended the B-reliability importance to a general case in which system output performance, denoted by function $g(\mathbf{p})$, is related to not only the reliability attributes $\mathbf{p}$ of components but also their nominal performance rates. The systems reliability is a measure of the ability of the system to meet the required performance level (i.e., demand). For example, in power engineering, it is the ability of the system to provide an adequate supply of electrical energy. To estimate the impact of each component on system output performance $g(\mathbf{p})$, the generalized B-reliability importance of component i is defined as

$$\frac{\partial g(\mathbf{p})}{\partial p_i}.$$

The system output performance $g(\mathbf{p})$ possibly represents system availability, mean system output performance, or mean unsupplied demand during an operating period, and so on. Straightforwardly, the generalized B-reliability importance is a local sensitivity analysis of the prescribed system output performance.

18.1.2   The Multidirectional Sensitivity Measure

In addition to the local sensitivity based on the generalized B-reliability importance of components in Subsection 18.1.1, Gandini (1990) and Do Van et al. (2008a, 2008b) proposed a sensitivity methodology concerned with basic system parameters. The methodology uses Markov process models and is based on importance measures. It is efficient, especially for *intercomponents, functional dependencies* (standby redundancy, shared load, shared resources, and so on), and *repairable components* in dynamic Markovian systems. For example, in an active redundancy system (e.g., in Figure 2.3), the failure rates of the other components are switched to higher values when one component fails because of a shared load.

In the context of discrete event dynamic Markov process models, Do Van et al. (2008a, 2008b) proposed a multidirectional sensitivity measure (MDSM), which provides a tool to investigate not only the importance of a given component but also the importance of a group of components, the importance of a system state, and, more generally, the effect of the simultaneous change of several design parameters that are related to the components or the system state.

For an irreducible homogeneous Markov process with a finite m-state space, let $\boldsymbol{T} = [t_{\ell j}]$ denote the transition rate matrix. A perturbation on one or a group of parameters of a Markov process is equivalent to a perturbation in the transition rate matrix $\boldsymbol{T}$. Then, the transition rate matrix $\boldsymbol{T}$ changes to

$$\boldsymbol{T}_\delta = \boldsymbol{T} + \delta \boldsymbol{Q}, \tag{18.1}$$

where δ is a real number and $\boldsymbol{Q} = [q_{\ell j}]$ is a matrix representing the *direction of perturbation*: $q_{\ell j} = 0$ indicates that the matrix entry $t_{\ell j}$ is not perturbed and a nonzero $q_{\ell j}$ indicates that the matrix entry $t_{\ell j}$ is perturbed by an amount $\delta q_{\ell j}$. The only condition on the structure of $\boldsymbol{Q}$ is that the matrix $\boldsymbol{T}_\delta$ is also a transition rate matrix. That is, the sum of each row of $\boldsymbol{Q}$ equals zero, $\boldsymbol{Q}\mathbf{e} = \mathbf{0}$ with $\mathbf{e}$ being a column vector with all elements of 1.

A general type of performance measure (including the availability of the system but excluding mean times like the MTBF and MTTR) is a linear function of the state probabilities. Let an m-dimensional row vector $\pi(t)$ represent the probabilities of m operational states and an m-dimensional column vector $\boldsymbol{\alpha}$ be associated with the performance of the system in each state. Then, the performance measure of the Markov process is

$$\eta(t) = \pi(t)\boldsymbol{\alpha}. \tag{18.2}$$

For example, taking $\alpha_j = 1$ if the system functions in state j and $\alpha_j = 0$ otherwise, the performance measure is the system availability. With respect to the general perturbation and performance measure, the MDSM is defined to study the sensitivity of the performance of the Markovian system for its steady state, its transient state, and a given time of period of interest.

Steady state

At the steady state of the Markov process, the vector of state probabilities and the performance measure are time-independent and are denoted simply by π and η, respectively, and $\eta = \pi\boldsymbol{\alpha}$. The stationary performance measure of the perturbed Markov process (i.e., the Markov process with transition rate matrix T_δ) is denoted as $\eta_\delta = \pi_\delta\boldsymbol{\alpha}$. According to Chapman-Kolmogorov equations, at steady state,

$$\pi T = 0 \quad \text{and} \quad \pi_\delta T_\delta = 0.$$

Definition 18.1.1 *The MDSM at steady state, denoted by $I_{\text{MDSM}}(Q; T)$, is defined as the derivative of η in direction Q. Mathematically,*

$$I_{\text{MDSM}}(Q; T) = \frac{d\eta}{dQ} = \lim_{\delta \to 0} \frac{\eta_\delta - \eta}{\delta}.$$

The following example from Do Van et al. (2008b)[*] interprets these concepts, especially the perturbation, as well as the B-importance of components.

Example 18.1.2 Consider a parallel system consisting of two independent components with constant failure rates λ_1 and λ_2 and constant repair rates μ_1 and μ_2. The transition rate matrix of this system is given by

$$T = \begin{pmatrix} -\lambda_1 - \lambda_2 & \lambda_1 & \lambda_2 & 0 \\ \mu_1 & -\mu_1 - \lambda_2 & 0 & \lambda_2 \\ \mu_2 & 0 & -\mu_2 - \lambda_1 & \lambda_1 \\ 0 & \mu_2 & \mu_1 & -\mu_1 - \mu_2 \end{pmatrix}.$$

The system availability at steady state equals

$$A = 1 - (1 - A_1)(1 - A_2),$$

[*]Reprinted from Do Van P, Barros A, and Berenguer C. 2008. Reliability importance analysis of Markovian systems at steady state using perturbation analysis. *Reliability Engineering and Systems Safety* **93**, 1605–1615, with permission from Elsevier Ltd.

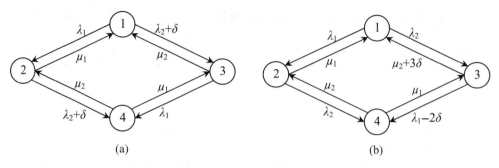

Figure 18.1   (a) Permutation on λ_2 and (b) permutation on the exit transition rates of state 3

where A_i is the steady-state availability of component i, $i = 1, 2$. When time t approaches infinity, it is written as

$$A_i = \lim_{t \to \infty} \left(\frac{\mu_i}{\lambda_i + \mu_i} + \frac{\lambda_i}{\lambda_i + \mu_i} e^{-(\lambda_i + \mu_i)t} \right) = \frac{\mu_i}{\lambda_i + \mu_i}.$$

In this case, the B-importance of the two components is

$$I_B(1; T) = \frac{\partial A}{\partial A_1} = 1 - A_2 \quad \text{and} \quad I_B(2; T) = \frac{\partial A}{\partial A_2} = 1 - A_1.$$

For two different types of perturbations, Figure 18.1 sketches the state diagram of this system: in state 1, components 1 and 2 function; in state 2, component 1 functions and component 2 fails; in state 3, component 1 fails and component 2 functions; and in state 4, components 1 and 2 fail. Figure 18.1(a) shows the case of a perturbation on one specific *parameter*, λ_2, which corresponds to the directional perturbation matrix $\mathbf{Q}(\lambda_2)$:

$$\mathbf{Q}(\lambda_2) = \begin{pmatrix} -1 & 0 & 1 & 0 \\ 0 & -1 & 0 & 1 \\ 0 & 0 & 0 & 0 \\ 0 & 0 & 0 & 0 \end{pmatrix}.$$

$I_{\text{MDSM}}(\mathbf{Q}(\lambda_2); T)$ quantifies the importance of the parameter λ_2, which gives the impact of a component failure on system performance, and at steady state,

$$I_{\text{MDSM}}(\mathbf{Q}(\lambda_2); T) = \frac{d\eta}{d\mathbf{Q}(\lambda_2)} = \frac{\partial \eta}{\partial \lambda_2}.$$

More generally, the partial derivative of η with respect to parameter λ_i corresponds to a particular $I_{\text{MDSM}}(\mathbf{Q}(\lambda_i); T)$ calculated with the perturbation matrix $\mathbf{Q}(\lambda_i) = [q_{\ell j}]$ where $q_{\ell j}$ equals the coefficient of the parameter λ_i in the entry $t_{\ell j}$ of the matrix $\mathbf{T}$.

Figure 18.1(b) presents the case of a perturbation on the exit transition rates of one specific *state*, state 3. This perturbation corresponds to the directional perturbation matrix $\mathbf{Q}_1$:

$$\mathbf{Q}_1 = \begin{pmatrix} 0 & 0 & 0 & 0 \\ 0 & 0 & 0 & 0 \\ 3 & 0 & -1 & -2 \\ 0 & 0 & 0 & 0 \end{pmatrix}.$$

On the contrary, $I_{\text{MDSM}}(\boldsymbol{Q}_1; \boldsymbol{T})$ quantifies the importance of the transition rate from one state to another. It is a kind of conditional importance measure that quantifies the sensitivity to the failure rate of component 2, knowing that the system is in state 3. It is more closely connected to an importance measure at the system state level than at the component level.

Take the system availability as a performance function, $\eta = A$. The link of the MDSM at the component level (e.g., in the case of $\boldsymbol{Q}(\lambda_2)$) with the B-importance is directly established using the chain rule:

$$I_{\text{MDSM}}(\boldsymbol{Q}(\lambda_2); \boldsymbol{T}) = \frac{\partial A}{\partial \lambda_2} = \frac{\partial A}{\partial A_2}\frac{\partial A_2}{\partial \lambda_2} = I_{\text{B}}(2; \boldsymbol{T})\frac{\partial A_2}{\partial \lambda_2}.$$

In the case of independent components, then

$$I_{\text{MDSM}}(\boldsymbol{Q}(\lambda_2); \boldsymbol{T}) = \frac{\lambda_1}{\lambda_1 + \mu_1}\frac{-\mu_2}{(\lambda_2 + \mu_2)^2}.$$

However, in the case of a system with dependent components, as in the standby system in Figure 2.11, the link between the MDSM and B-importance cannot be directly established. However, an MDSM at the component level gives the similar kind of information as the B-importance since the MDSM can provide partial derivatives with respect to any system parameters.

At steady state, the particular structure of the Chapman-Kolmogorov equations and the linearity of the performance measure lead to the following expression of the measure derivatives (Cao and Chen 1997):

$$\frac{\mathrm{d}\eta}{\mathrm{d}\boldsymbol{Q}} = -\pi\boldsymbol{Q}\boldsymbol{T}^*\boldsymbol{\alpha}, \tag{18.3}$$

where $\boldsymbol{T}^*$ is the *inverse group* of $\boldsymbol{T}$ defined as $\boldsymbol{T}^* = (\boldsymbol{T} - \mathbf{e}\pi)^{-1} - \mathbf{e}\pi$. This provides a way for calculating the exact value of the MDSM at steady state.

According to Equation (18.3), if a perturbation matrix $\boldsymbol{Q}$ is a linear function of the elementary perturbation matrices, it is possible to evaluate the MDSM at steady state in direction $\boldsymbol{Q}$ on the basis of the MDSM in the directions of the elementary perturbation matrices, as shown in Theorem 18.1.3 (Do Van et al. 2008b).

Theorem 18.1.3   *If a joint perturbation matrix with a different derivative direction for each parameter in a group of k parameters is defined as*

$$\boldsymbol{Q}(c_1\lambda_1, c_2\lambda_2, \ldots, c_k\lambda_k) = \sum_{i=1}^{k} c_i\boldsymbol{Q}(\lambda_i),$$

where $c_1, c_2, \ldots, c_k$ are constants, then at steady state

$$I_{\text{MDSM}}(\boldsymbol{Q}(c_1\lambda_1, c_2\lambda_2, \ldots, c_k\lambda_k); \boldsymbol{T}) = -\pi(c_1\boldsymbol{Q}(\lambda_1) + c_2\boldsymbol{Q}(\lambda_2) + \cdots + c_k\boldsymbol{Q}(\lambda_k))\boldsymbol{T}^*\boldsymbol{\alpha}$$

$$= \sum_{i=1}^{k} c_i I_{\text{MDSM}}(\boldsymbol{Q}(\lambda_i); \boldsymbol{T}).$$

For instance, in Example 18.1.2, for the joint sensitivity of the group of parameters (λ_1, λ_2), it is possible to define a joint perturbation matrix:

$$Q(\lambda_1, \lambda_2) = Q(\lambda_1) + Q(\lambda_2),$$

where the matrix entry $q_{\ell j}(\lambda_i)$ equals the coefficient of the parameter λ_i in the entry $t_{\ell j}$ of the matrix T. Then,

$$\begin{aligned}
I_{\mathrm{MDSM}}(Q(\lambda_1, \lambda_2); T) &= \frac{\partial \eta}{\partial Q(\lambda_1, \lambda_2)} \\
&= -\pi Q(\lambda_1, \lambda_2) T^* \alpha = -\pi (Q(\lambda_1) + Q(\lambda_2)) T^* \alpha \\
&= I_{\mathrm{MDSM}}(Q(\lambda_1)) + I_{\mathrm{MDSM}}(Q(\lambda_2)).
\end{aligned}$$

In this way, the joint MDSM of a group of parameters can be the sum of the MDSM of each parameter in the group. This property of the joint MDSM is concerned with perturbation directions but not the components. In some particular cases, that is, when the MDSM related to one component can be defined by only one perturbation direction, the joint MDSM of a group of components can be expressed as the sum of the MDSM of the components. But it is not true as a rule, especially for systems under stochastic dependencies. As a conclusion, if the MDSM related to elementary perturbation directions are calculated, the joint MDSM related to any linear combination of these directions requires no additional calculations.

Transient state

Do Van et al. (2008a) extended the MDSM to conduct a reliability importance analysis in the transient state. Let the column vector π_0 be the initial vector of state probabilities. The system of the first-order Chapman-Kolmogorov equations applied to the homogeneous Markov process is

$$\frac{\mathrm{d}\pi(t)}{\mathrm{d}t} = T\pi(t). \tag{18.4}$$

The solution of Equation (18.4) can be expressed as

$$\pi(t) = e^{Tt}\pi_0 = W_T(t)\pi_0,$$

where $W_T(t) = e^{Tt}$ is the matrix exponential. The perturbations in the transition rate matrix affect the transient solution $\pi(t)$, which becomes $\pi_\delta(t)$. A necessary condition that is classically respected in perturbation analysis is assumed:

$$\pi_\delta(0) = \pi(0).$$

Then $\pi_\delta(t)$ satisfies

$$\frac{\mathrm{d}\pi_\delta(t)}{\mathrm{d}t} = T_\delta \pi_\delta(t).$$

Definition 18.1.4 *The MDSM at the transient state, denoted by* $I_{\text{MDSM}}(Q; t, T)$, *is defined as the derivative of* $\eta(t)$ *in direction* Q. *Mathematically,*

$$I_{\text{MDSM}}(Q; t, T) = \frac{d\eta(t)}{dQ} = \lim_{\delta \to 0} \frac{\eta_\delta(t) - \eta(t)}{\delta} = \left[\lim_{\delta \to 0} \frac{\pi_\delta(t) - \pi(t)}{\delta} \right] \alpha.$$

The MDSM at the transient state can be expressed as (Do Van et al. 2008a)

$$I_{\text{MDSM}}(Q; t, T) = \int_0^t W_T(t - s) Q W_T(s) \pi_0 \alpha ds, \tag{18.5}$$

which may be evaluated by a numerical integration method or directly by making a suitable expansion of matrix exponentials with, for example, the uniformization method (Neuts 1995). Equation (18.5) allows for the evaluation of the system sensitivity (e.g., availability) in any direction of interest at time t in the transient state. Gandini (1990) showed a similar formula evaluation by using the generalized perturbation theory method in the context of Markovian systems. However, the direction Q in Gandini (1990) is limited to the direction of a single parameter.

Average on a finite time horizon

Considering the system performance averaged over a given period $[0, \bar{t}]$, denoted by $\bar{\eta}(\bar{t})$, then the MDSM on a finite time horizon can be defined as follows (Do Van et al. 2008a). This MDSM allows the calculation of the average sensitivity analysis during a given period $[0, \bar{t}]$ in any direction Q.

Definition 18.1.5 *The MDSM on a time horizon* $[0, \bar{t}]$, *denoted by* $I_{\text{MDSM}}(Q; \bar{t}, T)$, *is defined as the derivative of* $\bar{\eta}(\bar{t})$ *in direction* Q. *Mathematically,*

$$I_{\text{MDSM}}(Q; \bar{t}, T) = \frac{d\bar{\eta}(\bar{t})}{dQ} = \frac{1}{\bar{t}} \frac{d\bar{\pi}(\bar{t})}{dQ} \alpha = \frac{1}{\bar{t}} \left[\int_0^{\bar{t}} W_T(\bar{t} - s) Q \bar{\pi}(s) ds \right] \alpha,$$

where $\bar{\pi}(\bar{t}) = \int_0^{\bar{t}} \pi(s) ds$ *and* $\bar{\eta}(\bar{t}) = \int_0^{\bar{t}} \eta(s) ds = \bar{\pi}(\bar{t}) \alpha / \bar{t}$.

Applications

The MDSM-based sensitivity analysis can rank the effect of the change in any direction of one parameter or, more generally, the simultaneous change of several design parameters, as Q can represent the change of one parameter or a group of parameters. For example, in a repairable system, the MDSM of the repair rates can be used to evaluate the importance of the maintenance. In the context of Markovian systems, Do Van et al. (2008a) applied the MDSM to the study of the production capacity of multistate production systems such as manufacturing, production lines, and power generation, in which the performance can settle at different levels depending on the operational conditions of the constitutive components. The production capacity differs at each state. Both availability and expected production capacity criteria are considered for the steady state of a Markovian system, for its transient state, and also for a given time period of interest.

18.1.3   The Multidirectional Differential Importance Measure and Total Order Importance

In the context of Markov reliability models at steady state, Do Van et al. (2010) extended the first-order DIM (i.e., DIM^I) and the highest-order TOI (TOI^O) using multidirectional derivatives in the directions specified by perturbation matrices. This subsection continues to use the perturbation representation (18.1) and performance measure (18.2) and assumes the steady state of the system as described in Subsection 18.1.2. Particularly, consider the case in which the transition rate matrix is perturbed in n different directions, $Q_1, Q_2, \ldots, Q_n$. The perturbed transition rate matrix is then

$$T_\delta = T + \delta_1 Q_1 + \delta_2 Q_2 + \cdots + \delta_n Q_n, \tag{18.6}$$

where $\delta_1, \delta_2, \ldots, \delta_n$ are the amounts of variations in directions $Q_1, Q_2, \ldots, Q_n$, respectively. Let η_{δ_Σ} represent a variation of the system performance η corresponding to the variations in transition rate matrix described in Equation (18.6).

The DIM^I

If the changes of parameters in Equation (18.6) are small enough, η_{δ_Σ} can be then approximated by the first-order contribution $\eta_{\delta_\Sigma}^I$:

$$\eta_{\delta_\Sigma}^I = \sum_{i=1}^n \delta_i \frac{\mathrm{d}\eta}{\mathrm{d}Q_i} = \sum_{i=1}^n \delta_i I_{\mathrm{MDSM}}(Q_i; T),$$

where $I_{\mathrm{MDSM}}(Q_i; T) = \mathrm{d}\eta/\mathrm{d}Q_i$ is the MDSM in direction Q_i and $\delta_i I_{\mathrm{MDSM}}(Q_i; T) = \delta_i(\mathrm{d}\eta/\mathrm{d}Q_i)$ is the first-order contribution of the change in transition rate matrix with respect to the direction Q_i and amount δ_i.

Referring to Definition 7.2.1, the DIM^I can be extended by using directional derivatives and the MDSM at steady state. As the direction Q_i can relate to a component, a group of components, a state, or a group of states, the DIM^I in direction Q_i represents then the corresponding contribution on the total variation of the system performance.

Definition 18.1.6   *The DIM^I at steady state in direction Q_i, denoted by $I_{\mathrm{DIM}^I}(Q_i; T)$, is defined as*

$$I_{\mathrm{DIM}^I}(Q_i; T) = \frac{\delta_i \frac{\mathrm{d}\eta}{\mathrm{d}Q_i}}{\sum_{j=1}^n \delta_j \frac{\mathrm{d}\eta}{\mathrm{d}Q_j}} = \frac{\delta_i I_{\mathrm{MDSM}}(Q_i; T)}{\eta_{\delta_\Sigma}^I}.$$

The same as in Subsection 7.2.1, the DIM^I has the additivity property, that is,

$$I_{\mathrm{DIM}^I}(Q_1, Q_2, \ldots, Q_s; T) = I_{\mathrm{DIM}^I}(Q_1; T) + I_{\mathrm{DIM}^I}(Q_2; T) + \cdots + I_{\mathrm{DIM}^I}(Q_s; T).$$

This property can be used to calculate the DIM^I of a group of components, given the DIM^I of each individual component in the group. In addition, the sum of the DIM^I of all directions equals unity, that is, $\sum_{i=1}^{n} I_{\text{DIM}^I}(Q_i; T) = 1$.

Applying Equation (18.3), the DIM^I at steady state can be expressed as

$$I_{\text{DIM}^I}(Q_i; T) = \frac{\delta_i \pi Q_i T^* \alpha}{\pi \left(\sum_{j=1}^{n} \delta_j Q_j \right) T^* \alpha},$$

which can be used to calculate the exact value of the DIM^I.

The TOI^O

If the transition rate matrix is perturbed by the perturbation matrix δQ as in Equation (18.1), the variation of the system performance is proved to be (Do Van et al. 2010)

$$\eta_\delta - \eta = (\pi_\delta - \pi)\alpha = -\delta \pi Q T^* (I + \delta Q T^*)^{-1} \alpha, \qquad (18.7)$$

where I denotes the $n \times n$ identity matrix. The total variation of the system performance can be calculated for any value of δ, that is, for any magnitude of change, with Equation (18.7).

Referring to Theorem 7.3.3, the TOI^O can be extended using directional derivatives and Equation (18.7).

Definition 18.1.7 *The TOI^O at steady state in direction Q_i, denoted by $I_{\text{TOI}^O}(Q_i; T)$, is defined as*

$$I_{\text{TOI}^O}(Q_i; T) = \frac{\eta_{\delta_i} - \eta}{\eta_{\delta_\Sigma}} = \frac{\delta_i \pi Q_i T^* (I + \delta_i Q_i T^*)^{-1} \alpha}{\pi \left(\sum_{j=1}^{n} \delta_j Q_j \right) T^* \left(I + \left(\sum_{j=1}^{n} \delta_j Q_j \right) T^* \right)^{-1} \alpha}.$$

Do Van et al. (2010) demonstrated the use of the DIM^I and TOI^O in a simple power generation system of three units.

18.1.4   Perturbation Analysis

Perturbation analysis in operations research is to analyze the effect of perturbed input parameters on the system output and to approximate sensitivity measures based on partial derivatives or gradients. It is known as sensitivity analysis in the statistics literature (Golub and Zha 1994), because it derives the sensitivity information of the steady-state performance of the system. These sensitivity measures and information may then be used as the basis for computing more specific importance measures such as the B-importance, MDSM, DIM, and TOI.

In a broader sense, perturbation analysis deals with performance evaluation, such as performance gradient and sensitivity by means of simulation. Ho and other researchers have done a significant amount of work on the perturbation analysis of a discrete event dynamic system (Dai and Ho 1995; Gong and Ho 1987; Ho 1992; Ho and Cao 1991), such as queueing networks (Ho and Cao 1983, 1985) and flexible manufacturing systems (Cao and Ho 1987; Ho and Cao 1985). A Markov process is the most fundamental model for many discrete event systems. In this context, a parameter change can generate propagable perturbations in the timing of events in the sample path of a dynamic system and the perturbation analysis evaluates the perturbed performance from nominal by means of simulation. The state-event sequence together with the time of the occurrence of each and every event constitutes a *trajectory* (i.e., sample path) of the discrete event dynamic system. If the deviations caused by the parameter perturbations between the nominal and the perturbed trajectories are small and only temporary, then the perturbation analysis can give an unbiased estimate for the derivative of the sample performance. Ho (1988, 1992) has explained the basic concepts and algorithms of perturbation analysis. Subsections 18.1.2 and 18.1.3 present importance measures in the context of Markov process models, which can be estimated using perturbation analysis.

Perturbation analysis is effective for estimating the sensitivity-related importance measures of the parameters of a discrete event dynamic system with respect to its performance measure using only one sample path or Monte Carlo experiment of the system. It is particularly applicable to cases that lack an explicit general analytical form of performance measure in a stochastic environment. Ordinarily, unless closed-form formulas are available, performance sensitivity is calculated by brute force using two different experiments with only the parameter values being different between them. This can, of course, be very time-consuming, expensive, and numerically difficult. Perturbation analysis, in effect, is a method of reconstructing the perturbed performance value from the nominal experiment or sample path without the need to actually carry out the perturbed experiment. It makes no difference whether or not the experiment deals with steady-state or transient performance; thus, perturbation analysis can be used for the determination of either steady-state or transient performance gradients.

Perturbation analysis and its variants, infinitesimal perturbation analysis, perturbation realization, perturbation-theory-based methods, and so on, have been proposed based on the use of a stochastic gradient that can be estimated without change of model parameters in a single sample path of simulation. The single sample path based perturbation analysis saves a great deal of computation in simulation for system optimization and can be applied to online performance optimization of real systems, where changing the values of parameters is not feasible.

In the framework of Markovian systems at steady state, Do Van et al. (2008b) used perturbation analysis and particularly, perturbation realization (Cao and Chen 1997), to estimate the MDSM defined in Subsection 18.1.2. They also presented several numerical examples with the asymptotic availability as the performance measure. The MDSM in combination with perturbation realization as an estimation method offers a promising tool for steady-state sensitivity analysis of Markov processes in reliability studies.

Along with the work in perturbation analysis for the MDSM, Do Van et al. (2010) used perturbation analysis, that is, a single sample path of a Markov process, to estimate the DIM^I and TOI^O defined in Subsection 18.1.3 and showed that the estimated results are closely related to the analytical results. Marseguerra and Zio (2004) applied the perturbed Monte Carlo simulation to evaluate the DIM^I of components using the differential sensitivity measures of perturbed parameters characterizing the stochastic behavior of the components.

In a different but related study, Gandini (1990) used the generalized perturbation theory method within the Markov chain representation to estimate the differential sensitivity measures with respect to a parameter, which, in turn, can be directly related to the B-importance, criticality importance, and BP importance.

18.2 Global Sensitivity and Uncertainty Analysis

Unlike the local sensitivity analysis in Section 18.1, another approach to sensitivity is global sensitivity analysis, which does not specify the input $\mathbf{x}$ at a nominal point. The global sensitivity approach considers the mathematical model $f(\mathbf{x})$ over the entire feasible region, which specifies the range of $\mathbf{x}$, and studies the impact of individual input variables or groups of the variables on the output $f(\mathbf{x})$. Consequently, global sensitivity measures are defined and used for estimating the impact of individual variables or groups of variables, while all other variables are varied as well. Thus, they account for interactions among variables and do not depend on the choice of a nominal point. Global sensitivity measures can help identify significant variables whose uncertainty most affects output uncertainty; they are thus also referred to as uncertainty importance measures (Borgonovo 2007). Global sensitivity measures should be regarded as a tool for studying a mathematical model rather than its specified solution.

The definitions of global sensitivity measures make sense only if they can be effectively estimated. Therefore, one direction of the research in this area relates to computation methods, for example, the design of Monte Carlo or quasi-Monte Carlo methods for the evaluation of integrals. However, this section focuses on the concept of the sensitivity measures themselves and provides a brief introduction of computation methods without details.

Subsections 18.2.1, 18.2.2, and 18.2.3 present three most prevalent types of global sensitivity measures—analysis-of-variances (ANOVA) decomposition-based, elementary-effect-based, and derivative-based sensitivity measures. Then, Subsections 18.2.4 and 18.2.5 give the relationships between the ANOVA-decomposition-based and derivative-based measures and their applications in the case of random input variables, respectively. Subsection 18.2.6 presents moment-independent global sensitivity measures of random input variables and the extension to fuzzy inputs.

18.2.1 ANOVA-decomposition-based Global Sensitivity Measures

On the basis of the ANOVA-decomposition of the model output $f(\mathbf{x})$, Sobol (1993, 2001) defined a global sensitivity measure. This ANOVA-decomposition-based global sensitivity measure (AGSM) estimates the ratio of output uncertainty caused by an individual input or a group of inputs over the total output uncertainty. Before presenting the definition of the AGSM, first introduce the ANOVA decomposition. Assume that $f(\mathbf{x})$ is an integrable function defined in the n-dimensional unit hypercube $[0, 1]^n$. For simplification, all of the integrals have each integration variable vary independently from zero to one unless stated otherwise, and introduce Lebesgue measure $d\mathbf{x} = dx_1 dx_2 \ldots dx_n$ and similar notation directly.

Definition 18.2.1 *The representation of $f(\mathbf{x})$ in a form*

$$f(\mathbf{x}) = f_0 + \sum_{k=1}^{n} \sum_{1 \leq i_1 < i_2 < \ldots < i_k \leq n} f_{i_1 i_2 \ldots i_k}(x_{i_1}, x_{i_2}, \ldots, x_{i_k}) \tag{18.8}$$

is called ANOVA decomposition if

$$\int f_{i_1 i_2 \ldots i_k}(x_{i_1}, x_{i_2}, \ldots, x_{i_k}) \, dx_{i_1} dx_{i_2} \ldots dx_{i_k} = 0. \tag{18.9}$$

Equation (18.8) is unique and means that

$$f(\mathbf{x}) = f_0 + \sum_i f_i(x_i) + \sum_{i<j} f_{ij}(x_i, x_j) + \cdots + f_{12\ldots n}(x_1, x_2, \ldots, x_n),$$

and the total number of summands in Equation (18.8) is 2^n. Following Equation (18.9), the members in Equation (18.8) are orthogonal and can be expressed as integrals of $f(\mathbf{x})$. Indeed,

$$\int f(\mathbf{x})d\mathbf{x} = f_0,$$

$$\int f(\mathbf{x}) \prod_{s\neq i} dx_s = f_0 + f_i(x_i),$$

$$\int f(\mathbf{x}) \prod_{s\neq i,j} dx_s = f_0 + f_i(x_i) + f_j(x_j) + f_{ij}(x_i, x_j),$$

and so on. The last member, $f_{12\ldots n}(x_1, x_2, \ldots, x_n)$, is defined by Equation (18.8).

Assuming that $f(\mathbf{x})$ is square integrable, then all of the $f_{i_1 i_2 \ldots i_k}$ in Equation (18.8) are also square integrable. Define the variances as the constants

$$V_{i_1 i_2 \ldots i_k} = \int f^2_{i_1 i_2 \ldots i_k}(x_{i_1}, x_{i_2}, \ldots, x_{i_k})\, dx_{i_1} dx_{i_2} \ldots dx_{i_k}.$$

Squaring both sides of Equation (18.8) and integrating over $[0, 1]^n$, the total variance is

$$V = \int f^2(\mathbf{x})d\mathbf{x} - f_0^2 = \sum_{k=1}^{n} \sum_{1 \leq i_1 < i_2 < \ldots < i_k \leq n} V_{i_1 i_2 \ldots i_k}. \tag{18.10}$$

If $\mathbf{x}$ is a random point uniformly distributed in $[0, 1]^n$, then $f(\mathbf{x})$ and $f_{i_1 i_2 \ldots i_k}(x_{i_1}, x_{i_2}, \ldots, x_{i_k})$ are random variables with variances V and $V_{i_1 i_2 \ldots i_k}$, respectively. In other words, the variances V and $V_{i_1 i_2 \ldots i_k}$ show the variability of $f(\mathbf{x})$ and $f_{i_1 i_2 \ldots i_k}(x_{i_1}, x_{i_2}, \ldots, x_{i_k})$, respectively. Thus, the basic type of AGSM is defined using V and $V_{i_1 i_2 \ldots i_k}$ as follows (Sobol 1993).

Definition 18.2.2   *The basic AGSM of k variables $x_{i_1}, x_{i_2}, \ldots, x_{i_k}$ are defined as the ratio of variances:*

$$I_A(i_1, i_2, \ldots, i_k) = \frac{V_{i_1 i_2 \ldots i_k}}{V}.$$

All of the $I_A(i_1, i_2, \ldots, i_k)$ values are nonnegative, and their sum is

$$\sum_{k=1}^{n} \sum_{1 \leq i_1 < i_2 < \ldots < i_k \leq n} I_A(i_1, i_2, \ldots, i_k) = 1.$$

For a piecewise continuous function $f(\mathbf{x})$, $I_A(i_1, i_2, \ldots, i_k) = 0$ means that $f_{i_1 i_2 \ldots i_k}(x_{i_1}, x_{i_2}, \ldots, x_{i_k}) = 0$. Thus, the functional structure of $f(\mathbf{x})$ can be investigated by estimating $I_A(i_1, i_2, \ldots, i_k)$.

Let $K = \{i_1, i_2, \ldots, i_k\}$ be a subset of variable indices and K^c the set of $n-k$ complementary variables. The variance corresponding to subset K is

$$V_K^{\text{var}} = \sum_{s=1}^{k} \sum_{(i_1 < i_2 < \ldots < i_s) \in K} V_{i_1 i_2 \ldots i_s},$$

where the sum is taken over all groups of $(i_1, i_2, \ldots, i_s)$ in which all of the elements are in K. The variance $V_{K^c}^{\text{var}}$ is similarly defined for subset K^c. Then the total variance corresponding to K is

$$V_K^{\text{tot}} = V - V_{K^c}^{\text{var}} = \sum_{s=1}^{k} \sum_{\{i_1 < i_2 < \ldots < i_s\} \cap K \neq \emptyset} V_{i_1 i_2 \ldots i_s},$$

where the sum is taken over all groups of $(i_1, i_2, \ldots, i_s)$ in which at least one element is in K. Another two types of the AGSM for subset K are then defined using V_K^{var} and V_K^{tot} as follows (Homma and Saltelli 1996; Sobol 2001).

Definition 18.2.3 *The var and tot AGSM of k variables $x_{i_1}, x_{i_2}, \ldots, x_{i_k}$ with indices in $K = \{i_1, i_2, \ldots, i_k\}$ are defined, respectively, as*

$$I_{A\text{var}}(K) = \frac{V_K^{\text{var}}}{V} = \sum_{s=1}^{k} \sum_{(i_1 < i_2 < \ldots < i_s) \in K} I_A(i_1, i_2, \ldots, i_s),$$

$$I_{A\text{tot}}(K) = \frac{V_K^{\text{tot}}}{V} = \sum_{s=1}^{k} \sum_{\{i_1 < i_2 < \ldots < i_s\} \cap K \neq \emptyset} I_A(i_1, i_2, \ldots, i_s).$$

Clearly, $I_{A\text{tot}}(K) = 1 - I_{A\text{var}}(K^c)$ and $0 \leq I_{A\text{var}}(K) \leq I_{A\text{tot}}(K) \leq 1$. Note that $I_{A\text{var}}(K) = I_{A\text{tot}}(K) = 0$ means that $f(\mathbf{x})$ does not depend on the variables in K; $I_{A\text{var}}(K) = I_{A\text{tot}}(K) = 1$ means that $f(\mathbf{x})$ depends only on the variables in K.

The integer k in Definitions 18.2.2 and 18.2.3 is the order of the AGSM. When only one variable x_i is considered in K, it is referred to as the first-order $I_{A\text{var}}(i)$ and $I_{A\text{tot}}(i)$. The first-order AGSM, especially $I_{A\text{tot}}(i)$, is often used for ranking the impact of individual variables on the model output. The following example (Sobol 2001) illustrates the basic concepts of AGSM.

Example 18.2.4 Consider a g-function with separated variables as

$$g = \prod_{i=1}^{n} \varphi_i(x_i),$$

where $\varphi_i(t) = (|4t - 2| + a_i)/(1 + a_i)$ depends on a nonnegative parameter a_i. If $a_i = 0$, the multiplier $\varphi_i(t)$ varies from 0 to 2, and the variable x_i is important. If $a_i = 3$, the $\varphi_i(t)$ varies from 0.75 to 1.25, and the corresponding x_i is insignificant. Let $n = 8$, $a_1 = a_2 = 0$, and

$a_3 = a_4 = \ldots = a_8 = 3$. Then, the significance of the first two variables can be seen from the AGSM. The first-order basic AGSM values are $I_A(1) = I_A(2) = 0.329$, while $I_A(3) = I_A(4) = \cdots = I_A(8) = 0.021$. The second-order basic AGSM values are $I_A(1, 2) = 0.110$; $I_A(i, j) = 0.007$ if one of the indices is 1 or 2; and $I_A(i, j) = 0.0004$ if both i and j correspond to insignificant variables. The largest third-order basic AGSM value is $I_A(1, 2, i) = 0.002$ for $i = 3, 4, \ldots, 8$; the other third-order basic AGSM values do not exceed 0.00014.

Assume now that $n = 3$ and $K = \{1, 2\}$. Then $K^c = \{3\}$, $I_{A\text{var}}(K) = I_A(1) + I_A(2) + I_A(1, 2)$, and $I_{A\text{tot}}(K) = I_A(1) + I_A(2) + I_A(1, 2) + I_A(1, 3) + I_A(2, 3) + I_A(1, 2, 3) = 1 - I_A(3)$.

Applications in nonlinear mathematical models

Sobol (2001) presented three types of problems in nonlinear mathematical models that can be studied with the aid of the AGSM. First, the AGSM can rank variables in $f(\mathbf{x})$ as shown in Example 18.2.4. Second, it can be used to fix insignificant variables in $f(\mathbf{x})$ (Sobol et al. 2007). Assuming that $I_{A\text{tot}}(K^c) \ll 1$, then $f(\mathbf{x})$ depends mainly on the variables in K, and an approximation $\tilde{f}(\mathbf{x}) = f(\mathbf{x}^K, \mathbf{x}_0^{K^c})$ with some fixed $\mathbf{x}_0^{K^c} \in [0, 1]^{n-k}$ can be suggested. The following theorem (Sobol 1993) shows that the approximation error $\delta(\mathbf{x}_0^{K^c}) = (\int [f(\mathbf{x}) - \tilde{f}(\mathbf{x})]^2 d\mathbf{x})/V$ depends on $I_{A\text{tot}}(K^c)$.

Theorem 18.2.5 *For an arbitrary $\mathbf{x}_0^{K^c} \in [0, 1]^{n-k}$, $\delta(\mathbf{x}_0^{K^c}) \geq I_{A\text{tot}}(K^c)$. But if $\mathbf{x}_0^{K^c}$ is a random point uniformly distributed in $[0, 1]^{n-k}$, then for an arbitrary $\varepsilon > 0$*

$$\Pr\left\{\delta(\mathbf{x}_0^{K^c}) < \left(1 + \frac{1}{\varepsilon}\right) I_{A\text{tot}}(K^c)\right\} \geq 1 - \varepsilon.$$

For example, selecting $\varepsilon = 0.5$, then the probability that $\delta(\mathbf{x}_0^{K^c}) < 3 I_{A\text{tot}}(K^c)$ exceeds 0.5.

Third, the approximation of $f(\mathbf{x})$ can be constructed by deleting high-order members in Equation (18.8). Quite often in mathematical models, the low-order interactions of input variables have the main impact upon the output. For such models the following approximation can be used:

$$\tilde{f}_L(\mathbf{x}) = f_0 + \sum_{s=1}^{L} \sum_{1 \leq i_1 < i_2 < \ldots < i_s \leq n} f_{i_1 i_2 \ldots i_s}(x_{i_1}, x_{i_2}, \ldots, x_{i_s})$$

with $L < n$. Sobol (2001) showed that the corresponding approximation error is

$$1 - \sum_{s=1}^{L} \sum_{1 \leq i_1 < i_2 < \ldots < i_s \leq n} I_A(i_1, i_2, \ldots, i_s).$$

Monte Carlo approaches

Numerically, the AGSM can be efficiently computed by Monte Carlo or quasi-Monte Carlo methods (Sobol 2001; Sobol and Kucherenko 2005; Sobol and Levitan 1999; Sobol and Shukhman 2007) for rather complex mathematical models, utilizing values of $f(\mathbf{x})$ at special random or quasi-random points. For an arbitrary set $K = \{i_1, i_2, \ldots, i_k\}$, denote $\mathbf{y} = \mathbf{x}^K = (x_{i_1}, x_{i_2}, \ldots, x_{i_k})$ and $\mathbf{z} = \mathbf{x}^{K^C}$ for the sake of notation simplification; thus, $\mathbf{x} = (\mathbf{y}, \mathbf{z})$. The Monte Carlo methods are based on the following results (Sobol 2001)

$$V_K^{\text{var}} = \int f(\mathbf{x})f(\mathbf{y}, \mathbf{z}')\mathrm{d}\mathbf{x}\mathrm{d}\mathbf{z}' - f_0^2, \tag{18.11}$$

$$V_K^{\text{tot}} = \frac{1}{2} \int [f(\mathbf{y}, \mathbf{z}) - f(\mathbf{y}', \mathbf{z})]^2 \mathrm{d}\mathbf{x}\mathrm{d}\mathbf{y}'. \tag{18.12}$$

Thus, for computing $I_{A^{\text{var}}}$ and $I_{A^{\text{tot}}}$ four integrals must be estimated:

$$\int f(\mathbf{x})\mathrm{d}\mathbf{x}, \quad \int f^2(\mathbf{x})\mathrm{d}\mathbf{x}, \quad \int f(\mathbf{x})f(\mathbf{y}, \mathbf{z}')\mathrm{d}\mathbf{x}\mathrm{d}\mathbf{z}', \quad \text{and} \quad \int f(\mathbf{x})f(\mathbf{y}', \mathbf{z})\mathrm{d}\mathbf{x}\mathrm{d}\mathbf{y}'.$$

Then, a Monte Carlo or quasi-Monte Carlo method can be constructed. Consider two independent random points $(\boldsymbol{\eta}, \boldsymbol{\zeta})$ and $(\boldsymbol{\eta}', \boldsymbol{\zeta}')$ uniformly distributed in $[0, 1]^n$. Each Monte Carlo trial requires three computations of the model: $f(\boldsymbol{\eta}, \boldsymbol{\zeta})$, $f(\boldsymbol{\eta}, \boldsymbol{\zeta}')$, and $f(\boldsymbol{\eta}', \boldsymbol{\zeta})$. After N trials, crude Monte Carlo estimates are obtained:

$$\frac{1}{N}\sum_{j=1}^{N} f(\boldsymbol{\eta}_j, \boldsymbol{\zeta}_j) \to^p f_0, \quad \frac{1}{N}\sum_{j=1}^{N} f^2(\boldsymbol{\eta}_j, \boldsymbol{\zeta}_j) \to^p V + f_0^2,$$

$$\frac{1}{N}\sum_{j=1}^{N} f(\boldsymbol{\eta}_j, \boldsymbol{\zeta}_j)f(\boldsymbol{\eta}_j, \boldsymbol{\zeta}_j') \to^p V_K^{\text{var}} + f_0^2, \quad \frac{1}{N}\sum_{j=1}^{N} f(\boldsymbol{\eta}_j, \boldsymbol{\zeta}_j)f(\boldsymbol{\eta}_j', \boldsymbol{\zeta}_j) \to^p V_{K^C}^{\text{var}} + f_0^2,$$

$$\tag{18.13}$$

$$\frac{1}{2N}\sum_{j=1}^{N}[f(\boldsymbol{\eta}_j, \boldsymbol{\zeta}_j) - f(\boldsymbol{\eta}_j, \boldsymbol{\zeta}_j')]^2 \to^p V_{K^C}^{\text{tot}}, \quad \frac{1}{2N}\sum_{j=1}^{N}[f(\boldsymbol{\eta}_j, \boldsymbol{\zeta}_j) - f(\boldsymbol{\eta}_j', \boldsymbol{\zeta}_j)]^2 \to^p V_K^{\text{tot}}.$$

$$\tag{18.14}$$

The stochastic convergence $(\to^p)$ is implied by the absolute convergence of the four integrals that follows from the square integrability of $f(\mathbf{x})$. The two Monte Carlo estimates in (18.13) are derived from (18.11) and alternatively used with the two in (18.14) that are derived from (18.12).

Research has been conducted on the AGSM, mainly on its computational aspects. Homma and Saltelli (1996) discussed the Monte Carlo computation and the related issues of random points generation, error estimates, and random transformation. They also tested applications of the AGSM on a few analytical and computer models and concluded that the higher-order terms in the AGSM can affect the variable ranking even when the sum of the first-order terms is not far from unity. Corresponding to the same Monte Carlo algorithm, Sobol and Kucherenko (2005) applied different quasi-Monte Carlo algorithms whose efficiencies may differ even when their constructive dimensions are equal and the same quasi-random points are used. Sobol and Kucherenko (2005) also confirmed that the Brownian bridge is superior to the standard approximation of a Wiener integral. The AGSM can also be estimated using an extended version of the fourier amplitude sensitivity test (Homma and Saltelli 1996; Saltelli 2002), especially the low-order AGSM. Sobol (2001) provided detailed computation algorithms. Sobol and Shukhman (2007) studied the bias in the AGSM produced by random errors in model evaluation.

18.2.2   Elementary-effect-based Global Sensitivity Measures

The elementary effect method proposed by Morris (1991) can be regarded as global, as the final measure is obtained by averaging local measures (i.e., the elementary effects attributable to each input variable). Let model $f(\mathbf{x})$ be defined in the unit hypercube $[0, 1]^n$. For a given value of $\mathbf{x}$, the so-called elementary effect of variable x_i is defined as

$$\frac{1}{\delta}[f(x_1, x_2, \ldots, x_{i-1}, x_i + \delta, x_{i+1}, \ldots, x_n) - f(\mathbf{x})], \qquad (18.15)$$

where $\mathbf{x}$ is taken on an n-dimensional p-level grid, in which each x_i may take on values from $\{0, 1/(p-1), 2/(p-1), \ldots, 1\}$ such that $x_i + \delta \leq 1$ and δ is the predetermined multiple of $1/(p-1)$. The final measure is composed of individually "one-factor-at-a-time" experiments and uses random sampling of points from the fixed p-level grid for averaging elementary effects as in Equation (18.15), which are calculated as finite differences with the increment δ comparable with the range of uncertainty. For this reason, it cannot correctly account for the effects with characteristic dimensions much less than δ.

The distribution of elementary effects, F_i, is obtained by randomly sampling points from $[0, 1]^n$. Two sensitivity measures are evaluated for each variable x_i:

$I_{E\mu}(i) = $ an estimate of the mean of the distribution F_i, and
$I_{E\sigma}(i) = $ an estimate of the standard deviation of F_i.

A high value of $I_{E\mu}(\cdot)$ indicates a variable with an important overall impact on the output. A high value of $I_{E\sigma}(\cdot)$ indicates a variable involved in interaction with other variables or whose effect is nonlinear. Morris (1991) proposed efficient sampling methods. Campolongo et al. (2007) also provided a more effective sample strategy, which allows a better exploration of the space of uncertain inputs.

A revised elementary effect method

Campolongo et al. (2007) presented a revised version of the elementary effect method, which is

$I_{E\tilde{\mu}}(i) = $ an estimate of the mean of the distribution of the absolute values
of elementary effects.

When the model is nonmonotonic, $I_{E\widetilde{\mu}}(\cdot)$ gives a better estimate of the order of importance than $I_{E\mu}(\cdot)$ because of the canceling effect. That is, if the model is nonmonotonic, then the distribution F_i contains both positive and negative elements, and some elements may thus cancel each other out, producing a low $I_{E\mu}(\cdot)$ value even for an important variable. Thus, the original elementary effect method can produce inaccurate measures for nonmonotonic functions (Kucherenko et al. 2009).

The drawback of $I_{E\widetilde{\mu}}(\cdot)$ is the loss of information on the sign of the effect. Nevertheless, this information can be recovered by the simultaneous examination of $I_{E\mu}(\cdot)$ and $I_{E\widetilde{\mu}}(\cdot)$, as an estimate of $I_{E\mu}(\cdot)$ comes at no extra computational cost. If $I_{E\mu}(i)$ and $I_{E\widetilde{\mu}}(i)$ are both high, the sign of the effect is always the same, that is, the output function is monotonic with respect to variable x_i. If, in contrast, $I_{E\mu}(i)$ is low while $I_{E\widetilde{\mu}}(i)$ is high, variable x_i carries the effects of different signs, depending on the values assumed by the other variables.

In contrast to $I_{E\mu}(\cdot)$, $I_{E\widetilde{\mu}}(\cdot)$ can be adjusted to work with a group of variables. The idea is to move all variables in the same group simultaneously. The definition of the elementary effect in Equation (18.15) cannot be extended straightforwardly to cases in which more than one variable is moved at the same time, as two variables may have been changed in opposite directions, that is, one increased and another one decreased by δ. However, if using $I_{E\widetilde{\mu}}(\cdot)$, this problem is overcome, as the focus is not on the elementary effect itself but on its absolute value. While the conceptual design is unchanged and the evaluated points must be in the p-level grid, from a computational point of view, the strategy has been slightly modified to allow variables in the same group to be changed simultaneously in opposite directions.

The total computational cost of the elementary effect methods is considerably cheaper than that of the AGSM (Kucherenko et al. 2009). However, the elementary effect sensitivity measures $I_{E\mu}(\cdot)$, $I_{E\sigma}(\cdot)$, and $I_{E\widetilde{\mu}}(\cdot)$ lack the ability of the AGSM in providing information about main effects (contribution of individual variables to uncertainty without considering the interactions) and cannot distinguish between low- and high-order interactions.

18.2.3 Derivative-based Global Sensitivity Measures

By modifying the elementary effect method, a series of derivative-based global sensitivity measures (DGSM) were proposed by averaging local derivatives using Monte Carlo or quasi-Monte Carlo sampling methods. Consider a differentiable function $f(\mathbf{x})$ over the unit hypercube $[0, 1]^n$. For a given value of $\mathbf{x}$, the local sensitivity measure of variable x_i is partial derivative:

$$\frac{\partial f(\mathbf{x})}{\partial x_i}$$

for $i = 1, 2, \ldots, n$. The dependency of the local sensitivity on a nominal point can be overcome by averaging $\partial f(\mathbf{x})/\partial x_i$ over the variable space $[0, 1]^n$. Using this idea, Kucherenko et al. (2009) defined four DGSM as in expressions (18.16)–(18.19); Sobol and Kucherenko (2009) defined one DGSM as in expression (18.20); and Sobol and Kucherenko (2010) defined one DGSM for individual variables as in expression (18.21) and for a group of variables $x_{i_1}, x_{i_2}, \ldots, x_{i_k}$ as in expression (18.22).

Definition 18.2.6 *Integrating over $[0, 1]^n$, the following types of the DGSM of variable x_i are defined:*

$$I_{D\mu}(i) = \int \frac{\partial f(\mathbf{x})}{\partial x_i} d\mathbf{x} \tag{18.16}$$

$$I_{D\tilde{\mu}}(i) = \int \left| \frac{\partial f(\mathbf{x})}{\partial x_i} \right| d\mathbf{x} \tag{18.17}$$

$$I_{D\sigma}(i) = \left[\int \left(\frac{\partial f(\mathbf{x})}{\partial x_i} - I_{D\mu}(i) \right)^2 d\mathbf{x} \right]^{1/2} = \left[\int \frac{\partial f(\mathbf{x})}{\partial x_i}^2 d\mathbf{x} - I_{D\mu}^2(i) \right]^{1/2} \tag{18.18}$$

$$I_{D\tilde{\sigma}}(i) = \left[\int \left(\left| \frac{\partial f(\mathbf{x})}{\partial x_i} \right| - I_{D\tilde{\mu}}(i) \right)^2 d\mathbf{x} \right]^{1/2} \tag{18.19}$$

$$I_{D^2}(i) = \int \left(\frac{\partial f(\mathbf{x})}{\partial x_i} \right)^2 d\mathbf{x} \tag{18.20}$$

$$I_{D_A^2}(i) = \int \left(\frac{\partial f(\mathbf{x})}{\partial x_i} \right)^2 \frac{1 - 3x_i + 3x_i^2}{6} d\mathbf{x}. \tag{18.21}$$

In addition, for a group of variables $x_{i_1}, x_{i_2}, \ldots, x_{i_k}$

$$I_{D_A^2}(i_1, i_2, \ldots, i_k) = \sum_{s=1}^{k} \int \left(\frac{\partial f(\mathbf{x})}{\partial x_{i_s}} \right)^2 \frac{1 - 3x_{i_s} + 3x_{i_s}^2}{6} d\mathbf{x}. \tag{18.22}$$

Measures (18.16) and (18.18) represent the mean and standard deviation of $\partial f(\mathbf{x})/\partial x_i$, respectively. To avoid the canceling effect (due to the positive and negative values of $\partial f(\mathbf{x})/\partial x_i$), measures (18.17) and (18.19) are based on the absolute value of $|\partial f(\mathbf{x})/\partial x_i|$, similarly to $I_{E\tilde{\mu}}(i)$ within the framework of the elementary effect method. Note that $I_{D\tilde{\mu}}(i)$ is the limiting value of the elementary effect sensitivity measure $I_{E\tilde{\mu}}(i)$. The canceling effect does not affect measures (18.20)–(18.22). Kucherenko et al. (2009) proposed a normalized version of $I_{D^2}(i)$ over all of the variables. From a practical point of view, $I_{D\tilde{\mu}}(i)$ and $I_{D^2}(i)$ can be evaluated by the same numerical algorithm and are linked by relations

$$I_{D\tilde{\mu}}(i) \le \sqrt{I_{D^2}(i)} \quad \text{and} \quad I_{D^2}(i) \le C I_{D\tilde{\mu}}(i) \quad \text{if} \quad \left| \frac{\partial f(\mathbf{x})}{\partial x_i} \right| \le C.$$

Measures (18.21) and (18.22) are motivated by the Taylor approximation of V_K^{tot}; Sobol and Kucherenko (2010) showed that $I_{D_A^2}(i)$ is a slight improvement of $I_{D^2}(i)$.

Except for $I_{D_A^2}(\cdot)$, the DGSM do not originally include a grouping algorithm, while Kiparissides et al. (2009) presented a technique to group variables with the DGSM, adapting the results for grouping variables with the revised elementary effect method in Subsection 18.2.2. These make it possible to study the effects of groups of variables on model output.

Computation of the DGSM is based on the evaluation of integrals in (18.16)–(18.22) using the Monte Carlo and quasi-Monte Carlo integration methods. This integration method is more accurate than the elementary effect method because the elementary effects are evaluated as strict local derivatives with small increments compared to the variable uncertainty ranges, while local derivatives are evaluated at randomly or quasi-randomly selected points in the entire range of uncertainty rather than at points from a fixed grid as in the elementary effect method. It has been shown (Kucherenko et al. 2009) numerically that the Monte Carlo and quasi-Monte Carlo integration methods can be much faster and more accurate than the elementary effect method and that they are especially efficient if automatic calculation of derivatives is used. However, in many instances, especially when the number of uncertain variables is high and/or the model is expensive to compute, the elementary effect method may still be considered as a good compromise between accuracy and efficiency. The DGSM saves impressive amounts of time while proving to be a robust screening technique (Campolongo et al. 2007).

18.2.4 Relationships between the ANOVA-decomposition-based and the Derivative-based Sensitivity Measures

The following theorem establishes links between the first-order $I_{A^{tot}}(i)$ and the derivative $\partial f(\mathbf{x})/\partial x_i$ (Sobol and Kucherenko 2009), which use the limiting values of $|\partial f(\mathbf{x})/\partial x_i|$ (related to $I_{D\widetilde{\mu}}(i)$) and the mean value of $(\partial f(\mathbf{x})/\partial x_i)^2$ (related to $I_{D^2}(i)$), as well as links between the kth-order $I_{A^{tot}}(K)$ and $I_{D_A^2}(i_1, i_2, \ldots, i_k)$ (Sobol and Kucherenko 2010).

Theorem 18.2.7 *Let function $f(\mathbf{x})$ is integrable and differentiable in the n-dimensional unit hypercube $[0, 1]^n$.*

(i) *Assuming that $c \leq |\partial f(\mathbf{x})/\partial x_i|$, then $c^2/(12V) \leq I_{A^{tot}}(i) \leq C^2/(12V)$, and the constant factor 12 cannot be improved.*

(ii) *Assuming that $\partial f(\mathbf{x})/\partial x_i$ is square integrable, then $I_{A^{tot}}(i) \leq \int (\partial f(\mathbf{x})/\partial x_i)^2 \, d\mathbf{x}/(\pi^2 V) = I_{D^2}(i)/(\pi^2 V)$.*

(iii) *For a subset of variable indices $K = \{i_1, i_2, \ldots, i_k\}$, $I_{A^{tot}}(K) \leq 24 I_{D_A^2}(i_1, i_2, \ldots, i_k)/(\pi^2 V)$.*

(iv) *If $f(\mathbf{x})$ is linear with respect to variables $x_{i_1}, x_{i_2}, \ldots, x_{i_k}$, then $I_{A^{tot}}(K) = I_{D_A^2}(i_1, i_2, \ldots, i_k)/V$.*

Theorem 18.2.7 shows that small values of $I_{D^2}(i)$ or $I_{D_A^2}(i)$ imply small values of $I_{A^{tot}}(i)$. Thus, Theorem 18.2.7 allows identification of a set of insignificant variables (usually defined by a condition of the type $I_{A^{tot}}(i) < \delta$). Theorem 18.2.7(iv) shows that if the model $f(\mathbf{x})$ is near to linear, the performance of $I_{D_A^2}(i)$ is close to the performance of $I_{A^{tot}}(i)$. However, for highly nonlinear functions, Sobol and Kucherenko (2009, 2010) showed through counterexamples that the rankings of variables based on $I_{D^2}(\cdot)$ and $I_{D_A^2}(\cdot)$ may be different from that of the AGSM.

The numerical experiments in Kucherenko et al. (2009) empirically showed that the DGSM ratio of $I_{D\widetilde{\mu}}(i)/I_{D\widetilde{\sigma}}(i)$ is a good proxy for the AGSM ratio $I_A(i)/I_{A^{tot}}(i)$; therefore, it can be used to assess the importance of higher-order interactions in the same way as

$I_A(i)/I_{A\text{tot}}(i)$. In terms of computational cost, for linear and quasi-linear output functions, the DGSM require fewer function evaluations and are faster, while for highly nonlinear functions, computation of the AGSM can be cheaper (Sobol and Kucherenko 2010). The AGSM is essentially composed of variance-based methods, which generally require a large number of function evaluations to achieve reasonable convergence and can become impractical for large engineering problems. In this regard, elementary effect methods and derivative-based methods generally require less computation cost for numerical evaluation. It is shown in Kucherenko et al. (2009) and Kiparissides et al. (2009) that the computation time required for Monte Carlo evaluation of the DGSM can be much shorter than that for estimation of the AGSM, possibly many orders of magnitude.

18.2.5   The Case of Random Input Variables

Global sensitivity analysis can be used to analyze the impact of the uncertainties of inputs on the uncertainty of model output. Uncertainty and sensitivity analysis is an important step in the model building process. Treat model inputs as independent random variables, $X_1, X_2, \ldots, X_n$ with distribution functions $F_1(x_1), F_2(x_2), \ldots, F_n(x_n)$, and then model output $f(\mathbf{X})$ is a random variable with a finite variance $V = \text{Var}(f(\mathbf{X}))$. The theory of global sensitivity measures can be easily generalized and applied in this case.

For example, in a systems reliability model, the systems unreliability is determined by the unreliability of components that is characterized as random variables q_i for $i = 1, 2, \ldots, n$, and $\mathbf{q} = (q_1, q_2, \ldots, q_n)$. Both the AGSM in Subsection 18.2.1 and the DGSM in Subsection 18.2.2 can be directly applied to analyze the impact of the uncertainties of each component unreliability, q_i (input), on the uncertainty of systems unreliability, $Q(\mathbf{q})$ (output). The reliability of components and system is in $(0, 1)$ and, in general, systems unreliability is a nonlinear function of the unreliability values of components. In using global sensitivity measures with the aim of reducing the uncertainty of systems unreliability, the process can be iterated until an acceptable uncertainty range of systems unreliability is achieved.

To use the AGSM, the total output variance for a model needs to be decomposed as ANOVA-decomposition. The ANOVA-decomposition in Equation (18.8) remains true, but requirement (18.9) should be replaced by corresponding expectations

$$\int_{-\infty}^{\infty} f_{i_1 i_2 \ldots i_k}(x_{i_1}, x_{i_2}, \ldots, x_{i_k}) dF_{i_1}(x_{i_1}) dF_{i_2}(x_{i_2}) \ldots dF_{i_k}(x_{i_k}) = 0$$

and

$$f_0 = \mathbb{E}(f(\mathbf{X})).$$

In this case, the variances $V_{i_1 i_2 \ldots i_k}$ are real variances,

$$V_{i_1 i_2 \ldots i_k} = \text{Var}(f_{i_1 i_2 \ldots i_k}(X_{i_1}, X_{i_2}, \ldots, X_{i_k})).$$

Functional relationships that include random variables are true with probability one.

The first part in the following theorem (Sobol and Kucherenko 2009) is generalized from the first part in Theorem 18.2.7. The second (Sobol and Kucherenko 2009) and third (Sobol and Kucherenko 2010) parts in the following theorem are similar to the second and third parts in Theorem 18.2.7, respectively, but they are not generalizations.

Theorem 18.2.8

(i) *Assume that $c \leq |\partial f(\mathbf{x})/\partial x_i| \leq C$ and that the variance of X_i is finite, $\mathrm{Var}(X_i) < \infty$. Then $c^2 \mathrm{Var}(X_i)/V \leq I_{A^{\mathrm{tot}}}(i) \leq C^2 \mathrm{Var}(X_i)/V$, and the constant factor $\mathrm{Var}(X_i)$ cannot be improved.*

(ii) *Assume that random variable X_i is normally distributed with variance $\mathrm{Var}(X_i)$ and that the integral in inequality (18.23) is finite. Then*

$$I_{A^{\mathrm{tot}}}(i) \leq \frac{\mathrm{Var}(X_i)}{V} \int_{\mathbb{R}^n} \left(\frac{\partial f(\mathbf{x})}{\partial x_i} \right)^2 \prod_{s=1}^{n} \mathrm{d}F_s(x_s). \tag{18.23}$$

The constant factor $\mathrm{Var}(X_i)$ cannot be reduced.

(iii) *Assume that random variables $X_1, X_2, \ldots, X_n$ are independent and normally distributed. Then for a subset of variable indices $K = \{i_1, i_2, \ldots, i_k\}$, $I_{A^{\mathrm{tot}}}(K) \leq 2I_{D_A^2}(i_1, i_2, \ldots, i_k)/V$.*

18.2.6 Moment-independent Sensitivity Measures

The aforementioned global sensitivity measures implicitly assume that the moments (e.g., expectation or variance) are sufficient to describe output variation. Indeed, a state of knowledge on an input variable or on a model output is represented by the entire uncertainty distribution. In this respect, moment-independent sensitivity measures refer to the entire distributions both of the input variables and of the output in a moment-independent fashion, thus reflecting the impact of distributional changes of input variables on the change of entire output distribution.

Borgonovo (2007) defined such a moment-independent global sensitivity measure, referred to as B-MGSM, by specifying a joint distribution of the input variables without requiring independence so that its computation is well posed in the presence of correlations among the variables. Let $Y = f(\mathbf{X}) : S \subseteq \mathbb{R}^n \mapsto \mathbb{R}$ be the functional relationship between output Y and input $\mathbf{X}$. Let $F_{\mathbf{X}}(\mathbf{x})$ ($f_{\mathbf{X}}(\mathbf{x})$) be the joint cumulative distribution (density) function of X_i, $i = 1, 2, \ldots, n$, that is, the analyst's (subjective) state of knowledge on $\mathbf{X}$. Then, $f_{X_i}(x_i) = \int f_{\mathbf{X}}(\mathbf{x}) \prod_{s \neq i} \mathrm{d}x_s$ is the marginal density of X_i. Let $F_Y(y)$ ($f_Y(y)$) be the cumulative distribution (density) function of the model output Y and $f_{Y|X_i}(y)$ be the conditional density function of Y, given that X_i assumes a fixed value.

Definition 18.2.9 *The B-MGSM of random input variable X_i with respect to output Y, denoted by $I_{\mathrm{BM}}(i)$, is defined as*

$$I_{\mathrm{BM}}(i) = \frac{1}{2} \mathbb{E}_{X_i}(s(X_i)),$$

where

$$s(X_i) = \int |f_Y(y) - f_{Y|X_i}(y)| \mathrm{d}y$$

represents the shift between $f_Y(y)$ and $f_{Y|X_i}(y)$, and

$$\mathbb{E}_{X_i}(s(X_i)) = \int f_{X_i}(x_i) \left[\int |f_Y(y) - f_{Y|X_i}(y)| \mathrm{d}y \right] \mathrm{d}x_i$$

is the expected shift since $s(X_i)$ is dependent on X_i and thus a random variable.

Letting $\overline{\mathbf{X}} = (X_{i_1}, X_{i_2}, \ldots, X_{i_k})$ be a group of random input variables, then the B-MGSM of $\overline{\mathbf{X}}$ is defined as

$$
\begin{aligned}
I_{\mathrm{BM}}&(i_1, i_2, \ldots, i_k) \\
&= \frac{1}{2}\mathbb{E}_{\overline{\mathbf{X}}}(s(\overline{\mathbf{X}})) \\
&= \int f_{X_{i_1},X_{i_2},\ldots,X_{i_k}}(x_{i_1}, x_{i_2}, \ldots, x_{i_k}) \left[\int |f_Y(y) - f_{Y|X_{i_1},X_{i_2},\ldots,X_{i_k}}(y)| \mathrm{d}y \right] \mathrm{d}x_{i_1} \mathrm{d}x_{i_2} \ldots \mathrm{d}x_{i_k},
\end{aligned}
$$

where

$$
f_{X_{i_1},X_{i_2},\ldots,X_{i_k}}(x_{i_1}, x_{i_2}, \ldots, x_{i_k}) = \int f_{\mathbf{X}}(\mathbf{x}) \prod_{s \neq i_1, i_2, \ldots, i_k} \mathrm{d}x_s.
$$

$I_{\mathrm{BM}}(i)$ represents the normalized expected shift in the distribution of Y provoked by X_i. Apparently, $0 \leq I_{\mathrm{BM}}(i) \leq 1$, and the B-MGSM of all input variables is unity. When the output is independent of a variable or group of variables, the corresponding B-MGSM equals zero (in fact, $f_{Y|X_i}(y) = f_Y(y)$).

Borgonovo (2007) proposed an algorithm for the computation of the B-MGSM. Numerical results from an application in PRA show that the AGSM and B-MGSM agree in identifying the less relevant variables with respect to output uncertainty. However, differences in the ranking of the more relevant variables emerge because of the different scope of the measures. That implies that the variable that most strongly affects variance is not necessarily the variable that most strongly affects the output distribution. Borgonovo (2006) obtained the same conclusion. In addition, Liu and Homma (2009) proposed a simple computational method for evaluating the B-MGSM based on a distance metric between the probability distribution of Y and the conditional probability distribution of Y given X_i.

There is another type of moment-independent sensitivity measure as described in Definition 18.2.10, referred to as CHT-MGSM (Chun et al. 2000).

Definition 18.2.10   Let y^i_α be the αth quantile of Y for the sensitivity case relative to X_i, y^0_α be the αth quantile of Y for the nominal case, and $\mathbb{E}(Y_0)$ be the expectation of Y for the nominal case. The CHT-MGSM of random input variable X_i, denoted by $I_{\mathrm{CM}}(i)$, is defined as

$$
I_{\mathrm{CM}}(i) = \frac{1}{\mathbb{E}(Y_0)} \left[\int (y^i_\alpha - y^0_\alpha)^2 \mathrm{d}\alpha \right]^{1/2}. \tag{18.24}
$$

$I_{\mathrm{CM}}(\cdot)$ takes into account the entire distribution of the model output Y and is expressed in terms of the cumulative distribution function of Y (F_Y). Intuitively, it represents the (square of the) area related to a shift in F_Y from the nominal case to the sensitivity case. By the sensitivity case, a recomputation of the model is meant when (i) the uncertainty in a variable is completely eliminated; (ii) the uncertainty range is changed; or (iii) the type of distribution is changed. All of the three cases reflect a change in the state of knowledge regarding the input variables. Therefore, $I_{\mathrm{CM}}(\cdot)$ can determine the variable that provokes the greatest change in the distribution of Y when, for example, the uncertainty in all variables is reduced by a same factor.

The main difference between the CHT-MGSM and B-MGSM is that the CHT-MGSM requires hypothesizing a sensitivity case, while the B-MGSM does not. The CHT-MGSM evaluates the importance of the variables on affecting output uncertainty when uncertainty in each of variables, one at a time, is under a hypothesized change. On the other hand, the B-MGSM represents the importance of the entire distribution of X_i with respect to the entire distribution of Y, given the current state of knowledge (joint distribution of input variables) without considering an artificially hypothesized change. Numerical examples in Borgonovo (2007) show that the CHT-MGSM and B-MGSM can produce different rankings of variables.

Extension to fuzzy input uncertainty

The fuzziness is distinguished from the randomness. Song et al. (2011) extended the B-MGSM to measure the influence of fuzzy input uncertainty on fuzzy structural response (i.e., fuzzy output uncertainty). Let x_i be a fuzzy-valued variable, $i = 1, 2, \ldots, n$, and $z = f(x_1, x_2, \ldots, x_n)$ be the structural response, which is a function of basic fuzzy-valued variables. The characteristic of fuzzy-valued variable x_i is represented by a membership function $\mu_{x_i}(\cdot)$, quantifying the degree of possibility that variable x_i takes value $\cdot$. Fuzzy-valued variables can be evaluated using λ-cuts; at each λ-cut for $0 \le \lambda \le 1$, the variation of fuzzy-valued variable x_i is defined by a lower and an upper bound $[x_i^L(\lambda), x_i^U(\lambda)]$.

Once the bounds of the cuts of the input fuzzy-valued variables are obtained by the corresponding membership functions, the bounds on response z at various λ-cuts can be evaluated by means of optimization technique, such as the local search, simulated annealing, genetic algorithm, and neural network. In turn, the membership function of response z, denoted by μ_z, can be obtained. Let $z_i' = f(x_1, x_2, \ldots, x_{i-1}, x_i^*, x_{i+1}, \ldots, x_n)$ be the fuzzy response when variable x_i is fixed at a value x_i^*, and let $\mu_{z_i'}$ be the corresponding membership function. Obviously, the difference between μ_z and $\mu_{z_i'}$ can be expressed as

$$\int_{-\infty}^{+\infty} |\mu_z - \mu_{z_i'}| \mathrm{d}z,$$

reflecting the effect of the fixed value of x_i on the response. On average, the effect of x_i can be measured by

$$\int_{-\infty}^{+\infty} \mu_{x_i} \left[\int_{-\infty}^{+\infty} |\mu_z - \mu_{z_i'}| \mathrm{d}z \right] \mathrm{d}x_i.$$

However, the fuzzy-valued variables and the response have the measure units. To eliminate the effect of the measure units, Song et al. (2011) proposed the following B-MGSM of fuzzy-valued variables.

Definition 18.2.11 *The B-MGSM of fuzzy-valued input variable x_i with respect to response z, denoted by $I_{\text{BM-Fuzzy}}(i)$, is defined as*

$$I_{\text{BM-Fuzzy}}(i) = \int_{-\infty}^{+\infty} \frac{\mu_{x_i}}{\int \mu_{x_i} dx_i} \left[\int_{-\infty}^{+\infty} \left| \frac{\mu_z}{\int \mu_z dz} - \frac{\mu_{z_i'}}{\int \mu_{z_i'} dz} \right| dz \right] dx_i.$$

The B-MGSM of a pair of fuzzy-valued variables x_i and x_j is defined in two ways as

$$I_{\text{BM-Fuzzy}}(i, j) = \int_{-\infty}^{+\infty} \frac{\mu_{x_i}}{\int \mu_{x_i} dx_i} \frac{\mu_{x_j}}{\int \mu_{x_j} dx_j} \left[\int_{-\infty}^{+\infty} \left| \frac{\mu_z}{\int \mu_z dz} - \frac{\mu_{z_{ij}'}}{\int \mu_{z_{ij}'} dz} \right| dz \right] dx_i dx_j$$

$$I_{\text{BM-Fuzzy}}(i|j) = \int_{-\infty}^{+\infty} \frac{\mu_{x_i}}{\int \mu_{x_i} dx_i} \frac{\mu_{x_j}}{\int \mu_{x_j} dx_j} \left[\int_{-\infty}^{+\infty} \left| \frac{\mu_{z_i'}}{\int \mu_{z_i'} dz} - \frac{\mu_{z_{ij}'}}{\int \mu_{z_{ij}'} dz} \right| dz \right] dx_i dx_j,$$

where $\mu_{z_{ij}'}$ is the membership function of fuzzy response z_{ij}' when variables x_i and x_j take the fixed values x_i^* and x_j^*, respectively.

Apparently, $I_{\text{BM-Fuzzy}}(i) \geq 0$. Similar to the B-MGSM of random input variables, when response z is independent on fuzzy variable x_i, the corresponding B-MGSM equals zero. Song et al. (2011) also showed that $I_{\text{BM-Fuzzy}}(i, j) \leq I_{\text{BM-Fuzzy}}(i) + I_{\text{BM-Fuzzy}}(i|j)$, and that if z is dependent on x_i but independent on x_j, then $I_{\text{BM-Fuzzy}}(i, j) = I_{\text{BM-Fuzzy}}(j|i) = I_{\text{BM-Fuzzy}}(i)$ and $I_{\text{BM-Fuzzy}}(i|j) = 0$.

Generally, the membership function of response z cannot be expressed by explicit form. It is determined by the lower and upper bounds at various λ-cuts. Thus, the integrals in Definition 18.2.11 need to be evaluated by numerical integrations, that is,

$$I_{\text{BM-Fuzzy}}(i) \simeq \sum_{k=1}^{N} \left\{ \Delta x_i \frac{\mu_{x_i}(x_{i,k})}{\int \mu_{x_i} dx_i} \sum_{m=1}^{M} \left[\Delta z \left| \frac{\mu_z(z_m)}{\int \mu_z dz} - \frac{\mu_{z_i'}(z_m)}{\int \mu_{z_i'} dz} \right| \right] \right\}$$

$$I_{\text{BM-Fuzzy}}(i, j) \simeq \sum_{k=1}^{N_i} \sum_{\ell=1}^{N_j} \left\{ \Delta x_i \Delta x_j \frac{\mu_{x_i}(x_{i,k})}{\int \mu_{x_i} dx_i} \frac{\mu_{x_j}(x_{j,\ell})}{\int \mu_{x_j} dx_j} \sum_{m=1}^{M} \left[\Delta z \left| \frac{\mu_z(z_m)}{\int \mu_z dz} - \frac{\mu_{z_{ij}'}(z_m)}{\int \mu_{z_{ij}'} dz} \right| \right] \right\}$$

$$I_{\text{BM-Fuzzy}}(i|j) \simeq \sum_{k=1}^{N_i} \sum_{\ell=1}^{N_j} \left\{ \Delta x_i \Delta x_j \frac{\mu_{x_i}(x_{i,k})}{\int \mu_{x_i} dx_i} \frac{\mu_{x_j}(x_{j,\ell})}{\int \mu_{x_j} dx_j} \sum_{m=1}^{M} \left[\Delta z \left| \frac{\mu_{z_i'}(z_m)}{\int \mu_{z_i'} dz} - \frac{\mu_{z_{ij}'}(z_m)}{\int \mu_{z_{ij}'} dz} \right| \right] \right\},$$

where Δx_i and Δz are the spans of isometric sampling, $x_{i,k}$ is the kth fixed value of fuzzy-valued variable x_i, z_m is the mth value of response z, and the other notation is defined similarly. Song et al. (2011) presented the details and four examples for the calculations of the B-MGSM of fuzzy-valued variables.

18.3 Systems Reliability Subject to Uncertain Component Reliability

One difficulty in estimating the unreliability (or unavailability) of a system is insufficiency of component failure data. Consequently, component unreliability and systems unreliability are subject to uncertainty. It is desirable to know which components are most important for causing the uncertainty of the systems unreliability so that this uncertainty can be greatly reduced by improving the estimation of the unreliability of these components. This leads to studies of the uncertainty importance measures of components. Assume that the unreliability of component i is a random variable q_i with mean of $\overline{q}_i$ and variance of estimate of q_i, v_i, for $i = 1, 2, \ldots, n$. Then the systems unreliability $Q(\mathbf{q})$ is a random variable, denoting its mean as $\overline{Q}$ and its variance as V. In addition to the global sensitivity measures in Section 18.2, this section introduces three variance-based uncertainty importance measures, similar to which some measures have also been proposed in early work (Bhattacharyya and Ahmed 1982; Bier 1983; Hora and Iman 1989; Iman 1987; Iman and Hora 1990; Ishigami and Homma 1990; Rushdi 1985). These variance-based uncertainty importance measures are frequently and mainly used in uncertainty and sensitivity analysis for systems reliability.

Nakashima and Yamato (1982) proposed a first variance-based uncertainty importance that is defined for component i as

$$\frac{\partial V}{\partial v_i} \frac{v_i}{V}.$$

The $(\partial V / \partial v_i) v_i$ equals the absolute value of the change of system variance when the variance of component i varies from v_i to zero. Thus, components with larger importance values contribute more to the variance of systems unreliability. If reduce v_i to αv_i ($0 < \alpha < 1$) by collecting failure data of component i, then the change of V is $(1 - \alpha)(\partial V / \partial v_i) v_i$. Thus, for fixed α, the component with the largest value of this importance has the greatest effect in reducing the variance of systems unreliability, V.

Pan and Tai (1988) proposed another variance-based uncertainty importance that is defined for component i as

$$\frac{\partial V}{\partial v_i}.$$

This uncertainty importance can also be expressed as

$$\mathbb{E}\left(\left(\frac{\partial Q}{\partial q_i}\right)^2\right),$$

where $\partial Q / \partial q_i$ is the B-importance. Miman and Pohl (2006) extended this variance-based uncertainty importance to the repairable system and its components by substituting the availability for the reliability. To calculate these uncertainty importance measures, one must evaluate partial derivatives and V, either analytically or numerically by Monte Carlo simulation (Pan and Tai 1988).

In a fault tree analysis, Cho and Yum (1997) proposed an uncertainty importance measure that can be experimentally evaluated by a two-stage procedure utilizing the Taguchi tolerance design technique (Taguchi 1986). Under the assumption that all basic events are independent and that the probability of each basic event follows a lognormal distribution, this uncertainty importance measure is defined for basic event i as

$$\frac{\text{Uncertainty of } \ln Q \text{ due to the uncertainty of } \ln q_i}{\text{Total uncertainty of } \ln Q}$$

and for a pair of basic events i and j as

$$\frac{\text{Uncertainty of } \ln Q \text{ due to the uncertainty of } \ln q_i \text{ and } \ln q_j}{\text{Total uncertainty of } \ln Q}.$$

This uncertainty importance measure identifies the basic events that significantly contribute to the uncertainty of the top event probability. The one of a basic event assesses the main effect of uncertainty, while the one of a pair of basic events assesses the interaction effect. It can apparently be extended to more than two basic events. In the two-stage procedure, the first stage screens out the basic events with negligible main effects on the uncertainty of $\ln Q$, and the second stage conducts a more detailed analysis, including of interaction effects, with a substantially smaller number of basic events. This two-stage procedure calculates the contribution ratios, which are a general concept in Taguchi tolerance design, of basic event i and of the interaction between basic events i and j as

$$\frac{SS_i - DOF_i \times MSE}{SST} \quad \text{and} \quad \frac{SS_{ij} - DOF_{ij} \times MSE}{SST},$$

respectively, where SST, SS_i, SS_{ij}, DOF_i, DOF_{ij}, and MSE are the total sum of squares, sum of squares due to the main effect of basic event i, sum of squares due to the interaction between basic events i and j, degree of freedom associated with basic event i, degree of freedom associated with the interaction, and error mean squares, respectively. The contribution ratios are used to estimate the defined uncertainty importance measure of a basic event or of a pair of basic events.

In addition, Cho and Yum (1997) presented a measure to quantify the percentage reduction in the uncertainty of the log-transformed top event probability when the uncertainty of each basic-event probability is reduced. This measure has the advantage of numerical robustness but has the disadvantage of not easily converting back to Q.

Remark 18.3.1 In reliability analysis, uncertainty importance measures differ from the local importance measures (e.g., the B-importance), which are used for quantifying a component's contribution to system performance at a nominal point. These local importance measures assume that component (systems) reliability is expressed by the knowledge-based probability that a component (system) functions or in a particular state and thus are also referred to as *probabilistic importance measures* (Pan and Tai 1988). Uncertainty and global sensitivity measures are concerned with output uncertainty, while probabilistic (local) importance measures are concerned with the output itself. The uncertainty importance measures are for quantifying the contribution of the uncertainty of component reliability to the uncertainty of system performance. They are frequently used in risk assessment and safety analysis; Chapter 19 introduces more uncertainty importance measures in the context of the risk assessment and safety analysis.

18.3.1   Software Reliability

The development of the theory of importance measures for software systems is challenging because of the following fundamental differences with hardware systems (Fiondella and Gokhale 2008; Simmons et al. 1998). First, the architecture of a software system, which is given by the branching behavior among its components, is probabilistic, as opposed to a deterministic structure as in the case of hardware systems. The components in software reliability are the functional modules that are normally assumed to be logically independent. Second, these branching probabilities, which are determined by the *operational profile* of the applications, may not be known with complete certainty in the early phases of the software life cycle and are subject to the uncertainties in this profile (Musa et al. 1987). Since a software system can have a variety of applications, the operational profile (also known as the user profile) represents the probabilistic distribution of the utilization of the components in an operational environment.

In software reliability analysis, architecture-based reliability analysis (also known as the user-oriented reliability model) expresses systems reliability in terms of the failure behavior of its components and the probabilistic system architecture (also known as the component dependency graph). Using this approach, the reliability of the system is a function of both the deterministic properties of the structure of the program and the stochastic properties of the operational profile and component failure behavior. The control (program) flow graph can represent the structure of the system. This approach assumes sequential execution of components; parallelism is not considered.

Architecture-based reliability analysis assumes that the transfer of control among components is a Markov process. Considering a terminating application with n components, the architecture of the application is represented by its probabilistic control flow graph, which is mapped to a discrete time Markov chain. Without loss of generality, assume that the application begins with the execution of component 1 and terminates upon the execution of component n. Thus, state 1 of the Markov chain is designated as the initial state, state n as the final or absorbing state. Let $\mathbf{p}$ denote the vector of component reliability, where p_i is the reliability of component i for $i = 1, 2, \ldots, n$. The systems reliability $R(\mathbf{p})$ may be obtained in the following steps: (i) set $t_{n,n} = 0$, which is the (n, n)th entry of the transition probability matrix $\mathbf{W}$, (ii) compute $S = (\mathbf{I} - \mathbf{DW})^{-1}$ where $\mathbf{D}$ is the $n \times n$ diagonal matrix with entry $d_{i,i} = p_i$, and (iii) $R(\mathbf{p}) = s_{1,n}p_n$ where $s_{1,n}$ represents the probability that the absorbing state is reached from the initial state. The value of $s_{1,n}$ times the reliability of component n provides an estimate of application reliability.

Importance-measure-based sensitivity analysis can help identify critical components (modules) for further improvement. It can be applied early in the life cycle to detect potential problems and to enable decision-makers to make more responsive choices. Early detection of critical architecture elements (those that most affect the overall reliability of the system) is useful in distributing resources in later development phases. Quantitative importance measures are highly valuable when there is an attitude for reliability improvement subject to the limited resources at the components. With respect to the calculated software reliability $R(\mathbf{p})$, Cheung (1980) provided sensitivity analysis of a modular software system based on the B-importance of the components. The architecture-based reliability analysis and sensitivity analysis can be used to develop both an effective testing strategy to detect as many bugs as possible and a maintenance strategy to correct the critical errors. With limited testing resources, concentration should be put on the module with the largest B-importance value. On the other hand, for software reliability, resource distribution should depend on the effectiveness (i.e., the B-importance) and the marginal cost of improving the reliability of the module (e.g., size, logical complexity, and structure of the module; relationship with other modules; and the programmer's understanding of the module).

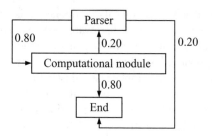

Figure 18.2   A simplified ESA architecture

For software systems, a significant level of uncertainty is associated with the reliability of the individual components and the probabilistic architecture, especially in the early design phases. To investigate how effectively component improvements decrease the variance of systems reliability, Fiondella and Gokhale (2008) assumed that increasing a component's expected reliability simultaneously decreases its variance; in turn, both the increase in component reliability and the decrease in its variance contribute not only to an increase in the mean systems reliability but also to a decrease in system-level variance. They defined an unbiased estimator for variance as

$$V(\mathbf{p}) = R(\mathbf{p})(1 - R(\mathbf{p})),$$

where $R(\mathbf{p})$ denotes the expected systems reliability and is evaluated using the architecture-based reliability analysis. Adapting the B-importance in Equation (4.10) and the improvement potential importance in Equation (4.18), the corresponding variance importance measures of component i can be defined as

$$I_{\mathrm{B}^V}(i) = \frac{\partial V(\mathbf{p})}{\partial p_i}$$

and

$$I_{\mathrm{B_s}^V}(i) = V(1_i, \mathbf{p}) - V(\mathbf{p}),$$

respectively.

Example 18.3.2   This example is a simplified application used by the European Space Agency (ESA) (Fiondella and Gokhale 2008) to configure an array of antennas. The application has 11,000 lines of code, and its architecture is shown in Figure 18.2. Upon receiving input, the application begins by parsing a configuration script. The application terminates if the parser encounters a syntax error within the script; otherwise, it proceeds to the computational module. Additional invocations of the parser may be recursive. Eventually, the computational module proceeds directly to the end of the application. The component reliability for the present version of the application based on testing data is $\mathbf{p} = (p_p, p_c, p_e) = (0.84, 0.83, 1.0)$.

The transition probability matrix of the ESA application is given by

$$W = \begin{pmatrix} 0 & t_{p,c} & t_{p,e} \\ t_{c,p} & 0 & t_{c,e} \\ 0 & 0 & 0 \end{pmatrix} = \begin{pmatrix} 0 & 0.8 & 0.2 \\ 0.2 & 0 & 0.8 \\ 0 & 0 & 0 \end{pmatrix},$$

Table 18.1   Results in Example 18.3.2

Components	I_B	I_{Bs}	I_{B^V}	I_{Bs^V}
Parser	0.9263	0.1518	−0.3545	−0.0812
computational module	0.7097	0.1238	−0.2716	−0.0627

where the absorbing state $t_{e,e}$ is set to 0. The expected reliability of the ESA application is

$$R(\mathbf{p}) = \frac{(t_{p,e} + p_c t_{p,c} t_{c,e}) p_p p_e}{1 - p_p p_c t_{p,c} t_{c,p}},$$

and the corresponding

$$V(\mathbf{p}) = \frac{(t_{p,e} + p_c t_{p,c} t_{c,e}) p_p p_e}{1 - p_p p_c t_{p,c} t_{c,p}} \left(1 - \frac{(t_{p,e} + p_c t_{p,c} t_{c,e}) p_p p_e}{1 - p_p p_c t_{p,c} t_{c,p}}\right).$$

Then, the expected application reliability and its variance are 0.6913 and 0.2134, respectively.

Table 18.1 shows the values of the importance measures. The parser consistently ranks higher than the computational module because within the context of the application architecture it is more likely that the parser is invoked. $I_B(\cdot)$ and $I_{Bs}(\cdot)$ are relative to the expectation of systems reliability, and particularly, $I_{Bs}(i)$ represents the maximum increase in expected systems reliability that would be achieved by improving component i. $I_{B^V}(\cdot)$ and $I_{Bs^V}(\cdot)$ are relative to the variance of systems reliability and are negative, indicating a decrease in the variance of systems reliability corresponding to the improvement of components.

Achieving a highly reliable software application is a difficult task, even when high quality, pretested, and trusted software components are composed together (Jin and Santhanam 2001). The main reason is that the transition from one component to another may fail. In the aforementioned methods including the architecture-based reliability analysis, the transition is failure free. Extending those methods to considering the failure of transitions, Yacoub et al. (2004) presented a scenario-based reliability analysis for component-based and distributed software systems to identify critical components, subsystems, and links between components. The software reliability problem is defined by a set of software components, a generic architecture for communicating components, and a set of possible execution scenarios. For each scenario that represents a sequence of component interactions, there is an associated scenario probability, that is, the execution frequency of the scenario with respect to all other scenarios. The sum of execution probabilities of all scenarios is unity. The component execution times and the transition execution probabilities are the average values over all scenarios weighted by the scenario probabilities. The component execution times are included because of the existence of loops between two or more components. By virtue of using the average execution time of the application (i.e., the average execution time of all scenarios weighted by the scenario probabilities) to terminate the depth traversal of the graph, these loops never lead to a deadlock.

Cheung (1980) and Yacoub et al. (2004) proposed the methods for estimation of the parameters used in the reliability models and construction of the probabilistic system architecture. Basically, these probabilistic parameters can be estimated by stochastic testing, that is, by selecting a representative sample of sets of valid input data, running the program, and measuring the relevant quantities.

18.4 Broad Applications

From local sensitivity analysis to uncertainty and global sensitivity analysis, the importance measures are desirable to be model independent, include multidimensional averaging, and be capable of evaluating the importance of groups of variables, as discussed in this chapter. For the local sensitivity analysis, Borgonovo and Peccati (2004) applied the DIM for investment project decisions, where the economic relevant models are expected net profit value, internal rate of return, and so on. Borgonovo (2008) extended the DIM in combination with comparative statics technique to inventory management, in which the models are implicit, and the problem is to determine the most influential parameter on an inventory policy.

The importance-measure-based uncertainty and global sensitivity have been widely used in various areas to analyze the relative importance of inputs to the model output, especially for nonlinear models, ranging from food safety (Frey 2002) to hurricane losses (Iman et al. 2005a, 2005b). For example, in structural reliability analysis, sensitivity and importance measures have been widely used (Hohenbichler and Rackwitz 1986). Gupta and Manohar (2004) represented and investigated uncertainties in the structural and load properties through a vector of random variables having specified joint probability distributions, and Haukaas and Kiureghian (2005) did work along this line. For studying the uncertainty response of laminated composite structures, António and Hoffbauer (2008) compared the first-order differential local sensitivity method and the Monte-Carlo-based first-order AGSM.

Campolongo et al. (2007) employed the revised elementary effect method to assess the sensitivity of a chemical reaction model for dimethylsulphide, a gas involved in climate change. Kiparissides et al. (2009) applied both the AGSM and DGSM (Equations (18.16)–(18.20)) in highly nonlinear dynamic models in the context of biological systems. Linking model-based design of experiments and model-based control and optimization into the actual industrial bioprocess, global sensitivity measures are used to identify significant variables that can be experimentally estimated, as well as insignificant variables that can be fixed at their literature values or approximated. They showed that the DGSM is the most time-efficient and robust alternative to the Monte-Carlo-based AGSM.

Saltelli and Tarantola (2002) tested another commonly used uncertainty importance measure that was proposed by Iman (1987) and is defined as follows.

Definition 18.4.1 *Let* $\mathbb{E}(\mathrm{Var}(f(\mathbf{X})|X_i))$ *be the conditional expected value of the variance of output* $f(\mathbf{X})$ *when* X_i *is known and* $\mathrm{Var}(\mathbb{E}(f(\mathbf{X})|X_i))$ *be the variance of the expected value of* $f(\mathbf{X})$ *when* X_i *is known. The I-uncertainty importance measure of random input variable* X_i, *denoted by* $I_{\mathrm{IU}}(i)$, *is defined as*

$$I_{\mathrm{IU}}(i) = \mathrm{Var}(f(\mathbf{X})) - \mathbb{E}(\mathrm{Var}(f(\mathbf{X})|X_i)) = \mathrm{Var}(\mathbb{E}(f(\mathbf{X})|X_i)). \tag{18.25}$$

The basis for the importance measure is $\mathrm{Var}(f(\mathbf{X})|X_i)$, but as this term is unknown, the reference becomes the expected value $\mathbb{E}(\mathrm{Var}(f(\mathbf{X})|X_i))$. The I-uncertainty importance represents the expected reduction in output variance that can be achieved if uncertainty in X_i

is eliminated (Aven and Nøkland 2010; Borgonovo 2007; Hora and Iman 1989; Ishigami and Homma 1990). Saltelli and Tarantola (2002) applied this first-order conditional variance measure and the first-order tot AGSM to judge the relative importance of input variables in a nonlinear, nonmonotonic level E model with or without correlation of input variables. The level E model predicts the radiologic dose to humans over geologic timescales due to the underground migration of radionuclides from a nuclear waste disposal site. Chapter 19 further presents the typical applications of uncertainty and sensitivity analyses in PRA and PSA.

References

António CC and Hoffbauer LN. 2008. From local to global importance measures of uncertainty propagation in composite structures. *Composite Structures* **85**, 213–225.

Aven T and Nøkland TE. 2010. On the use of uncertainty importance measures in reliability and risk analysis. *Reliability Engineering and System Safety* **95**, 127–133.

Bhattacharyya AK and Ahmed S. 1982. Establishing data requirements for plant probabilistic risk assessment. *Transactions of the American Nuclear Society* **43**, 477–478.

Bier VM. 1983. A measure of uncertainty importance for components in fault trees. *Transactions of the 1983 Winter Meeting of the American Nuclear Society* **45**(1), 384–385.

Borgonovo E. 2006. Measuring uncertainty importance: Investigation and comparison of alternative approaches. *Risk Analysis* **26**, 1349–1361.

Borgonovo E. 2007. A new uncertainty importance measure. *Reliability Engineering and System Safety* **92**, 771–784.

Borgonovo E. 2008. Differential importance and comparative statics: An application to inventory management. *International Journal of Production Economics* **111**, 170–179.

Borgonovo E and Peccati L. 2004. Sensitivity analysis in investment project evaluation. *International Journal of Production Economics* **90**, 17–25.

Campolongo F, Cariboni J, and Saltelli A. 2007. An effective screening design for sensitivity analysis of large models. *Environmental Modelling and Software* **22**, 1509–1518.

Cao XR and Chen HF. 1997. Perturbation realization, potentials, and sensitivity analysis of Markov processes. *IEEE Transactions on Automatic Control* **42**, 1382–1393.

Cao XR and Ho YC. 1987. Sensitivity estimate and optimization of throughput in a production line with blocking. *IEEE Transactions on Automatic Control* **32**, 959–967.

Cheung RC. 1980. A user-oriented software reliability model. *IEEE Transactions on Software Engineering* **6**, 118–125.

Cho JG and Yum BJ. 1997. Development and evaluation of an uncertainty importance measure in fault tree analysis. *Reliability Engineering and System Safety* **57**, 143–157.

Chun MH, Han SJ, and Tak NI. 2000. An uncertainty importance measure using a distance metric for the change in a cumulative distribution function. *Reliability Engineering and System Safety* **70**, 313–321.

Dai L and Ho YC. 1995. Structural infinitesimal perturbation analysis for derivative estimation of discrete-event dynamic systems. *IEEE Transactions on Automatic Control* **40**, 1154–1166.

Do Van P, Barros A, and Berenguer C. 2008a. Importance measure on finite time horizon and application to Markovian multistate production systems. *Proceedings of the IMechE, Part O: Journal of Risk and Reliability, Vol. 222*, pp. 449–461.

Do Van P, Barros A, and Berenguer C. 2008b. Reliability importance analysis of Markovian systems at steady state using perturbation analysis. *Reliability Engineering and Systems Safety* **93**, 1605–1615.

Do Van P, Barros A, and Berenguer C. 2010. From differential to difference importance measures for Markov reliability models. *European Journal of Operational Research* **24**, 513–521.

Fiondella L and Gokhale SS. 2008. Importance measures for a modular software system. *Proceedings of the 8th International Conference on Quality Software*, pp. 338–343.

Frey CH. 2002. Introduction to special section on sensitivity analysis and summary of NCSU/USDA workshop on sensitivity analysis. *Risk Analysis* **22**, 539–545.

Gandini A. 1990. Importance & sensitivity analysis in assessing system reliability. *IEEE Transactions on Reliability* **39**, 61–70.

Golub GH and Zha H. 1994. Perturbation analysis of the canonical correlations of matrix pairs. *Linear Algebra and its Applications* **210**, 3–28.

Gong WB and Ho YC. 1987. Smoothed (conditional) perturbation analysis of discrete event dynamical systems. *IEEE Transactions on Automatic Control* **32**, 858–866.

Gupta S and Manohar CS. 2004. An improved response surface method for the determination of failure probability and importance measures. *Structural Safety* **26**, 123–139.

Haukaas T and Kiureghian AD. 2005. Parameter sensitivity and importance measures in nonlinear finite element reliability analysis. *Journal of Engineering Mechanics* **131**, 1013–1026.

Ho YC. 1988. Perturbation analysis explained. *IEEE Transactions on Automatic Control* **33**, 761–763.

Ho YC. 1992. Perturbation analysis: Concepts and algorithms. In *Proceedings of the 1992 Winter Simulation Conference* (ed. Swain JJ, Goldsman D, Crain RC and Wilson JR), pp. 231–240.

Ho YC and Cao XR. 1983. Perturbation analysis and optimization of queueing networks. *Journal of Optimization Theory and Applications* **40**, 559–582.

Ho YC and Cao XR. 1985. Performance sensitivity to routing changes in queueing networks and flexible manufacturing systems using perturbation analysis. *IEEE Journal of Robotics and Automation* **RA-I**, 165–172.

Ho YC and Cao XR. 1991. *Perturbation Analysis of Discrete Event Dynamic Systems*. Kluwer Academic Publishers, Boston.

Hohenbichler M and Rackwitz R. 1986. Sensitivity and importance measures in structural reliability. *Civil Engineering Systems* **3**, 203–209.

Homma T and Saltelli A. 1996. Importance measures in global sensitivity analysis of nonlinear models. *Reliability Engineering and System Safety* **52**, 1–17.

Hora SC and Iman RL. 1989. Expert opinion in risk analysis: The NUREG-1150 methodology. *Nuclear Science and Engineering* **102**, 323–331.

Iman RL. 1987. A matrix-based approach to uncertainty and sensitivity analysis for fault trees. *Risk Analysis* **7**, 21–23.

Iman RL and Hora SC. 1990. A robust measure of uncertainty importance for use in fault tree system analysis. *Risk Analysis* **10**, 401–406.

Iman RL, Johnson ME, and Watson CCJ. 2005a. Sensitivity analysis for computer model projections of hurricane losses. *Risk Analysis* **25**, 1277–1297.

Iman RL, Johnson ME, and Watson CCJ. 2005b. Uncertainty analysis for computer model projections of hurricane losses. *Risk Analysis* **25**, 1299–1312.

Ishigami T and Homma T. 1990. An importance quantification technique in uncertainty analysis for computer models. *Proceedings of the 1st International Symposium on Uncertainty Modelling and Analysis*, pp. 398–403.

Jin H and Santhanam P. 2001. An approach to higher reliability using software components. *Proceedings of the 12th IEEE International Symposium Software Reliability Engineering (ISSRE)*, pp. 1–11.

Kiparissides A, Kucherenko SS, Mantalaris A, and Pistikopoulos EN. 2009. Global sensitivity analysis challenges in biological systems modeling. *Industrial and Engineering Chemistry Research* **48**, 7168–7180.

Kucherenko S, Rodriguez-Fernandez M, Pantelides C, and Shah N. 2009. Monte Carlo evaluation of derivative-based global sensitivity measures. *Reliability Engineering and System Safety* **94**, 1135–1148.

Levitin G and Lisnianski A. 1999. Importance and sensitivity analysis of multi-state systems using the universal generating function method. *Reliability Engineering and System Safety* **65**, 271–282.

Liu Q and Homma T. 2009. A new computational method of a moment-independent uncertainty importance measure. *Reliability Engineering and System Safety* **94**, 1205–1211.

Marseguerra M and Zio E. 2004. Monte Carlo estimation of the differential importance measure: Application to the protection system of a nuclear reactor. *Reliability Engineering and System Safety* **86**, 11–24.

Miman M and Pohl EA. 2006. Uncertainty assessment for availability: Importance measures. *Proceedings of Annual Reliability and Maintainability Symposium*, pp. 222 –227.

Morris MD. 1991. Factorial sampling plans for preliminary computational experiments. *Technometrics* **33**, 161–174.

Musa J, Iannino A, and Okumoto K. 1987. *Software Reliability: Measurement, Prediction, Application*. Series of Software Engineering and Technology. McGraw-Hill, New York.

Nakashima K and Yamato K. 1982. Variance-importance of system components. *IEEE Transactions on Reliability* **R-31**, 99–100.

Neuts MF. 1995. *Algorithm Probability: A Collection of Problems*. Chapman and Hall, London.

Pan ZJ and Tai YC. 1988. Variance importance of system components by Monte Carlo. *IEEE Transactions on Reliability* **37**, 421–423.

Rushdi AM. 1985. Uncertainty analysis of fault-tree outputs. *IEEE Transactions on Reliability* **R-34**, 458–462.

Saltelli A and Tarantola S. 2002. On the relative importance of input factors in mathematical models: Safety assessment for nuclear waste disposal. *Journal of American Statistical Association* **97**, 702–709.

Simmons D, Ellis N, Fujihara H, and Kuo W. 1998. *Software Measurement: A Visualization Toolkit for Project Control and Process Improvement*. Prentice Hall, New Jersey. 459 pages and CD software.

Sobol I. 1993. Sensitivity estimates for nonlinear mathematical models. *Mathematical Modeling and Computational Experiment* **1**, 407–414.

Sobol IM. 2001. Global sensitivity indices for nonlinear mathematical models and their Monte Carlo estimates. *Mathematics and Computers in Simulation* **55**, 271–280.

Sobol IM and Kucherenko SS. 2005. On global sensitivity analysis of quasi-Monte Carlo algorithms. *Monte Carlo Methods and Applications* **11**, 83–92.

Sobol IM and Kucherenko SS. 2009. Derivative based global sensitivity measures and their link with global sensitivity indices. *Mathematics and Computers in Simulation* **79**, 3009–3017.

Sobol IM and Kucherenko SS. 2010. A new derivative based importance criterion for groups of variables and its link with the global sensitivity indices. *Computer Physics Communications* **181**, 1212–1217.

Sobol IM and Levitan YL. 1999. On the use of variance reducing multipliers in Monte Carlo computations of a global sensitivity index. *Computer Physics Communications* **117**, 52–61.

Sobol IM and Shukhman BV. 2007. On global sensitivity indices: Monte Carlo estimates affected by random errors. *Monte Carlo Methods Applications* **13**, 89–97.

Sobol IM, Tarantola S, Gatelli D, Kucherenko S, and Mauntz W. 2007. Estimating the approximation error when fixing unessential factors in global sensitivity analysis. *Reliability Engineering and System Safety* **92**, 957–960.

Song S, Lu Z, and Cui L. 2012. A generalized Borgonovo's importance measure for fuzzy input uncertainty. *Fuzzy Sets and Systems* **189**, 53–62 .

Taguchi G. 1986. *Introduction to Quality Engineering*. Asian Productivity Association, Tokyo, Japan.

Yacoub S, Cukic B, and Ammar HH. 2004. A scenario-based reliability analysis approach for component-based software. *IEEE Transactions on Reliability* **53**, 465–480.

19

Risk and Safety in Nuclear Power Plants

Importance measures are commonly applied to identify the risk and safety significant contributors of basic events in PRA and PSA where people care about the risk and safety of the system, for example, in nuclear power plants. The B-importance, criticality importance for system failure, FV importance, RAW, and RRW have been traditionally used in PRA and PSA. Most recently, the DIM and TOI were proposed for PSA, and their influence in PRA and PSA has been growing. On the other hand, uncertainty and global sensitivity measures are used to identify uncertainty drivers of parameters to the uncertainty of the system performance such as risk metrics.

Section 19.1 introduces the fundamental descriptions of PRA and PSA. Section 19.2 presents the main probabilistic importance measures in PRA and PSA, and Section 19.3 is for the global importance measures in PRA and PSA. Following that, Section 19.4 presents a case study, and Section 19.5 gives a brief review on various applications of importance measures in PRA and PSA. Finally, Section 19.6 discusses the practical implementations of fault diagnosis and maintenance policy.

19.1 Introduction to Probabilistic Risk Analysis and Probabilistic Safety Assessment

PRA and PSA are methodologies that produce numerical estimates for a number of risk metrics for complex technological systems. The common risk metrics of interest could be the system unavailability; the top event frequency; the core damage frequency; the large early release frequency; the radioactive release; and the health effects. Among them, core damage frequency is probably the most commonly used in nuclear power plants. Note that there is no intent to distinguish PRA and PSA, since the importance measures and the methods based on them are normally interchangeable in these contexts. PSA normally deals with issues of safety, while PRA may be used to deal with nonsafety issues.

Importance Measures in Reliability, Risk, and Optimization: Principles and Applications, First Edition.
Way Kuo and Xiaoyan Zhu. © 2012 John Wiley & Sons, Ltd. Published 2012 by John Wiley & Sons, Ltd.

In PRA and PSA, there often emerges the need to distinguish between two types of uncertainties. Aleatory uncertainty is the uncertainty about the occurrence of a random event, and epistemic uncertainty or "state-of-knowledge" uncertainty is the uncertainty in the values of the probabilities of such occurrences (Borgonovo 2008). Unlike aleatory uncertainties, epistemic uncertainties are associated with nonobservable quantities, for example, the parameters of models such as failure rates. Normally, the uncertainty and global sensitivity measures are used in the presence of epistemic uncertainties to drive the reduction of uncertainties, but probabilistic importance measures have been used to analyze both aleatory and epistemic uncertainties (Remark 18.3.1).

Note that the presence of epistemic uncertainties also affects the calculation and comparisons of probabilistic importance measures of components. Baraldi et al. (2009) incorporated the epistemic uncertainties of parameters in calculating the importance measures, and the resulting importance values of components are random variables following certain distributions. Then, the difference of the random importance values of two components i and j, denoted by Δ_{ij}, is a random variable. The importance ranking of the two components depends on the distribution of Δ_{ij}. If the probability or possibility of $\Delta_{ij} > 0$ is greater than an upper threshold value, then component i is more important than component j; if it is less than a lower threshold value, then component i is less important than component j; if it is between the two threshold values, then component i is equally important to component j, and in this case, additional information (e.g., costs, times, impacts on public opinion, and so on) can further guide the ranking.

The generic risk metric can be written as a function of the frequencies of initiating events (i.e., events that disturb the normal operation of the facility, such as failure of emergency generators under flooding in Example 1.0.1) and of conditional probabilities of failure modes of structures, systems, and components:

$$Risk(\boldsymbol{\zeta}, \mathbf{q}), \tag{19.1}$$

where synthetically, $\boldsymbol{\zeta} = \{\zeta_i\}$, $i = 1, 2, \ldots, \ell$, is the set of the frequencies of initiating events with ℓ as the total number of initiating events included in the PSA model, and $\mathbf{q} = \{q_j\}$, $j = 1, 2, \ldots, m$, is the set of basic event probabilities with m as the total number of basic events. Equation (19.1) relates the risk metric to the basic events and hence is referred to as the representation of *Risk at the basic event level*. The basic event representation (19.1) should be used when conducting the local sensitivity analysis (with respect to the events) of the nominal risk. The estimate of *Risk* at a nominal point is denoted by $Risk_0$ hereafter.

The risk metric can also be expressed as a function of fundamental parameters. For example, the failure time of a component is usually assumed to follow an exponential distribution with a failure rate λ. In the case that the component is renewed every τ units of time, its average (over time) unavailability is

$$q_j = \frac{\lambda_j \tau}{2}, \tag{19.2}$$

where basic event j corresponds to the failure of the component. These input parameters may be uncertain and are expressed using epistemic probability distributions. Then, the propagation of these distributions produces the epistemic probability distribution of *Risk*. Epistemic dependencies and conditional dependencies are not captured by the basic event representation

(19.1) of *Risk*. The risk metric can be written as a function of the PSA model parameters $x_1, x_2, \ldots, x_n$:

$$Risk(\mathbf{x}) = g(x_1, x_2, \ldots, x_n). \tag{19.3}$$

This representation (19.3) of *Risk at the parameter level* is needed when conducting the uncertainty and global sensitivity analysis.

Importance measures are strongly affected by the scope and quality of the PRA and PSA. For example, incomplete assessments of risk contributions from low-power and shutdown operations, fires, and human performance will distort the importance measures (Powers 2000).

19.2   Probabilistic (Local) Importance Measures

This section discusses the definitions and properties, at both the basic event and parameter levels, of the probabilistic (local) importance measures (Remark 18.3.1) that are commonly used in PRA and PSA. These importance measures (e.g., the B-importance, improvement potential importance, FV importance, RAW, RRW, and DIM) are local measures and are computed with basic event probabilities/parameters fixed at their nominal values. They are used to identify the basic events or initiating events that contribute most to risk and safety, and their results provide information about the significance of the elements fixed at one (nominal) point of the uncertainty space, representing the most possibly normal operation.

The B-reliability importance in Section 4.1 is a partial derivative of systems reliability with respect to component reliability. It can easily be adjusted to risk indices such as the expected number of fatalities by expressing the risk indices as functions of a set of event probabilities. Two related measures that are commonly used in PRA and PSA are the criticality importance described in Subsections 4.1.2 and 5.1.1 and the improvement potential importance described in Subsection 4.1.1. Both of them are proportional to the B-importance.

The FV importance of a basic event in PRA and PSA is defined as (Cheok et al. 1998b)

$$I_{\text{FV}}(E_j; \boldsymbol{\zeta}, \mathbf{q}) = \frac{\text{Fr}(\bigcup_{C \in \overline{\mathscr{C}}_j} C)}{Risk_0},$$

where E_j stands for basic/initiating event j, and $\text{Fr}(\bigcup_{C \in \overline{\mathscr{C}}_j} C)$ is the frequency of the union of all minimal cuts containing event j. The definition of the FV importance can be extended at the parameter level (Borgonovo and Apostolakis 2001) as

$$I_{\text{FV}}(x_j; \mathbf{x}_0) = \frac{\sum_j G_R^{x_j}(\mathbf{x}_0)}{Risk_0},$$

where $G_R^{x_j}(\mathbf{x}_0)$ denotes the generic term in the parameter representation of risk metric (Equation (19.3)) that contains parameter x_j; the numerator is the sum over all terms in the expression of $Risk(\mathbf{x})$ that contain parameter x_j. $I_{\text{FV}}(x_j; \mathbf{x}_0)$ is the fraction of the risk that is associated with parameter x_j. In addition, the FV importance in the field of risk and safety analysis is often interpreted as

$$\frac{Risk_0 - Risk_j^-}{Risk_0},$$ (19.4)

where $Risk_j^-$ is the new risk that is produced when basic event j is assumed not to happen. This is consistent with the definition of the FV importance in Subsection 16.1.3 but is different from the original definition of the FV importance as presented in Sections 4.2 and 5.2. The following papers reviewed the FV importance as in Equation (19.4): Levitin et al. (2003, Section 3.3), Zio et al. (2004, Section 2.5), van der Borst and Schoonakker (2001, Section 2), Espiritu et al. (2007, Section 2), Ramirez-Marquez and Coit (2005, Section III), as well as others.

The RAW and the RRW stand for the risk achievement worth and the risk reduction worth in PRA and PSA. Vesely and Davis (1985) and Zio and Podofillini (2006) discussed the interpretations and applications of these two importance measures in PRA. The RAW is defined for basic events as in Equation (7.15) or consistently as

$$I_{RAW}(E_j; \zeta, \mathbf{q}) = \frac{Risk_j^+}{Risk_0},$$

where $Risk_j^+$ is the new risk that is produced when basic event j is assumed to have happened. Similarly, the RRW is defined as in Equation (7.16) or as

$$I_{RRW}(E_j; \zeta, \mathbf{q}) = \frac{Risk_0}{Risk_j^-}.$$

However, neither the RAW nor the RRW can be used to compute the importance of parameters. Kim et al. (2005) proposed a balancing method and reviewed other methods to estimate the RAW of components from the RAW of basic events of common cause failure, which are acquired as PSA in the risk informed regulation and applications.

In PRA and PSA, the DIM^I can be defined for both basic/initiating events and model parameters by treating component i in Definition 7.2.1 as basic/initiating event i and parameter x_i, respectively, and by using the risk metric (19.1) at the basic event level and risk metric (19.3) at the parameter level, respectively. The DIM^I of a basic/initiating event or a parameter is the fraction of the local change in risk metric $Risk$ that is due to a change in the event or the parameter, respectively. Thus, the DIM^I measures the risk significance of basic/initiating events or parameters. As in Subsection 7.2.1, the formats and computation of DIM^I depend on assumptions regarding the way events or parameters are affected by the changes. The DIM^I possesses the additivity property (i.e., the DIM^I of a group of events or parameters is equal to the sum of the DIM^I of the individual events or parameters in the group) and thus can evaluate the importance of a group of basic/initialing events or parameters that, however, cannot be easily realized by the B-importance, FV importance, RAW, or RRW.

Borgonovo et al. (2003) demonstrated the significance of the DIM^I as in Section 19.4, and Vinod et al. (2003) presented an application of the DIM^I in a risk-informed in-service inspection of a nuclear power plant. The DIM^I and DIM^{II} have many applications in PRA and PSA to evaluate the importance of basic/initiating events and of model parameters (Borgonovo and Apostolakis 2001; Zio and Podofillini 2006).

19.3 Uncertainty and Global Sensitivity Measures

Dealing with uncertainty is one of the challenges of many quantitative risk assessment problems. Uncertainty and global sensitivity measures (e.g., the ones in Section 18.2) are often used in PRA and PSA to study how uncertainties in input parameters of a model affect the uncertainty of the model output, or equivalently, how uncertainty of the output can be apportioned to different sources of uncertainties in the input parameters. To utilize global sensitivity techniques in the context of uncertainty analysis in PRA and PSA, the entire epistemic uncertainty in system risk and in the parameters should be taken into account. Identifying the significant parameters for the output uncertainty could direct the efforts to reducing the epistemic uncertainty in these parameters so as to achieve the great reduction in the uncertainty of risk metric *Risk*. Cheok et al. (1998b) and van der Borst and Schoonakker (2001) have called attention to the uncertainty of models and corresponding importance measures in PRA and PSA.

Borgonovo et al. (2003) emphasized that in the presence of epistemic uncertainty, the uncertainty and global sensitivity analysis must be performed on Equation (19.3), the parameter level, and not on Equation (19.1), the basic event level, because the different basic events (components) may have the distributions of the correlated parameters due to epistemic uncertainty. Thus, the uncertainty and global importance measures in Section 18.2 are properly defined for parameters and not for basic events.

Referring to the AGSM in Subsection 18.2.1, the variance of risk metric *Risk* is decomposed as Equation (18.10), that is,

$$V_R = \sum_i V_i + \sum_{i<j} V_{ij} + \sum_{i<j<k} V_{ijk} + \cdots + V_{12\ldots n},$$

where n denotes the number of the uncertain parameters. Then, the basic and tot AGSM are defined accordingly. It is worth mentioning that the AGSM is able to identify the parameters that individually (first-order AGSM) or as groups (higher-order AGSM) contribute most to the uncertainty in risk metric *Risk* without stating any assumption on the type of the dependence of *Risk* on the individual parameters. The first-order basic AGSM of parameter x_i represents the expected percentage reduction in V_R that is obtained when uncertainty in parameter x_i is eliminated. The first-order tot AGSM of parameter x_i represents the expected percentage of variance that remains if all parameters are known except for parameter x_i. This gain in information may lead to a cost of the increase in computational effort. The AGSM is tested in the case study in Section 19.4.

19.4 A Case Study

The PSA model

In this case study (Borgonovo et al. 2003)*, the reference PSA model is a large loss of coolant accident (LLOCA) sequence of the advanced test reactor. Two major safety systems

---

* Reprinted from Borgonovo E, Apostolakis GE, Tarantola S, and Saltelli A. 2003. Comparison of global sensitivity analysis techniques and importance measures in PSA. *Reliability Engineering and System Safety* **79**, 175–185, with permission from Elsevier Ltd.

are involved in an LLOCA, namely, the scram system and the firewater injection system (FIS). Failure of the scram system leads directly to core damage. The scram system failure is dominated by the common cause failure event, "failure to insert four safety rods," since a system failure due to a series of independent events is very unlikely. If the scram system is successful, then the FIS must also be successful to assure that no core damage results. The FIS cools the core after the LLOCA. Water is injected into the core by four injection lines. During an LLOCA, one of these lines is assumed to have failed. Failure of the other three lines is necessary to fail the system. If the FIS fails, core damage results.

The number of basic events for this model is 45, for a total of 289 minimal cuts. Of these, ten minimal cuts contain only one basic event, 32 are formed by two basic events, and the remaining 247 are given by three basic events. The exponential failure model as in Equation (19.2) is assumed for the basic events.

The risk metric used in this case is the core damage frequency that is associated with the LLOCA initiating event (CDF^{LLOCA}). CDF^{LLOCA} is written under the rare-event approximation as a function of the event probabilities as

$$CDF^{LLOCA}(\zeta, \mathbf{q}) = \zeta_{LLOCA}\left[\sum_{j=1}^{10} p(E_1^j) + \sum_{j=1}^{32}\prod_{i=1}^{2} p(E_i^j) + \sum_{j=1}^{247}\prod_{i=1}^{3} p(E_i^j)\right], \qquad (19.5)$$

where ζ_{LLOCA} is the frequency of the LLOCA initiating event; $\sum_{j=1}^{10} p(E_1^j)$ is the sum of the probabilities of the ten minimal cuts that contain one basic event; $\sum_{j=1}^{32}\prod_{i=1}^{2} p(E_i^j)$ is the sum of the probabilities of the 32 minimal cuts that contain two basic events; and $\sum_{j=1}^{247}\prod_{i=1}^{3} p(E_i^j)$ is the sum of the probabilities of the 247 minimal cuts that contain three basic events.

The total number of parameters is 31, and no correlation occurs among the parameters. The number of parameters is smaller than the number of basic events due to epistemic dependence. For example, the three deep-well pumps are considered identical, and the same parameters thus characterize their failure modes. With these data an uncertainty distribution of the CDF^{LLOCA} can be approximated by a lognormal.

Note that *Risk* in Equation (19.1) is linear in ζ and $\mathbf{q}$, once the logical expression of the minimal cuts is expanded and the rare-event approximation is considered (Borgonovo and Apostolakis 2001). Thus, as shown in Equation (19.5), the CDF^{LLOCA} is linear in individual basic event probabilities. On the other hand, in the presence of epistemic uncertainty, the CDF^{LLOCA} becomes a polynomial function of the parameters of the PSA model because the same parameters can characterize several basic event probabilities that are in the same minimal cut. In this case study, a parameter (e.g., the rate of independent failures while running of the three pumps) can appear at most three times in a minimal cut and the CDF^{LLOCA}; thus, the parameter appears at most in the third power in the polynomial. Using regression analysis, this CDF^{LLOCA} model is confirmed to be nonlinear at the parameter level.

The FV importance, RAW, and DIM^I results

According to three local importance measures (the FV importance, RAW, and DIM^I) at the basic event level, the most important event is the initiating event, the LLOCA. The FV importance and DIM^I produce the same rankings. Using the local importance measures (the FV importance and DIM^I) at the parameter level, the most relevant parameter is the frequency of the initiating event, ζ_{LLOCA}. This parameter is also associated with the most relevant event. However, there is now a discrepancy between the rankings of the FV importance and DIM^I, showing that the model is nonlinear at the parameter level, as anticipated.

The first-order AGSM results

The first-order AGSM (see Subsection 18.2.1) is chosen as the global sensitivity measure, and the extended Fourier amplitude sensitivity test (Homma and Saltelli 1996; Saltelli 2002) is used to simultaneously compute the first-order basic and tot AGSM at the parameter level. The results show that the sum of the first-order basic AGSM ($I_A(\cdot)$) over all parameters is equal to 0.23. This means that the portion of the variance explainable in terms of individual parameters is about 23% and that parameter groups and interactions among parameters account for 77% of V_R. Thus, the CDF^{LLOCA} at the parameter level is highly nonlinear. In this case, the first-order tot AGSM ($I_{A^{tot}}(\cdot)$) is better for evaluating the apportionment of uncertainty.

The AGSM-based global sensitivity analysis qualitatively interprets how uncertainty is partitioned between the two safety systems by tracking the parameters to the basic events of two safety systems. In this PSA model, uncertainty in the CDF^{LLOCA} is unequally distributed between the two safety systems, with FIS being responsible for most of it.

Suppose that further data are collected for the parameters with the largest values of $I_A(\cdot)$ and that, in turn, these parameters are fixed to their true values. The interpretation of $I_A(i)$ and $I_{A^{tot}}(i)$ is that the percentage reduction of V_R is between $I_A(i)$ and $I_{A^{tot}}(i)$ when parameter x_i is fixed. The actual reduction is obtained by reconducting the entire uncertainty analysis, with assumed x_i known. Fixing ten parameters in this case, V_R is reduced by almost two orders of magnitude (from 10^{-15} to 10^{-17}) with the error factor falling from 12 to 9.94.

Risk significant contributors versus uncertainty drivers

To understand whether risk significant contributors (i.e., the ones that are important to system risk) identified by the local sensitivity analysis are uncertainty drivers identified by the uncertainty and global sensitivity analysis, it is necessary to compare local importance measure results to the AGSM global sensitivity results. Such comparisons are sometimes not direct because the local importance measures are produced at the basic event level by most standard PSA software tools such as SAPHIRE (Borgonovo and Smith 1999, 2000), while the AGSM is only defined and computed at the parameter level. In this case study, the FV importance and RAW results at the basic event level are indirectly compared to the AGSM results, and the FV importance and DIM^I results at the parameter level are directly compared to the AGSM results. Note that the RAW is not defined for parameters.

To compare a parameter ranking to a basic/initiating event ranking, each parameter is first associated with the corresponding event(s). Then, the savage scores (Campolongo and Saltelli 1997) are used to compare two rankings produced by different importance measures. The degree of agreement of the rankings of two importance measures is quantified by the correlation coefficient of the savage scores of the two measures. In general, the closer the correlation coefficient is to unity, the higher the agreement in the rankings according to the two measures. A negative value correlation coefficient means that the two measures tend to give opposite rankings, that is, parameters that are ranked high by one technique tend to be ranked low by the other.

The comparison of risk contributors and uncertainty drivers at the basic event level produces intermediate values of the correlation coefficients of savage scores for the FV importance and AGSM (0.49) and for the RAW and AGSM (0.51). The comparison at the parameter level produces again intermediate values of the correlation coefficients of savage scores for the FV importance and AGSM (0.62) and for the DIM^I and AGSM (0.64). These results indicate that uncertainty drivers are not necessarily risk significant contributors. However, the initiating event, the LLOCA, is ranked first by all measures. This means that getting information to reduce the uncertainty in the initiating event frequency (ζ_{LLOCA}) would allow a reduction in the system risk while effectively reducing the uncertainty in the CDF^{LLOCA}.

The analyst can then utilize this information to allocate resources and prioritize information and data collection for the model. The local importance measures can help identify the basic/initiating events that contribute most to system risk. The AGSM global sensitivity technique can provide useful analytical capabilities that respond to the need for improving uncertainty analysis of PRA and PSA models. In this respect, the AGSM provides more detailed information at the possible cost of increasing computational effort.

19.5 Review of Applications

Abundant work has been conducted on the probabilistic importance measures in PRA and PSA. Borgonovo (2007a) discussed the computation of the B-importance, criticality importance, FV importance, RAW, and DIM for individual basic events, basic event groups, and components. Findings show that the estimation of the B-importance or criticality importance of a group of basic events or components is not possible. Kim and Han (2009) developed a new quantification algorithm to calculate the FV importance in a fire PRA model. Youngblood (2001) applied the FV importance of basic/initiating events to an example of nuclear power plant risk models where the output is the top event frequency. Johnson and Apostolakis (2009) applied the FV importance, RAW, and RRW to the PRA of the prism reactor design. Reinert and Apostolakis (2006) presented a methodology using the RAW to identify the basic events in the PRA model that have the potential to change the decision. Borgonovo (2008) introduced a methodical approach based on the FV importance and DIM^I to categorizing PSA model parameters in the presence of epistemic uncertainty and tested the approach on the system described in Section 19.4. Borgonovo and Smith (2010) discussed when interactions do matter in complex PSA models and how they relate to uncertainty through the application of the TOI in a space PSA model.

Recently, Aven and Nøkland (2010) discussed the use of the probabilistic importance measures (the B-importance and improvement potential importance) and the I-uncertainty importance in reliability and risk analysis, using a small example of three components and focusing on the link between the probabilistic and uncertainty importance measures. They also tested a nonparametric technique of uncertainty analysis by evaluating the Pearson correlation coefficient between each input and the output. Similar to the case study in Section 19.4, the probabilistic and uncertainty importance measures generate different rankings since they are concerned with the different problems and aspects of a system.

Borgonovo (2007b) illustrated the application of the B-MGSM to the PRA model, in which the probability of the top event (output) fits into the lognormal distribution well. This PRA model was also utilized by Iman (1987) and Chun et al. (2000) to illustrate the applications of the I-uncertainty importance in Equation (18.25) and the CHT-MGSM in Equation (18.24), respectively.

Risk-informed decision-making means making use of information that is derived from PRA and PSA models. For the interpretation of the results of PRA and PSA, it is important to have measures not only to identify the basic events and/or groups of basic events that contribute most to the system performance (e.g., the top event frequency) but also to identify basic events and/or their groups that are the main drivers to the uncertainty in system performance. Pörn (1997) developed an uncertainty importance within a decision theory framework, thereby providing an indication of the basic event on which it would be the most valuable, from the decision-making point of view, to procure more information. Vinod et al. (2003) studied the importance ranking of piping segments from process and safety systems with an initiating event of a medium loss of cooling accident. Reinert and Apostolakis (2006) presented a methodology using the RAW to determine how much error in the basic event probabilities would be necessary in order to impact the decision. They found that the decision is fairly insensitive to uncertainties in these basic events. For example, the basic event probabilities would be between two and four orders of magnitude larger than modeled in the PRA model before they would become important to the decision. The basic events in risk-informed applications could be component failures, human errors, common cause failures, and so on.

Vaurio (2010) addressed the accuracy of various alternative methods used for quantifying accident sequence probabilities when negations or success branches of event trees are involved. They developed criteria for selecting truncation limits and cutoff errors so that importance measures (e.g., the B-importance and DIM^I) can be estimated reliably and risk-informed decision-making is robust. A truncation process aims to determine among the set of minimal cuts produced by a PSA model which of them are significant. Duflot et al. (2009) proposed a truncation process to estimate from a single set of minimal cuts the importance measure of any basic event with the desired accuracy level and tested the method on a complete level one PSA model of a 900 MWe nuclear power plant. They showed that to reach the same level of accuracy, their proposed method produces a set of minimal cuts whose size is significantly reduced. Contini and Matuzas (2011) focused on computing the B-importance of initiating and enabling events in fault tree analysis.

The general goal of risk-informed applications is to make requirements on operation and maintenance activities more risk-effective and less costly by better focusing on what is risk-important. To this end, importance measures can be used for the risk-informed optimization of system design and management. Zio and Podofillini (2007) presented an optimization approach in which the information provided by importance measures is incorporated in the formulation of a multiobjective optimization problem to drive the design toward a solution which, besides being optimal from the points of view of economics and safety, is also balanced in the sense that all components have similar importance values. The importance measure used is the generalized FV importance for a multistate system. The approach allows identifying design systems with test/maintenance activities calibrated according to the components' importance ranking. The approach is tested against the risk-informed optimization problem of the technical specifications of a safety system in a nuclear power plant, and genetic algorithms are used for the optimization.

Podofillini and Zio (2008) further investigated on which importance measure among the B-importance, FV importance, and RAW is most suitable as the balancing objective in the test/maintenance optimization problem in Zio and Podofillini (2007). The RAW is found inappropriate because it relates to the defense of the system against the failure of components, which is independent on frequency of the component testing, whereas the use of the B-importance or FV importance in allocating test/maintenance activities to components is in agreement with the principle of the risk-informed philosophy of avoiding unnecessary regulatory burdens and defining more efficient inspection and maintenance activities.

In practice, the various categories of risk significance are determined by defining threshold values for the importance measures. For example, in some applications, a component or a subsystem is in the "high" risk-significant category when the FV importance value is greater than 0.005 and the RAW value is greater than 2.0, or the FV importance value is greater than 0.1, or the RAW value is greater than 100 (Powers 2000). In other applications, the numerical values are different.

As mentioned earlier, there has been considerable research in the importance-measure-based PRA and PSA of nuclear power plants, including additional examples such as un-certainty analysis on core damage frequency (Borgonovo et al. 2003); uncertainty analysis on nuclear water disposal (Saltelli and Tarantola 2002); risk and safety contributor analy-sis (with respect to the core damage frequency) based on probabilistic importance measures (Borgonovo et al. 2003; Cho and Jung 1989; Reinert and Apostolakis 2006); analysis of com-ponent contributions to system risk in nuclear waste repositories (Eisenberg and Sagar 2000); and reliability/availability analysis in a nuclear reactor protection system (Marseguerra and Zio 2004).

Although related, this section does not intend to give a detailed illustration or review of the applications and computation of importance measures in PRA and PSA. Some review papers have been presented regarding this aspect (Borgonovo 2007a; Cheok et al. 1998b; van der Borst and Schoonakker 2001; Vasseur and Llory 1999; Vaurio 2011; Vinod et al. 2003). For example, Cheok et al. (1998b) discussed the use of importance measures to analyze PRA results and some issues that potentially limit the usefulness of importance measures. They recognized that the most frequently used importance measures in PRA include the B-importance, criticality importance for system failure, RAW, and RRW, and discussed their extensions to multiple basic events. van der Borst and Schoonakker (2001) discussed the use of importance measures in PSA to optimize the performance of a nuclear power plant as well as the limitations of importance measures. They pointed out that in this context, the B-importance, FV importance, and RAW are the most commonly used. Interested readers are referred to the papers mentioned in this section and the resources of Borgonovo and Smith (2011), Cheok et al. (1998a), Park and Ahn (1994), Vesely (1998), Vesely et al. (1994), and their numerous references.

19.6   System Fault Diagnosis and Maintenance

As in Example 1.0.11, when a system fails, a systems analyst needs to find the components that caused the system failure and then repair them to restore the system as soon as possible.

In one way or another, the analyst has to assign priorities to the components in the system and then check and repair the components in the order of these priorities. Using the appropriate importance measures or ratios of importance measures to cost factors (Gupta and Kumar 2010), these priorities can be assigned so that the systems analyst can generate a repair checklist to sequence the components that have likely failed and caused or contributed to the system failure. The components with large importance measure values are checked first in the fault diagnosis. Different types of importance measures have been suggested for this purpose.

The BP importance can be used in fault diagnosis (Barlow and Proschan 1975; Lambert 1975b) because the BP importance reveals the relative extent to which a component is critical for system failure. The particular expression of the BP importance that is used to rank components on a checklist depends on when system failure is detected. If system failure is observed at some instant of time, for example, a continuously operating system that fails during operation, then a component must have caused the system to fail precisely at that time. Then, analogous to Equation (5.21), the components can be ranked according to

$$\frac{\left[R(1_i, \overline{\mathbf{F}}(t)) - R(0_i, \overline{\mathbf{F}}(t))\right] f_i(t)}{\sum_{i=1}^{n} \left[R(1_i, \overline{\mathbf{F}}(t)) - R(0_i, \overline{\mathbf{F}}(t))\right] f_i(t)},$$

which is the probability that component i causes system failure, given that the system fails precisely at time t. On the other hand, if system failure can be observed only at the end of some time interval $[0, t]$, for example, a passive standby safety system that is checked at the end of the testing interval $[0, t]$, then components can cause the system failure at any time in $[0, t]$. In this case, the components should be ranked according to the BP TDL importance, as in Equation (5.21), which is the probability that component i has caused system to fail by time t, given that the system has failed by time t.

Alternatively, Lambert (1975b) suggested that components contained in minimal cuts of size two or higher can be ranked according to

$$\frac{\sum_{j:j \neq i} \int_0^t \left[R(0_i, 1_j, \overline{\mathbf{F}}(u)) - R(0_i, 0_j, \overline{\mathbf{F}}(u))\right] F_i(u) dF_j(u)}{1 - R(\overline{\mathbf{F}}(t))}, \tag{19.6}$$

where the sum over j includes only those components that appear in at least one minimal cut containing component i. The expression (19.6) is the probability that the failure of component i is a factor for system failure when another component j causes the system to fail, given that the system fails.

The criticality TDL importance for system failure defined in Subsection 5.1.1 may be another indicator; it is the probability that component i has failed by time t and that component i is critical for the system at time t, given that the system has failed by time t. The c-FV importance in Sections 4.2 and 5.2 can also be suggested for diagnosing system failure, since it evaluates the contribution of a component to system failure no matter whether the component is critical for the system or not. The Bayesian reliability importance defined in Subsection 4.1.3 may also be applied because it is the probability that component i fails, given that the system fails.

The ordering of components on the checklist should be done on a conditional basis. The ordering should reflect the knowledge gained about the system as each component in the checklist is examined. For example, if the first component on the checklist has not failed or if repairing the first component given that it has failed does not restore the system to functioning, then the second component on the checklist should be the most critical to system failure given that the first component has not failed. In general, the kth component on the checklist is the most important to system failure given that the first $k-1$ components on the checklist have not failed, supposing that any component among the first $k-1$ components that has failed is already repaired. Lambert (1975a) applied the aforementioned checklist generation scheme to diagnose and maintain a low-pressure injection system, which is a standby safety system that forms part of the emergency core cooling system in a nuclear power plant.

References

Aven T and Nøkland TE. 2010. On the use of uncertainty importance measures in reliability and risk analysis. *Reliability Engineering and System Safety* **95**, 127–133.

Baraldi P, Zio E, and Compare M. 2009. A method for ranking components importance in presence of epistemic uncertainties. *Journal of Loss Prevention in the Process Industries* **22**, 582–592.

Barlow RE and Proschan F. 1975. Importance of system components and fault tree events. *Statistic Processes and Their Applications* **3**, 153–172.

Borgonovo E. 2007a. Differential, criticality and Birnbaum importance measures: An application to basic event, groups and SSCs in event trees and binary decision diagrams. *Reliability Engineering and System Safety* **92**, 1458–1467.

Borgonovo E. 2007b. A new uncertainty importance measure. *Reliability Engineering and System Safety* **92**, 771–784.

Borgonovo E. 2008. Epistemic uncertainty in the ranking and categorization of probabilistic safety assessment model elements: Issues and findings. *Risk Analysis* **28**, 983–1001.

Borgonovo E and Apostolakis GE. 2001. A new importance measure for risk-informed decision making. *Reliability Engineering and System Safety* **72**, 193–212.

Borgonovo E and Smith C. 1999. A case study: Two components in parallel with epistemic uncertainty, Part I. *SAPHIRE Facets INEEL 1999*.

Borgonovo E and Smith C. 2000. A case study: Two components in parallel with epistemic uncertainty, Part II. *SAPHIRE Facets INEEL 2000*.

Borgonovo E and Smith C. 2010. Total order reliability importance in space PSA. *Procedia Social and Behavioral Sciences* **2**, 7617–7618.

Borgonovo E and Smith C. 2011. A study of interactions in the risk assessment of complex engineering systems: An application to space PSA, *Operations Research* **59**, 1461–1476.

Borgonovo E, Apostolakis GE, Tarantola S, and Saltelli A. 2003. Comparison of global sensitivity analysis techniques and importance measures in PSA. *Reliability Engineering and System Safety* **79**, 175–185.

Campolongo F and Saltelli A. 1997. Sensitivity analysis of an environmental model: An application of different analysis methods. *Reliability Engineering and System Safety* **57**, 49–69.

Cheok MC, Parry GW, and Sherry RR. 1998a. Response to "Supplemental viewpoints on the use of importance measures in risk-informed regulatory applications". *Reliability Engineering and System Safety* **60**, 261–261.

Cheok MC, Parry GW and Sherry RR. 1998b. Use of importance measures in risk-informed regulatory applications. *Reliability Engineering and System Safety* **60**, 213–226.

Cho NZ and Jung WS. 1989. A new method for evaluating system-level importance measures based on singular value decomposition. *Reliability Engineering and System Safety* **25**, 197–205.

Chun MH, Han SJ, and Tak NI. 2000. An uncertainty importance measure using a distance metric for the change in a cumulative distribution function. *Reliability Engineering and System Safety* **70**, 313–321.

Contini S and Matuzas V. 2011. New methods to determine the importance measures of initiating and enabling events in fault tree analysis. *Reliability Engineering and System Safety* **96**, 775–784.

Duflot N, Bérenguer C, Dieulle L, and Vasseur D. 2009. A min cut-set-wise truncation procedure for importance measures computation in probabilistic safety assessment. *Reliability Engineering and System Safety* **94**, 1827–1837.

Eisenberg NA and Sagar B. 2000. Importance measures for nuclear waste repositories. *Reliability Engineering and System Safety* **70**, 217–239.

Espiritu JF, Coit DW, and Prakash U. 2007. Component criticality importance measures for the power industry. *Electric Power Systems Research* **77**, 407–420.

Gupta S and Kumar U. 2010. Maintenance resource prioritization in a production system using cost-effective importance measure. *Proceedings of the 1st International Workshop and Congress on eMaintenance*, pp. 196–204.

Homma T and Saltelli A. 1996. Importance measures in global sensitivity analysis of nonlinear models. *Reliability Engineering and System Safety* **52**, 1–17.

Iman RL. 1987. A matrix-based approach to uncertainty and sensitivity analysis for fault trees. *Risk Analysis* **7**, 21–23.

Johnson BC and Apostolakis GE. 2009. The limited exceedance factor importance measure: An application to the prism reactor design. *Proceedings of the 17th International Conference on Nuclear Engineering (ICONE17)*, pp. 349–353.

Kim K and Han S. 2009. A study on importance measures and a quantification algorithm in a fire PRA model. *Reliability Engineering and System Safety* **94**, 969–972.

Kim K, Kang D, and Yang J. 2005. On the use of the balancing method for calculating component RAW involving CCFs in SSC categorization. *Reliability Engineering and System Safety* **87**, 233–242.

Lambert HE. 1975a. *Fault Trees for Decision Making in System Safety and Availability*. PhD Thesis, University of California, Berkeley.

Lambert HE. 1975b. Measure of importance of events and cut sets in fault trees. In *Reliability and Fault Tree Analysis* (eds. Barlow RE, Fussell JB, and Singpurwalla ND). Society for Industrial and Applied Mathematics, Philadelphia, pp. 77–100.

Levitin G, Podofillini L, and Zio E. 2003. Generalised importance measures for multi-state elements based on performance level restrictions. *Reliability Engineering and System Safety* **82**, 287–298.

Marseguerra M and Zio E. 2004. Monte Carlo estimation of the differential importance measure: Application to the protection system of a nuclear reactor. *Reliability Engineering and System Safety* **86**, 11–24.

Park CK and Ahn KI. 1994. A new approach for measuring uncertainty importance and distributional sensitivity in probabilistic safety assessment. *Reliability Engineering and System Safety* **46**, 253–261.

Podofillini L and Zio E. 2008. Designing a risk-informed balanced system by genetic algorithms: Comparison of different balancing criteria. *Reliability Engineering and System Safety* **93**, 1842–1852.

Pörn K. 1997. A decision-oriented measure of uncertainty importance for use in PSA. *Reliability Engineering and System Safety* **56**, 17–27.

Powers DA. 2000. Importance measures derived from probabilistic risk assessments. Technical report, U.S. Nuclear Regulatory Commission. http://www.nrc.gov/reading-rm/doc-collections/acrs/letters/2000/4691872.html.

Ramirez-Marquez JE and Coit DW. 2005. Composite importance measures for multi-state systems with multi-state components. *IEEE Transactions on Reliability* **54**, 517–529.

Reinert JM and Apostolakis GE. 2006. Including model uncertainty in risk-informed decision making. *Annals of Nuclear Energy* **33**, 354–369.

Saltelli A. 2002. Making best use of model evaluations to compute sensitivity indices. *Computer Physics Communications* **145**, 280–297.

Saltelli A and Tarantola S. 2002. On the relative importance of input factors in mathematical models: Safety assessment for nuclear waste disposal. *Journal of American Statistical Association* **97**, 702–709.

van der Borst M and Schoonakker H. 2001. An overview of PSA importance measures. *Reliability Engineering and System Safety* **72**, 241–245.

Vasseur D and Llory M. 1999. International survey on PSA figures of merit. *Reliability Engineering and System Safety* **66**, 261–274.

Vaurio JK. 2010. Ideas and developments in importance measures and fault-tree techniques for reliability and risk analysis. *Reliability Engineering and System Safety* **95**, 99–107.

Vaurio JK. 2011. Importance measures in risk-informed decision making: Ranking, optimisation and configuration control. *Reliability Engineering and System Safety* **96**, 1426–1436.

Vesely WE. 1998. Supplemental viewpoints on the use of importance measures in risk-informed regulatory applications. *Reliability Engineering and System Safety* **60**, 257–259.

Vesely WE and Davis TC. 1985. Two measures of risk importance and their applications. *Nuclear Technology and Radiation Protection* **68**, 226–234.

Vesely WE, Belhadj M, and Rezos JT. 1994. PRA importance measures for maintenance prioritization applications. *Reliability Engineering and System Safety* **43**, 307–318.

Vinod G, Kushwaha HS, Verma AK, and Srividy A. 2003. Importance measures in ranking piping components for risk informed in-service inspection. *Reliability Engineering and System Safety* **80**, 107–113.

Youngblood RW. 2001. Risk significance and safety significance. *Reliability Engineering and System Safety* **73**, 121–136.

Zio E and Podofillini L. 2006. Accounting for components interactions in the differential importance measure. *Reliability Engineering and System Safety* **91**, 1163–1174.

Zio E and Podofillini L. 2007. Importance measures and genetic algorithms for designing a risk-informed optimally balanced system. *Reliability Engineering and System Safety* **92**, 1435–1447.

Zio E, Podofillini L, and Levitin G. 2004. Estimation of the importance measures of multi-state elements by Monte Carlo simulation. *Reliability Engineering and System Safety* **86**, 191–204.

Afterword

Few people can doubt the myriad benefits that recent developments in technology have brought to countless people around the world in terms of health, industry, communication, social networking and entertainment, to name but a few.

However, the efficiency and convenience that have advanced electronic and bio products such as smart phones, tablets, and so forth are not enough. For society to solve the innumerable problems it faces, we need to be more creative and self-reflective in the way we think, act, and behave.

Therefore, design for quality and reliability becomes a critical element of the process if we are to ensure that the technological revolution continues to spearhead positive change.

Allaying concerns

Rapid economic development has aggravated the global energy crisis, and hence, the growth of nuclear power in the past half a century. Today, nuclear reactors provide 17% of the total power supply in the world, but the 9.0-magnitude earthquake and the ensuing tsunami that hit the northeast coast of Japan in March 2011 have many people looking over their shoulders at what happened in Chernobyl in Ukraine in 1986 and on Three Mile Island in the United States in 1979.

Although today nuclear energy is, according to many, the world's safest energy source, we need reliability studies and the use of probabilistic risk analysis for problems both nuclear-related and others. People are anxious to know more about the chances of similar accidents occurring in the future and their potential impact.

Reliability awareness

For the WASH-1400 published in 1975 in the USA and the State-of-the-Art Reactor Consequence Analysis published by the Nuclear Regulatory Commission in 2007, reliability was the mode of assessment. Computer simulations were used to illustrate potential accidents in nuclear power plants (Kuo 2011).

Importance Measures in Reliability, Risk, and Optimization: Principles and Applications, First Edition.
Way Kuo and Xiaoyan Zhu. © 2012 John Wiley & Sons, Ltd. Published 2012 by John Wiley & Sons, Ltd.

In the manufacturing industry, reliability has a close relationship with product warranty. It includes elements such as product cycle analysis and risk evaluation. As far as nuclear power is concerned, reliability is achieved by assessing every link of the whole process for possible failures. However, even if nuclear power plants are built on a sound scientific and engineering basis, the biggest uncertainty is the human factor. Both the Chernobyl and Three Mile Island nuclear accidents were caused by human error. The Fukushima accident was aggravated by management, maintenance, and human issues.

In all kinds of systems, the interaction of humans with the system is crucial to safety. In a complex system, we must locate bottlenecks and implement radical reforms where necessary. Therefore, whereas the nuclear industry and nuclear power plants are enhancing their safety measures, they also need to identify the critical elements using importance measures as presented in this book.

These measures are vital for effective decision-making in the areas of energy generation, energy saving, economic growth, safety, reliability and sustainability.

Orderings and ranges of component reliability

As discussed, the performance of importance-measure-based heuristics is also related to the system structure and the ranges of component reliability. Even though component reliability might not be accurately estimated, knowing the ranges of component reliability allows a good choice of the heuristics to be made on the basis of the results and discussion for the CAP. The accuracy and efficiency of the heuristics allow multiple trials using different estimates of component reliability. The structure importance measures, such as the permutation importance, internal and external domination importance measures, and so on, can also be used in determining potentially important components. These structure importance measures have close relationships with the B-importance and can provide decisions based only on the orderings or ranges of component reliability. This characteristic is attractive, especially when component reliability cannot be well estimated.

As for $Con/k/n$ systems, the patterns of importance measures of components (positions) exist. If the component reliability values are unknown or partially known, these patterns can at least be used to avoid poor-quality solutions, for example, for the CAP. Intuitively, the positions of the large importance values should be assigned the high reliable components; thus, any arrangement that is against the discovered patterns may be abandoned. In addition, these patterns could be incorporated in designing the heuristics for effectively addressing the reliability optimization problems.

Importance-measure-based B&B method

Many reliability optimization problems are NP-hard, either combinatorial optimization or integer programming, for example, CAP and redundancy allocation problems. There is little work on exact optimal solution methodologies, except possibly the enumeration method. Taking an example of the CAP, the exact enumeration consumes intolerably long computation time, and even well-developed commercial solvers such as GAMS/CoinBonmin cannot solve the CAP to optimality in most cases. The importance-measure-based B&B method for solving general integer problems could be specified for finding the optimal solutions of CAP, redundancy allocation problems, or other discrete optimization problems. By using the valuable information from the importance measures in B&B node selection and a branching strategy in conjunction with branching variable selection, the B&B method is expected to be effective in solving a particular optimization problem, for example, the CAP in reliability.

Multidimensional approach

Consistent with the review of Kuo and Zhu (2012a, 2012b) and Zhu and Kuo (2012), in summary we can say the following:

(i) Reliability importance measures are fundamental to addressing complex reliability problems. Importance measures for components or groups of components provide insight on system structures and constitutive components. Almost all of the importance measures are defined through (minimal) cuts and (minimal) paths, which characterize the system structure.

(ii) Importance measures have been designed from the known component reliability over a fixed mission time to the stochastic versions of the known lifetime distributions over a long or infinite mission time. They enter various stages of system design, upgrading, diagnosis, maintenance, and so on, by means of including design requirements, cost parameters, failure rates, maintenance policies, repair characteristics, and so on. It is a challenge to incorporate these types of information into one entity in a systematic way to facilitate decision-making. A common way is to incorporate all of the information into the objective function and then to partition it to each object (e.g., component), which often results in complexity of the mathematical expressions. Procedures need to be developed to avoid the complicated math and vague interpretations, for example, the multistage procedures, and importance measures using vectors in which each element relates to a type of information. Importance measures do not need to be a scalar. They could be in any formats, for example, a vector or a random value. If an importance measure is random, then the comparisons of two components should be based on the stochastic orderings of their importance values.

(iii) Importance measures have also been developed from individual components to groups of input factors, since the decisions are often made on the basis of more than one factor and these factors are often correlated. The existing importance measures of pairs or groups of components do not explicitly address how decisions are made, since some factors are conditional on other factors.

(iv) Importance measures have been well developed for independent components or factors, while some are extended to dependent cases, for example, the B-importance, BP importance, JRI, L_1 importance, yield importance, and link importance, with limited results. For a complex system, dependence, such as common cause failures and load sharing, should be addressed. A recent research in fault tree analysis extends the importance measures to phased missions (Vaurio 2011) in which a component can have different roles in disjoint phases and component failures are mutually exclusive rather than i.i.d.

(v) Different importance measures have their own probabilistic interpretations. However, these importance measures should be tested in real applications, such as those used in nuclear power plants. The identifications of problems and the corresponding solution methodologies are some of the future developments in the area of importance measures. The existing importance measures may be appropriate in some applications, whereas variations or even new importance measures need to be derived for a better fit for potential applications.

Practical decision-making

Even though some of the applications presented in this book achieve only preliminary results, they are luminous and can be elaborated more and more assertively as future research develops more feasible and effective methods based on importance measures for solving open-ended problems that exist in the practical work environment, including problems in uncertain environment, for multistate systems, for continuum systems of degrading and deteriorating components, and for measures of pairs or groups of components.

Inspired by these applications and the corresponding results, importance-measure-based methods deserve investigation, since they are probably the most practical decision-making tools. The concept and methods of importance measures can be propagated to many fields such as information theory.

In a globalized environment, what happens in the region will affect local and global economies as well as political and social stability, making it all the more necessary for us to identify critical elements and bottlenecks and look for overall optimization of the benefits under consideration. Risk management is imperative in managing commercial, political, or nonprofit organizations. Importance measures help identify the uncertainties, safeguard stability, and optimize the final objectives.

Reliability, risk, and optimization

This book provides a comprehensive view on modeling and the combination of importance measures with other design tools such as simulation, optimization, and risk analysis. It also provides insight into improving simulation, optimization, and risk analysis. Some solution methods are specifically included that require only orderings or ranges of component reliability. With respect to this feature, some importance measures are designed under the assumption of unknown component reliability or performance; some others are designed for different ranges of component reliability or performance measures. In the book, we have identified many difficult problems that need to be further explored by readers.

This book is an integral part of reliability modeling (Kuo and Zuo 2003) and reliability design (Kuo et al. 2006). In broader research, importance-measure-based methods for solving various difficult problems in the fields of reliability, risk, and mathematical programming deserve special attention.

References

Kuo W. 2011. Reliability and nuclear power. *IEEE Transactions on Reliability* **60**, 365–367.

Kuo W, Prasad VR, Tillman FA, and Hwang CL. 2006. *Optimal Reliability Design: Fundamentals and Applications*, 2nd edn. Cambridge University Press, Cambridge, UK.

Kuo W and Zhu X. 2012a. Relations and generalizations of importance measures in reliability. *IEEE Transactions on Reliability* **61**.

Kuo W and Zhu X. 2012b. Some recent advances on importance measures in reliability. *IEEE Transactions on Reliability* **61**.

Kuo W and Zuo MJ. 2003. *Optimal Reliability Modeling: Principles and Applications*. John Wiley & Sons, New York.

Vaurio JK. 2011. Importance measures for multi-phase missions. *Reliability Engineering and System Safety* **96**, 230–235.

Zhu X and Kuo W. 2012. Importance measures in reliability and mathematical programming. *Annals of Operations Research*, Accepted.

Appendix

Proofs

A.1 Proof of Theorem 8.2.7

The following proof is based on Zhu et al. (2011).

Proof. When $s = 0$, parts (i) and (ii) come directly from Theorem 8.2.3(iv)b and (v), respectively. The following proves the statement for the case of $s = k - 3 \geq 1$ (i.e., $p \geq 1/3$). The other cases of $1 \leq s \leq k - 4$ can be proved similarly. In this proof, let $I_n(i; p)$ denote the B-i.i.d. importance of component i with a same component reliability p in a $Lin/Con/k/n$:F system.

Part (i): For a $Lin/Con/k/n$:F system, $R_L(1_n, p) = R_L(k, n - 1)$, $R_L(0_n, p) = R_L(k, n - 1) - pq^{k-1}R_L(k, n - k - 1)$ and then

$$R_L(k, n) = pR_L(1_n, p) + qR_L(0_n, p) = R_L(k, n - 1) - pq^k R_L(k, n - k - 1). \tag{A.1}$$

Using Equations (8.1) and (A.1), for $n \geq 2k + 1$,

$$I_n(k + 1; p) - I_n(k - 2; p)$$

$$= \frac{R_L(k, k)R_L(k, n - k - 1) - R_L(k, n)}{q} - \frac{R_L(k, k - 3)R_L(k, n - k + 2) - R_L(k, n)}{q}$$

$$= \frac{1}{q}\left[(1 - q^k)R_L(k, n - k - 1) - R_L(k, n - k + 2)\right]$$

$$= \frac{1}{q}[(1 - q^k)R_L(k, n - k - 1) - R_L(k, n - k - 1)$$

$$+ pq^k\left(R_L(k, n - 2k - 1) + R_L(k, n - 2k) + R_L(k, n - 2k + 1)\right)]$$

$$= q^{k-1}\left[p\left(R_L(k, n - 2k - 1) + R_L(k, n - 2k) + R_L(k, n - 2k + 1)\right) - R_L(k, n - k - 1)\right],$$

where the third equation is obtained by using Equation (A.1) iteratively three times. Furthermore, Equation (A.1) implies that $R_L(k, n)$ decreases as n increases; thus, each of

---

Importance Measures in Reliability, Risk, and Optimization: Principles and Applications, First Edition.
Way Kuo and Xiaoyan Zhu. © 2012 John Wiley & Sons, Ltd. Published 2012 by John Wiley & Sons, Ltd.

$R_L(k, n - 2k - 1), R_L(k, n - 2k)$, and $R_L(k, n - 2k + 1)$ is greater than $R_L(k, n - k - 1)$. Therefore, $p \geq 1/3$ implies that

$$I_n(k + 1; p) > I_n(k - 2; p). \tag{A.2}$$

Part (ii): For $i \geq k + 1$ and $0 < j - i < k$,

$$I_n(j; p) - I_n(i; p) = pq^k \sum_{r=1}^{j-i} \left[I_{n-k-r}(i; p) - I_{n-k-r}(j - k - r; p) \right]. \tag{A.3}$$

Letting $j = i + 1$ in Equation (A.3), for $i \geq k + 1$,

$$I_n(i + 1; p) - I_n(i; p) = pq^k \left[I_{n-k-1}(i; p) - I_{n-k-1}(i - k; p) \right].$$

Thus, to show that for $n \geq 4k + s + 2 = 5k - 1$ and $k + 1 \leq i \leq k + s + 1 = 2k - 2$, $I_n(i + 1; p) > I_n(i; p)$, it suffices to show that $I_{n-k-1}(i; p) > I_{n-k-1}(i - k; p)$.

First, by Theorem 8.2.3(iv)a and inequality (A.2),

$$I_{n-k-1}(k + 1; p) > I_{n-k-1}(i - k; p) \tag{A.4}$$

for $n - k - 1 > 2k + 1$ and $2 \leq i - k \leq k - 2$, which are satisfied when $n \geq 5k - 1$ and $k + 2 \leq i \leq 2k - 2$.

Second, for $n \geq 5k - 1$ and $k + 2 \leq i \leq 2k - 2$, by Equation (A.3),

$$I_{n-k-1}(i; p) - I_{n-k-1}(k + 1; p) = pq^k \sum_{r=1}^{i-k-1} \left[I_{n-2k-r-1}(k + 1; p) - I_{n-2k-r-1}(i - k - r; p) \right],$$

from which

$$I_{n-k-1}(i; p) > I_{n-k-1}(k + 1; p) \tag{A.5}$$

because $I_{n-2k-r-1}(k + 1; p) > I_{n-2k-r-1}(i - k - r; p)$. To show that, by inequality (A.2), it suffices to show $I_{n-2k-r-1}(k - 2; p) > I_{n-2k-r-1}(i - k - r; p)$, which is verified by Theorem 8.2.3(iv)a and noting that $n - 2k - r - 1 \geq n - k - i \geq 2k + 1$ and $1 \leq i - k - r \leq k - 3$ for $1 \leq r \leq i - k - 1$.

Inequalities (A.5) and (A.4) result in $I_{n-k-1}(i; p) > I_{n-k-1}(i - k; p)$ for $n \geq 5k - 1$ and $k + 2 \leq i \leq 2k - 2$. By Theorem 8.2.3(iv)b, $I_{n-k-1}(i; p) > I_{n-k-1}(i - k; p)$ also holds for $n \geq 5k - 1$ and $i = k + 1$. This completes the proof.

A.2   Proof of Theorem 10.2.10

The following proof is based on Natvig (1979).

Proof. Part (i): For a series system, the result follows immediately from Equation (10.15).

Part (ii): For a parallel system, the numerator in Equation (10.16) reduces to

$$-\frac{\lambda_i}{\alpha}\int_0^\infty u^{1/\alpha}\, e^{(-\lambda_i u)}\prod_{j\neq i}(1-e^{(-\lambda_j u)})du = -\frac{\lambda_i}{\alpha}\left[\lambda_i^{-\beta}\Gamma(\beta)-\sum_{j\neq i}(\lambda_i+\lambda_j)^{-\beta}\Gamma(\beta)\right.$$

$$\left.+\sum_{\substack{j<k\\ j,k\neq i}}(\lambda_i+\lambda_j+\lambda_k)^{-\beta}\Gamma(\beta)-\ldots+(-1)^{n-1}(\lambda_1+\lambda_2+\ldots+\lambda_n)^{-\beta}\Gamma(\beta)\right],$$

which gives expression (10.20).

Part (iii): For a two-component series system, the results follow immediately from Equation (10.19). For a two-component parallel system, Equation (10.20) reduces to

$$I_{L_1}(i;\overline{\mathbf{F}}) = \frac{\lambda_i^{-1/\alpha}-\lambda_i(\lambda_1+\lambda_2)^{-\beta}}{\lambda_1^{-1/\alpha}+\lambda_2^{-1/\alpha}-(\lambda_1+\lambda_2)^{-1/\alpha}},\quad i=1,2,$$

from which the result in Equation (10.21) follows immediately. Now for $i\neq j$

$$\frac{\partial I_{L_1}(i;\overline{\mathbf{F}})}{\partial\lambda_j}=\{\lambda_1\lambda_2(\alpha+1)[(\lambda_1^{1/\alpha}+\lambda_2^{1/\alpha})(\lambda_1+\lambda_2)^{1/\alpha}-\lambda_1^{1/\alpha}\lambda_2^{1/\alpha}]+[(\lambda_1+\lambda_2)^\beta-\lambda_1^\beta]$$

$$\times[(\lambda_1+\lambda_2)^\beta-\lambda_2^\beta]\}\Big/\{\alpha\lambda_i^{1/\alpha}\lambda_j^\beta(\lambda_1+\lambda_2)^{2\beta}[\lambda_1^{-1/\alpha}+\lambda_2^{-1/\alpha}-(\lambda_1+\lambda_2)^{-1/\alpha}]\}^2,$$

the numerator of which reduces to

$$\lambda_1\lambda_2\alpha\left[(\lambda_1^{1/\alpha}+\lambda_2^{1/\alpha})(\lambda_1+\lambda_2)^{1/\alpha}-\lambda_1^{1/\alpha}\lambda_2^{1/\alpha}\right]$$

$$+(\lambda_1+\lambda_2)^{1/\alpha}\left\{2\lambda_1\lambda_2(\lambda_1+\lambda_2)^{1/\alpha}+\lambda_1^2[(\lambda_1+\lambda_2)^{1/\alpha}-\lambda_1^{1/\alpha}]+\lambda_2^2[(\lambda_1+\lambda_2)^{1/\alpha}-\lambda_2^{1/\alpha}]\right\}.$$

This expression is obviously positive. Hence, $I_{L_1}(i;\overline{\mathbf{F}})$ is decreasing in $\overline{F}_j(t)$. Because $I_{L_1}(i;\overline{\mathbf{F}})+I_{L_1}(j;\overline{\mathbf{F}})=1$ by Theorem 10.2.4, $I_{L_1}(j;\overline{\mathbf{F}})$ is decreasing in $F_i(t)$, and the proof is complete.

A.3   Proof of Theorem 10.2.17

The following proof is based on Natvig (1985).

Proof. Suppose that components i and j are in series with the rest of the system. Note that $F_i(t)\geq F_j(t)$ for $t\geq 0$ is equivalent to $\lambda_i\geq\lambda_j$. From Equation (10.29),

$$\frac{\partial\mathbb{E}(T_\phi)}{\partial\lambda_i^{-1}}=\lambda_i^2\int_0^\infty h(t)\exp(-(\lambda_i+\lambda_j)h(t))R(1_i,1_j,\overline{\mathbf{F}}(t)^{(ij)})dt,$$

which is obviously no less than $\partial\mathbb{E}(T_\phi)/\partial\lambda_j^{-1}$, and the proof of the series connection case is complete. Here $(1_i,1_j,\overline{\mathbf{F}}(t)^{(ij)})$ is a vector with 1 in positions i and j, and $\overline{\mathbf{F}}(t)^{(ij)}$ is the vector obtained by deleting $\overline{F}_i(t)$ and $\overline{F}_j(t)$ from $\overline{\mathbf{F}}(t)$.

Now, suppose that components i and j are in parallel with the rest of the system. First note that

$$I_B(i; \overline{\mathbf{F}}(t)) = 1 - R(0_i, \mathbf{F}(t)) = 1 - [\overline{F}_j(t)R(0_i, 1_j, \overline{\mathbf{F}}(t)^{(ij)}) + F_j(t)R(0_i, 0_j, \overline{\mathbf{F}}(t)^{(ij)})]$$
$$= 1 - \overline{F}_j(t) - F_j(t)R(0_i, 0_j, \overline{\mathbf{F}}(t)^{(ij)}) = F_j(t)(1 - R(0_i, 0_j, \overline{\mathbf{F}}(t)^{(ij)})).$$

Similarly,

$$I_B(j; \overline{\mathbf{F}}(t)) = F_i(t)[1 - R(0_i, 0_j, \overline{\mathbf{F}}(t)^{(ij)})].$$

From Equation (10.29), it is easy to verify that

$$\frac{\partial \mathbb{E}(T_\phi)}{\partial \lambda_i^{-1}} - \frac{\partial \mathbb{E}(T_\phi)}{\partial \lambda_j^{-1}} = \int_0^\infty \frac{(1 - R(0_i, 0_j, \overline{\mathbf{F}}(t)^{(ij)}))F_i(t)F_j(t)}{h(t)}$$
$$\times \left[\frac{\overline{F}_i(t)(-\ln \overline{F}_i(t))^2}{F_i(t)} - \frac{\overline{F}_j(t)(-\ln \overline{F}_j(t))^2}{F_j(t)} \right] dt.$$

Introducing the functions $g(x) = x(\ln x)^2/(1 - x)$, $0 \le x \le 1$, and $\psi(t) = g(\overline{F}_i(t)) - g(\overline{F}_j(t))$, $t \ge 0$, it is easy to see that $g(x)$ obtains a single maximum in $(0, 1)$ and that $g(0) = g(1) = 0$. By the assumptions that $\overline{F}_i(t) \ge \overline{F}_j(t)$ for $t \ge 0$ and that both functions are decreasing in t, there exists $t_0 \ge 0$ such that $\psi(t) \le 0$ for $0 \le t \le t_0$ and $\psi(t) \ge 0$ for $t \ge t_0$. Meanwhile, $[1 - R(0_i, 0_j, \overline{\mathbf{F}}(t)^{(ij)})]h(t)/h'(t)$ is increasing in t; hence,

$$\frac{\partial \mathbb{E}(T_\phi)}{\partial \lambda_i^{-1}} - \frac{\partial \mathbb{E}(T_\phi)}{\partial \lambda_j^{-1}} \ge \int_0^{t_0} [1 - R(0_i, 0_j, \overline{\mathbf{F}}(t_0)^{(ij)})] \frac{h(t_0)}{h'(t_0)} F_i(t)F_j(t) \frac{h'(t)}{(h(t))^2} \psi(t)dt$$
$$+ \int_{t_0}^\infty [1 - R(0_i, 0_j, \overline{\mathbf{F}}(t_0)^{(ij)})] \frac{h(t_0)}{h'(t_0)} F_i(t)F_j(t) \frac{h'(t)}{(h(t))^2} \psi(t)dt$$
$$= [1 - R(0_i, 0_j, \overline{\mathbf{F}}(t_0)^{(ij)})] \frac{h(t_0)}{h'(t_0)} \int_0^\infty F_i(t)F_j(t) \frac{h'(t)}{(h(t))^2} \psi(t)dt.$$

Note that $h'(t) > 0$ for $t \ge 0$. By substituting $u = h(t)$ in the integral, the right-hand side of this expression is reduced to zero, and the proof of the parallel connection case is complete.

For the case of the two-component parallel system, similar to the proof for the parallel connection case, the results are obtained by noting that $R(0_i, 0_j, \overline{\mathbf{F}}(t)^{(ij)}) = 0$ when $n = 2$.

A.4 Proof of Theorem 10.3.11

The following proof is based on Xie and Bergman (1991).

Proof. Part (i): It follows immediately from

$$I_Y(i; \overline{\mathbf{F}}) - I_Y(j; \overline{\mathbf{F}}) = \frac{1}{EY'} \int_0^\infty Y'(t) \left(I_B(i; \overline{\mathbf{F}}(t))f_i(t) - I_B(j; \overline{\mathbf{F}}(t))f_j(t) \right) dt$$

because the integrand is positive by the assumption.

Parts (ii) and (iii): To prove part (ii), let $F_r(t)$ be the lifetime distribution of the rest of the system. Then, in the parallel connection case, $F_\phi(t) = F_i(t)F_j(t)F_r(t)$, and

$$I_B(i; \overline{\mathbf{F}}(t))f_i(t) - I_B(j; \overline{\mathbf{F}}(t))f_j(t)$$

$$= \frac{\partial \overline{F}_\phi}{\partial \overline{F}_i}f_i(t) - \frac{\partial \overline{F}_\phi}{\partial \overline{F}_j}f_j(t) = \frac{\partial F_\phi}{\partial F_i}f_i(t) - \frac{\partial F_\phi}{\partial F_j}f_j(t)$$

$$= F_j(t)F_r(t)f_i(t) - F_i(t)F_r(t)f_j(t) = F_r(t)\left(f_i(t)F_j(t) - f_j(t)F_i(t)\right).$$

Then, part (ii) follows from part (i). The proof of part (iii) is similar.

Part (iv): Let component i be in series with a module containing component j. Denote by $G(t)$ the lifetime distribution of this module, and let $\overline{F}_M(t) = \overline{F}_i(t)\overline{G}(t)$. It follows that

$$I_B(i; \overline{\mathbf{F}}(t)) = \frac{\partial \overline{F}_\phi}{\partial \overline{F}_i} = \frac{\partial \overline{F}_\phi}{\partial \overline{F}_M}\frac{\partial \overline{F}_M}{\partial \overline{F}_i} = \frac{\partial \overline{F}_\phi}{\partial \overline{F}_M}\overline{G}(t),$$

$$I_B(j; \overline{\mathbf{F}}(t)) = \frac{\partial \overline{F}_\phi}{\partial \overline{F}_j} = \frac{\partial \overline{F}_\phi}{\partial \overline{F}_M}\frac{\partial \overline{F}_M}{\partial \overline{F}_j} = \frac{\partial \overline{F}_\phi}{\partial \overline{F}_M}\overline{F}_i(t)\frac{\partial \overline{G}}{\partial \overline{F}_j}.$$

By part (i), it suffices to show that $I_B(i; \overline{\mathbf{F}}(t))f_i(t) \geq I_B(j; \overline{\mathbf{F}}(t))f_j(t)$. From the earlier two equations, this condition can then be reduced to

$$\overline{G}(t)\overline{F}_i(t)r_i(t) \geq \overline{F}_i(t)\frac{\partial \overline{G}}{\partial \overline{F}_j}\overline{F}_j(t)r_j(t).$$

Because it is assumed that $r_i(t) \geq r_j(t)$ for $t \geq 0$, it reduces to show that $\overline{G}(t) - (\partial \overline{G}/\partial \overline{F}_j)\overline{F}_j(t) \geq 0$, which follows from Equation (5.6).

A.5   Proof of Theorem 10.3.15

The following proof is based on Xie and Bergman (1991).

Proof. Let ϕ denote the original system and ψ denote the system in which component i in system ϕ is replaced with a module of m i.i.d. components in parallel. Then, system ψ can be treated as system ϕ, whose component i is a supercomponent of lifetime distribution $F_i^m(t)$. The difference of the lifetime distributions of systems ϕ and ψ is

$$\overline{F}_\psi(t) - \overline{F}_\phi(t) = (1 - F_i^m(t))\frac{\overline{F}_\phi(t)}{\overline{F}_i(t)} - \overline{F}_\phi(t) = \frac{F_i(t)(1 - F_i^{m-1}(t))}{\overline{F}_i(t)}\overline{F}_\phi(t) \qquad (A.6)$$

for $t \geq 0$. Considering

$$E_\phi Y' = \int_0^\infty Y'(t) dF_\phi(t) = -Y'(\infty) + \int_0^\infty \overline{F}_\phi(t) dY'(t) \text{ and}$$

$$E_\psi Y' = -Y'(\infty) + \int_0^\infty \overline{F}_\psi(t) dY'(t),$$

if $Y'(t)$ is an increasing (decreasing) function, then $E_\psi Y' \geq (\leq) E_\phi Y'$ because $\overline{F}_\psi(t) \geq \overline{F}_\phi(t)$ for $t \geq 0$ from Equation (A.6).

If $Y'(t)$ is an increasing function, by Equation (10.42)

$$\begin{aligned}
I_Y(i; \overline{\mathbf{F}}_\phi) - I_Y(i_1, \overline{\mathbf{F}}_\psi) &= \frac{1}{E_\phi Y'} \int_0^\infty Y'(t) \overline{F}_\phi(t) r_i(t) dt - \frac{1}{E_\psi Y'} \int_0^\infty Y'(t) \overline{F}_\psi(t) r_i(t) dt \\
&\geq \frac{1}{E_\phi Y'} \int_0^\infty Y'(t) \overline{F}_\phi(t) r_i(t) dt - \frac{1}{E_\phi Y'} \int_0^\infty Y'(t) \overline{F}_\psi(t) r_i(t) dt \\
&= \frac{1}{E_\phi Y'} \int_0^\infty Y'(t) \frac{F_i(t)(1 - F_i^{m-1}(t))}{\overline{F}_i(t)} \overline{F}_\phi(t) r_i(t) dt.
\end{aligned}$$

The second inequality holds because $E_\psi Y' \geq E_\phi Y'$. If $Y'(t)$ is a decreasing function, the second inequality changes the direction. The last equality holds due to Equation (A.6). For the BP TIL importance, $Y(t) = t$, $E_\psi Y' = E_\phi Y' = 1$, and thus, the second inequality holds equality.

References

Natvig B. 1979. A suggestion of a new measure of importance of system component. *Stochastic Processes and Their Applications* **9**, 319–330.

Natvig B. 1985. New light on measures of importance of system components. *Scandinavian Journal of Statistics* **12**, 43–54.

Xie M and Bergman B. 1991. On a general measure of component importance. *Journal of Statistical Planning and Inference* **29**, 211–220.

Zhu X, Yao Q, and Kuo W. 2012. Patterns of Birnbaum importance in linear consecutive-*k*-out-of-*n* systems. *IIE Transactions* **44**, 277–290.

Bibliography

Abouammoh AM and Al-Kadi MA. 1991. On measures of importance for components in multistate coherent systems. *Microelectronics and Reliability* **31**, 109–122.

Achterberg T, Kocha T, and Martinb A. 2005. Branching rules revisited. *Operations Research Letters* **33**, 42–54.

Agrawal A and Barlow R. 1984. A survey of network reliability and domination theory. *Operations Research* **32**, 478–492.

Akhavein A and Fotuhi-Firuzabad M. 2011. A heuristic-based approach for reliability importance assessment of energy producers. *Energy Policy* **39**, 1562–1568.

Andrews JD and Beeson S. 1983. On the *s*-importance of elements and prime implicants of noncoherent systems. *IEEE Transactions on Reliability* **R-32**, 21–25.

Andrews JD and Beeson S. 2003. Birnbaum's measure of component importance for noncoherent systems. *IEEE Transactions on Reliability* **52**, 213–219.

António CC and Hoffbauer LN. 2008. From local to global importance measures of uncertainty propagation in composite structures. *Composite Structures* **85**, 213–225.

Archer KJ and Kimes RV. 2008. Empirical characterization of random forest variable importance measures. *Computational Statistics and Data Analysis* **52**, 2249–2260.

Armstrong MJ. 1995. Joint reliability-importance of components. *IEEE Transactions on Reliability* **44**, 408–412.

Armstrong MJ. 1997. Reliability-importance and dual failure-mode components. *IEEE Transactions on Reliability* **46**, 212–221.

Aven T. 1986. On the computation of certain measures of importance of system components. *Microelectronics and Reliability* **26**, 279–281.

Aven T and Jensen U. 1999. *Stochastic Models in Reliability*. Springer, New York.

Aven T and Nøkland TE. 2010. On the use of uncertainty importance measures in reliability and risk analysis. *Reliability Engineering and System Safety* **95**, 127–133.

Aven T and Ostebo R. 1986. Two new component importance measures for a flow network system. *Reliability Engineering* **14**, 75–80.

Bailey MP and Kulkarni VG. 1986. A recursive algorithm for computing exact reliability measures. *IEEE Transactions on Reliability* **R-35**, 36–40.

Ball MO. 1980. Complexity of network reliability computations. *Networks* **10**, 153–165.

Ball MO. 1986. Computational complexity of network reliability analysis: An overview. *IEEE Transactions on Reliability* **R-35**, 230–239.

Banzhaf JF. 1965. Weighted voting doesn't work: A mathematical analysis. *Rutges Law Review* **19**, 317–343.

Barabady J and Kumar U. 2007. Availability allocation through importance measures. *International Journal of Quality and Reliability Management* **24**, 643–657.

Baraldi P, Zio E, and Compare M. 2009. A method for ranking components importance in presence of epistemic uncertainties. *Journal of Loss Prevention in the Process Industries* **22**, 582–592.

Barlow R and Iyer S. 1988. Computational complexity of coherent systems and the reliability polynomial. *Probability in the Engineering and Informational Sciences* **2**, 461–469.

Barlow R and Wu A. 1978. Coherent systems with multistate components. *Mathematics of Operations Research* **3**, 275–281.

Barlow RE and Proschan F. 1965. *Mathematical Theory of Reliability*. John Wiley & Sons, New York.

Barlow RE and Proschan F. 1975. *Statistical Theory of Reliability and Life Testing Probability Models*. Holt, Rinehart and Winston, New York.

Barlow RE and Proschan F. 1975a. Importance of system components and fault tree events. *Statistic Processes and Their Applications* **3**, 153–172.

Barlow RE and Proschan F. 1975b. *Statistical Theory of Reliability and Life Testing Probability Models*. Holt, Rinehart and Winston, New York.

Bartlett LM and Andrews JD. 2001. An ordering heuristic to develop the binary decision diagram based on structural importance. *Reliability Engineering and System Safety* **72**, 31–38.

Baxter LA. 1984. Continuum structures I. *Journal of Applied Probability* **21**, 802–815.

Baxter LA. 1986. Continuum structures II. *Mathematical Proceedings of the Cambridge Philosophical Society* **99**, 331–338.

Baxter LA. 1988. On the theory of cannibalization. *Journal of Mathematical Analysis and Applications* **136**, 290–297.

Baxter LA and Harche F. 1992a. Note: On the greedy algorithm for optimal assembly. *Naval Research Logistics* **39**, 833–837.

Baxter LA and Harche F. 1992b. On the optimal assembly of series-parallel systems. *Operations Research Letters* **11**, 153–157.

Baxter LA and Kim C. 1986. Bounding the stochastic performance of continuum structure functions I. *Journal of Applied Probability* **23**, 660–669.

Baxter LA and Kim C. 1987. Bounding the stochastic performance of continuum structure functions II. *Journal of Applied Probability* **24**, 609–618.

Baxter LA and Lee SM. 1989. Further properties of reliability importance for continuum structure functions. *Probability in the Engineering and Informational Sciences* **3**, 237–246.

Bazaraa MS, Jarvis JJ, and Sherali HD. 1990. *Linear Programming and Network Flows*, 2nd edn. John Wiley & Sons, Hoboken, New Jersey.

Beale EML. 1979. Branch and bound methods for mathematical programming systems. *Annals of Discrete Mathematics* **5**, 201–219.

Beeson S and Andrews JD. 2003. Importance measures for non-coherent system analysis. *IEEE Transactions on Reliability* **52**, 301–310.

Bergman B. 1985a. On reliability theory and its applications. *Scandinavian Journal of Statistics* **12**, 1–41.

Bergman B. 1985b. On some new reliability importance measures. *Proceedings of IFAC SAFECOMP'85* (ed. Quirk WJ), pp. 61–64.

Bhattacharyya AK and Ahmed S. 1982. Establishing data requirements for plant probabilistic risk assessment. *Transactions of the American Nuclear Society* **43**, 477–478.

Bier VM. 1983. A measure of uncertainty importance for components in fault trees. *Transactions of the 1983 Winter Meeting of the American Nuclear Society* **45**(1), 384–385.

Birnbaum ZW. 1969. On the importance of different components in a multicomponent system. In *Multivariate Analysis, Vol. 2* (ed. Krishnaiah PR). Academic Press, New York, pp. 581–592.

Block, HW and Savits TH. 1982. A decomposition for multistate monotone system. *Journal of Applied Probability* **19**, 391–402.

Block HW and Savits TH. 1984. Continuous multistate structure functions. *Operations Research* **32**, 703–714.

Boland PJ and El-Neweihi E. 1995. Measures of component importance in reliability theory. *Computers and Operations Research* **22**, 455–463.

Boland PJ, El-Neweihi E, and Proschan F. 1988. Active redundancy allocation in coherent systems. *Probability in the Engineering and Informational Sciences* **2**, 343–353.

Boland PJ, El-Neweihi E, and Proschan F. 1991. Redundancy importance and allocation of spares in coherent systems. *Journal of Statistical Planning and Inference* **29**, 55–65.

Boland PJ, El-Neweihi E, and Proschan F. 1992. Stochastic order for redundancy allocations in series and parallel systems. *Advances in Applied Probability* **24**, 161–171.

Boland PJ and Proschan F. 1983. The reliability of *k* out of *n* systems. *Annals of Probability* **11**, 760–764.

Boland PJ and Proschan F. 1984. Optimal arrangement of systems. *Naval Research Logistics* **31**, 399–407.

Boland PJ, Proschan F, and Tong YL. 1989. Optimal arrangement of components via pairwise rearrangements. *Naval Research Logistics* **36**, 807–815.

Borgonovo E. 2006. Measuring uncertainty importance: Investigation and comparison of alternative approaches. *Risk Analysis* **26**, 1349–1361.

Borgonovo E. 2007a. A new uncertainty importance measure. *Reliability Engineering and System Safety* **92**, 771–784.

Borgonovo E. 2007b. Differential, criticality and Birnbaum importance measures: An application to basic event, groups and SSCs in event trees and binary decision diagrams. *Reliability Engineering and System Safety* **92**, 1458–1467.

Borgonovo E. 2008a. Differential importance and comparative statics: An application to inventory management. *International Journal of Production Economics* **111**, 170–179.

Borgonovo E. 2008b. Epistemic uncertainty in the ranking and categorization of probabilistic safety assessment model elements: Issues and findings. *Risk Analysis* **28**, 983–1001.

Borgonovo E. 2010. The reliability importance of components and prime implicants in coherent and non-coherent systems including total-order interactions. *European Journal of Operational Research* **204**, 485–495.

Borgonovo E and Apostolakis GE. 2001. A new importance measure for risk-informed decision making. *Reliability Engineering and System Safety* **72**, 193–212.

Borgonovo E, Apostolakis GE, Tarantola S, and Saltelli A. 2003. Comparison of global sensitivity analysis techniques and importance measures in PSA. *Reliability Engineering and System Safety* **79**, 175–185.

Borgonovo E and Peccati L. 2004. Sensitivity analysis in investment project evaluation. *International Journal of Production Economics* **90**, 17–25.

Borgonovo E and Smith C. 1999. A case study: Two components in parallel with epistemic uncertainty, Part I. *SAPHIRE Facets INEEL 1999*.

Borgonovo E and Smith C. 2000. A case study: Two components in parallel with epistemic uncertainty, Part II. *SAPHIRE Facets INEEL 2000*.

Borgonovo E and Smith C. 2010. Total order reliability importance in space PSA. *Procedia Social and Behavioral Sciences* **2**, 7617–7618.

Borgonovo E and Smith C. 2011. A study of interactions in the risk assessment of complex engineering systems: An application to space PSA. *Operations Research* **59**, 1461–1476.

Boros E, Crama Y, Ekin O, Hammer P, Ibaraki T, and Kogan A. 2000. Boolean normal forms, shellability, and reliability computations. *SIAM Journal on Discrete Mathematics* **13**, 212–226.

Bossche A. 1987. Calculation of critical importance for multi-state components. *IEEE Transactions on Reliability* **R-36**, 247–249.

Brown M and Proschan F. 1983. Imperfect repair. *Journal of Applied Probability* **20**, 851–859.

Brunelle RD and Kapur KC. 1998. Continuous-state system reliability: An interpolation approach. *IEEE Transactions on Reliability* **47**, 181–187.

Brunelle RD and Kapur KC. 1999. Review and classification of reliability measures for multistate and continuum models. *IIE Transactions* **31**, 1171–1180.

Bueno VC. 1989. On the importance of components for multistate monotone systems. *Statistics and Probability Letters* **7**, 51–59.

Bueno VC. 2000. Component importance in a random environment. *Statistics and Probability Letters* **48**, 173–179.

Butler DA. 1977. An importance ranking for system components based upon cuts. *Operations Research* **25**, 874–879.

Butler DA. 1979. A complete importance ranking for components of binary coherent systems with extensions to multi-state systems. *Naval Research Logistics* **4**, 565–578.

Cai J. 1994. Reliability of a large consecutive-k-out-of-n:F system with unequal component-reliability. *IEEE Transactions on Reliability* **43**, 107–111.

Campolongo F, Cariboni J, and Saltelli A. 2007. An effective screening design for sensitivity analysis of large models. *Environmental Modelling and Software* **22**, 1509–1518.

Campolongo F and Saltelli A. 1997. Sensitivity analysis of an environmental model: An application of different analysis methods. *Reliability Engineering and System Safety* **57**, 49–69.

Cao XR and Chen HF. 1997. Perturbation realization, potentials, and sensitivity analysis of Markov processes. *IEEE Transactions on Automatic Control* **42**, 1382–1393.

Cao XR and Ho YC. 1987. Sensitivity estimate and optimization of throughput in a production line with blocking. *IEEE Transactions on Automatic Control* **32**, 959–967.

Carot V and Sanz J. 2000. Criticality and sensitivity analysis of the components of system. *Reliability Engineering and System Safety* **68**, 1147–1152.

Chacko VM and Manoharan M. 2011. Joint importance measures in network system. *Reliability: Theory & Applications #04 (23)* **2**, 129–139.

Chadjiconstantinidis S and Koutras MV. 1999. Measures of component importance for Markov chain imbeddable reliability structures. *Naval Research Logistics* **46**, 613–639.

Chang GJ, Cui L, and Hwang FK. 1999. New comparisons in Birnbaum importance for the consecutive-*k*-out-of-*n* system. *Probability in the Engineering and Informational Sciences* **13**, 187–192.

Chang GJ, Hwang FK, and Cui L. 2000. Corrigenda on "New comparisons in Birnbaum importance for the consecutive-*k*-out-of-*n* system". *Probability in the Engineering and Informational Sciences* **14**, 405.

Chang GJ, Chen RJ, and Hwang FK. 2002. The structural Birnbaum importance of consecutive-*k* systems. *Journal of Combinatorial Optimization* **6**, 183–197.

Chang HW and Hwang FK. 1999. Existence of invariant series consecutive-*k*-out-of-*n*:G systems. *IEEE Transactions on Reliability* **R-48**, 306–308.

Chang HW and Hwang FK. 2002. Rare-event component importance for the consecutive-*k* system. *Naval Research Logistics* **49**, 159–166.

Chang YR, Amari SV, and Kuo SY. 2004. Computing system failure frequencies and reliability importance measures using OBDD. *IEEE Transactions on Computers* **53**, 54–68.

Chang YR, Amari SV, and Kuo SY. 2005. OBDD-based evaluation of reliability and importance measures for multistate systems subject to imperfect fault coverage. *IEEE Transactions on Dependable and Secure Computing* **2**, 336–347.

Chao MT and Lin GD. 1984. Economical design of large consecutive-*k*-out-of-*n*:F systems. *IEEE Transactions on Reliability* **R-33**, 411–413.

Cheok MC, Parry GW, and Sherry RR. 1998a. Response to "Supplemental viewpoints on the use of importance measures in risk-informed regulatory applications". *Reliability Engineering and System Safety* **60**, 261–261.

Cheok MC, Parry GW, and Sherry RR. 1998b. Use of importance measures in risk-informed regulatory applications. *Reliability Engineering and System Safety* **60**, 213–226.

Chern MS. 1992. On the computational complexity of reliability redundancy allocation in a series system. *Operations Research Letters* **11**, 309–315.

Cheung RC. 1980. A user-oriented software reliability model. *IEEE Transactions on Software Engineering* **6**, 118–125.

Chiang DT and Chiang RF. 1986. Relayed communication via consecutive-*k*-out-of-*n*:F system. *IEEE Transactions on Reliability* **35**, 65–67.

Chiang DT and Niu SC. 1981. Reliability of consecutive-*k*-out-of-*n*:F system. *IEEE Transactions on Reliability* **R-30**, 87–89.

Chiu CC, Yeh YS, and Chou JS. 2001. An effective algorithm for optimal *k*-terminal reliability of distributed systems. *Malaysian Journal of Library & Information Science* **6**, 101–118.

Cho JG and Yum BJ. 1997. Development and evaluation of an uncertainty importance measure in fault tree analysis. *Reliability Engineering and System Safety* **57**, 143–157.

Cho NZ and Jung WS. 1989. A new method for evaluating system-level importance measures based on singular value decomposition. *Reliability Engineering and System Safety* **25**, 197–205.

Chun MH, Han SJ, and Tak NI. 2000. An uncertainty importance measure using a distance metric for the change in a cumulative distribution function. *Reliability Engineering and System Safety* **70**, 313–321.

Contini S and Matuzas V. 2011. New methods to determine the importance measures of initiating and enabling events in fault tree analysis. *Reliability Engineering and System Safety* **96**, 775–784.

Coyle D, Buxton MJ, and O'Brien BJ. 2003. Measures of importance for economic analysis based on decision modeling. *Journal of Clinical Epidemiology* **56**, 989–997.

Cui LR and Hawkes AG. 2008. A note on the proof for the optimal consecutive-*k*-out-of-*n*:G line for $n \leq 2k$. *Journal of Statistical Planning and Inference* **138**, 1516–1520.

Dai L and Ho YC. 1995. Structural infinitesimal perturbation analysis for derivative estimation of discrete-event dynamic systems. *IEEE Transactions on Automatic Control* **40**, 1154–1166.

Dakin RJ. 1965. A tree search algorithm for mixed integer programming problems. *Computer Journal* **8**, 250–255.

Davis L. 1991. *Handbook of Genetic Algorithms*. Van Nostrand Reinhold, New York.

Derman C, Lieberman GJ, and Ross SM. 1972. On optimal assembly of system. *Naval Research Logistics* **19**, 569–574.

Derman C, Lieberman GJ, and Ross SM. 1974. Assembly of systems having maximum reliability. *Naval Research Logistics* **21**, 1–12.

Derman C, Lieberman GJ, and Ross SM. 1982. On the consecutive-*k*-of-*n*:F system. *IEEE Transactions on Reliability* **R-31**, 57–63.

Doulliez P and Jamoulle E. 1972. Transportation networks with random arc capacities. *RAIRO* **3**, 45–60.

Do Van P, Barros A, and Berenguer C. 2008a. Importance measure on finite time horizon and application to Markovian multistate production systems. *Proceedings of the IMechE, Part O: Journal of Risk and Reliability*, Vol. 222, pp. 449–461.

Do Van P, Barros A, and Berenguer C. 2008b. Reliability importance analysis of Markovian systems at steady state using perturbation analysis. *Reliability Engineering and Systems Safety* **93**, 1605–1615.

Do Van P, Barros A, and Berenguer C. 2010. From differential to difference importance measures for Markov reliability models. *European Journal of Operational Research* **24**, 513–521.

Du DZ and Hwang FK. 1985. Optimal consecutive-2 systems of lines and cycles. *Networks* **15**, 439–447.

Du DZ and Hwang FK. 1986. Optimal consecutive 2-out-of-*n* systems. *Mathematics of Operations Research* **11**, 187–191.

Du DZ and Hwang FK. 1987. Optimal assignments for consecutive-2 graphs. *SIAM Journal on Algebraic and Discrete Methods* **8**, 510–518.

Du D, Hwang FK, Jung Y, and Ngo HQ. 2001. Optimal consecutive-*k*-out-of-$(2k + 1)$:G cycle. *Journal of Global Optimization* **19**, 51–60.

Du D, Hwang FK, Jia X, and Ngo HQ. 2002. Optimal consecutive-*k*-out-of-*n*:G cycle for $n \leq 2k + 1$. *SIAM Journal on Discrete Mathematics* **15**, 305–316.

Duflot N, Bérenguer C, Dieulle L, and Vasseur D. 2009. A min cut-set-wise truncation procedure for importance measures computation in probabilistic safety assessment. *Reliability Engineering and System Safety* **94**, 1827–1837.

Dutuit Y and Rauzy A. 2001. Efficient algorithms to assess component and gate importance in fault tree analysis. *Reliability Engineering and System Safety* **72**, 213–222.

Eisenberg NA and Sagar B. 2000. Importance measures for nuclear waste repositories. *Reliability Engineering and System Safety* **70**, 217–239.

El-Neweihi E. 1980. A relationship between partial derivatives of the reliability function of a coherent system and its minimal path (cut) sets. *Mathematics of Operations Research* **5**, 553–555.

El-Neweihi E, Proschan F, and Sethuraman J. 1986. Optimal allocation of components in parallel-series and series-parallel systems. *Journal of Applied Probability* **23**, 770–777.

El-Neweihi E, Proschan F, and Sethuraman J. 1987. Optimal assembly of systems using Schur functions and majorization. *Naval Research logistics* **34**, 705–712.

El-Neweihi E and Sethuraman J. 1993. Optimal allocation under partial ordering of lifetimes of components. *Advances in Applied Probability* **25**, 914–925.

Espiritu JF, Coit DW, and Prakash U. 2007. Component criticality importance measures for the power industry. *Electric Power Systems Research* **77**, 407–420.

Finkelstein MS. 1994. Once more on measures of importance of system components. *Microelectronics and Reliability* **34**, 1431–1439.

Fiondella L and Gokhale SS. 2008. Importance measures for a modular software system. *Proceedings of the 8th International Conference on Quality Software*, pp. 338–343.

Foldes S and Hammer PL. 2005. Submodularity, supermodularity, and higher-order monotonicities of pseudo-Boolean functions. *Mathematics of Operations Research* **30**, 453–461.

Freisleben B and Merz P. 1996. A genetic local search algorithm for solving symmetric and asymmetric traveling salesman problems. In *Proceedings of the 1996 IEEE International Conference on Evolutionary Computation* (ed. Grefenstette JJ). IEEE Press, New York, pp. 616–621.

Freixas J and Pons M. 2008a. Identifying optimal components in a reliability system. *IEEE Transactions on Reliability* **57**, 163–170.

Freixas J and Pons M. 2008b. The influence of the node criticality relation on some measures of component importance. *Operations Research Letters* **36**, 557–560.

Freixas J and Puente MA. 2002. Reliability importance measures of the components in a system based on semivalues and probabilistic values. *Annals of Operations Research* **109**, 331–342.

Frey CH. 2002. Introduction to special section on sensitivity analysis and summary of NCSU/USDA workshop on sensitivity analysis. *Risk Analysis* **22**, 539–545.

Fussell JB. 1975. How to hand-calculate system reliability and safety characteristics. *IEEE Transactions on Reliability* **R-24**, 169–174.

Gámiz ML and Martínez Miranda MD. 2010. Regression analysis of the structure function for reliability evaluation of continuous-state system. *Reliability Engineering and System Safety* **95**, 134–142.

Gandini A. 1990. Importance & sensitivity analysis in assessing system reliability. *IEEE Transactions on Reliability* **39**, 61–70.

Gao X, Cui L, and Li J. 2007. Analysis for joint importance of components in a coherent system. *European Journal of Operational Research* **182**, 282–299.

Garey M and Johnson D. 1979. *Computers and Intractability: A Guide to the Theory of NP-Completeness.* Freeman, San Francisco.

Gebre BA and Ramirez-Marques JE. 2007. Element substitution algorithm for general two-terminal network reliability analyses. *IIE Transactions* **39**, 265–275.

Gertsbakh I and Shpungin Y. 2008. Network reliability importance measures: Combinatorics and Monte Carlo based computations. *WSEAS Transactions on Computers* **7**, 216–227.

Golub GH and Zha H. 1994. Perturbation analysis of the canonical correlations of matrix pairs. *Linear Algebra and its Applications* **210**, 3–28.

Gong WB and Ho YC. 1987. Smoothed (conditional) perturbation analysis of discrete event dynamical systems. *IEEE Transactions on Automatic Control* **32**, 858–866.

Grabisch M, Marichal JL, and Roubens M. 2000. Equivalent representations of set functions. *Mathematics of Operations Research* **25**, 157–178.

Grefenstette JJ, Gopal R, Rosmaita B, and Cucht DV. 1985. Genetic algorithms for the traveling salesman problem. *Proceedings of the 1st International Conference on Genetic Algorithms and Their Applications.* Erlbaum, Hillsdale, NJ, pp. 160–168.

Griffith W. 1986. On consecutive k-out-of-n:F failure system and their generations. In *Reliability and Quality Control* (ed. Basu AP). Elsevier (North-Holland), New York, pp. 157–165.

Griffith WS. 1980. Multistate reliability models. *Journal of Applied Probability* **17**, 735–744.

Griffith WS and Govindarajulu Z. 1985. Consecutive k-out-of-n failure systems: Reliability, availability, component importance, and multistate extensions. *American Journal of Mathematical and Management Sciences* **5**, 125–160.

Gupta S and Kumar U. 2010. Maintenance resource prioritization in a production system using cost-effective importance measure. *Proceedings of the 1st International Workshop and Congress on eMaintenance*, pp. 196–204.

Gupta S and Manohar CS. 2004. An improved response surface method for the determination of failure probability and importance measures. *Structural Safety* **26**, 123–139.

Ha C and Kuo W. 2005. Multi-path approach for reliability-redundancy allocation using a scaling method. *Journal of Heuristics* **11**, 201–217.

Habib A and Szántai T. 1997. An algorithm evaluating the exact reliability of a consecutive k-out-of-r-from-n:F system. *Proceedings of 1997 IMACS Conference*, pp. 421–425.

Habib A and Szántai T. 2000. New bounds on the reliability of the consecutive k-out-of-r-from-n:F system. *Reliability Engineering and System Safety* **68**, 97–104.

Habib A, Al-Seedy R, and Radwan T. 2007. Reliability evaluation of multi-state consecutive k-out-of-r-from-n:G system. *Applied Mathematical Modelling* **31**, 2412–2423.

Hagstrom JN. 1990. Redundancy, substitutes and complements in system reliability. Technical report, College of Business Administration, University of Illinois, Chicago.

Hagstrom JN and Mak KT. 1987. System reliability analysis in the presence of dependent component failures. *Probability in the Engineering and Informational Sciences* **1**, 425–440.

Haukaas T and Kiureghian AD. 2005. Parameter sensitivity and importance measures in nonlinear finite element reliability analysis. *Journal of Engineering Mechanics* **131**, 1013–1026.

Higashiyama Y. 2004. A method for exact reliability of consecutive 2-out-of-(r, r)-from-(n, n):F systems. *International Transactions in Operational Research* **11**, 217–224.

Higashiyama Y, Ariyoshi H, and Kraetzl M. 1995. Fast solutions for consecutive 2-out-of-r-from-n:F system. *IEICE Transactions on Fundamentals* **E-87A**, 680–684.

Higashiyama Y, Kraetzl M, and Caccetta L. 1996. Efficient algorithms for a consecutive 2-out-of-r-from-n:F system. *Australasian Journal of Combinatorics* **14**, 31–36.

Higashiyama Y, Kraetzl M, and Caccetta L. 1999. Formulas for the reliability of a consecutive k-out-of-r-from-n:F system. *Proceedings of SCI'99*, pp. 131–134.

Higashiyama Y, Ohkura T, and Rumchev VG. 2006. Recursive method for reliability evaluation of circular consecutive 2-out-of-(r, r)-from-(n, n):F systems. *International Journal of Reliability, Quality and Safety Engineering* **13**, 355–363.

Hilber P and Bertling L. 2007. Component reliability importance indices for electrical networks. *Proceedings of the 8th International Power Engineering Conference, IPEC, Singapore.*

Ho YC. 1988. Perturbation analysis explained. *IEEE Transactions on Automatic Control* **33**, 761–763.

Ho YC. 1992. Perturbation analysis: Concepts and algorithms. In *Proceedings of the 1992 Winter Simulation Conference* (ed. Swain JJ, Goldsman D, Crain RC and Wilson JR), pp. 231–240.

Ho YC and Cao XR. 1983. Perturbation analysis and optimization of queueing networks. *Journal of Optimization Theory and Applications* **40**, 559–582.

Ho YC and Cao XR. 1985. Performance sensitivity to routing changes in queueing networks and flexible manufacturing systems using perturbation analysis. *IEEE Journal of Robotics and Automation* **RA-I**, 165–172.

Ho YC and Cao XR. 1991. *Perturbation Analysis of Discrete Event Dynamic Systems.* Kluwer Academic Publishers, Boston.

Ho YC, Zhao QC, and Jia QS. 2007. *Ordinal Optimization: Soft Optimization for Hard Problems.* Springer, New York.

Hohenbichler M and Rackwitz R. 1986. Sensitivity and importance measures in structural reliability. *Civil Engineering Systems* **3**, 203–209.

Homma T and Saltelli A. 1996. Importance measures in global sensitivity analysis of nonlinear models. *Reliability Engineering and System Safety* **52**, 1–17.

Hong JS, Koo HY, and Lie CH. 2000. Computation of joint reliability importance of two gate events in a fault tree. *Reliability Engineering and System Safety* **68**, 1–5.

Hong JS, Koo HY, and Lie CH. 2002. Joint reliability-importance of k-out-of-n systems. *European Journal of Operational Research* **142**, 539–547.

Hong JS and Lie CH. 1993. Joint reliability-importance of two edges in an undirected network. *IEEE Transactions on Reliability* **42**, 17–33.

Hong Y and Lee L. 2009. Reliability assessment of generation and transmission systems using fault-tree analysis. *Energy Conversion and Management* **50**, 2810–2817.

Hora SC and Iman RL. 1989. Expert opinion in risk analysis: The NUREG-1150 methodology. *Nuclear Science and Engineering* **102**, 323–331.

Hsu SJ and Yuang MC. 1999. Efficient computation of marginal reliability importance for reducible networks in network management. *Proceedings of the 1999 IEEE International Conference on Communications*, pp. 1039–1045.

Huseby AB and Natvig B. 2010. Advanced discrete simulation methods applied to repairable multistate systems. In *Reliability, Risk and Safety: Theory and Applications, Vol. 1* (eds. Bris R, Guedes Soares C, and Martorell S). CRC Press, London, pp. 659–666.

Hwang FK. 1982. Fast solutions for consecutive k-out-of-n:F systems. *IEEE Transactions on Reliability* **R-31**, 447–448.

Hwang FK. 1989a. Invariant permutations for consecutive-k-out-of-n cycles. *IEEE Transactions on Reliability* **R-38**, 65–67.

Hwang FK. 1989b. Optimal assignment of components to a two-stage k-out-of-n system. *Mathematics of Operations Research* **14**, 376–382.

Hwang FK. 2001. A new index of component importance. *Operations Research Letters* **28**, 75–79.

Hwang FK. 2005. A hierarchy of importance indices. *IEEE Transactions on Reliability* **54**, 169–172.

Hwang FK, Cui LR, Chang JC, and Lin WD. 2000. Comments on "Reliability and component importance of a consecutive-k-out-of-n systems" by Zuo. *Microelectronics and Reliability* **40**, 1061–1063.

Hwang FK and Pai CK. 2000. Sequential construction of a circular consecutive-2 system. *Information Processing Letters* **75**, 231–235.

Hwang FK and Shi D. 1987. Redundant consecutive-k-out-of-n:F systems. *Operations Research Letters* **6**, 293–296.

Iman RL. 1987. A matrix-based approach to uncertainty and sensitivity analysis for fault trees. *Risk Analysis* **7**, 21–23.

Iman RL and Hora SC. 1990. A robust measure of uncertainty importance for use in fault tree system analysis. *Risk Analysis* **10**, 401–406.

Iman RL, Johnson ME, and Watson CCJ. 2005a. Sensitivity analysis for computer model projections of hurricane losses. *Risk Analysis* **25**, 1277–1297.

Iman RL, Johnson ME, and Watson CCJ. 2005b. Uncertainty analysis for computer model projections of hurricane losses. *Risk Analysis* **25**, 1299–1312.

Inagaki T and Henley EJ. 1980. Probabilistic evaluation of prime implicants and top-events for non-coherent systems. *IEEE Transactions on Reliability* **R-29**, 361–367.

Isbell J. 1956. A class of majority games. *Quarterly Journal of Mathematics, Oxford Series* **7**, 183–187.

Ishigami T and Homma T. 1990. An importance quantification technique in uncertainty analysis for computer models. *Proceedings of the 1st International Symposium on Uncertainty Modelling and Analysis*, pp. 398–403.

Iyer S. 1992. The Barlow-Proschan importance and its generalizations with dependent components. *Stochastic Processes and Their Applications* **42**, 353–359.

Jain SP and Gopal K. 1985. Reliability of k-to-ℓ-out-of-n systems. *Reliability Engineering* **12**, 175–179.

Jalali A, Hawkes AG, Cui LR, and Hwang FK. 2005. The optimal consecutive-k-out-of-n:G line for $n \leq 2k$. *Journal of Statistical Planning and Inference* **128**, 281–287.

Jan S and Chang HW. 2006. Joint reliability importance of k-out-of-n systems and series-parallel systems. *Proceedings of PDPTA'06*, pp. 395–398.

Jenelius E. 2010. Redundancy importance: Links as rerouting alternatives during road network disruptions. *Procedia Engineering* **3**, 129–137.

Jensen PA and Barnes JW. 1980. *Network Flow Programming*. John Wiley & Sons, New York.

Jin H and Santhanam P. 2001. An approach to higher reliability using software components. *Proceedings of the 12th IEEE International Symposium Software Reliability Engineering (ISSRE)*, pp. 1–11.

Johnson BC and Apostolakis GE. 2009. The limited exceedance factor importance measure: An application to the prism reactor design. *Proceedings of the 17th International Conference on Nuclear Engineering (ICONE17)*, pp. 349–353.

Kao SC. 1982. Computing reliability from warranty. *Proceedings of the 1982 American Statistical Association, Section on Statistical Computing*, pp. 309–312.

Khachiyan L, Boros E, Elbassioni K, Gurvich V, and Makino K. 2007. Enumerating disjunctions and conjunctions of paths and cuts in reliability theory. *Discrete Applied Mathematics* **155**, 137–149.

Kim C and Baxter LA. 1987a. Axiomatic characterizations of continuum structure functions. *Operations Research Letters* **6**, 297–300.

Kim C and Baxter LA. 1987b. Reliability importance for continuum structure functions. *Journal of Applied Probability* **24**, 779–785.

Kim C and Baxter LA. 1997. Approximation of reliability importance for continuum structure functions. *Kangweon-Kyungki Mathematical Journal* **5**, 55–60.

Kim K and Han S. 2009. A study on importance measures and a quantification algorithm in a fire PRA model. *Reliability Engineering and System Safety* **94**, 969–972.

Kim K, Kang D, and Yang J. 2005. On the use of the balancing method for calculating component RAW involving CCFs in SSC categorization. *Reliability Engineering and System Safety* **87**, 233–242.

Kiparissides A, Kucherenko SS, Mantalaris A, and Pistikopoulos EN. 2009. Global sensitivity analysis challenges in biological systems modeling. *Industrial and Engineering Chemistry Research* **48**, 7168–7180.

Kochar S, Mukerjee H, and Samaniego FJ. 1999. The "signature" of a coherent system and its application to comparisons among systems. *Naval Research Logistics* **46**, 507–523.

Kontoleon JM. 1979. Optimal link allocation of fixed topology networks. *IEEE Transactions on Reliability* **28**, 145–147.

Koutras MV, Papadopoylos G, and Papastavridis SG. 1994. Note: Pairwise rearrangements in reliability structures. *Naval Research Logistics* **41**, 683–687.

Kucherenko S, Rodriguez-Fernandez M, Pantelides C, and Shah N. 2009. Monte Carlo evaluation of derivative-based global sensitivity measures. *Reliability Engineering and System Safety* **94**, 1135–1148.

Kuo W, Prasad VR, Tillman FA, and Hwang CL. 2006. *Optimal Reliability Design: Fundamentals and Applications*, 2nd edn. Cambridge University Press, Cambridge, UK.

Kuo W and Wan R. 2007. Recent advances in optimal reliability allocation. *IEEE Transactions on Systems, Man, and Cybernetics, Series A* **37**, 143–156.

Kuo W, Zhang W, and Zuo MJ. 1990. A consecutive k-out-of-n:G: The mirror image of a consecutive k-out-of-n:F system. *IEEE Transactions on Reliability* **R-39**, 244–253.

Kuo W and Zhu X. 2012a. Relations and generalizations of importance measures in reliability. *IEEE Transactions on Reliability* **61**.

Kuo W and Zhu X. 2012b. Some recent advances on importance measures in reliability. *IEEE Transactions on Reliability* **61**.

Kuo W and Zuo MJ. 2003. *Optimal Reliability Modeling: Principles and Applications*. John Wiley & Sons, New York.

Lambert HE. 1975a. *Fault Trees for Decision Making in System Safety and Availability*. PhD Thesis, University of California, Berkeley.

Lambert HE. 1975b. Measure of importance of events and cut sets in fault trees. In *Reliability and Fault Tree Analysis* (eds. Barlow RE, Fussell JB, and Singpurwalla ND). Society for Industrial and Applied Mathematics, Philadelphia, pp. 77–100.

Land AH and Powell S. 1979. Computer codes for problems of integer programming. *Annals of Discrete Mathematics* **5**, 221–269.

Lee HS, Lie CH, and Hong JS. 1997. A computation method for evaluating importance-measures of gates in a fault tree. *IEEE Transactions on Reliability* **46**, 360–365.

Levitin G. 2002. Optimal allocation of elements in a linear multi-state sliding window system. *Reliability Engineering and System Safety* **76**, 245–254.

Levitin G. 2003a. Element availability importance in generalized k-out-of-r-from-n systems. *IIE Transactions* **35**, 1125–1131.

Levitin G. 2003b. Linear multi-state sliding window systems. *IEEE Transactions on Reliability* **52**, 263–269.

Levitin G. 2004a. Consecutive k-out-of-r-from-n system with multiple failure criteria. *IEEE Transactions on Reliability* **53**, 394–400.

Levitin G. 2004b. Protection survivability importance in systems with multilevel protection. *Quality and Reliability Engineering International* **20**, 727–738.

Levitin G. 2005. Reliability of linear multistate multiple sliding window systems. *Naval Research Logistics* **52**, 212–223.

Levitin G and Lisnianski A. 1999. Importance and sensitivity analysis of multi-state systems using the universal generating function method. *Reliability Engineering and System Safety* **65**, 271–282.

Levitin G, Podofillini L, and Zio E. 2003. Generalised importance measures for multi-state elements based on performance level restrictions. *Reliability Engineering and System Safety* **82**, 287–298.

Liaw A and Wiener M. 2002. Classification and regression by ramdomforest. *Resampling Methods in R: The Boot Package* **2**, 18–22.

Lim MH, Yuan Y, and Omatu S. 2000. Efficient genetic algorithms using simple genes exchange local search policy for the quadratic assignment problem. *Computational Optimization and Applications* **15**, 249–268.

Lim MH, Yuan Y, and Omatu S. 2002. Extensive testing of a hybrid genetic algorithm for solving quadratic assignment prolems. *Computational Optimization and Applications* **23**, 47–64.

Lin FH and Kuo W. 2002. Reliability importance and invariant optimal allocation. *Journal of Heuristics* **8**, 155–171.

Lin FH, Kuo W, and Hwang F. 1999. Structure importance of consecutive-k-out-of-n systems. *Operations Research Letters* **25**, 101–107.

Lin MS, Chang MS, and Chen DJ. 2000. A generalization of consecutive k-out-of-n:G systems. *IEICE Transactions on Information and Systems* **E83-D**, 1309–1313.

Lin MS and Chen DJ. 1997. The computational complexity of the reliability problem on distributed systems. *Information Processing Letters* **64**, 143–147.

Lisnianski A. 2001. Estimation of boundary points for continuum-state system reliability measures. *Reliability Engineeging and System Safety* **74**, 81–88.

Liu Q and Homma T. 2009. A new computational method of a moment-independent uncertainty importance measure. *Reliability Engineering and System Safety* **94**, 1205–1211.

Lu L and Jiang J. 2007. Joint failure importance for noncoherent fault trees. *IEEE Transactions on Reliability* **56**, 435–443.

MacGregor M, Grover WD, and Maydell UM. 1993. Connectability: A performance metric forreconfigurability transport networks. *IEEE Journal on Selected Areas in Communications* **11**, 1461–1468.

Makri FS and Psillakis ZM. 1996. Bounds for reliability of k-within two-dimensional consecutive-r-out-of-n failure systems. *Microelectronics Reliability* **36**, 341–345.

Malon DM. 1984. Optimal consecutive 2-out-of-n:F component sequencing. *IEEE Transactions on Reliability* **R-33**, 414–418.

Malon DM. 1985. Optimal consecutive k-out-of-n:F component sequencing. *IEEE Transactions on Reliability* **R-34**, 46–49.

Malon DM. 1990. When is greedy module assembly optimal?. *Naval Research Logistics* **37**, 847–854.

Marinacci M and Montrucchio L. 2005. Ultramodular functions. *Mathematics of Operations Research* **30**, 311–332.

Marseguerra M and Zio E. 2004. Monte Carlo estimation of the differential importance measure: Application to the protection system of a nuclear reactor. *Reliability Engineering and System Safety* **86**, 11–24.

Marshall A and Olkin I. 1979. *Inequalities: Theory of Majorization and Its Applications*. Academic Press, New York.

Meng FC. 1993a. Component-relevancy and characterization results in multistate systems. *IEEE Transactions on Reliability* **42**, 478–483.

Meng FC. 1993b. On selecting components for redundancy in coherent systems. *Reliability Engineering and System Safety* **41**, 121–126.

Meng FC. 1994. Comparing criticality of nodes via minimal cut (path) sets for coherent systems. *Probability in the Engineering and Informational Sciences* **8**, 79–87.

Meng FC. 1995. Some further results on ranking the importance of system components. *Reliability Engineering and System Safety* **47**, 97–101.

Meng FC. 1996a. Comparing the importance of system components by some structural characteristics. *IEEE Transactions on Reliability* **45**, 59–65.

Meng FC. 1996b. More on optimal allocation of componenets in coherent systems. *Journal of Applied Probability* **33**, 548–556.

Meng FC. 2000. Relationships of Fussell-Vesely and Birnbaum importance to structural importance in coherent systems. *Reliability Engineering and System Safety* **67**, 55–60.

Meng FC. 2004. Comparing Birnbaum importance measure of system components. *Probability in the Engineering and Informational Sciences* **18**, 237–245.

Merz P and Freisleben B. 1997. A genetic local search approach to the quadratic assignment problem. In *Proceedings of the 7th International Conference on Genetic Algorithms* (ed. Grefenstette JJ). Morgan Kaufmann, San Francisco, CA, pp. 465–472.

Mi J. 1998. Bolstering components for maximizing system life. *Naval Research Logistics* **45**, 497–509.

Mi J. 1999. Optimal active redundancy in *k*-out-of-*n* system. *Journal of Applied Probability* **36**, 927–933.

Mi J. 2003. A unified way of comparing the reliability of coherent systems. *IEEE Transactions on Reliability* **52**, 38–43.

Miles Jr. EP. 1960. Generalized Fibonacci numbers and associated metrices. *The American Mathematical Monthly* **67**, 745–752.

Miman M and Pohl EA. 2006. Uncertainty assessment for availability: Importance measures. *Proceedings of Annual Reliability and Maintainability Symposium*, pp. 222 –227.

Misra RB and Agnihotri G. 1979. Peculiarities in optimal redundancy for a bridge network. *IEEE Transactions on Reliability* **R-28**, 70–72.

Montero J, Tejada J, and Yáñez J. 1990. Structural properties of continuum systems. *European Journal of Operational Research* **45**, 231–240.

Morris MD. 1991. Factorial sampling plans for preliminary computational experiments. *Technometrics* **33**, 161–174.

Musa J, Iannino A, and Okumoto K. 1987. *Software Reliability: Measurement, Prediction, Application*. Series of Software Engineering and Technology. McGraw-Hill, New York.

Nakashima K and Yamato K. 1982. Variance-importance of system components. *IEEE Transactions on Reliability* **R-31**, 99–100.

Naoki S and Hiroshi M. 2004. Wireless charging system by microwave power transmission for electric motor vehicles. *IEICE Transactions on Electronics* **J87-C**, 433–443.

Natvig B. 1979. A suggestion of a new measure of importance of system component. *Stochastic Processes and Their Applications* **9**, 319–330.

Natvig B. 1982a. On the reduction in remaining system lifetime due to the failure of a specific component. *Journal of Applied Probability* **19**, 642–652.

Natvig B. 1982b. Two suggestions of how to define a multistate coherent system. *Advances in Applied Probability* **14**, 434–455.

Natvig B. 1985a. New light on measures of importance of system components. *Scandinavian Journal of Statistics* **12**, 43–54.

Natvig B. 1985b. Recent developments in multistate reliability theory. In *Probabilistic Methods in the Mechanics of Solids and Structures* (eds. Eggwertz S and Lind NC). Springer Verlag, Berlin, pp. 385–393.

Natvig B. 2011. *Multistate Systems Reliability Theory with Applications*. Wiley Series in Probability and Statistics. John Wiley & Sons, West Sussex, UK.

Natvig B, Eide KA, Gåsemyr J, Huseby AB, and Isaksen SL. 2009. Simulation based analysis and an application to an offshore oil and gas production system of the Natvig measures of component importance in repairable systems. *Reliability Engineering and System Safety* **94**, 1629–1638.

Natvig B and Gåsemyr J. 2009. New results on the Barlow-Proschan and Natvig measures of component importance in nonrepairable and repairable systems. *Methodology and Computing in Applied Probability* **11**, 603–620.

Natvig B, Huseby AB, and Reistadbakk MO. 2011. Measures of component importance in repairable multistate systems—a numerical study. *Reliability Engineering and System Safety* **96**, 1680–1690.

Navarro J, Spizzichino F, and Balakrishnan N. 2010. Average systems and their role in the study of coherent systems. *Journal of Multivariate Analysis* **101**, 1471–1482.

Nelson JB. 1978. Minimal-order models for false-alarm calculations on sliding windows. *IEEE Transactions on Aerospace Electron System* **ASE-14**, 351–363.

Nemhauser GL and Wolsey LA. 1988. *Integer and Combinatorial Optimization.* Wiley-Interscience Series in Discrete Mathematics and Optimization. John Wiley & Sons, New York.

Neuts MF. 1995. *Algorithm Probability: A Collection of Problems.* Chapman and Hall, London.

Norros I. 1986a. A compensator representation of multivariate life length distributions, with applications. *Scandinavian Journal of Statistics* **13**, 99–112.

Norros I. 1986b. Notes on Natvig's measure of importance of system components. *Journal of Applied Probability* **23**, 736–747.

Pörn K. 1997. A decision-oriented measure of uncertainty importance for use in PSA. *Reliability Engineering and System Safety* **56**, 17–27.

Page LB and Perry JE. 1994. Reliability polynomials and link importance in networks. *IEEE Transactions on Reliability* **43**, 51–58.

Pan ZJ and Tai YC. 1988. Variance importance of system components by Monte Carlo. *IEEE Transactions on Reliability* **37**, 421–423.

Papastavridis S. 1987. The most important component in a consecutive-k-out-of-n:F system. *IEEE Transactions on Reliability* **R-36**, 266–268.

Papastavridis S. 1990. m-consecutive-k-out-of-n:F systems. *IEEE Transactions on Reliability* **39**, 386–388.

Papastavridis S and Hadzichristos I. 1988. Formulas for the reliability of a consecutive-k-out-of-n:F system. *Journal of Applied Probability* **26**, 772–779.

Papastavridis S and Koutras MV. 1993. Bounds for reliability of consecutive-k within-m-out-of-n systems. *IEEE Transactions on Reliability* **R-42**, 156–160.

Papastavridis SG and Sfakianakis M. 1991. Optimal-arrangement and importance of the components in a consecutive-k-out-of-r-from-n:F system. *IEEE Transactions on Reliability* **R-40**, 277–279.

Park CK and Ahn KI. 1994. A new approach for measuring uncertainty importance and distributional sensitivity in probabilistic safety assessment. *Reliability Engineering and System Safety* **46**, 253–261.

Pham H. 1991. Optimal design for a class of noncoherent systems. *IEEE Transactions on Reliability* **40**, 361–363.

Podofillini L and Zio E. 2008. Designing a risk-informed balanced system by genetic algorithms: Comparison of different balancing criteria. *Reliability Engineering and System Safety* **93**, 1842–1852.

Powers DA. 2000. Importance measures derived from probabilistic risk assessments. Technical report, U.S. Nuclear Regulatory Commission. http://www.nrc.gov/reading-rm/doc-collections/acrs/letters/2000/4691872.html.

Prasad VR, Aneja YP, and Nair KPK. 1991a. A heuristic approach to optimal assignment of components to parallel-series network. *IEEE Transactions on Reliability* **40**, 555–558.

Prasad VR and Kuo W. 2000. Reliability optimization of coherent systems. *IEEE Transactions on Reliability* **49**, 323–330.

Prasad VR, Nair KPK, and Aneja YP. 1991b. Optimal assignment of components to parallel-series and series-parallel systems. *Operations Research* **39**, 407–414.

Prasad VR and Raghavachari M. 1998. Optimal allocation of interchangeable components in a series-parallel system. *IEEE Transactions on Reliability* **R-47**, 255–260.

Räde L. 1989. Expected time to failure of reliability systems. *Mathematical Scientist* **14**, 24–37.

Ramirez-Marquez JE and Coit DW. 2005. Composite importance measures for multi-state systems with multi-state components. *IEEE Transactions on Reliability* **54**, 517–529.

Ramirez-Marquez JE and Coit DW. 2007. Multi-state component criticality analysis for reliability improvement in multi-state systems. *Reliability Engineering and System Safety* **92**, 1608–1619.

Ramirez-Marquez JE, Rocco CM, Gebre BA, Coit DW, and Tortorella M. 2006. New insights on multi-state component criticality and importance. *Reliability Engineering and System Safety* **91**, 894–904.

Reinert JM and Apostolakis GE. 2006. Including model uncertainty in risk-informed decision making. *Annals of Nuclear Energy* **33**, 354–369.

ReliaSoft. 2003. Using reliability importance measures to guide component improvement efforts. Available at http://www.maintenanceworld.com/Articles/reliasoft/usingreliability.html.

Rocco C, Ramirez-Marquez JE, Salazar D, and Zio E. 2010. A flow importance measure with application to an Italian transmission power system. *International Journal of Performability Engineering* **6**, 53–61.

Rushdi AM. 1985. Uncertainty analysis of fault-tree outputs. *IEEE Transactions on Reliability* **R-34**, 458–462.

Ryabinin IA. 1994. A suggestion of a new measure of system components importance by means of a boolean difference. *Microelectronics and Reliability* **34**, 603–613.

Saltelli A. 2002. Making best use of model evaluations to compute sensitivity indices. *Computer Physics Communications* **145**, 280–297.

Saltelli A and Tarantola S. 2002. On the relative importance of input factors in mathematical models: Safety assessment for nuclear waste disposal. *Journal of American Statistical Association* **97**, 702–709.

Salvia AA and Lasher WC. 1990. 2-dimensional consecutive-k-out-of-n:F models. *IEEE Transactions on Reliability* **R-39**, 382–385.

Samaniego FJ. 2007. *System Signatures and Their Applications in Engineering Reliability*. Springer, New York.

Santha M and Zhang Y. 1987. Consecutive-2 systems on trees. *Probability in the Engineering and Informational Sciences* **1**, 441–456.

Schwender H, Bowers K, Fallin M, and Ruczinski I. 2011. Importance measures for epistatic interactions in case-parent trios. *Annals of Human Genetics* **75**, 122–132.

Separstein B. 1973. On the occurrence of n successes within n Bernouli trials. *Technometrics* **15**, 809–818.

Separstein B. 1975. Note on a clustering problem. *Journal of Applied Probability* **12**, 629–632.

Sfakianakis M. 1993. Optimal arrangement of components in a consecutive k-out-of-r-from-n:F system. *Microelectronics and Reliability* **33**, 1573–1578.

Sfakianakis M, Kounias S, and Hillaris A. 1992. Reliability of a consecutive-k-out-of-r-from-n:F system. *IEEE Transactions on Reliability* **R-41**, 442–447.

Shaked M and Shanthikumar JG. 1992. Optimal allocation of resources to nodes of parallel and series systems. *Advances in Applied Probability* **24**, 894–914.

Shapley LS and Shubik M. 1954. A method for evaluating the distribution of power in a committee system. *American Political Science Review* **48**, 787–792.

Shen J and Zuo MJ. 1994. Optimal design of series consecutive-k-out-of-n:G system with age-dependent minimal repair. *Reliability Engineering and System Safety* **45**, 277–283.

Shen K and Xie M. 1989. The increase of reliability of k-out-of-n systems through improving a component. *Reliability Engineering and System Safety* **26**, 189–195.

Shen K and Xie M. 1990. On the increase of system reliability by parallel redundancy. *IEEE Transactions on Reliability* **R-39**, 607–611.

Shen K and Xie M. 1991. The effectiveness of adding standby redundancy at system and component levels. *IEEE Transactions on Reliability* **40**, 53–55.

Shen ZJM, Coullard C, and Daskin MS. 2003. A joint location-inventory model. *Transportation Science* **37**, 40–55.

Shingyoch K, Yamamoto H, Tsujimura Y, and Akiba T. 2010. Proposal of simulated annealing algorithms for optimal arrangement in a circular consecutive-k-out-of-n:F system. *Quality Technology and Quantitative Management* **7**, 395–405.

Shingyoch K, Yamamoto H, Tsujimura Y, and Kambayashi Y. 2009. Improvement of ordinal representation scheme for solving optimal component arrangement problem of circular consecutive-k-out-of-n:F system. *Quality Technology and Quantitative Management* **6**, 11–22.

Shohat JA and Tamarkin JD. 1943. *The Problem of Moments*. American Mathematical Society, Providence, RI.

Shrestha A, Xing L, and Coit DW. 2010. An efficient multistate multivalued decision diagram-based approach for multistate system sensitivity analysis. *IEEE Transactions on Reliability* **59**, 581–592.

SIMLAB. 2007. SIMLAB *Reference Manual* POLIS-JRC ISIS. Joint Research Center of the European Community.

Simmons D, Ellis N, Fujihara H, and Kuo W. 1998. *Software Measurement: A Visualization Toolkit for Project Control and Process Improvement*. Prentice Hall, New Jersey. 459 pages and CD software.

Singh H and Misra N. 1994. On redundancy allocations in systems. *Journal of Applied Probability* **31**, 1004–1014.

Singpurwalla ND. 2006. *Reliability and Risk: A Bayesian Perspective*. Wiley Series in Probability and Statistics. John Wiley & Sons, New York.

Smith CL and Borgonovo E. 2007. Decision making during nuclear power plant incidents – a new approach to the evaluation of precursor events. *Risk Analysis* **27**, 1027–1042.

Sobol I. 1993. Sensitivity estimates for nonlinear mathematical models. *Mathematical Modeling and Computational Experiment* **1**, 407–414.

Sobol IM. 2001. Global sensitivity indices for nonlinear mathematical models and their Monte Carlo estimates. *Mathematics and Computers in Simulation* **55**, 271–280.

Sobol IM and Kucherenko SS. 2005. On global sensitivity analysis of quasi-Monte Carlo algorithms. *Monte Carlo Methods and Applications* **11**, 83–92.

Sobol IM and Kucherenko SS. 2009. Derivative based global sensitivity measures and their link with global sensitivity indices. *Mathematics and Computers in Simulation* **79**, 3009–3017.

Sobol IM and Kucherenko SS. 2010. A new derivative based importance criterion for groups of variables and its link with the global sensitivity indices. *Computer Physics Communications* **181**, 1212–1217.

Sobol IM and Levitan YL. 1999. On the use of variance reducing multipliers in Monte Carlo computations of a global sensitivity index. *Computer Physics Communications* **117**, 52–61.

Sobol IM and Shukhman BV. 2007. On global sensitivity indices: Monte Carlo estimates affected by random errors. *Monte Carlo Methods Applications* **13**, 89–97.

Sobol IM, Tarantola S, Gatelli D, Kucherenko S, and Mauntz W. 2007. Estimating the approximation error when fixing unessential factors in global sensitivity analysis. *Reliability Engineering and System Safety* **92**, 957–960.

Song J and Der Kiureghian A. 2005. Component importance measures by linear programming bounds on system reliability. *Proceedings of the 9th International Conference on Structural Safety and Reliability*, pp. 19–23.

Song S, Lu Z, and Cui L. 2012. A generalized Borgonovo's importance measure for fuzzy input uncertainty. *Fuzzy Sets and Systems* **189**, 53–62.

Soofi ES, Retzer JJ, and Yasai-Ardekani M. 2000. A framework for measuring the importance of variables with applications to management research and decision models. *Decision Sciences* **31**, 595–625.

Spielberg K. 1979. Enumerative methods in integer programming. *Annals of Discrete Mathematics* **5**, 139–183.

Strobl C, Boulesteix AL, Zeileis A, and Hothorn T. 2007. Bias in random forest variable importance measures: Illustrations, sources and a solution. *BMC Bioinformatics* **8**, 1–25.

Taguchi G. 1986. *Introduction to Quality Engineering*. Asian Productivity Association, Tokyo, Japan.

Taylor A and Zwicker W. 1999. *Simple Games: Desirability Relations, Trading, and Pseudoweightings*. Princeton Universtiy Press, Princeton, NJ.

Tillman FA, Hwang CL, and Kuo W. 1977. Determining component reliability and redundancy for optimum system reliability. *IEEE Transactions on Reliability* **R-26**, 162–165.

Tomlin JA. 1970. Branch and bound methods for integer and non-convex programming. In *Integer and Nonlinear Programming* (ed. Abadie J). Elsevier (North-Holland), New York, pp. 437–450.

Tong YL. 1985. A rearrangement inequality for the longest run, with an application to network reliability. *Journal of Applied Probability* **22**, 386–393.

Tong YL. 1986. Some new results on the reliability of circular consecutive-k-out-of-n:F system. In *Reliability and Quality Control* (ed. Basu AP). Elsevier (North-Holland), New York, pp. 395–400.

Triantafyllou IS and Koutras MV. 2008. On the signature of coherent systems and applications. *Probability in the Engineering and Informational Sciences* **22**, 19–35.

Ulder NLJ, Aarts EHL, Bandelt HJ, van Laarhoven PJM, and Pesch E. 1991. Genetic local search algorithms for the traveling salesman problem. In *Parallel Problem Solving from Nature I* (ed. Schwefel H and Manner R). Springer-Verlag, Berlin, pp. 109–116.

Upadhyaya SJ and Pham H. 1993. Analysis of noncoherent systems and an architecture for the computation of the system reliability. *IEEE Transactions on Computers* **42**, 484–493.

Valiant LG. 1979. The complexity of enumeration and reliability problems. *SIAM Journal on Computing* **8**, 410–421.

van der Borst M and Schoonakker H. 2001. An overview of PSA importance measures. *Reliability Engineering and System Safety* **72**, 241–245.

Vasseur D and Llory M. 1999. International survey on PSA figures of merit. *Reliability Engineering and System Safety* **66**, 261–274.

Vaurio JK. 2010. Ideas and developments in importance measures and fault-tree techniques for reliability and risk analysis. *Reliability Engineering and System Safety* **95**, 99–107.

Vaurio JK. 2011a. Importance measures for multi-phase missions. *Reliability Engineering and System Safety* **96** 230–235.

Vaurio JK. 2011b. Importance measures in risk-informed decision making: Ranking, optimisation and configuration control. *Reliability Engineering and System Safety* **96**, 1426–1436.

Vesely WE. 1970. A time dependent methodology for fault tree evaluation. *Nuclear Engineering and Design* **13**, 337–360.

Vesely WE. 1998. Supplemental viewpoints on the use of importance measures in risk-informed regulatory applications. *Reliability Engineering and System Safety* **60**, 257–259.

Vesely WE, Belhadj M, and Rezos JT. 1994. PRA importance measures for maintenance prioritization applications. *Reliability Engineering and System Safety* **43**, 307–318.

Vesely WE and Davis TC. 1985. Two measures of risk importance and their applications. *Nuclear Technology and Radiation Protection* **68**, 226–234.

Vinod G, Kushwaha HS, Verma AK, and Srividy A. 2003. Importance measures in ranking piping components for risk informed in-service inspection. *Reliability Engineering and System Safety* **80**, 107–113.

Wang W, Loman J, and Vassiliou P. 2004. Reliability importance of componetns in complex system. *Proceedings of Annual Reliability and Maintainability Symposium*, pp. 6–8.

Wei VK, Hwang FK, and Sös VT. 1983. Optimal sequencing of items in a consecutive-2-out-of-*n* system. *IEEE Transactions on Reliability* **R-32**, 30–33.

Whitson JC and Ramirez-Marquez JE. 2009. Resiliency as a component importance measure in network reliability. *Reliability Engineering and System Safety* **94**, 1685–1693.

Wu S. 2005. Joint importance of multistate systems. *Computers and Industrial Engineering* **49**, 63–75.

Wu S and Chan LY. 2003. Performance utility-analysis of multi-state systems. *IEEE Transactions on Reliability* **52**, 14–21.

Xie M. 1987. On some importance measures of system components. *Stochastic Processes and Their Applications* **25**, 273–280.

Xie M. 1988. A note on the Natvig measure. *Scandinavian Journal of Statistics* **15**, 211–214.

Xie M and Bergman B. 1991. On a general measure of component importance. *Journal of Statistical Planning and Inference* **29**, 211–220.

Xie M and Lai CD. 1996. Exploiting symmetry in the reliability analysis of coherent systems. *Naval Research Logistics* **43**, 1025–1034.

Xie M and Shen K. 1989. On ranking of system components with respect to different improvement actions. *Micro-electronics and Reliability* **29**, 159–164.

Xie M and Shen K. 1990. On the increase of system reliability due to some improvement at component level. *Reliability Engineering and System Safety* **28**, 111–120.

Yacoub S, Cukic B, and Ammar HH. 2004. A scenario-based reliability analysis approach for component-based software. *IEEE Transactions on Reliability* **53**, 465–480.

Yang K and Xue J. 1996. Continuous state reliability analysis. *Processings of Annual Reliability and Maintainability Symposium*, pp. 251–257.

Yao Q, Zhu X, and Kuo W. 2012. Importance-measure based genetic local search for component assignment problems. *Annals of Operations Research*, In review.

Yao Q, Zhu X, and Kuo W. 2011. Heuristics for component assignment problems based on the Birnbaum importance. *IIE Transactions* **43**, 1–14.

Youngblood RW. 2001. Risk significance and safety significance. *Reliability Engineering and System Safety* **73**, 121–136.

Zakaria RS, David HA, and Kuo W. 1992. A counter-intuitive aspect of component importance measures in linear consecutive-*k*-out-of-*n* systems. *IIE Transactions* **24**, 147–154.

Zhang C, Ramirez-Marquez JE, and Sanseverino CMR. 2011. A holistic method for reliability performance assessment and critical components detection in complex networks. *IIE Transactions* **43**, 661–675.

Zhang Q and Mei Q. 1985. Element importance and system failure frequency of a 2-state system. *IEEE Transactions on Reliability* **R-34**, 308–313.

Zhang Z, Kulkarni A, Ma X, and Zhou Y. 2009. Memory resource allocation for file system prefetching: From a supply chain management perspective. *Proceedings of the EuroSys Conference*, pp. 75–88.

Zhu X and Kuo W. 2008. Comments on "A hierarchy of importance indices". *IEEE Transactions on Reliability* **57**, 529–531.

Zhu X and Kuo W. 2012. Importance measures in reliability and mathematical programming. *Annals of Operations Research*, Accepted.

Zhu X, Yao Q, and Kuo W. 2011. Patterns of Birnbaum importance in linear consecutive-*k*-out-of-*n* systems. *IIE Transactions* **44**, 277–290.

Zio E, Marella M, and Podofillini L. 2007. Importance measures-based prioritization for improving the performance of multi-state systems: Application to the railway industry. *Reliability Engineering and System Safety* **92**, 1303–1314.

Zio E and Podofillini L. 2003a. Importance measures of multi-state components in multi-state systems. *International Journal of Reliability Quality and Safety Engineering* **10**, 289–310.

Zio E and Podofillini L. 2003b. Monte-Carlo simulation analysis of the effects on different system performance levels on the importance on multi-state components. *Reliability Engineering and System Safety* **82**, 63–73.

Zio E and Podofillini L. 2006. Accounting for components interactions in the differential importance measure. *Reliability Engineering and System Safety* **91**, 1163–1174.

Zio E and Podofillini L. 2007. Importance measures and genetic algorithms for designing a risk-informed optimally balanced system. *Reliability Engineering and System Safety* **92**, 1435–1447.

Zio E, Podofillini L, and Levitin G. 2004. Estimation of the importance measures of multi-state elements by Monte Carlo simulation. *Reliability Engineering and System Safety* **86**, 191–204.

Zio E, Podofillini L, and Zille V. 2006. A combination of Monte Carlo simulation and cellular automata for computing the availability of complex network systems. *Reliability Engineering and System Safety* **91**, 181–190.

Zuo MJ. 1993a. Reliability and component importance of a consecutive-k-out-of-n system. *Microelectronics and Reliability* **33**, 243–258.

Zuo MJ. 1993b. Reliability and design of 2-dimensional consecutive-k-out-of-n systems. *IEEE Transactions on Reliability* **R-42**, 488–490.

Zuo MJ and Kuo W. 1988. Reliability design for high system utilization. In *Advances in Reliability and Quality Control* (ed. Hamza MH). ACTA Press, Zurich, pp. 53–56.

Zuo MJ and Kuo W. 1990. Design and performance analysis of consecutive k-out-of-n structure. *Naval Research Logistics* **37**, 203–230.

Zuo MJ and Shen J. 1992. System reliability enhancement through heuristic design. *1992 OMAE Volume II, Safety and Reliability, ASME*, pp. 301–304.

Index

Importance Measures in Reliability, Risk, and Optimization: Principles and Applications, First Edition.
Way Kuo and Xiaoyan Zhu. © 2012 John Wiley & Sons, Ltd. Published 2012 by John Wiley & Sons, Ltd.